International Series in Operations Research & Management Science

Volume 169

For further volumes:
http://www.springer.com/series/6161

Cynthia Barnhart · Barry C. Smith
Editors

Quantitative Problem Solving Methods in the Airline Industry

A Modeling Methodology Handbook

Springer

Cynthia Barnhart
Department of Civil and Environmental Engineering
Massachusetts Institute of Technology
Massachusetts Avenue 77
Cambridge, MA 02139-4307
USA
e-mail: barnhart@mit.edu

Barry C. Smith
Barry C. Smith LLC
6286 Willowgate
Dallas, TX 75230
USA
e-mail: barrycsmith@mac.com

ISSN 0884-8289
ISBN 978-1-4614-1607-4
e-ISBN 978-1-4614-1608-1
DOI 10.1007/978-1-4614-1608-1
Springer New York Dordrecht Heidelberg London

Library of Congress Control Number: 2011940035

Cover design: eStudio Calamar, Berlin/Figueres

Printed on acid-free paper

Springer is part of Springer Science+Business Media (www.springer.com)

Preface

Airlines and the practice of Operations Research have had a long, mutually beneficial relationship. As part of a regulated industry until 1978, US domestic airlines had the luxury of supporting significant research and development efforts in operations research. Since then, in a largely de-regulated industry, airlines continue to be major consumers, and developers, of Operations Research theory, methodology and applications. As a result, Operations Research has become a critical component of the problem solving culture at airlines, and is credited with enabling advanced capabilities leading to tremendous benefits.

The airline industry has been a consistent source of Operations Research opportunity. Airline applications have motivated many developments in Operations Research methodology in areas such as large-scale optimization, forecasting and customer modeling. Some of Operations Research's major success stories are from the airline industry, in particular crew scheduling, fleet assignment, operations management and revenue management. As airlines work to gain competitive advantage, they experiment with new technology; if it works, they use it. Today, no major airline operates without significant Operations Research support. Due to the size of the industry, even modest measureable improvements can have significant impact.

Even with this long history, the airline industry and related Operations Research opportunities continue to evolve. In this book, we review Operations Research theory, applications and practice in seven major areas of airline planning and operations. In each area, a team of academic and industry experts provide an overview of the business and technical landscape, a view of current best practices, a summary of open research questions and suggestions for relevant future research. There are several common themes in current airline Operations Research efforts. First is a growing focus on the customer in terms of: (1) what they want; (2) what they are willing to pay for services and (3) how they are impacted by planning, marketing and operational decisions. Second, as algorithms improve and computing power increases, the scope of modeling applications expands, often re-integrating processes that had been broken into smaller parts in order to solve them in the past. Finally, there is a growing awareness of the uncertainty in many airline planning and operational processes and decisions. Airlines now recognize the need to develop 'robust' solutions that effectively cover many possible outcomes, not just the best case, "blue sky" scenario. The topics by chapter include:

1. *Customer Modeling*. Airline success depends on creating and delivering products and services valued by customers. Historically airlines have used multiple models to analyze the effects of schedule, price and competition. As a result, different functional areas within an airline often had inconsistent views of their customers. For example, demand models used for scheduling are generally different and inconsistent with those used in revenue management. Airlines are currently adopting modeling approaches based on customer preferences and intent toward travel, combining the various aspects of customer behavior. These models are not only improving the accuracy of current applications, they offer the potential to integrate customer modeling and decision making across multiple application areas. Chapter 1 covers the various methodologies for customer modeling as well as the current and emerging applications.
2. *Airline Planning and Schedule Development*. The schedule is one of the most visible aspects of airline product design. Airline scheduling has been the focus of significant Operations Research work in the past. The full scheduling problem is so large and complex that it has been broken into a sequence of smaller problems to allow solutions to be generated for specific components, such as fleet assignment and aircraft routing. Advances in modeling and IT are allowing components to be re-integrated, resulting in improved airline performance through the generation of schedules that more accurately model customer demand, pricing and revenue management, crew scheduling and operations. Airline scheduling continues to be an area of active research. Chapter 2 also describes many remaining open research questions.
3. *Revenue Management*. There is a long history of modeling success and well-documented benefits in revenue management. Industry business practices continue to evolve in areas such as restriction-free pricing andi product de-bundling. Revenue Management requirements and opportunities evolve as well. Its scope, once limited to overbooking and discount allocation, is expanding into airline pricing and product design. Chapter 3 provides a view of current business and technical landscapes, as well as suggested areas for future research.
4. *Airline Distribution*. With the development of computerized reservation systems and global distribution systems, the airline industry had a sophisticated electronic commerce infrastructure long before the dot-com age. Pioneering work in the 1980s measured the impact of screen placement on airline bookings sold through travel agencies. Since then, the number of distribution channels has increased, along with opportunities for new Operations Research applications. These applications apply to channel providers as well as to airlines. Chapter 4 provides a comprehensive overview of this newly emerging area.
5. *Crew Management Information Systems*. Developing efficient crew schedules has been one of the key applications areas for Operations Research. Chapter 5 describes the development of crew management optimization algorithms

including recent advances using the volume algorithm, primal-dual methods and constraint-based approaches. Chapter 5 also reviews the development of information systems that facilitate the integration of crew management modeling with airline planning and operations.

6. *Airline Operations*. Airlines manage the flows of aircraft, crew and passengers, choreographing thousands of components to provide efficient, reliable transportation. Due to the size and complexity of airline operations, Operations Research methodologies often fail to solve enough of the problem, fast enough, to be useful in daily practice. Chapter 6 describes recent advances and successes in solving the airline operations problem. While the use of Operations Research in airline operations is becoming more typical, many research questions remain.
7. *Air Traffic Flow Management*. Flow management focuses on maintaining safe and efficient flow of aircraft through controlled airspace. Historically, flow management was a constraint for airlines. Now, however, the relationship between airlines and air traffic control is evolving to provide airlines with greater participation in flow management decisions. Chapter 7 describes the modeling environment and opportunities for both Air Traffic Flow Management and the airlines.

This book is the result of hard work and collaboration; we would like to thank the many contributors to this effort. First, we would like to thank the authors. Each chapter was written by a team of experts, drawing on decades of both academic and industrial experience. They provided great insights into how the business and technical environments shaped current modeling approaches as well as an expert view of the future. Next we would like to thank Fred Hillier for his vision and for giving us the opportunity to take on this project. We hope this is a small addition to his substantial contribution to the Operations Research profession. Finally, we would like to thank Matthew Amboy and Springer for their support in producing and distributing this book.

Cynthia Barnhart
Barry C. Smith

Contents

Chapter 1
Customer Modeling

Laurie Garrow

1.1 Introduction

Understanding demand for products and services is an integral part of many fields, including transportation, economics, marketing, public policy, and operations research. The methods used to model demand are as diverse as the fields in which they appear. Within the aviation industry, demand has typically been modeled using time series, averaging methods, or simple probability distributions. As an example, consider the first generation revenue management systems designed in the 1990s. These systems initially modeled demand for each booking class on a flight using a Poisson or other statistical distribution. Booking classes were clearly defined and represented one or more products with similar prices and attributes. That is, each booking class corresponded to a distinct product offering that was defined by "fences" such as advance purchase, Saturday night stay, minimum stay, and other requirements. The more the requirements associated with a product, the lower the price. Due to the distinct nature of the products, demand was assumed to be independent across booking classes and future demand for a booking class was predicted using historical demand for that booking class.

In practice, independence across booking classes was a fairly reasonable assumption because product restrictions effectively helped to prevent (or "fence") business customers from purchasing lower fares designed for leisure customers. This is because product restrictions were explicitly designed to segment business and leisure customers, the former of which tend to purchase close to flight departure, stay fewer days away from home, and not travel over a Saturday night. Consequently, the underlying types of customers purchasing specific booking classes were distinct and modeling demand using time series or other averaging

L. Garrow (✉)
Georgia Institute of Technology, Atlanta, Georgia, USA
e-mail: laurie.garrow@ce.gatech.edu

C. Barnhart and B. C. Smith, LLC (eds.), *Quantitative Problem Solving Methods in the Airline Industry*, International Series in Operations Research & Management Science 169,
DOI: 10.1007/978-1-4614-1608-1_1,

methods that assumed independence among booking classes worked well. Indeed, American Airlines credits their first-generation revenue management system for creating an additional $1.4 billion in revenues over a three-year period—a feat that won them the 1991 Edelman Prize (INFORMS 2010).

However, today the airline industry is very different from when the first-generation decision-support systems were built for revenue management, scheduling, marketing/planning, distribution, and other areas. According to the Air Transport Association, in the first 30 years after passenger deregulation (which in the US occurred in 1978), domestic airline prices fell 41.2% in real terms (2010). Numerous factors have contributed to this decrease in fares and more competitive environment, most notably the increased market penetration of low cost carriers (LCCs) and the increased use of the internet as a major distribution channel. In turn, this more competitive environment has increased the importance of more accurately modeling customers' behavior, particularly how customers make tradeoffs among different product attributes.

The increased market penetration of LCCs is important in the context of modeling demand because LCCs typically use different pricing models from legacy carriers, most notably one-way (versus round-trip) fares. One-way fares eliminate the ability to segment demand based on minimum stay and Saturday night stay requirements, i.e., without knowledge of when a customer is returning, length of stay and specific departure, and return days of week cannot be determined. Some LCCs such as JetBlue further market themselves as an "instant purchase" airline, implying that they have a number of seats they will sell at a certain price. When this number is sold, they will increase the price. The characteristics of these "products" are similar, differing only by price. Further, under this pricing model, no explicit price increases are associated with advance purchase deadlines.

Consequently, LCCs have transformed the competitive landscape by making it much more difficult for the industry as a whole to segment business and leisure travelers. Product characteristics such as minimum stay and Saturday night restrictions have effectively become obsolete. In turn this has led to booking classes that are less clearly defined and that contain a mix of customers with potentially quite different willingness to pay. Corresponding assumptions, such as independence across booking classes, are no longer valid, raising the question as to whether other types of demand models, such as those based on discrete choice methods, may be more appropriate in revenue management applications.

A second major factor that has driven an interest in using discrete choice methods to model demand is the increased use of the internet as a major distribution channel. For example, a study based on data from Continental Airlines finds that in 1998, only about 1% of domestic leisure flight tickets were sold through the internet, but in 2005, this number was 35% and expected to grow to 50% by 2008 (Brunger and Perelli 2008). Use of the internet to make bookings is even higher among some LCCs. For example, in 2009 Southwest Airlines was the largest US domestic carrier by total passengers, carrying over 101.3 million passengers; 81% of these passengers made their bookings via southwest.com (Southwest Airlines 2009a, b).

Industry surveys by SITA and PhoCusWright also find that more than half of all leisure travel purchases are made online (Aaron 2007) and approximately 60% of online leisure travelers purchase the lowest fare they can find (Harteveldt et al. 2004; PhoCusWright 2004). The increased use of the internet and ability to easily compare fares across multiple carriers means that customers are more affected by price than in the past (and more capable of finding the lowest fares).

Combine the increasing market penetration of LCCs (that eliminated traditional product restrictions and effectively destroyed the ability to segment business and leisure travelers into distinct booking classes) with the increasing market penetration of the internet (that enabled customers to easily compare price and product characteristics across multiple carriers), and it becomes clear why the airline industry has been expressing interest in using discrete choice methods to model customer demand. Unlike time-series methods or probability distributions based on historical information, discrete choice models explicitly model *how customers make trade-offs among product attributes*. That is, discrete choice models consider not only the attributes of the products (or alternatives) that the customer *chose*, but also the attributes of other alternatives presented to the customer that were *not chosen*. In discrete choice models, the probability of purchasing a particular product (or alternative) is driven by the characteristics of the chosen product as well as how similar (or dissimilar) these characteristics are compared to other alternatives. A demand modeling framework grounded in discrete choice theory that captures how passengers are making decisions and tradeoffs among different alternatives is intuitively appealing, particularly in a competitive environment in which passengers can easily identify and compare alternatives across one or more carriers.

The subsequent chapters present numerous examples of how discrete choice models are being used in scheduling, distribution, revenue management, sales/marketing and product design. Given the prevalence of these models throughout the airline industry today, one cannot help but ask why the airline industry did not initially gravitate towards using discrete choice models to predict demand. Simply stated, the data needed to estimate discrete choice model parameters were not available. As noted earlier, discrete choice models predict how customers make tradeoffs among alternatives. To make these predictions, discrete choice models need information about the alternative the customer chose, as well as information about alternatives the customer *did not* choose. Historically, information about non-chosen alternatives has been difficult to obtain. Even if this data were available, it was often not viable to store or use it in forecasting applications that required frequent (often daily) updates.

For example, to predict the number of customers that will no show for a flight using a time-series method, one needs to have information about historical *flight-level* no show rates. A major airline with thousands of daily flight departures would need to store information about the total number of customers who showed (or did not show) for each flight. Information collected over a minimum time period (e.g., 60 days) for each flight would enable time-series forecasts that incorporated day of week, time of day, and equipment effects. But … would no show forecasts be more accurate if, instead of using flight-level averages as an input to these time-series models, we could use cabin-level averages? Intuitively, yes but this potentially doubles (if not triples) data storage requirements depending on how many cabins of service an airline offers (first, business, and/or coach). Similarly, would no show forecasts be more accurate if, instead of using cabin-level averages as input to time-series models, we used booking-class averages instead? Intuitively, yes, but data requirements increase again (the number of booking classes an airline uses can easily exceed 20). And now the ultimate question: would no show forecasts be more accurate if, instead of using information simply about the booking class, we could incorporate information associated with individual bookings, e.g., when the booking was made, how it was purchased, whether the customer making the booking was a frequent flyer on the carrier of interest? That is, could discrete choice models be used to model no show probabilities based on characteristics of the booking and the individual making the booking? Yes, but data requirements to predict a no show rate for each individual booking is now an order or two higher in magnitude than the initial requirement, which was to forecast no show rates using historical flight-level averages. Computational requirements to create these forecasts would also increase when using passenger-level data versus flight- or cabin-level data.

This example serves to illustrate several recurring themes that help to explain why the airline industry did not initially use discrete choice models to forecast demand. First, the sheer volume of flights, passengers, and data an airline handles is large. At the time initial decision-support systems were being developed, it was often not viable to estimate data-intensive discrete choice models or use them to produce frequent forecasts. Second, as with the revenue management demand forecasting example, if bookings associated with a particular booking class represent similar customers and attributes (e.g., most frequent flyers purchase booking classes close to booking departure), then historical booking-class averages would lead to accurate forecasts using time-series methods. However, as customers become less segmented, capturing individual booking and passenger detail directly into the forecasts becomes more important.

The no show example also helps explain why particular applications of discrete choice models are included in the text, whereas other applications common to other disciplines are omitted or are noted as future research directions. In the no show example, the choices available to a customer are easy to define: the customer either shows or no shows. Further, every customer has the option to show or no show. Consequently, creating the set of chosen and non-chosen alternatives is very straightforward and numerous airlines have reported how they have used passenger-level data (often based on discrete choice models) to forecast no show rates. These include Continental (Pastor 2000; Gorin et al. 2006), United (Garrow and Koppelman 2004a, b), Air Canada (Dupuis et al. 2010), and Lufthansa (Kalka and Weber 2000).

To date, the majority of discrete choice modeling implementations have focused on areas where it has been relatively straightforward to identify the alternatives that individuals consider during the choice process. In addition, the majority of discrete choice modeling implementations have focused on areas in which it would be relatively easy for airlines to replace an existing module (e.g., a no show forecast) that is part of a much larger decision-support system (e.g., a revenue management system). It is no coincidence that numerous airlines have published articles describing how passenger-level data has been used to produce no show forecasts, but no airline has successfully implemented the next-generation of choice-based revenue management systems. The latter is a much more challenging problem, and one that requires fundamental research in how discrete choice models can be used with censored data.

Conceptually, choice-based revenue management methods use data that effectively track individuals' purchase decisions, as well as the menus of choices they viewed prior to purchase (i.e., the alternatives they did not chose). That is, in contrast to traditional booking data, online shopping data provide a detailed snapshot of the products available for sale at the time an individual was searching for fares, as well as information on whether the search resulted in a purchase (or booking). These data effectively enable firms to replace revenue management demand models based on probability and time-series models with models grounded in discrete choice theory. To date, several theoretical papers on choice-based revenue management techniques have appeared in the research community and a few empirical studies based on a limited number of markets and/or departure dates have also been reported (e.g., see Besbes and Zeevi 2006; Bodea et al. 2009; Bront et al. 2007; Gallego and Sahin 2006; Hu and Gallego 2007; Talluri and van Ryzin 2004; van Ryzin and Liu 2004; van Ryzin and Vulcano 2008a, b; Vulcano et al. 2008; Zhang and Cooper 2005).

However, in contrast to the no show problem, the choice-based revenue management problem is much more complex and numerous questions remain to be answered: How should the choice set be defined for different distribution channels? Should only the online products shown for a particular carrier be included in the choice set, or should online products returned from a general search engine be used? What alternatives do individuals actually consider when making their purchase decision and how important is inclusion of (and forecasting of future) competitor products to forecasting accuracy? If search and purchase occur at different periods, how should these dynamics be modeled, i.e., is it important to track customers through multiple visits to an online site to understand how search and alternatives shown early in a process ultimately affect choice? Do factors such as diversity in prices shown across alternatives affect purchase probabilities? Are there ways to make online menus more attractive to the consumers, thereby helping to shift market share to a particular airline?

These questions highlight many of the challenging questions researchers and practitioners are facing as they contemplate how to design the next-generation of decision-support systems. At a fundamental level, this complexity is being driven by questions related to how to model alternatives that were not chosen by the individual, particularly when data are censored in the sense that an airline only has detailed information for its own bookings. Joint consideration of the "market size," or the number of customers who purchase on the airline of interest or one of its competitors and "market share," or the number of customers purchasing on the airline of interest (and distribution of products purchased), is a problem that has not been examined in depth by the discrete choice modeling community. That is, using booking data for a single airline and estimating a "no purchase alternative" to represent market size is an open research question being driven by the unique data and decision-support requirements of the airline industry. Solving this problem will likely benefit not only the airline industry, but also marketing (through better representation of stock-outs) and urban travel demand modeling (through better representation of latent demand for inter-city or intra-city trips). However, estimation of choice-based revenue management models, and specifically how the "no purchase" alternative in the choice set should be represented, requires much more fundamental research before it will be ready for implementation (e.g., see Newman et al. 2010).

Due to the complexity that is often encountered when trying to integrate discrete choice models within a revenue management (or other decision-support system) framework, finding ways to integrate "the best" of consumer behavior modeling with "the best" of existing systems has also been an active research stream. For example, the revenue management chapter describes a hybrid demand forecasting model that estimates the fraction of "priceable" customers who always purchase the lowest fare with the fraction of customers who are "yieldable" or purchase higher fares even when lower fares are available. This example represents ways in which researchers are creatively integrating important customer behavior factors into complex decision-support systems. The comparison of no show and choice-based revenue management examples are meant to highlight that it is important to keep in mind the complexity of the problem and ease with which non-chosen alternatives can be defined when reading subsequent chapters. That is, these factors help explain why certain discrete choice modeling applications are described in detail, whereas others are mentioned as ideas or current research areas, in subsequent chapters.

What is common to both examples, though, is their potential for generating additional revenue for the airline industry. In the airline industry, even small improvements (or deteriorations) in the forecasting accuracy of key decision-support models—such as no show and demand forecasting systems—may lead to millions of dollars of annual revenue gains (or losses) for an airline. For example, Stefan Polt, affiliated with Lufthansa German Airlines, stated that "as a rule of thumb, a 10% improvement in (demand) forecasting accuracy translates to a 1–2% revenue increase" (Polt 2002). Further, the forecasting improvements associated with models that incorporate more realistic assumptions relating to customer behavior can be impressive, as seen in the context of itinerary choice modeling.

Itinerary choice models, covered in detail in Chap. 2, are similar to revenue management models in the sense that they predict demand for air travel. Itinerary choice models are used to help support long-range scheduling decisions (e.g., where to fly, what equipment to fly, what flights to market using code-share or joint venture agreements, etc.). The initial models used to predict air travel were based on Quality of Service (QSI) indices. QSI indices were developed in 1957 and predicted how demand would shift among carriers as a function of flight frequency, level of service (e.g., nonstop, single-connection, double-connection), and equipment type (Civil Aeronautics Board 1970). At the time, the airline industry was regulated, fares and service levels were set by the government, and load factors were about 50% (e.g., see Ben-Yosef 2005). Competition was based primarily on marketing promotion and image.

One of the first published studies of itinerary share choice models based on a discrete choice framework was published by United Airlines in 2003 (Coldren, Koppelman, Kasturirangan, and Mukherjee). Multinomial logit (MNL) model parameters were estimated from a single month of itineraries (January 2000) and validated on monthly flight departures in 1999 in addition to selected months in 2001 and 2002. Using quarterly market sizes adjusted by a monthly seasonality factor, validation was undertaken at the flight segment level for the carrier's segments. That is, the total number of forecasted passengers on each segment was obtained by summing passengers on each itinerary using the flight segment. These forecasts were compared to onboard passenger count data. Errors, defined as the mean absolute percentage deviation, were averaged across segments for regional entities and compared to predictions from the original QSI model. Regional entities are defined by time zone for each pair of continental time zones in the US (e.g., East–East, East-Central, East-Mountain, East–West, ..., West–West). The MNL forecasts were consistently superior to the QSI model, with the magnitude of errors reduced on the order of 10–15% of the QSI errors. Further, forecasts were stable across months, including months that occurred after September 11, 2001 (Coldren et al. 2003).

This example is impressive not only because of the dramatic improvements in forecasting accuracy (10–15%), but because of what the MNL model did not include. Itinerary share choice modeling is one of the first areas that adopted a discrete choice modeling framework, but this was done using aggregate data, i.e., counts of the total number of customers booking an itinerary. Fare information was also aggregated, representing a quarterly average of all bookings sold for a particular carrier. Yet, despite the limitations, the MNL model dominated the performance of United's QSI model. Further gains in forecasting accuracy are expected if online purchase data could be used (which would capture the alternatives and prices actually available at the time of purchase).

In addition to improved forecasting accuracy, the itinerary choice application example also highlights some of the additional benefits of using discrete choice models to represent demand. Compared to traditional methods (such as QSI or time-series methods), the parameter estimates of discrete choice models tend to be more stable over time, resulting in less frequent model calibration updates. Conceptually, this is because customer preferences, or willingness to pay for different product attributes, tend to be stable with time. Further, once these customer preferences for different product attributes are quantified, it is relatively straightforward to predict how demand will shift when underlying products change or become unavailable. That is, discrete choice models are particularly useful in applications in which the underlying alternatives being chosen change—either in their availability or underlying characteristics. Examples include experimentation with new product designs (such as product debundling), shifting aircraft capacities across different markets, and predicting recapture rates when fares become unavailable. In this sense, then, discrete choice models are helpful for modeling demand during periods of turmoil as they are better able to capture underlying changes. Applications that combine discrete choice models (that capture how customers are making trade-offs across product attributes) with time-series methods (that capture longer term market trends) have also been investigated and also show strong potential to improve forecasting accuracy.

To summarize, discrete choice models are becoming a fundamental tool in revenue management, pricing, schedule planning, marketing, and product development. Depending on the application area, significant challenges remain to be solved; however, the potential revenue implications are significant and likely to be on the order of hundreds of millions of dollars per year for a major airline. Modeling customer behavior and demand is more important now than in the past due to the transparency of product offerings and creation of new product offerings.

The balance of this chapter is organized as follows. The next section provides an overview of how the US airline industry has evolved over the last decade. Section 1.3 provides an overview of the most common discrete choice models used in airline practice; specific applications are covered in subsequent chapters. The chapter concludes with a discussion of research opportunities specific to airline customer behavior modeling and discrete choice modeling.

1.2 Background Context

This section provides a useful background context for the structural market factors that have helped drive the interest in developing better models of customer demand (and revenue). The first decade of the twentieth century was especially challenging for major US airlines. Indeed, faced with increased market penetration of LCCs, unprecedented fuel costs, continued security threats post 9/11, health outbreaks (SARS, HN1N), economic recessions, and the global financial crisis, it is no surprise that in the first decade of the twenty-first century, the seven largest US network carriers collectively lost $35.1 billion[1] (US DOT 2010). From 2001 to 2005, four out of six legacy carriers went into bankruptcy (Delta, Northwest, United, and US Airways) and from 2005 to 2010 eight major US carriers went through mergers/acquisitions (America West and US Airways in 2005; Delta and

[1] We define largest using total number of passengers carried in 2006. The seven largest network carriers (Alaska, American, Continental, Delta, Northwest, United, US Airways) lost $35.1B over this time period. However, the seven largest low cost carriers (AirTran, American West, ATA, Frontier, Jet Blue, Southwest and Spirit) earned $4.9B and the seven largest regional carriers (American Eagle, Atlantic Southeast, Comair, Expressjet, Mesa, Pinnacle, and Skywest) earned $5.3B over this time period.

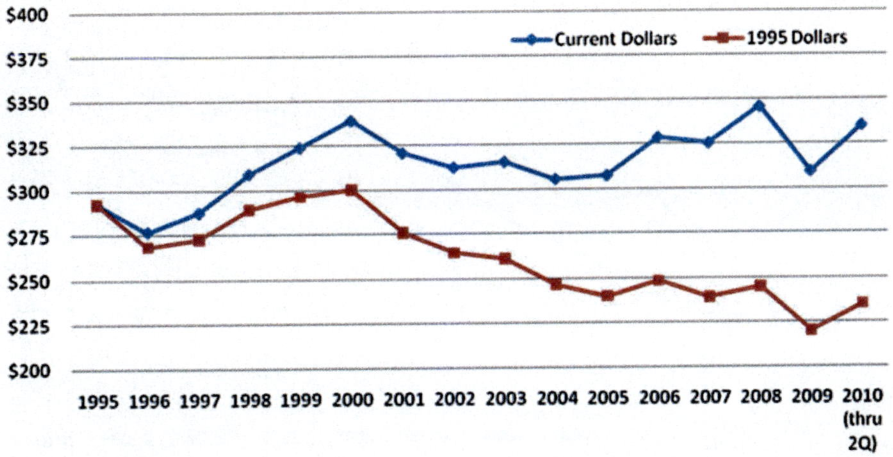

Fig. 1.1 Annual US domestic average itinerary fare—current and constant dollars. *Source* Bureau of Transportation Statistics data. Rate calculated using Bureau of Labor Statistics Consumer Price Index

Northwest in 2008; Continental and United in 2010; Southwest and AirTran in 2010).

The intense pressure that many airlines faced to reduce cost while simultaneously competing with low cost carriers is reflected in several measures. Figure 1.1 shows that in constant dollar terms, domestic US fares continued to decline in the 2000s. The average domestic fare in 2010 is 27% lower in constant dollars than that in 2001, reflecting both the increased availability of lower fares as well as the inability of major carriers to price discriminate using traditional product restrictions. Lower fares were accompanied by an increase in load factors, as shown in Fig. 1.2. In addition, many legacy carriers reduced their domestic schedules and/or shifted a larger proportion of their capacity to (less competitive) international markets. This trend is portrayed in Fig. 1.3. That is, although it is true that from 2000 to 2009, the total domestic available seat miles (ASMs) by US passenger carriers dropped by 2.7%, it is important to recognize that domestic ASMs fell by 7.2% whereas international ASMs grew by 12.1% (US DOT. Further, network carriers—Alaska , American , Continental, Delta, Northwest, United and US Airways—moved much of their capacity from domestic to international markets, reducing their domestic capacity by 23.8% and increased their international capacity by 7.4%. In contrast, low cost carriers—AirTran, Frontier, JetBlue, Southwest and Spirit—increased their domestic capacity by 103% and began to provide international service (US DOT 2010). As of 2009, the major network carriers concentrated 62.8% of their ASMs in domestic markets (compared to 70.4% in 2000) whereas the low cost carriers concentrated 96.5% of their ASMs in domestic markets (compared to 99.1% in 2000). Looking ahead, pricing pressures and competition can be expected to increase as Southwest expands into

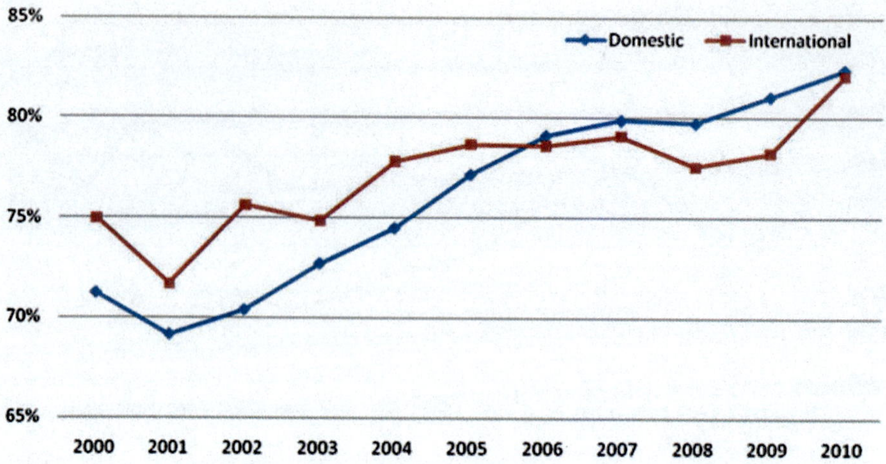

Fig. 1.2 Load factors. *Source* Bureau of Transportation Statistics data. Load factors are expressed as passenger-miles as a proportion of available seat-miles in percent (%)

international markets (and takes over AirTran's international markets due to the merger).

However, higher load factors, reduced domestic schedules, and expanded international schedules were not enough for legacy carriers to earn a profit due to increasing cost pressures. Higher operating costs (driven by labor and fuel) led many legacy airlines to reduce the number of employees, as shown in Fig. 1.4 (the increase in 2009 and 2010 for Delta is due to its merger with Northwest). Labor costs were also reduced voluntarily or during bankruptcies. However, while cost-cutting measures could be implemented for labor, the same was not true for fuel, as seen in Fig. 1.5. Fuel prices expressed as dollars per gallon (without adjusting for inflation) more than doubled from 2000 to 2010. Although the actual dollars per gallon spent by an airline will be influenced by many factors (such as aircraft type and stage length), the trend towards higher fuel prices is clear. The advantage Southwest enjoyed with fuel hedges is also clear.

Given the increasing thin profit margins airlines faced at the end of the twentieth century, along with their limited ability to raise fares or further reduce labor and other costs, airlines started creating new products and ancillary revenue streams. That is, new ancillary revenue sources including checked baggage fees and seat reservation fees were introduced in the late 2000s. In addition, many existing ancillary fees including fees for redeeming mileage award tickets, day of departure standby fees, agent-assisted ticketing fees, domestic and international ticketing exchange fees, on-board checked pet fees, unaccompanied minor fees, in-flight entertainment fees, food for sale, and amenity packages such as blankets and pillows for sale on red-eye flights were increased (see Garrow et al. 2010 for a review of recent de-bundling trends in the US).

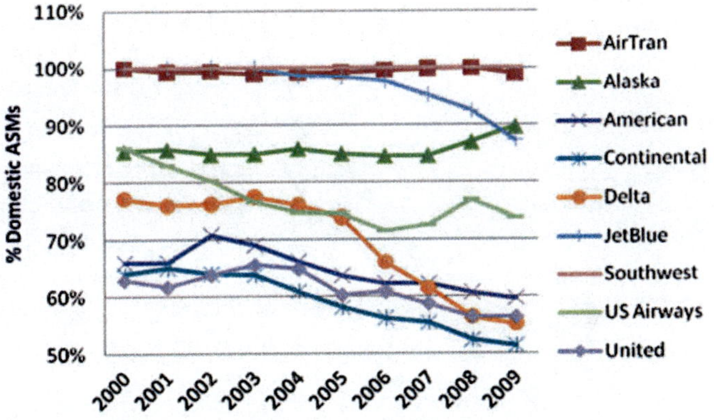

Fig. 1.3 Percent of available seat miles flown domestically. *Source* Bureau of Transportation Statistics T-100 Segment data

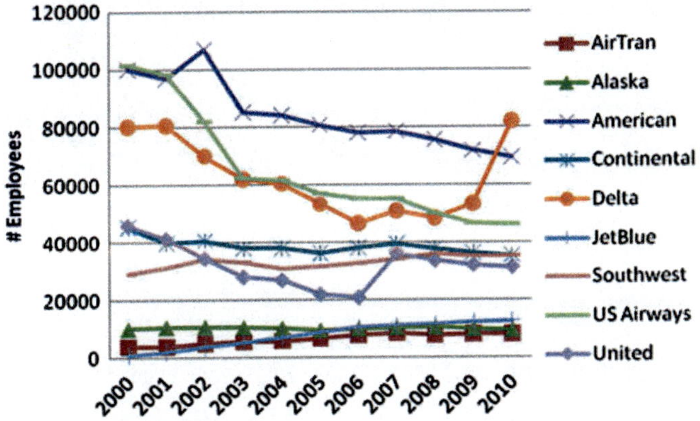

Fig. 1.4 Number of employees by airline. *Source* Bureau of Transportation Statistics data. Number of employees represents sum of full-time and part-time employees during the month of October for each year shown

There are several points of interest with respect to ancillary fees. Baggage fees and refund/exchange fees comprise the largest percentage of ancillary revenue streams. During the first three quarters of 2010, more than $2.57 billion was generated in baggage fees by US airlines (compared to $2.73 billion for the entire year of 2009 and $464 million for 2007). However, because baggage and ticketing exchange fees tend to be similar across airlines (e.g., each charges $50 for the first checked bag), these fees comprise a larger percentage of revenues for LCCs competing against Southwest Airlines (that does not charge fees for the first two checked bags or exchange fees). This relationship is seen in Fig. 1.6 which shows exchange and cancellation revenues for the major and LCC carriers listed in Table 1.1.

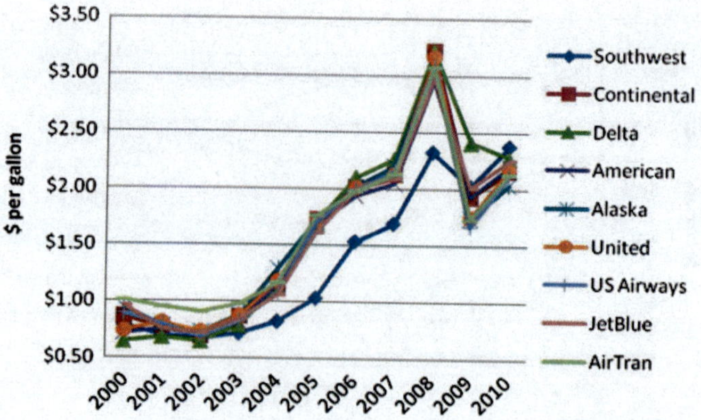

Fig. 1.5 Average dollars per gallon for fuel. *Source* Bureau of Transportation Statistics data. Data is based on total scheduled service from form P-12A

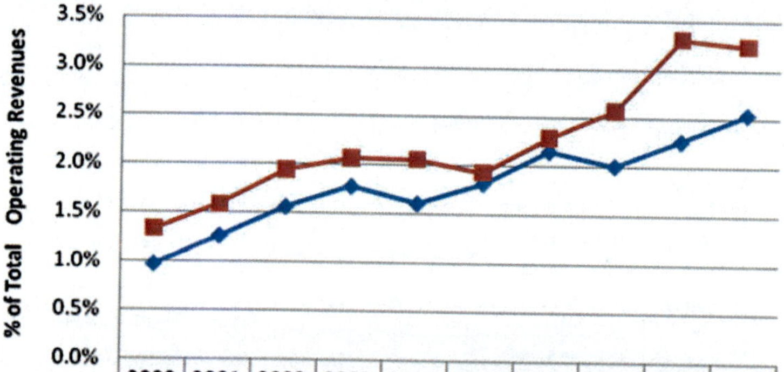

	2000	2001	2002	2003	2004	2005	2006	2007	2008	2009
Major	1.0%	1.3%	1.6%	1.8%	1.6%	1.8%	2.2%	2.0%	2.3%	2.5%
LCC	1.3%	1.6%	1.9%	2.1%	2.1%	1.9%	2.3%	2.6%	3.3%	3.2%

Fig. 1.6 Exchange and cancellation revenues as a percentage of total operating revenues. *Source* Bureau of Transportation Statistics data

Table 1.1 also highlights another trend seen with many ancillary revenue streams, i.e., fees are often waived for certain fare classes and/or loyal customers. Implementing new products (tied directly to observable customer segmentation characteristics such as status in a carrier's frequent flyer program) is another recent trend emerging in the industry.

Looking ahead, the "ancillary revenue" phenomenon is likely to continue in the US market, among both low cost and network carriers. Numerous airlines are reporting the importance of ancillary fees in their investor reports and/or listing generation of these fees as a top priority (e.g., see AirTran Holdings Inc. 2009).

Table 1.1 Ticketing/agent assisted and exchange fees

	Ticketing/agent assisted fees	Waivers for ticketing/agent assisted fees	Domestic exchange fees
Network carriers			
Delta	$20 ($35)	Diamond, Platinum, Gold	$150
American	$20 ($30)	Executive Platinum	$150
United	$25 ($30)	1 K, Global Services	$150
Continental	$25	Platinum	$150
US Airways	$25 ($35)	Preferred	$150
Alaska	$15 ($25)	None	$75/$100[a]
Low cost carriers			
Southwest	N/A	N/A	N/A
AirTran	$15	Elites	$75[b]
JetBlue	$15	None	$100
Frontier	$25	Elites	$50 Classic fares
		Classic/Classic Plus fares	$100 Economy fares
Virgin Am.	$15	None	$75/$100[c]

KEY: Call center fee (airport agent/city ticket office fee)
[a]Some web tickets may not be exchanged. If fee applies, charge is $75 online, $100 otherwise
[b]Only business fares are refundable. Elites purchasing Y,B,M fares receive free exchanges/refunds
[c]If fee applies (based on ticketing class) fee is $75 online, $100 otherwise

In their March 26, 2010 presentation to Barclay Capital High Yield Bond and Syndicated Loan Conference, Continental stated that one of its five top priorities in 2010 was to "increase ancillary revenues through customer choice" (Continental 2010). Continental elaborated, stating that they plan to grow these ancillary revenue streams through product unbundling (baggage fees, booking fees, in-flight amenities) and product enhancements including day-of-departure upgrades, preferred coach seating, premium wines and liquors, PetSafe, and "many others to come" (Continental 2010).

In summary, the past decade has seen tremendous change in the airline industry. Airlines operate in a more competitive market with thinner margins than they did a decade ago and are able to make money only if they match supply and demand. In turn, this underscores the need to better understand customer demand and revenue generation. The increased transparency of this competitive environment means that inefficient carriers have had no place to hide and scales of economy have become even more critical. This is reflected in the large number of bankruptcies and mergers/acquisitions. Demand is more affected than price than in the past, which has led to the creation of new products and ancillary revenue sources. Modeling customer behavior, and demand for these new products, is more important than in the past due to the need to understand how to generate revenue in a competitive market environment with transparent prices and new product offerings. Discrete choice models offer one way in which to model customer behavior because they explicitly incorporate multiple products and capture how individuals make trade-offs among price and product attributes.

1.3 Discrete Choice Modeling Methodology

Discrete choice models are commonly used to model customer behavior. Discrete choice models are used to predict the probability a decision-maker will choose one alternative among a finite set of mutually exclusive and collectively exhaustive alternatives. Discrete choice models relate to demand models in the sense that the total demand for a specific good (or alternative) is represented as the collection of choices made by individuals. For example, a binary logit model can be used to predict the probability that an airline passenger will no show (versus show) for a flight. The total demand expected to no show for a flight can be obtained by adding the no show probabilities for all passengers booked on the flight. This approach is distinct from statistical techniques traditionally used by airlines to model flight, itinerary, origin-destination, market, and other aggregate demand quantities. Probability and time-series methodologies that directly predict aggregate demand quantities based on archival data are commonly used in airline practice (e.g., demand for booking classes on a flight arrives according to a Poisson process, cancellations are binomially distributed, the no show rate for a flight is a weighted average of flight-level no show rates for the previous two months). In general, probability and time-series models are easier to implement than discrete choice models, but the former are limited because they do not capture or explain how individual airline passengers make decisions.

This section presents the "fundamentals" of discrete choice modeling. In particular, the four fundamental elements of choice theory are reviewed, followed by the general derivation of choice probabilities. Specific models are then presented, including the multinomial logit (MNL), nested logit (NL), generalized nested logit (GNL), mixed logit, and latent class models. This section draws heavily on Garrow (2010) and sections of that text have been reproduced with permission of Ashgate Publishing.

1.3.1 Four Fundamental Elements of the Choice Process

Discrete choice modeling theory is based on conceptual decision-making framework. Following the framework of Domencich and McFadden (1975), it is common to characterize the choice process by four elements: a decision-maker, the alternatives available to the decision-maker, attributes of these alternatives, and a decision rule. Each of these elements is discussed below.

1.3.1.1 Decision-maker

A decision-maker can represent an individual (e.g., an airline passenger), a group of individuals (e.g., a family traveling for leisure), a corporation (e.g., a travel agency), a government agency, etc. Identifying the appropriate decision-making unit of analysis may be a complex task. For example, airlines often offer discounts to large corporate customers. As part of the discount negotiation process, airline sales representatives assess the ability of the corporation to shift high-yield trips from competitors to their airline. On one hand, the corporation's total demand is the result of thousands of independent travel decisions made by its employees. Employee characteristics (e.g., their membership and level in airlines' loyalty programs) and preferences (e.g., their preferences for aircraft equipment types, departure times, etc.) will influence the choice of an airline. In this sense, the decision-making unit of analysis is the individual employee. However, employees must also comply with their corporation's travel policies. In this sense, the corporation is also a decision-maker because it influences the choice of an airline through establishing and enforcing travel policies. Thus, failure to consider the potential interactions between employee preferences and corporate travel policies may lead the sales representative to overestimate (in the case of weakly enforced travel policies) or underestimate (in the case of strongly enforced travel policies) the ability of the corporation to shift high-yield trips to a selected airline.

1.3.1.2 Alternatives

Each decision-maker is faced with a choice of selecting one alternative from a finite set of mutually exclusive and collectively exhaustive alternatives. Although alternatives may be discrete or continuous, the primary focus of this chapter is on describing methods applicable to selection of discrete alternatives. The finite set of all alternatives is defined as the *universal choice set*, C. However, individual n may select from only a subset of these alternatives, defined as the *choice set*, C_n. In an itinerary choice application, the universal choice set could be defined to include all reasonable itineraries in US markets that depart from cities in the eastern time zone and serve cities in the western time zone, whereas the choice set for an individual traveling from Boston to Portland would contain only the subset of itineraries between these two city pairs. In practice, the universal choice set is often defined to contain only reasonable alternatives. In itinerary choice applications, distance-based circuitry logic can be used to eliminate unreasonable itineraries and minimum and maximum connection times can be used to ensure that unrealistic connections are not allowed.

There are several subtle concepts related to the construction of the universal choice set. First, the assumptions that alternatives are mutually exclusive and collectively exhaustive are generally not restrictive. For example, assume there are two shops in an airport concourse, a dining establishment and a newsstand, and an airport manager is interested in knowing the probability an airline passenger will make a purchase at one or both of these stores. The choice set cannot be defined using simply two alternatives, as they are not mutually exclusive, i.e., the passenger can choose to shop in both stores. Mutual exclusivity can be obtained using three alternatives: "purchase only at dining establishment," "purchase only at newsstand," and "purchase both at dining establishment and newsstand." To make the choice set exhaustive, a fourth alternative representing customers who "do not purchase" can be included.

Also, the way in which the universal choice set is defined can lead to different interpretations. Consider a situation in which the analyst wants to predict the probability an individual will select one of five itineraries serving a market. The universal choice set is defined to contain these five itineraries, $C_1 \in \{I_1, I_2, I_3, I_4, I_5\}$, and a discrete choice model calibrated using *actual booking data* is used to predict the probability that one of these alternatives is selected. Compare this to a situation in which the analyst has augmented the universal choice set to include a no purchase option, $C_2 \in \{I_1, I_2, I_3, I_4, I_5, NP\}$, and calibrates the choice model using *booking requests* that are assumed to be independent. The first model will predict the probability an individual will select a particular itinerary given that the individual has decided to book an itinerary. The second will predict both the probability that an individual requesting itinerary information will purchase an itinerary, $1 - \Pr(NP)$, and, if so, which one will be purchased. The probability that itinerary one will be chosen out of all booking requests is given as $\Pr(I_1)$ and the probability that itinerary one will be chosen out of all bookings is $\Pr(I_1)/\{1 - \Pr(NP)\}$. This example demonstrates how different interpretations can arise from seemingly subtle changes in the universal choice set. It also illustrates how data availability can influence the construction of the universal choice set.

1.3.1.3 Attributes of the Alternatives

The third element in the choice process defined by Domencich and McFadden (1975) refers to attributes of the alternatives. Attributes are characteristics of the alternative that individuals consider during the choice process. Attributes can represent both deterministic and stochastic quantities. Scheduled flight time is deterministic whereas the variance associated with on-time performance is stochastic. In itinerary choice applications, attributes include schedule quality (non-stop, direct, single connection, double connection), connection time, departure and/or arrival times, aircraft type, airline, average fare, etc. In practice, the attributes used in scheduling, revenue management, pricing, and other applications that support day-to-day airline operations are derived from revealed preference data. Revealed preference data are based on the actual, observed behavior of passengers. By definition, revealed preference data reflect passenger behavior under existing or historical market conditions. Internal airline data rarely contain gender, age, income, marital status or other socio-demographic information—although with the recent Transportation Security Administration (TSA) requirement that age and gender be provided at the time of booking to enable watch list matching, some demographic information may become more common in the future (Federal Register 2008). Passenger information is generally limited to that collected to support operations. This includes information about the passenger's membership and status in the airline's loyalty program as well as any special service requests (e.g., wheelchair assistance, infant-in-arms, unaccompanied minor, special meal request).

Model enhancements arc often driven by the need to include additional attributes to support or evaluate new business processes. For example, prior to the use of code-shares, there was no need to distinguish between the marketing carrier who sold a ticket and the carrier who operated the flight, as these were the same carrier. In order to predict incremental revenue associated with an airline entering into different code-share agreements, it was necessary to model how itineraries marketed as code-shares differed from those marketed and flown by the operating carrier. When prioritizing model enhancements, a balance needs to be obtained between making models complex enough to capture factors essential for accurately supporting and evaluating different "what-if scenarios" while making these models simple enough to be understood by users and flexible enough to incorporate new attributes that were not envisioned when the model was first developed.

1.3.1.4 Decision Rule

The final element of the choice process is the decision rule. Numerous decision rules can be used to model rational behavior. The concept of utility is used in the context of discrete choice models. Utility is a scalar index of value that is a function of attributes and/or individual characteristics. In contrast to the other decision rules, utility represents the "value" an individual places on different attributes and captures how individuals make trade-offs among different attributes. Individuals are assumed to select the alternative that has the maximum utility. Alternative i is chosen if the utility individual n obtains from alternative i, U_{ni}, is greater than the utility for all other alternatives. Formally, alternative i is chosen *iff* $U_{ni} > U_{nj} \quad \forall j \neq i$. The utility for alternative i and individual n, U_{ni}, has an observed component, V_{ni}, and an unobserved component, commonly referred to as an "error term," ε_{ni}, but is more precisely referred to as the "stochastic term." Formally, $U_{ni} = V_{ni} + \varepsilon_{ni}$, where $V_{ni} = \beta' x_{ni}$. The observed component is often called the systematic or representative component of utility. The observed component is typically assumed to be a linear-in-parameters function of attributes that vary across individuals and alternatives (e.g. price, flight duration, gender). Note the assumption that $\boldsymbol{\beta}$ is linear-in-parameters *does not* imply that attributes like price must have linear relationships, i.e., x_{ni} can take on different functional forms, such as price, log (price), price^2; a linear-in-parameters assumption means that just the coefficient, β, associated with x_{ni} must be linear.

The error component is a random term that represents the unobserved and/or unknown (to the analyst) portion of the utility function. The distribution of random terms may be influenced by several factors, including measurement errors, omitting attributes from the utility function that are important to the choice process but that cannot be measured and/or are not known, incorrectly specifying the functional form of attributes that are included in the model (e.g., using a linear relationship when the “true” relationship is nonlinear), etc. There is an implicit relationship between the attributes included in the model and the distribution of error terms. That is, by including different attributes and/or by changing how attributes are included in the model, the distribution of error terms may change. Conceptually, this is similar in spirit to the situation where an analyst specifies a linear regression model and then examines the distribution of residual errors using visual plots and/or statistical tests to ensure that homoscedasticity and other assumptions embedded in the linear regression model are maintained. However, because choice models predict the probabilities associated with multiple, discrete outcomes, the ability to visually assess the appropriateness of error distribution assumptions is limited. Consequently, discrete choice modeling relies on statistical tests to identify violations in assumptions related to error distributions (e.g., see Train 2003: 53-4 for an extensive discussion and review of these tests). In addition, it is common to estimate different models (derived from different assumptions on the error terms) as part of the modeling process and assess which model fits the data the best.

1.3.2 Derivation of Choice Probabilities

The derivation of choice probabilities for discrete choice models uses the fact that individuals are assumed to select the alternative that has the maximum utility. Specifically, the utility associated with alternative i for individual n is given as $U_{ni} = V_{ni} + \varepsilon_{ni}$ and the probability the individual selects the alternative i from all J alternatives in the choice set C_n is given as:

$$\begin{aligned} P_{ni} &= P(U_{ni} \geq U_{nj} \quad \forall j \neq i) \\ &= P(V_{ni} + \varepsilon_{ni} \geq V_{nj} + \varepsilon_{nj} \quad \forall j \neq i) \\ &= P(\varepsilon_{nj} - \varepsilon_{ni} \leq V_{ni} - V_{nj} \quad \forall j \neq i) \\ &= P(\varepsilon_{nj} \leq V_{ni} - V_{nj} + \varepsilon_{ni} \quad \forall j \neq i) \\ &= \int_{\varepsilon_i=-\infty}^{+\infty} \int_{\varepsilon_j=-\infty\,\forall j\neq i}^{V_i-V_j+\varepsilon_i} f(\varepsilon)\mathrm{d}\varepsilon_j, \ldots, \mathrm{d}\varepsilon_{i+1}, \mathrm{d}\varepsilon_i. \end{aligned}$$

This derivation above is general in the sense that no assumptions have been made on the distribution of error terms; these assumptions are required in order to derive choice probabilities for specific models. Specific choice probabilities for different discrete choice models are obtained by imposing different assumptions on the distribution of these error terms. The assumption that unobserved error components are independently and identically distributed (iid) and follow a Gumbel distribution with mode zero and scale one, $\varepsilon \sim \text{iid } G(0, 1)$, results in the binary logit (in the case of two alternatives) or the multinomial logit model (in the case of more than two alternatives) (McFadden 1974). The assumption that error terms follow a multivariate normal distribution leads to the multinomial probit (MNP) model (Daganzo 1979). Unlike the MNL, the probit model is more general in that it allows flexible substitution patterns, correlation among unobserved factors, heteroscedasticity, and random taste variation. However, the choice probabilities can no longer be expressed analytically in closed-form and must be numerically evaluated. In practical terms, it has been difficult to use the probit in applications that require the numerical evaluation of more than approximately ten integrals.

Conceptually, MNL and MNP models can be loosely thought of as the end-points on a spectrum of discrete choice models. On one end is the MNL, a restrictive model that has a closed-form probability expression that is computationally simple. On the other end is the MNP, a flexible model that has a probability expression that must be numerically evaluated. Since the 1970s, advances in discrete choice models have generally focused on either relaxing the substitution restriction of the MNL while maintaining a closed-form expression for the choice probabilities (such as the NL and GNL model) or reducing the computational requirements of open-form models and further expanding the spectrum of open-form models to include more general formulations (such as the mixed logit model). The next sections review those models that have been applied to airline revenue management, scheduling, and distribution problems. However, it should be noted that dozens of discrete choice models exist and those reviewed herein reflect only a small percentage of those that have been developed since the 1970s. For a more thorough review of discrete choice models and how they have been applied to the airline industry, see Garrow (2010).

1.3.3 Multinomial Logit Models

The multinomial logit (MNL) model is used to describe how an individual chooses among multiple discrete alternatives. MNL probabilities are derived from the assumption that error terms are distributed Gumbel with mode zero and scale one (which implies a variance of $\pi^2/6$). The MNL probabilities are given as

$$P_{ni} = \frac{\exp(V_{ni})}{\sum_{j \in C_n} \exp\left(V_{nj}\right)}$$

where $V_{ni} = \beta' x_{ni}$.

From an estimation perspective, the problem of interest is to solve for the parameters $\boldsymbol{\beta}^*$ given a random sample of observations (extensions to other types of samples may require different estimators than those reviewed here). Estimators based on maximum likelihood estimation are most commonly used. Although other, more complex estimators can be used such as those based on the method of moments or the method of scores (e.g., see Train 2003), the focus of this discussion is on maximum likelihood estimators. Maximum likelihood estimation solves for the values of $\boldsymbol{\beta}$ that maximize the likelihood function:

$$L(\beta) = \prod_{n=1}^{N} \prod_{i \in C_n} P(i|x_{ni,}\beta)^{d_{ni}}$$

where

N	is the number of individuals in the random sample,
$i \in C_n$	are alternatives in the choice set C for individual n,
x_{ni}	is the vector of attributes associated with alternative i and individual n,
d_{ni}	is an indicator variable equal to 1 if individual n selects alternative i, and 0 otherwise,
$P(i\|x_{ni}, \beta)$	is the probability of selecting alternative i given a sample of attributes x_{ni} and estimates $\boldsymbol{\beta}$. Earlier discussions defined this probability as P_{ni}. When the conditional form is used in this section, it is used to emphasize the fact that the probability is dependent on characteristics of the sampling distribution related to attributes and estimates.

Computationally, it is easier to maximize the logarithm of the likelihood function, i.e., the log likelihood (LL) function, or:

$$\mathrm{LL} = \sum_{n=1}^{N} \sum_{i \in C_n} d_{ni} \ln P_{ni}.$$

The $\boldsymbol{\beta}$ parameter estimates are obtained by using optimization algorithms that maximize the log likelihood function. In the case of the binary logit and multinomial logit models, the log likelihood function is globally concave. This can be verified by examining its first and second derivatives with respect to $\boldsymbol{\beta}$. Given:

$$LL = \sum_{n=1}^{N} \sum_{i \in C_n} d_{ni} \ln \left(\frac{e^{V_{ni}}}{\sum_{j \in C_n} e^{V_{nj}}} \right)$$

$$LL = \sum_{n=1}^{N} \sum_{i \in C_n} d_{ni} \left(\beta' x_{ni} - \ln \sum_{j \in C_n} e^{V_{nj}} \right),$$

the derivative of the log likelihood function with respect to the kth parameter is given as:

$$\frac{\partial \mathrm{LL}}{\partial \beta_k} = \sum_{n=1}^{N} \sum_{i \in C_n} d_{ni} \left(x_{nik} - \frac{\sum_{j \in C_n} e^{\beta' x_{nj}} \cdot x_{njk}}{\sum_{j \in C_n} e^{\beta' x_{nj}}} \right), \quad k = 1, \ldots, K.$$

Noting that d_{ni} is an indicator variable equal to 1 if individual n selects alternative i, and 0 otherwise, gives:

$$\frac{\partial \mathrm{LL}}{\partial \beta_k} = \sum_{n=1}^{N} \sum_{i \in C_n} d_{ni} \left(x_{nik} - \frac{e^{\beta' x_{ni}} \cdot x_{nik}}{\sum_{j \in C_n} e^{\beta' x_{nj}}} \right), \quad k = 1, \ldots, K$$

$$\frac{\partial \mathrm{LL}}{\partial \beta_k} = \sum_{n=1}^{N} \sum_{i \in C_n} (d_{ni} - P_{ni}) \cdot x_{nik}, \quad k = 1, \ldots, K.$$

For reference, the second derivatives of the log likelihood function are:

$$\frac{\partial \mathrm{LL}}{\partial \beta_k \partial \beta_l} = -\sum_{n=1}^{N} \sum_{i \in C_n} P_{ni} \cdot \left[x_{nik} - \sum_{j \in C_n} x_{njk} \cdot P_{nj} \right] \cdot \left[x_{nil} - \sum_{j \in C_n} x_{njl} \cdot P_{nj} \right].$$

The maximum of the log likelihood function is obtained when $\partial \mathrm{LL}/\partial \beta = 0$. Further, since the second derivative is negative semi-definite, the log likelihood function is globally concave (which implies there is one unique solution for $\boldsymbol{\beta}$ that maximizes the log likelihood function).

Although the log likelihood function for the binary logit and MNL is globally concave, the same is not true for more complex models, such as the nested logit and mixed logit models discussed below. Solution of these models requires non-linear optimization methods. Three of the most popular algorithms include the Newton–Rhapson method, BFGS, and BHHH. The BFGS algorithm is named after Broyden, Fletcher, Goldfarb, and Shanno and the BHHH algorithm is named after Berndt, Hall, Hall, and Hausman. Additional information on these algorithms can be found in Ruud (2000), Dennis and Schnabel (1996), and Nocedal and Wright (1999).

Among all of the different discrete choice models, MNL models have been most frequently used in practice. Applications covered in this text include itinerary choice selection for online low fare search results, determination of optimal dynamic markups for opaque airline products, and content management (Chap. 4), itinerary choice selection for long- and mid-term schedule planning (Chap. 2), and recapture rate modeling and choice-based revenue management (Chap. 3). Additional applications of MNL models to airline applications include no show modeling (Garrow and Koppelman 2004a, b). Thus, although the MNL is the most restrictive MNL model (in terms of the assumptions it imposes on the interpretation of $\boldsymbol{\beta}$ parameters and substitution patterns among alternatives), it is also the most commonly used model due to the simplicity of choice probabilities and the ease with which alternatives can be added or removed from choice sets. To clarify, the MNL is limited in the sense that: (1) it does not incorporate random taste variation (i.e., the $\boldsymbol{\beta}$ parameters represent the average preference of individuals in the sample vs. a distribution of individuals' preferences); and, (2) it exhibits the independence of irrelevant alternatives (IIA) property. The IIA property states that the ratio of choice probabilities P_{ni}/P_{nj} for $i, j \in C_n$ is independent of the attributes of any other alternative. In terms of substitution patterns, this means a change or improvement in the utility of one alternative will draw share proportionately from all other alternatives. In many applications, this may not be a realistic assumption.

For example, in itinerary choice model applications, one may expect the 10 a.m. departure to compete more with flights departing close to 10 a.m. The nested logit model, discussed below, partially helps to relax the IIA assumption.

1.3.4 Nested Logit Models

The nested logit (NL) model (Williams 1977; McFadden 1978), is a discrete choice model that relaxes the assumption that errors are independently distributed (and, correspondingly, relaxes the IIA property) by grouping alternatives into M nests, i.e., $i \in A_m$ $m = 1,2,\ldots, M$. An alternative belongs to one and only one nest. The NL utility function can be expressed as follows (suppressing the index for individual n for notational convenience):

$$U_{im} = V_i + \varepsilon_m + \varepsilon_i.$$

Under the assumptions that $(\varepsilon_i + \varepsilon_m)$ is identically distributed $G(0,\gamma)$ and that ε_i is distributed $G(0,\gamma)/\mu_m$, $i \in A_m$, $m = 1,2,\ldots, M$, the probability that individual n selects alternative i is given as:

$$P_i = \frac{e^{V_i/\mu_m}\left(\sum_{j\in A_m} e^{V_j/\mu_m}\right)^{\mu_m - 1}}{\sum_{l=1}^{M}\left(\sum_{j\in A_l} e^{V_j/\mu_l}\right)^{\mu_l}}, \quad 0 < \mu_m \leq 1.$$

A more intuitive expression for the NL choice probability can be derived as the product of a conditional and marginal probability (this derivation is provided in Train 2003, p. 90). This formulation is particularly helpful when extending NL models to include additional levels of nests.

$$P_i = P_{i|m} \times P_m = \frac{e^{\left(\frac{V_i}{\mu_m}\right)}}{\sum_{j_m} e^{\left(\frac{V_j}{\mu_m}\right)}} \times \frac{e^{V_m + \mu_m \Gamma_m}}{\sum_{l=1}^{M} e^{V_l + \mu_l \Gamma_l}}, \quad \Gamma_m = \ln\left(\sum_{j\in A_m} e^{\left(\frac{V_j}{\mu_m}\right)}\right),$$
$$0 < \mu_m \leq 1.$$

The first component of the product is the probability of selecting alternative i among all j alternatives in nest m, conditional on the choice of m, and the second product is the probability of selecting nest m among all nests. Γ_m is often called the "log-sum term" because it is the log of a sum (this terminology should not to be confused with μ_m, the "logsum" or "logsum parameter"). The logsum parameter, μ_m, is a measure of the degree of correlation and substitution among alternatives in nest m. Higher values of μ_m imply less, and lower values imply more, correlation among alternatives in the nest. In turn, higher correlation leads to greater competition effects among alternatives in the nest.

From an interpretation perspective, the incorporation of positive covariance among alternatives that share a common nest also leads to increased substitution

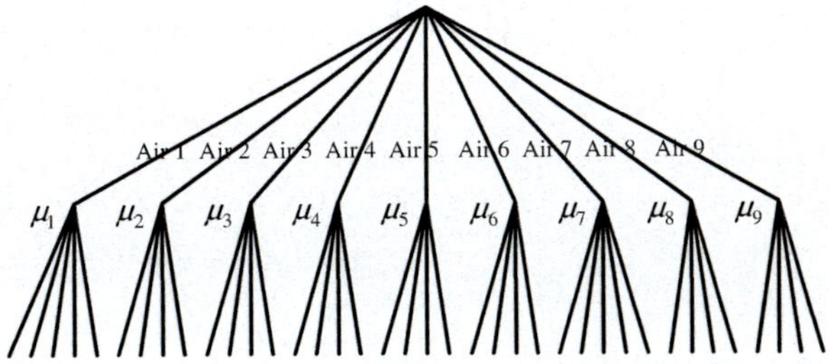

Fig. 1.7 Two-level NL carrier model structure

among these alternatives. In forecasting applications, this means that an improvement in an alternative will draw proportionately more share from alternatives that belong to the same nest than from alternatives that belong to different nests. As an example, consider an itinerary choice application in which nests are created that group itineraries by operating carrier as shown in Fig. 1.7. If Airline 3 improves one of its itineraries (e.g., by changing the schedule so that it operates at a time that is more desirable to passengers), the NL model will predict that the increased share associated with this itinerary is due to drawing proportionately more passengers from existing flights operated by Airline 3 than from those of its competitors. This is the main advantage associated with using a NL model over a MNL model. However, in this example, one can easily hypothesize that there may be different ways to group alternatives (or represent the underlying competition among itineraries). These alternative groupings may be by time of day, level of service (i.e., non-stop vs. connecting), fare class (if available), etc. The generalized nested logit model helps overcome this limitation by allowing alternatives to be allocated to more than one nest, i.e., competition among alternatives may occur along multiple dimensions. For this reason, the NL and GNL models have primarily been used to date in itinerary choice applications, where multiple dimensions of competition among alternatives "naturally" occurs (e.g., see Coldren 2005; Coldren and Koppelman 2005a, b).

1.3.5 Generalized Nested Logit Models

The MNL and NL models are special cases of the generalized nested logit (GNL) model (Wen and Koppelman 2001). The GNL is more "general" in the sense that its nesting structures are not restrictive, i.e., alternatives may be allocated to multiple nests. An example of a GNL model is shown in Fig. 1.8 for two train alternatives (one economy and one premium) and two air alternatives (one economy and one premium). The GNL model in Fig. 1.8 contains four nests. The first and second nests incorporate increased competition among the train and air alternatives, respectively; the third and fourth nests incorporate increased competition among the economy and premium alternatives, respectively.

Fig. 1.8 Generalized nested logit model

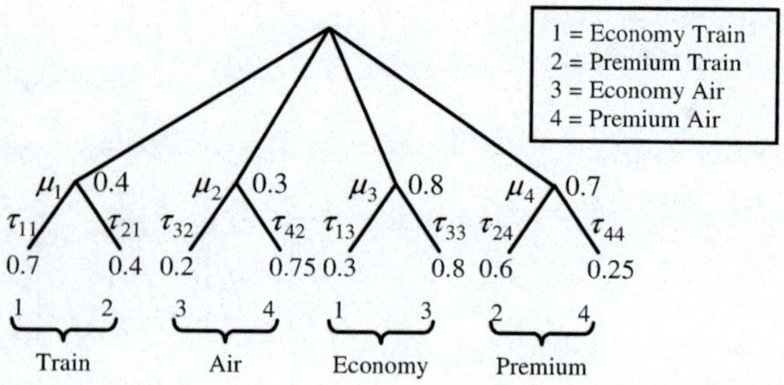

The GNL probability is given as:

$$P_i = \sum_m P_{i|m} \times P_m$$

$$= \sum_m \left[\frac{(\tau_{im} e^{V_i})^{\frac{1}{\mu_m}}}{\sum_{j \in A_m} (\tau_{jm} e^{V_j})^{\frac{1}{\mu_m}}} \times \frac{\left(\sum_{j \in A_m} (\tau_{jm} e^{V_j})^{\frac{1}{\mu_m}}\right)^{\mu_m}}{\sum_l \left(\sum_{j \in A_l} (\tau_{jl} e^{V_j})^{\frac{1}{\mu_l}}\right)^{\mu_l}} \right],$$

$$0 < \mu_m, \quad \mu_l \leq 1, \quad \sum_m \tau_{im} = 1 \, \forall i.$$

where

- $j \in A_m$ is the set of all alternatives that belong to nest m,
- m, l are indices used to sum over all nests,
- τ_{im} are unknown allocation parameters that characterize the portion of alternative i assigned to a nest. Allocation parameters are non-negative, i.e., $\tau_{im} \geq 0$ and must sum to one for every alternative, which is equivalent to $\sum_m \tau_{im} = 1, \forall i$
- μ_m is the logsum coefficient associated with nest m, $0 < \mu_m \leq 1, \forall m$.

Note that in the GNL model, both allocation and logsum parameters are estimated and alternatives may be allocated to one or more nests (the only condition is that the allocations associated with alternative i across all nests sum to one). For example, in Fig. 1.8, 70% of alternative one (representing economy train) is allocated to the nest one (the train nest) and 30% is allocated to nest three (the economy nest), or $\tau_{11} + \tau_{13} = 1$. Similarly, 40% of alternative two (representing premium train) is allocated to nest one (the train nest) and 60% is allocated to nest four (the premium nest), or $\tau_{21} + \tau_{24} = 1$.

From an interpretation perspective, both allocation and logsum parameters provide information on the amount of competition among alternatives. In Fig. 1.8, those nests with the smaller logsum parameters exhibit greater substitution, e.g., economy and premium train classes will compete more with each other ($\mu_1 = 0.4$) than economy train and economy air ($\mu_3 = 0.8$). Conceptually, larger allocations also lead to higher substitutions, e.g., 75% of alternative four, representing premium air, is allocated to the air competition nest, implying higher substitution between premium air and economy air ($\tau_{42} = 0.75$) than premium air and premium train ($\tau_{44} = 0.25$). Of course, as seen in GNL probability equation, the exact

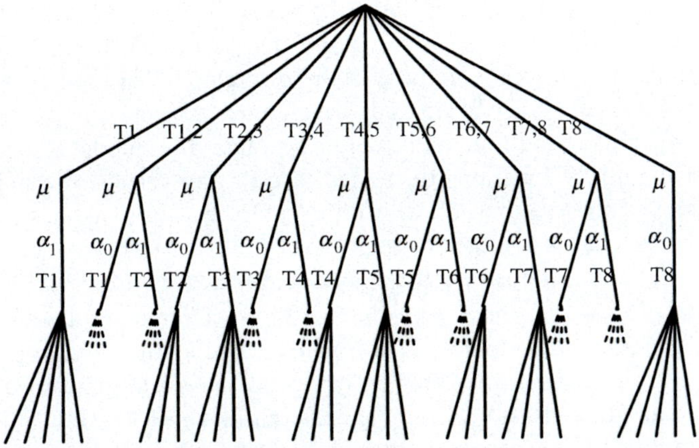

Note: T1=5-6:59 AM; T2=7-8:59 AM; T3=9-10:59 AM; T4=11-12:59PM; T5=1-2:59PM; T6=3-4:59PM; T7=5-6:59PM; T8=7-11:59PM

Fig. 1.9 OGEV model structure

value for the amount of competition or substitution between two alternatives is ultimately given as a function of allocation and logsum parameters, which may not be straightforward to calculate.

Figure 1.9 illustrates one of the "simplest" GNL models that has been used in the context of itinerary choice. Specifically, an ordered generalized extreme value (OGEV) model (Small 1987) is shown that allocates alternatives to multiple nests based on time periods. In this sense, the alternatives are first "ordered" by time and the nests are created using predefined time periods. In this OGEV model, there is an implicit assumption that itineraries in the same period compete most strongly, followed by itineraries in adjacent periods. For itineraries outside of adjacent periods, a MNL-level of competition applies. Note as well that in contrast to the general GNL formulation, all logsum coefficients are constrained to be equal (and thus the distinct name for this model). While theoretically, the logsums could be allowed to vary, in a practical sense it is common to constrain or impose relationships among logsums (and/or allocation parameters) to prevent empirical identification problems, i.e., there may not be a sufficient number of observations in the dataset that were chosen to enable estimation of all parameters.

To date, the only applications of GNL models the author is aware of have occurred in the context of itinerary choice models.

1.3.6 Mixed Logit Models

The MNL, NL, and GNL models discussed above all assume that parameters of the utility functions are deterministic, i.e., they represent the "average" value that individuals associate with an attribute. The mixed logit model, which first appeared in the early 1980s (e.g., see Boyd and Mellman 1980; Cardell and Dunbar 1980) and became predominate in the discrete choice modeling community in the late 1990s (e.g., see Train 2003) overcomes this and other assumptions. Conceptually, the mixed logit model is identical to the MNL model except that the parameters of the utility functions for mixed models can vary across individuals, alternatives, and/or observations. However, this added flexibility comes at a cost—choice probabilities can no longer be expressed in closed-form. Under a random parameters formulation, the utility that individual n obtains from alternative i is given as $U_{ni} = \beta' x_{ni} + \varepsilon_{ni}$ where $\boldsymbol{\beta}$ is the vector of parameters associated with attributes x_{ni}, and ε_{ni} is a random error component. Unlike the MNL model, the $\boldsymbol{\beta}$ parameters are no longer fixed values that represent "average" population values, but rather are random realizations from the density function $f(\boldsymbol{\beta})$. Thus, mixed logit choice probabilities are expressed as the integral of logit probabilities evaluated over the density of distribution parameters, or

$$P_{ni} = \int L_{ni}(\beta) f(\beta|\eta) \mathrm{d}\beta \tag{1.1}$$

where

P_{ni} is the probability individual n chooses alternative i,

$L_{ni}(\boldsymbol{\beta})$ is a logit probability evaluated at the vector of parameter estimates $\boldsymbol{\beta}$ that are random realizations from the density function $f(\boldsymbol{\beta})$,

$\boldsymbol{\eta}$ is a vector of parameter estimates associated with the density function $f(\boldsymbol{\beta})$.

In a mixed model, L_{ni} takes the MNL form. For example, for a particular realization of $\boldsymbol{\beta}$, the mixed MNL logit probability is:

$$L_{ni}(\beta) = \frac{\exp(V_{ni})}{\sum_{j \in C_n} \exp(V_{nj})}$$

where

C_n is the set of alternatives available in the choice set for individual n.

The problem of interest is to solve for the vector of distribution parameters $\boldsymbol{\eta}$ associated with the $\boldsymbol{\beta}$ coefficients given a random sample of observations from the population. Distinct from the formulation of the MNL, NL, and GNL models, some or all of the $\boldsymbol{\beta}$ coefficients are assumed to vary in an unspecified, therefore "random," pattern.

From a modeling perspective, the analyst begins with the assumption that individuals' "preferences" for an attribute, say cost, follow a specific distribution, in this case a normal. In contrast to the MNL and other discrete choice models discussed thus far, the use of a distribution allows the analyst to investigate the hypothesis that some individuals (facing the same product choices in the market and/or exhibiting similar socio-demographic characteristics) are more price-conscious than other individuals. That is, whereas the MNL and other discrete choice models belonging to the GNL families capture the *average* price sensitivity across the population or clearly defined market segment, the mixed MNL provides information on the *distribution* of individuals' price sensitivities.

Formally, maximum likelihood estimators can be used to solve simultaneously for the fixed $\boldsymbol{\beta}$ coefficients and distribution parameters $\boldsymbol{\eta}$ associated with the random $\boldsymbol{\beta}$ coefficients. Because the integral in Eq. 1.1 cannot be evaluated analytically, numerical approximation is used to maximize the simulated maximum likelihood function. The average probability that individual n selects alternative i is calculated by noting that for a particular realization of $\boldsymbol{\beta}$, the logit probability is known. Formally, the average simulated probability is given as:

$$\hat{P}_{ni} = \hat{P}(i|x_{ni}, \beta, \eta) = \frac{1}{R}\sum_{r=1}^{R} L_{ni}(\beta_r)$$

where

R is the number of draws or support points used to evaluate the integral,

$\hat{P}_{ni}$ is the average probability that individual n selects alternative i given attributes x_{ni} and parameter estimates $\boldsymbol{\beta}$, which are random realizations of a density function. The parameters of this density function are given by $\boldsymbol{\eta}$,

$\boldsymbol{\beta}_r$ is the vector of parameter estimates associated with draw or support point r.

The corresponding simulated likelihood (SL) and simulated log likelihood (SLL) functions are:

$$SL(\beta) = \prod_{n=1}^{N}\prod_{i \in C_n} \hat{P}(i|x_{ni}, \beta, \eta)^{d_{ni}}$$

$$SLL(\beta) = \sum_{n=1}^{N}\sum_{i \in C_n} d_{ni} \ln \hat{P}(i|x_{ni}, \beta, \eta)$$

where

d_{ni} is an indicator variable equal to 1 if individual n selects alternative i and 0 otherwise.

To date, there have been very few applications that have used mixed logit models in applications that have used airline industry data. One study is the work of Garrow (2004) that estimated mixed logit models for no show applications; however, the model fit associated with the mixed logit models were similar to those obtained from MNL and NL models. With one notable exception covered in Sect. 1.3.7, all other applications this author is aware of have been isolated to the discrete choice modeling community using stated preference (or survey-based) data, often in the context of multi-airport choice problems and/or access mode problems. In the context of airline applications, it remains an open research question as to whether mixed logit models that incorporate distributions to represent customer preferences will outperform MNL, NL, and other discrete choice models that have closed-form probability expressions. Part of the difference is due to the lack of socio-demographic information and "variability" in consumers in the air transport market as compared to other markets (e.g., public transport serves both choice and captive riders). Moreover, in the context of implementation, the use of mixed logit models will pose additional challenges due to the need to approximate integrals (which may increase implementation times on the order of one or two magnitudes, i.e., implementation and estimation times tend to be linearly related to the number of support points used to approximate these integrals; the number of support points required for stable numerical approximation is highly application- and data-specific). Unless a substantial improvement in model fit and forecasting accuracy is observed with a mixed logit model, the added complexity of using mixed logit models in practice will likely result in limited implementations. One exception, though, could lie in the use of latent class models to better represent customer segmentations.

1.3.7 Relationship Between Mixed Logit and Latent Class Models

Latent class models are a special case of mixed logit models. Both mixed logit models and latent class models allow the $\boldsymbol{\beta}$ coefficients to vary across individuals. However, in the mixed logit formulation, $\boldsymbol{\beta}$ was assumed to vary according to a known (continuous) density function $f(\boldsymbol{\beta})$ across the population. In contrast, latent class model allows $\boldsymbol{\beta}$ to vary across a discrete number of customer segments. The problem of interest for a latent class model involves determining how many segments are appropriate, the probability a customer belongs to a particular segment, and the $\boldsymbol{\beta}_s$ parameter estimates (which are estimated separately for each segment). In a latent class model, the probability alternative i is chosen by individual n is given as:

$$P(i|X_n) = \sum_{s=1}^{S} P(i|X_n, s) \times P(s|X_n)$$

where

$$P(i|X_n, s) = \frac{\beta_s' X_{ni}}{\sum_{s=1}^{S} \beta_s' X_{nj}}, \quad j \in C_n$$

is simply a MNL model (determining the probability that alternative i will be selected from all j alternatives in the universal choice set C_n) conditioned on the probability that individual n belongs to segment s. $P(s|X_n)$ is this conditional probability and is often referred to as the "class membership" model. Different models are often used to forecast class membership. For example, Wen and Lai

(2010) use a MNL model based on socioeconomic and trip characteristics (namely personal income, gender, age, group size, and trip purpose) to forecast class membership to predict carrier preference among airline passengers who have flown from Taipei to Tokyo or from Taipei to Hong Kong. Teichert et al. (2008) also use a class membership function based on how business customers ranked the importance of different product and service quality attributes (e.g., flight frequency, price, ticketing flexibility, ability to accrue miles, on-time punctuality, in-flight meal and beverage services, easy check-in services). These two examples highlight the more traditional use of latent class models, namely their use to support longer-term marketing decisions. That is, these examples are able to incorporate socioeconomic information and customer rankings/ratings data because they do not need to forecast these variables, i.e., these models are based on a stated preference survey that was designed to support marketing and product design objectives.

However, there is one noteworthy example of an application of latent class models that has been published in the context of revenue management. Carrier (2008) used information that could be obtained from booking records to predict airline itinerary and fare product choices. The variables he used were frequent flyer membership, outbound day of the week, and distribution channel. In this context, the use of a latent class model may be viable in an OR implementation setting as it would likely involve forecasting only one new variable: frequent flyer membership; the other two variables are typically included in most revenue management systems. This example highlights two factors associated with the viability of using latent class models in airline applications: (1) latent class models differ from traditional segmentation models in that they probabilistically assign individuals to different class memberships; (2) the ability to use latent class models for revenue management forecasting applications is contingent on the ability to forecast the variables used to assign individuals to these classes.

1.3.8 Summing it All Up: Which Discrete Choice Model to Use?

Given the dominate adoption of MNL models in airline applications, one question naturally arises: Are there substantial advantages associated with using advanced discrete choice models such as the NL, GNL, or mixed logit models? That is, although the NL and GNL models incorporate more realistic substitution patterns, their probability expressions are more complex and require additional information of which alternatives belong to which nest. Similarly, while the mixed logit model can incorporate random taste variation, its probability expression requires numerical approximation techniques. Thus, there is a tradeoff between using the "simple" MNL model and more complex NL, GNL, or mixed logit models.

In the context of itinerary choice models, the MNL model has been seen to offer dramatic improvements in model fit over QSI-based models at a major US airline

(Coldren et al. 2003); forecasting gains from using the more complex models are likely to be less (but have not yet been examined). Further, it is this author's opinion that in the context of itinerary choice models, using a MNL model and combining data from multiple sources (namely booking data and online shopping data) will lead to larger improvements in forecasting accuracy than developing more complex discrete choice models. This is because the on-line shopping data contain information on actual (and "disaggregate") prices and itineraries shown to the customers, not quarterly fare information. More broadly, among the discrete choice models reviewed here, the latent choice models may offer the largest promise for improvements in forecasting accuracy across a range of airline applications, particularly if individuals' class membership can be inferred from customers' search and/or purchase characteristics.

1.4 Future Research Directions

There are many future research directions related to the application of discrete choice models to airline passenger behavior. From the perspective of modeling airline customer behavior, the internet provides new opportunities to examine customers' search and purchase behaviors and investigate how these behaviors are influenced by menu designs, product displays, and product attributes such as discount levels. For example, using a dataset in which customers were offered discounts in exchange for accepting uncertainty in their travel dates, Lee et al. (2010) found that higher discounts result in greater visit search depths as well as nonlinear increases in revisit and purchase probabilities. In the context of evaluating promotion effectiveness, they found it was important to consider the timing of repeat visits, as failure to account for revisit activities lead to dramatic underestimation of the value of a promotion, as purchasing probabilities were three times higher for revisit (or latent demand) visitors versus first-time visitors. The example illustrates how firms can influence customer behavior and, more importantly, how different evaluation metrics can be unknowingly impacted by carriers' decisions.

A second key area of research relates to estimation of parameters for choice-based revenue management, and specifically development of methods to jointly estimate market share and market size. From the perspective of discrete choice modeling, this is a particularly unique problem as it involves censored data, i.e., only data for a single airline is known. Most applications of discrete choice models are based solely on observed data. Further research on data sampling methods that may be applicable to this censored data problem are needed.

Another area of research relates to evaluation of new product designs—both from the customer and systems perspectives. That is, ancillary revenue sources may be indirectly impacting customer choice in ways that decrease the forecasting accuracy of current decision-support systems. One example relates to seat fee reservations. By charging seat fees for extra legroom seats, some carriers may unknowingly be influencing customer preferences for certain flights (which impacts demand forecasting and recapture rate estimates). As an example, customers on a long-haul domestic flight may want an exit row seat, and select their flight from the subset of flights that have these particular seats available. Add to this the complication that some airlines currently make it very easy for customers to see their seat availabilities (e.g., by providing links next to the displayed prices), whereas other airlines only show their seat maps for a specific flight only after the customer selects a flight and provides her name, address, and other required booking information. Understanding the linkages among these multiple factors, and the downstream impacts they have on demand forecasts and recapture rates would be another area of research that would be interesting to investigate.

While the ultimate influence of ancillary revenues and other factors on customer behavior is unknown, one thing is certain—that the airline industry is very dynamic. A decade from now, it is almost certain that researchers will be challenged to solve revenue management, pricing, scheduling, marketing, and operations problems that need to incorporate factors that were not anticipated today. In this context, developing demand forecasting models that incorporate important elements of customer behavior while simultaneously enabling these decision-support systems to identify and respond to unanticipated market changes is an important research priority.

Acknowledgments The author wishes to express her gratitude to Georgia Tech students Susan Hotle and Stacey Mumbower who helped compile many of the statistics reported in this chapter.

References

Aaron K (2007) SITA survey: choice is key to encouraging self-service. Air Transp World 76–80

Air Transport Association (2010) Prices of air travel versus other goods and services. www.airlines.org/Economics/DataAnalysis/Pages/PriceofAirTravel-VersusOtherGoodsandServices.aspx Accessed on May 17, 2010

AirTran Holdings Inc (2009) Annual report to shareholders. http://investor.airtran.com/phoenix.zhtml?c=64267&p=irol-reportsAnnual Accessed on June 24, 2010

Ben-Yosef E (2005) The evolution of the US Airline Industry: theory, strategy, and policy. Springer, Dordrecht

Besbes O, Zeevi A (2006) Implications of choice model parameter estimation in a competitive environment. Working Paper, Stern School of Business, New York University

Bodea TD, Ferguson MA, Garrow LA (2009) Choice-based revenue management: data from a major hotel chain. Manuf Serv Oper Manag 11(2):356–361

Boyd J, Mellman J (1980) The effect of fuel economy standards on the U.S. automotive market: A hedonic demand analysis. Transp Res Part A 14:367–378

Bront JJM, Mendez-Diaz I, Vulcano G (2007) A column generation algorithm for choice-based network revenue management. Working Paper, Stern School of Business, New York University

Brunger WG, Perelli S (2008) The impact of the internet on airline fares: customer perspectives on the transition to internet distribution. J Revenue Pricing Manag 8(2/3):187–199

Cardell S, Dunbar F (1980) Measuring the societal impacts of automobile downsizing. Transp Res Part A 14:423–434

Carrier E (2008) Modeling the choice of an airline itinerary and fare product using booking and seat availability data. Ph.D. dissertation, Department of Civil and Environmental Engineering, Massachusetts Institute of Technology, Cambridge, MA

Civil Aeronautics Board (1970) Effect on total market traffic of changes in quality of service (QSI). Docket 21136 (box 1196). Exhibit BOR-R-300. In: Adley JF, Caridi CJ (eds) Exhibit Series Rebuttal Exhibits of the Bureau of Operating Rights

Coldren GM, Koppelman FS (2005a) Modeling the competition among air-travel itinerary shares: GEV model development. Transp Res Part A 39(4):345–365

Coldren GM, Koppelman FS (2005b) Modeling the proximate covariance property of air travel itineraries along the time-of-day dimension. Transp Res Rec 1915:112–123

Coldren GM, Koppelman FS, Kasturirangan K, Mukherjee A (2003) Modeling aggregate air-travel itinerary shares: Logit model development at a major U.S. airline. J Air Transp Manag 9(6):361–369

Coldren GM (2005) Modeling the competitive dynamic among air-travel itineraries with generalized extreme value models. Ph.D. dissertation, Department of Civil and Environmental Engineering, Northwestern University, Evanston, IL

Continental Airlines (2010). Barclays Capital High Yield Bond & Syndicated Loan Conference, March 26, 2010. http://www.continental.com/web/en-us/content/company/investor/docs/continental_presentation_barclays_20100326.pdf Accessed on June 25, 2010

Daganzo C (1979) Multinomial probit: the theory and its applications to demand forecasting. Academic Press, New York

Dennis JE, Schnabel RB (1996) Numerical methods for unconstrained optimization and nonlinear equations. Society for Industrial and Applied Mathematics, Philadelphia

Domencich TA, McFadden D (1975) Urban travel demand–a behavioral analysis. New York, Elsevier

Dupuis C, Gamache M, Page J-F (2010). Logical analysis of data for estimating passenger show rates in the airline industry. Working paper, École Polytechnique de Montréal. www.agifors.org/award/submissions2010 on June 30, 2010

Gallego G, Sahin O (2006) Inter-temporal valuations, product design and revenue management. Working Paper, Department of Industrial Engineering and Operations Research, Columbia University, New York

Garrow LA (2004) Comparison of choice models representing correlation and random taste variation: an application to airline passengers' rescheduling behavior. Ph.D. dissertation, Department of Civil and Environmental Engineering, Northwestern University, Evanston, IL

Garrow LA (2010) Discrete Choice Modelling and Air Travel Demand: Theory and Applications. Ashgate Publishing, Aldershot

Garrow LA, Koppelman FS (2004a) Multinomial and nested logit models of airline passengers' no show and standby behavior. J Revenue Pricing Manag 3(3):237–253

Garrow LA, Koppelman FS (2004b) Predicting air travelers' no show and standby behavior using passenger and directional itinerary information. J Air Transp Manag 10:401–411

Garrow LA, Jones SP, Parker RA (2007) How much airline customers are willing to pay: an analysis of price sensitivity in online distribution channels. J Revenue Pricing Manag 6(1): 1–20

Garrow LA, Hotle S, Mumbower SM (2010) Assessment of product de-bundling trends in the US airline industry: implications for customer service and the Airport and Airways Trust Fund. Working paper, Georgia Institute of Technology

Gorin T, Brunger WG, White M (2006) No show forecasting: a blended cost-based PNR-adjusted approach. J Revenue Pricing Manag 5(3):188–206

Harteveldt HH et al (2004) Why leisure travelers book at their favorite sites. Forrester Res: Trends

Hu M, Gallego G (2007) Dynamic pricing of perishable assets under competition. Department of Industrial Engineering and Operations Research, Columbia University, New York

INFORMS (2010) Edelman winners and finalists. www.scienceofbetter.org/Edelman/winners.htm Accessed on January 16, 2011
Kalka K, Weber K (2000) PNR-based No show Forecast. Presentation to the AGIFORS Reservation and Yield Management Study Group, New York
Lee M, Ferguson ME, Garrow LA, Post D (2010) The impact of leisure travelers' online search and purchase behaviors on promotion effectiveness. Working Paper, Georgia Institute of Technology
McFadden D (1974) Conditional logit analysis of qualitative choice behavior. In: Zarembka P (ed) Frontiers in econometrics. Academic Press, New York, pp 105–142
McFadden D (1978) Modeling the choice of residential location. In: Karlqvist A et al (eds) Spatial interaction theory and residential location. North-Holland, Amsterdam, pp 75–96
Newman JP, Ferguson ME, Garrow LA (2010) Estimation of choice-based models using sales data from a single firm. Working paper, Georgia Institute of Technology
Nocedal J, Wright SJ (1999) Numerical optimization. Springer, New York
Pastor J (2000) What exactly is data mining. Presentation to the AGIFORS Reservations and Yield Management Study Group, New York
PhoCusWright (2004) The PhoCusWright consumer travel trends survey. 2004. Downloaded from http://store.phocuswright.com/phcotrtrsusi.html
Polt S (2002) Revenue management tutorial. Presentation to the AGIFORS Reservations and Yield Management Study Group Meeting, Berlin
Ruud PA (2000) An introduction to classical econometric theory. Oxford University Press, Oxford
Small KA (1987) A discrete choice model for ordered alternatives. Econometrica 55(2):409–457
Southwest Airlines (2009) Southwest Airlines 2009 Filing 10-K, Part 2, Item 6
Southwest Airlines (2009) Southwest Airlines 2009 One Report. Accessed http://216.139.227.101/interactive/luv2009/luv2009ar.pdf on June 23, 2010
Talluri KT, van Ryzin G (2004) Revenue management under a general discrete choice model of consumer behavior. Manag Sci 50(1):15–33
Train KE (2003) Discrete choice methods with simulation. University Press, Cambridge
US Department of Transportation Research Innovative Technology Administration Bureau of Transportation Statistics (2010). 2010 Air Carrier Financial: Schedule P-12 (from 2000–2009). www.transtats.bts.gov on May 16, 2010
van Ryzin G, Liu Q (2004) On the choice-based linear programming model for network revenue management. Decision, Risk & Operations, Working Paper Series DRO-2004-04, Columbia Business School, New York
van Ryzin G, Vulcano G (2008a) Computing virtual nesting controls for network revenue management under customer choice behavior. Manuf Serv Oper Manag 10(3):448–467
van Ryzin G, Vulcano G (2008b) Simulation-based optimization of virtual nesting controls for network revenue management. Oper Res 56(4):865–880
Vulcano G, van Ryzin G, Chaar W (2006) Choice-based revenue management: An empirical study of estimation and optimization. Presentation to the Annual Meeting of INFORMS, Pittsburgh
Wen C-H, Koppelman F (2001) The generalized nested logit model. Transp Res Part B 35(7):627–641
Wen C-H, Lai S-C (2010) Latent class models of international air carrier choice. Transp Res Part E 46(2):211–221
Williams H (1977) On the formulation of travel demand models and economic evaluation measures of user benefits. Environ Plan A 9:285–344
Teichert T, Shehy E, von Wartburg I Customer segmentation revisited: The case of the airline industry. Transp Res Part A 42(1):227–242
Zhang D, Cooper W (2005) Revenue management for parallel flights with customer-choice behavior. Oper Res 53(3):415–431

Chapter 2
Airline Planning and Schedule Development

Timothy L. Jacobs, Laurie A. Garrow, Manoj Lohatepanont, Frank S. Koppelman, Gregory M. Coldren, and Hadi Purnomo

2.1 Introduction and Scope

Airlines have evolved over the past 70 years from simple contract mail carriers into sophisticated businesses. The current airline environment is very competitive and dynamic. Maintaining consistent profitability requires that appropriate trade-offs be made between the often competing objectives within planning, marketing and operations. Airlines have led other industries in the application of operations research and information technology to deal with these issues. The real-time solution of large-scale optimization models has played a significant role in shaping today's airline industry. This role will increase as the industry becomes more competitive and flight characteristics change due to the implementation of new technologies. Airline planning and scheduling represents an excellent example of the application of operations research and mathematical modeling to solve complex and real industry problems.

2.2 Overview of Airline Schedule Planning and Marketing

Planning and Marketing define an airline's products and determine how they will be sold. This is a continuous process which begins 5 or more years before a flight's departure and operates until the last passenger is boarded and the aircraft door is

T. L. Jacobs (✉) · L. A. Garrow · M. Lohatepanont · F. S. Koppelman ·
G. M. Coldren · H. Purnomo
e-mail: Tim.Jacobs@usairways.com

L. A. Garrow
e-mail: laurie.garrow@ce.gatech.edu

C. Barnhart and B. C. Smith, LLC (eds.), *Quantitative Problem Solving Methods in the Airline Industry*, International Series in Operations Research & Management Science 169,
DOI: 10.1007/978-1-4614-1608-1_2,

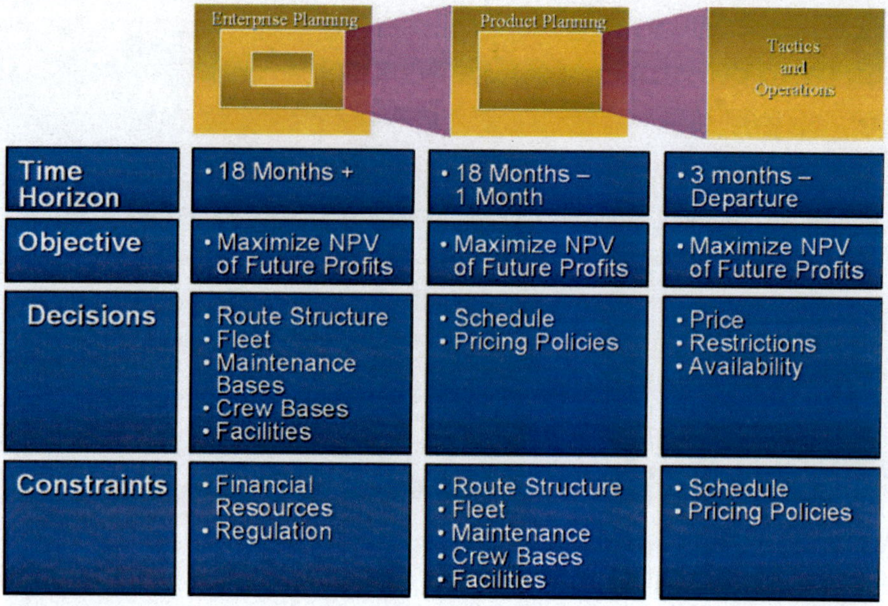

Fig. 2.1 Schedule planning and operation timeframe, objectives and constraints (Smith and Jacobs 1997)

closed. This process can be viewed as a series of overlapping sequential steps that include scheduling, marketing and distribution. This process requires an exchange of data and feedback between scheduling, pricing and revenue management and distribution. In addition, other considerations such as crew resources, maintenance and engineering and ground services help define the boundaries by which the airline schedule must operate and be managed (Fig. 2.1).

Scheduling determines where and when the airline will fly. Schedules are built to maximize long-term profitability. The revenue and cost associated with each schedule are based on very different views of the same information. Although the schedule is composed of individual flight legs between two cities, the airline's product and revenues are based on passenger origin and destination (O&D) markets. An O&D market is defined by a passenger's point of entry and exit from the airline system. The schedule is built to maximize its attractiveness to customers in a wide variety of O&D markets. The development of hub and spoke networks was based on providing maximum O&D coverage with a limited number of flight legs. The costs of operating the schedule depend on the flight legs, which drive the number and type of aircraft used. The schedule must consider the cost and availability of cabin and flight deck crews, as well as the requirement that aircraft cycle through maintenance bases at regular intervals. As a result, the schedule also determines the location and size of ground facilities, and the number and location of crew and maintenance bases.

Efficient schedules which match supply and demand are key to airline profitability. Profitable solutions require anticipation of general market conditions: the costs of capital, fuel and labor, as well as the level and nature of competition. Airlines address many scheduling issues (assigning aircraft and crews to flights, routing aircraft to maintenance bases) with large-scale combinatorial optimization techniques. The scale of today's airlines makes this increasingly difficult. For example, large U.S. domestic carriers operate more than 3,000 flights per day with 600 or more aircraft and can include more than 300 cities, serving over 10,000 unique O&D markets.

Marketing determines what specific products will be offered for sale and how many of each will be sold. The two primary components of airline marketing are pricing and yield management. Since deregulation of the U.S. domestic airline industry, both have evolved into very complex processes. Prior to deregulation, individual airlines served specific market segments. Scheduled carriers served the business traveler while charter carriers served the leisure market. Scheduled carriers flew with relatively low load factors but remained reasonably profitable due to the limited competition created by government regulation.

Just prior to deregulation, the scheduled carriers started to offer additional products to the leisure market segment to help fill some of the empty seats. There were two problems with this approach. First, airline revenue would have been severely diluted if the existing customer base switched from full fare to discounts. Restrictions were introduced to make discounts unappealing to existing business travelers (advance purchase, saturday night stay). Second, some flights were already full and discount sales would displace late booking, high value traffic. The yield management process was introduced shortly after deregulation to anticipate where discount sales would and would not be profitable. From this simple beginning, the pricing and yield management process has become very complex and dynamic. Their combined role is to help airlines fine tune demand and sales to meet the capacity provided by the schedule. Today, a U.S. domestic carrier's schedule can consist of up to 4 million fares. These fares and restrictions are adjusted frequently to match demand to supply; 100,000 fare changes per day is typical. As a result, revenue management departments must keep up with these changes. Every day the number of reservations available for sale is reviewed and adjusted in order to maximize total revenues on all future flights. This is another very large-scale optimization process that involves solving a highly stochastic and nonlinear optimization model. Controlling reservation availability for all future flights at a large U.S. carrier can represent a problem with approximately 500 million decision variables.

Distribution is the process of taking the airline products and putting them on the shelf for sale. The store front for the airline industry is primarily central reservation systems (CRS) and global distribution systems (GDS). A CRS allows an airline's reservation agents to book their own flights and fares. CRSs are relatively expensive to develop and maintain. Large airlines typically have their own CRS, while second and third tier carriers tend to rent space in another carrier's (often their competitor's) system. In the 1970s U.S. domestic carriers began to give travel agents access to their CRSs. This provided each airline with a method for selling products outside of the reservations office. Through the 1980s the number of travel agents with direct access to the major CRSs (Sabre, Apollo-United Airlines, Galileo-British Airways, World Span) grew substantially. Each CRS became a significant distribution outlet not only for the originating airline but for other participating airlines. Today the successors to these CRSs have become global in that an agent hooked into Sabre or Amadeus has access to the schedules, fares and reservation availability for most of the world's airlines. This gives any airline immediate access to a very significant distribution process. For example, Sabre contains schedules for over 700 airlines and agents can book tickets directly on 350 airlines. Sabre is installed in over 29,000 travel agency locations and processes more than 4,900 messages per second.

2.3 Chapter Outline

This chapter will focus on the application of forecasting and operations research techniques to airline scheduling problems and provide a brief overview of how airlines use these techniques to develop and evaluate schedules and business decisions. Section 2.2 of this chapter provides a brief overview of the forecasting process used to estimate passenger demand and determine the expected revenue, cost and profitability associated with a given schedule. This section will also provide insight as to how airlines use these techniques to evaluate incremental changes to market services such as frequency and/or aircraft assignment changes. Section 2.3 provides an overview of the fleeting process used by most network carriers. This section presents an introduction to the fleet assignment model (FAM) and some of the enhancements to the model better integrate the schedule development and fleeting processes with both operational and revenue management aspects of the airline business process. These enhancements will include the incorporation of O&D passenger effects (O&D FAM) and the inclusion of high-level maintenance and engineering (M&E) opportunities into the schedule development and fleeting process. Section 2.4 will provide a high-level overview of the aircraft routing process and its impact on other business units within the company such as M&E and flight and ground crew resources. Section 2.5 presents an overview of some new developments and directions in the operations research and forecasting and their application to the airline scheduling area. Section 2.6 provides a full list of references noted within the chapter for further study.

2.4 Forecasting Aspects and Methodologies for Schedule Planning

This section includes three major components. First, an overview of the data and major components of network-planning models is presented. Next, the two major types of market share models based on the Quality of Service Index (QSI) methodology or logit-based methodologies are reviewed. This is followed by a summary of the key experiences of a major U.S. airline that transitioned from using an itinerary choice model based on a QSI methodology to one based on a logit methodology. The section concludes with a discussion of one relatively new modeling technique: the use of continuous time-of-day functions (versus discrete time-of-day dummy variables or schedule delay functions). Part of the material in this section is reprinted from Garrow (2010, pp. 203–208, 228–229, 250) with permission from Ashgate Publishing. The material draws heavily on prior work from Coldren and Koppelman as well as information obtained via interviews with industry experts.

2.4.1 Introduction

Network-planning models (also called network-simulation or schedule profitability forecasting models) are used to forecast the profitability of airline schedules. These models support many important long- and intermediate-term decisions. For example, they aid airlines in performing merger and acquisition scenarios, route schedule analysis, code-share scenarios, minimum connection time studies, price-elasticity studies, hub location and hub buildup studies and equipment purchasing decisions. Conceptually, "network-planning models" refer to a collection of models that are used to determine how many passengers want to fly, which itineraries (defined as a flight or sequence of flights) they choose, and the revenue and cost implications of transporting passengers on their chosen flights.

Although various air carriers, aviation consulting firms and aircraft manufacturers own proprietary network-planning models, very few published studies exist describing them. Further, because the majority of academic researchers did not have access to the detailed ticketing and itinerary data used by airlines, the majority of published models are based on stated preference surveys and/or a high level of geographic aggregation. These studies provide limited insights into the range of scheduling decisions that network-planning models must support. Recent work by Coldren and Koppelman provides some of the first details into network-planning models used in practice (Coldren 2005; Coldren and Koppelman 2005a, b; Coldren et al. 2003; Koppelman et al. 2008).

2.4.2 Overview of Major Components of Network-Planning Models

As shown in Fig. 2.2, "network-planning models" refer to a collection of sub-models. First, an *itinerary generation algorithm* is used to build itineraries between each airport pair using leg-based air carrier schedule data obtained from a

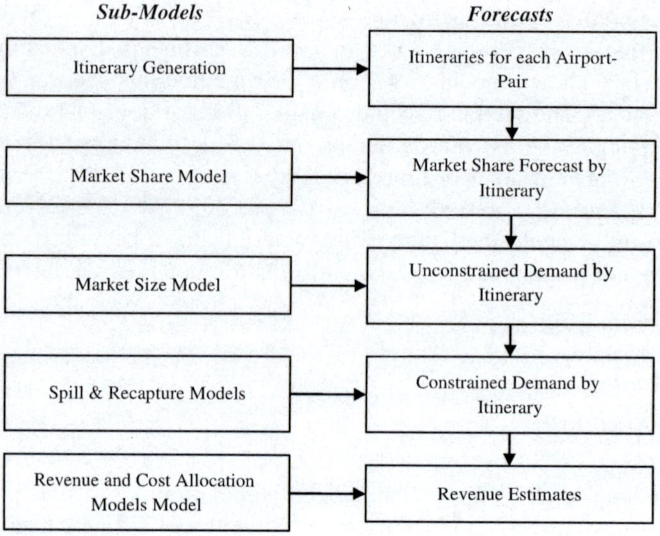

Fig. 2.2 Components and associated forecasts of a network-planning model

source such as the Official Airline Guide (OAG Worldwide Limited 2010). OAG data contain information for each flight including the operating airline, marketing airline (if a code-share leg), origin, destination, flight number, departure and arrival times, equipment, days of operation, leg mileage and flight time. Itineraries, defined as a flight or sequence of flights used to travel between the airport pair, are constructed from the OAG schedule. Itineraries are usually limited to those with a level-of-service that is either a non-stop, direct (a connecting itinerary not involving an airplane change), single-connect (a connecting itinerary with an airplane change) or double-connect (an itinerary with two connections). For a given day, an airport pair may be served by hundreds of itineraries, each of which offers passengers a potential way to travel between the airports. Although the logic used to build itineraries differs across airlines, in general itinerary generation algorithms include several common characteristics. These include distance-based circuitry logic to eliminate unreasonable itineraries and minimum and maximum connection times to ensure that unrealistic connections are not allowed. In addition, itineraries are typically generated for each day of the week to account for day-of-week differences in service offered.

An exception to the itinerary generation algorithm described above was developed by Boeing Commercial Airplanes for large-scale applications used to allocate weekly demand on a world-wide airline network. In this application, a weekly airline schedule involves the generation of 4.8 million paths across 280,000 markets that are served by approximately 950 airlines with 800,000 flights. Boeing's algorithms, outlined in Parker et al. (2005), integrate discrete choice theory into both the itinerary generation and itinerary selection. That is, the utility value of paths is explicitly considered as the paths are being generated; those paths with utility values "substantially lower" than the best path in a market are excluded from consideration.

After the set of itineraries connecting an airport pair is generated, a *market share model* is used to predict the percentage of travelers who select each itinerary in an airport pair. Different types of market share models are used in practice and can be generally characterized based on whether the underlying methodology uses a QSI or discrete choice (or logit-based) framework. Both types of market share models are discussed in this chapter.

Next, demand on each itinerary is determined by multiplying the percentage of travelers expected to travel on each itinerary by the forecasted *market size*, or the number of passengers traveling between an airport pair. However, because the demand for certain flights may exceed the available capacity, *spill and recapture models* are used to reallocate passengers from full flights to flights that have not exceeded capacity. Finally, *revenue and cost allocation models* are used to determine the profitability of an entire schedule (or a specific flight).

Market size and market share information can be obtained from ticketing data that provide information on the number of tickets sold across multiple carriers. In the U.S., ticketing data are collected as part of the U.S. Department of Transportation (U.S. DOT) *Origin and Destination Data Bank 1A or Data Bank 1B* (commonly referred to as DB1A or DB1B). The data are based on a 10% sample of flown tickets collected from passengers as they board aircraft operated by U.S. airlines. The data provide demand information on the number of passengers transported between origin–destination pairs, itinerary information (marketing carrier, operating carrier, class of service, etc.) and price information (quarterly fare charged by each airline for an origin–destination pair that is averaged across all classes of service). Although the raw DB datasets are commonly used in academic publications (after going through some cleaning to remove frequent flyer fares, travel by airline employees and crew, etc.), airlines generally purchase "Superset" data from the company Data Base Products (Data Base Products Inc. 2010). Superset data are a cleaned version of the DB data that are cross-validated against other data sources to provide a more accurate estimate of market sizes. See the websites of the Bureau of Transportation Statistics or Data Base Products for additional information.

The U.S. is the only country that requires airlines to report a 10% sample of used tickets. Thus, although ticketing information about domestic U.S. markets is publicly available, the same is not true for other markets. Two other sources of ticketing information include the Airline Reporting Corporation (ARC) and the Billing Settlement Plan (BSP), the latter of which is affiliated with the International Air Transport Association (IATA). ARC is the ticketing clearinghouse for many airlines in the U.S. and essentially keeps track of purchases, refunds and exchanges for participating airlines and travel agencies. Similarly, BSP is the primary ticketing clearinghouse for airlines and travel agencies outside the U.S.

Given an understanding of the major components of network-planning models and the OAG schedule, itinerary and ticketing data sources that are required to support the development of these models, the next sections provide a detailed description of QSI and logit-based market share models.

2.4.3 QSI Models

Market share models are used to estimate the probability a traveler selects a specific itinerary connecting an airport pair. Itineraries are the products that are ultimately purchased by passengers, and hence it is the characteristics of these itineraries that influence demand. In making their itinerary choices, travelers make tradeoffs among the characteristics that define each itinerary (e.g. departure time, equipment type(s), number of stops, route, carrier). Modeling these itinerary-level tradeoffs is essential to truly understand air-travel demand and is, therefore, one of the most important components of network-planning models.

The earliest market share models employed a demand allocation methodology referred to as QSI.[1] QSI models, developed by the U.S. government in 1957 in the era of airline regulation (Civil Aeronautics Board 1970) relate an itinerary's passenger share to its "quality" (and the quality of all other itineraries in its airport pair), where quality is defined as a function of various itinerary service attributes and their corresponding preference weights. For a given QSI model, these preference weights are obtained using statistical techniques and/or analyst intuition. Once the preference weights are obtained, the final QSI for a given itinerary is usually expressed as a linear or multiplicative function of its service characteristics and preference weights. For example, suppose a given QSI model measures itinerary quality along four service characteristics (e.g. number of stops, fare, carrier, equipment type) represented by independent variables X_1, X_2, X_3, X_4 and their corresponding preference weights β_1, β_2, β_3, β_4. The QSI for itinerary i, QSI_i, can be expressed as:

$$\text{QSI}_i = (\beta_1 X_1 + \beta_2 X_2 + \beta_3 X_3 + \beta_4 X_4), \text{ or}$$
$$\text{QSI}_i = (\beta_1 X_1)(\beta_2 X_2)(\beta_3 X_3)(\beta_4 X_4).$$

Other functional forms for the calculation of QSI's are also possible. For itinerary i, its passenger share is then determined by:

$$S_i = \frac{\text{QSI}_i}{\sum_{j \in J} \text{QSI}_j}$$

[1] QSI models described in this section are based on information in the Transportation Research Board's Transportation Research E-Circular E-C040 (Transportation Research Board 2002) and on the personal experiences of Gregory Coldren and Tim Jacobs.

where

S_i is the passenger share assigned to itinerary i,
QSI_i is the quality of service index for itinerary i,
$\sum_{j \in J} \mathrm{QSI}_j$ is the summation over all itineraries in the airport pair.

Theoretically, QSI models are problematic for two reasons. First, a distinguishing characteristic of these models is that their preference weights (or sometimes subsets of these weights) are usually obtained independently from the other preference weights in the model. Thus, QSI models do not capture interactions existing among itinerary service characteristics (e.g. elapsed itinerary trip time and equipment, elapsed itinerary trip time and number of stops). Second, QSI models are not able to measure the underlying competitive dynamic that may exist among air travel itineraries. This second inadequacy in QSI models can be seen by examining the cross-elasticity equation for the change in the passenger share of itinerary j due to changes in the QSI of itinerary i:

$$\eta_{\mathrm{QSI}_i}^{S_j} = \frac{\partial S_j}{\partial \mathrm{QSI}_i} \frac{\mathrm{QSI}_i}{S_j} = -S_i \mathrm{QSI}_i.$$

The expression on the right side of the equation is not a function of j. That is, changing the QSI (quality) of itinerary i will affect the passenger share of all other itineraries in its airport pair in the same proportion. This is not realistic since, for example, if a given itinerary (linking a given airport pair) that departs in the morning improves in quality, it is likely to attract more passengers away from the other morning itineraries than the afternoon or evening itineraries.

Thus, to summarize, because QSI models have a limited ability to capture the interactions between itinerary service characteristics or the underlying competitive dynamic among itineraries, other methodologies, such as those based on discrete choice models have emerged in the industry. A detailed overview of discrete choice models is provided in the Customer Modeling chapter. An overview of how discrete choice models have been applied to market share modeling is presented in the next section.

2.4.4 Application of Discrete Choice Models to Market Share Modeling

As presented in the Customer Modeling chapter, discrete choice (or "logit") models such as the multinomial logit (MNL) model are random utility maximizing models that describe how individuals choose one alternative among a finite set of mutually exclusive and collectively exhaustive alternatives. The individual chooses the alternative that has the maximum utility. The utility function for a random utility model is defined as

$$U_{ni} = \beta^* x_{ni}$$

where

U_{ni} is the total utility of alternative i for individual n.
β^* is the vector of parameters associated with attributes x. Utility is assumed to be a linear in parameters function of attributes x.
x_{ni} is the vector of attributes that vary across individuals and alternatives.

Because the utility the individual receives from each alternative is not known to the researcher, the utility function is assumed to have two components. The systematic or representative component contains observed variables that describe characteristics of the individuals and alternatives. The unobserved or error component is a random term that represents the unknown (to the researcher) portion of the individuals' utility function. The utility function is estimated using

$$V_{ni} = \beta x_{ni} + \varepsilon_{ni}$$

where

V_{ni} is the total observed utility of alternative i for individual n
β is the vector of estimates for β^*
x_{ni} is the vector of attributes for alternative i and individual n
ε_{ni} is an unobserved error component.

Different choice models are derived by imposing assumptions about the distribution of the error term and/or β. For example, the assumption that the error term is independent and identically distributed Gumbel[2] with mode[3] 0 and scale γ, iid G(0, γ,) leads to the multinomial logit (MNL) model (McFadden 1974). The MNL probability of selecting alternative i among all j alternatives in C_n, the choice set for individual n, can be expressed in closed-form as

$$P_{ni} = P(i|x_{ni}, \beta) = \frac{e^{\beta' x_{ni}}}{\sum_j e^{\beta' x_{nj}}}.$$

The main limitation of the MNL is exhibited in the independence of irrelevant alternatives (IIA) property which states that the ratio of choice probabilities P_{ni}/P_{nj} for $i, j \in C_n$ is independent of the attributes of any other alternative. The IIA property of the MNL model is also apparent by examining the cross-elasticity equation for the change in the probability of itinerary j due to changes in an attribute of itinerary i:

---

[2] An iid Gumbel distribution is also called Type I extreme value.

[3] Several publications incorrectly report the parameters describing the Gumbel distribution as the mean and scale. However, the shape and dispersion of the Gumbel distribution are formally defined by the mode and scale. Further, the mean of the Gumbel is given by the relationship mean = mode + $\{0.577/\gamma\}$.

$$\eta_{X_{ik}}^{P_j} = \frac{\partial P_j}{\partial X_{ik}} \frac{X_{ik}}{P_j} = -P_i \beta_k X_{ik}$$

Note that the expression on the right side is not a function of j. The IIA property of MNL is equivalent to the elasticity problem of the QSI model; that is, the cross-elasticity is undifferentiated across alternatives. In terms of substitution patterns, this means a change or improvement in the utility of one alternative will draw share proportionately from all other alternatives. In many applications, this may not be a realistic assumption. For example, in itinerary choice, the *unobserved factors* associated with the non-stop alternatives are expected to be correlated (since all non-stops are more convenient for passengers, may exhibit a decreased likelihood of lost luggage, etc.). Thus, the substitution between these alternatives is likely to be greater than between any of them and the connecting alternatives.

While the MNL model can be criticized for the restrictive substitution patterns it imposes, recent comparisons of itinerary choice models based on the MNL and QSI methodologies at a major U.S. airline clearly showed that the MNL outperformed their QSI model. In addition, many other discrete choice models (some developed specifically within the context of airline itinerary choice) can be used to incorporate flexible substitution patterns. Thus, the IIA property should not be viewed as a limitation of discrete choice models, as many other models (discussed extensively in Garrow 2010) can be used to relax this property. Nonetheless, it is interesting to note that in the context of itinerary choice models, even the simple MNL model dramatically outperformed the QSI model.

2.4.5 MNL and QSI Model Development at a Major U.S. Carrier

One of the first published studies modeling air-travel itinerary share choice based on a discrete framework was published in Coldren et al. (2003). MNL model parameters were estimated from a single month of itineraries (January 2000) and validated on monthly flight departures in 1999 in addition to selected months in 2001 and 2002. Using market sizes from the quarterly Superset data adjusted by a monthly seasonality factor, validation was undertaken at the flight segment level for the carrier's segments. That is, the total number of forecasted passengers on each segment was obtained by summing passengers on each itinerary using the flight segment. These forecasts were compared to onboard passenger count data. Errors, defined as the mean absolute percentage deviation, were averaged across segments for regional entities and compared to predictions from the original QSI model. Regional entities are defined by time zone for each pair of continental time zones in the U.S. (e.g. East–East, East-Central, East-Mountain, East–West,…,-West–West) in addition to one model for the Continental U.S. to Alaska/Hawaii and one model for Alaska/Hawaii to the Continental U.S. The MNL forecasts were consistently superior to the QSI model, with the magnitude of errors reduced on the order of 10–15% of the QSI errors. Further, forecasts were stable across months, including months that occurred after September 11, 2001. Additional validation details and estimation results are provided in Coldren et al. (2003).

The MNL itinerary share model reported in Coldren et al. (2003)—which represents a discrete choice model that was used to replace a major U.S. airlines' QSI model—includes variables for level-of-service, carrier presence, connection quality, aircraft size and type, fare and departure time-of-day. The representation of passengers' preferences for non-stop flights merits further discussion as it is unique from many other specifications found in practice and, more importantly, was found to be robust. Specifically, level-of-service (non-stop, direct, single-connect, or double-connect) is modeled using dummy variables to represent the level-of-service of the itinerary with *respect to the best level of service* available in the airport pair. That is, an itinerary with a double connection is much more onerous to passengers when the best level-of-service in the market is a non-stop than when the best level-of-service in the market is a single connection. Further, parameter estimates across 18 regional entities reveal that passengers have similar, but not identical responses, to changes in level-of-service across the entire domestic system. This is one example of the benefits of a "well-defined" utility function that captures the fundamental trade-offs of how passengers make itinerary choices; that is, parameter estimates are stable across datasets. Subsequently, this aids in transferability across different time periods and leads to better and more stable forecasting accuracy.

Another common industry practice reflected in the MNL itinerary share model of the major U.S. carrier is the inclusion of carrier presence variables. Numerous studies have found that increased carrier presence in a market leads to increased market share for that carrier (Algers and Beser 2001; Nako 1992; Proussaloglou and Koppelman 1999; Suzuki et al. 2001). In the MNL model, a "point of sale weighted airport presence" variable, used to represent carrier presence at both the origin and destination, is found to influence the value of itineraries.

Finally, it is important to note that in the major carrier's MNL itinerary share model, preferences for departure times are represented via the inclusion of time-of-day dummy variables for each hour of the day. In practice, there are other methods based on schedule delay formulations that are currently in use or are being explored in a research context to represent individuals' time-of-day preferences. Unfortunately, the terminology that has been used to describe the schedule delay functions is often referred to as a "nested logit model" within the airline community, which is incorrect. The next section clarifies the distinction between time-of-day preference and schedule delay functions. The Customer Choice chapter clarifies the definition of the nested logit model, as derived from discrete choice theory and discusses more advanced discrete choice models that have been applied to itinerary share and other airline applications.

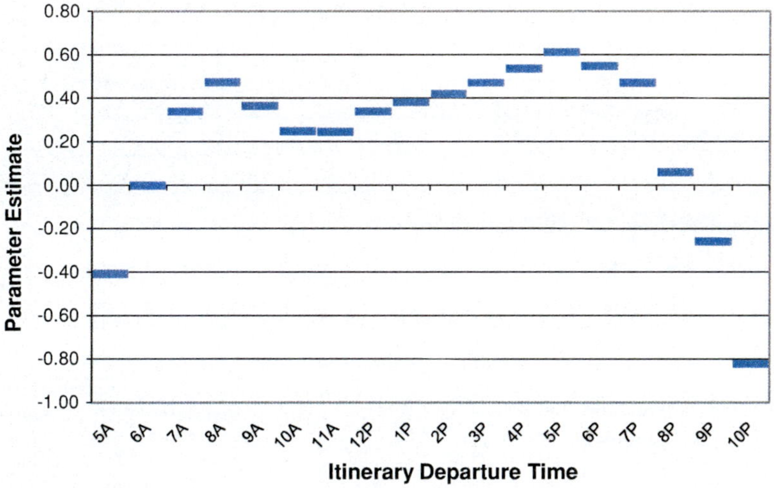

Fig. 2.3 Passenger departure time preferences from time period model. *Source* Reprinted from Koppelman et al. (2008) with permission of Elsevier

2.4.6 Time-of-Day Preferences Versus Schedule Delay Functions

Determining when to schedule flights is arguably one of the most important decisions made by airline scheduling departments. Scheduling flights during unpopular departure times will result in fewer passengers and/or lower average fares. As described in Koppelman et al. (2008), different approaches have been used to model air travelers' departure time preferences. The first is to include time-of-day dummy variables for each hour of the day. Figure 2.3 shows an example of time-of-day preferences for a month of continental U.S. departures based on all flights traveling westbound by one time zone. Parameter estimates based on a MNL formulation indicate passengers prefer to depart early in the morning (at 8 a.m.) or later in the afternoon (5 p.m.).

However, the use of discontinuous time periods poses interpretation problems in practice as a slight change in schedule (e.g. from 9:59 p.m. to 10:02 p.m.) can cause large and counter-intuitive shifts in the probabilities. Estimation of parameters for all time periods can also be subject to over-fitting problems. To address these deficiencies, Koppelman et al. (2008) propose an approach adopted by Zeid et al. (2006) in the context of urban travel activity models. The approach is to estimate weighting parameters for a series of sine and cosine curves to obtain an overall representation of the distribution of departure time preferences. The time-of-day preference for three sine and cosine curves is specified as:

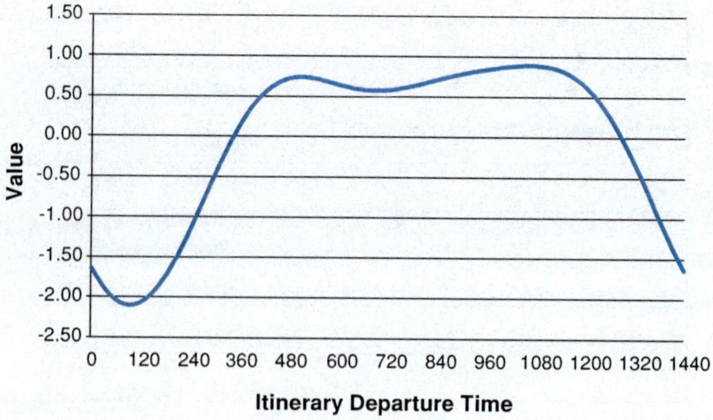

Fig. 2.4 Time of day utility curve from base sin–cos model. *Source* Reprinted from Koppelman et al. (2008) with permission of Elsevier

$$\text{Utility for alternative } i = \beta_1 \sin\left(\frac{2\pi t_i}{1440}\right) + \beta_2 \sin\left(\frac{4\pi t_i}{1440}\right) + \beta_3 \sin\left(\frac{6\pi t_i}{1440}\right) + \beta_4 \cos\left(\frac{2\pi t_i}{1440}\right) + \beta_5 \cos\left(\frac{4\pi t_i}{1440}\right) + \beta_6 \cos\left(\frac{6\pi t_i}{1440}\right)$$

where

t_i is the departure time of itinerary i expressed as minutes past midnight
1440 is the number of minutes in the day.

The final time-of-day value from this model is obtained by summing the six weighted trigonometric functions and is shown in Fig. 2.4. Statistical tests indicate that continuous specification is preferred over the time-of-day dummy variables.[4]

Carrier (2008) proposed a modification to this formulation to account for cycle lengths that are shorter than 24 hours. Formally, the equation:

$$\beta_1 \sin\left(\frac{2\pi h}{1440}\right) + \beta_2 \text{cosin}\left(\frac{2\pi h}{1440}\right) + \cdots$$

is replaced with

[4] Other trigonometric functions involving an additional offset parameter, such as those proposed by Gramming et al. (2005) were also estimated as part of the analysis. Results from these two approaches were virtually identical.

$$\beta_1 \sin\left\{\frac{2\pi(h-s)}{d}\right\} + \beta_2 \operatorname{cosin}\left\{\frac{2\pi(h-s)}{d}\right\} + \cdots$$
$$l - e \leq d \leq 24$$
$$0 \leq s \leq e$$

where e and l represent the departure times of the earliest and latest itineraries in the market, respectively, h represents the departure time, s represents the start time of the cycle (which is not uniquely identified and can be set to an arbitrary value) and d represents the cycle duration. The examples in this chapter use the 24-hours period, as Carrier's formulation leads to a nonlinear-in-parameters function, which he solved using a trial-and-error method. The trial-and-error method (often used by discrete choice modelers when they encounter nonlinear-in-parameters functions) essentially fixes d to different values and estimates the remaining parameters. The value of d that results in the best log likelihood value is the preferred model.

Figures 2.3 and 2.4 represent time-of-day preferences on a 24-hours cycle by measuring the relative value of a departure time relative to all other possible departure times. However, from a behavioral perspective, itinerary selection may be influenced by an individual's effort to depart as close as possible to his/her ideal departure time. The difference between an individual's desired departure time and actual departure time is defined as schedule delay. Formally, schedule delay for itinerary i, SD_i, can be expressed as:

$$SD_i = \sum_{\forall j} g(DT_i - TP_j) W_{TP_j}$$

where

DT_i is the departure time for itinerary i
TP_j is the start of each 15-min time period from 5:30 a.m. to 10:30 p.m.
$g()$ is a transformation function of the difference in minutes between the itinerary departure time and the time period
W_{TP_j} is the weight for time period j.

There are two key points to note about this formulation. First, the weights W_{TP_j} account for departure preferences and the distribution of observed passenger departures at different times-of-day. Second, the formulation is general in the sense that different schedule delay transformation functions are possible. Several functions, including linear, square root, squared, logistic, etc. were estimated. The logistic transformation shown in Fig. 2.5 was found to fit the data the best. Formally, this schedule delay transformation and time period weights are given as:

$$g(DT_i - TP_j) = \frac{1}{1 + \exp\left(\frac{\alpha_2 - |DT_i - TP_j|}{\alpha_1}\right)}$$

$$W_{TP_j} = \frac{\text{sincosValue}_{TP_j}}{\sum_j \text{sincosValue}_{TP_j}}$$

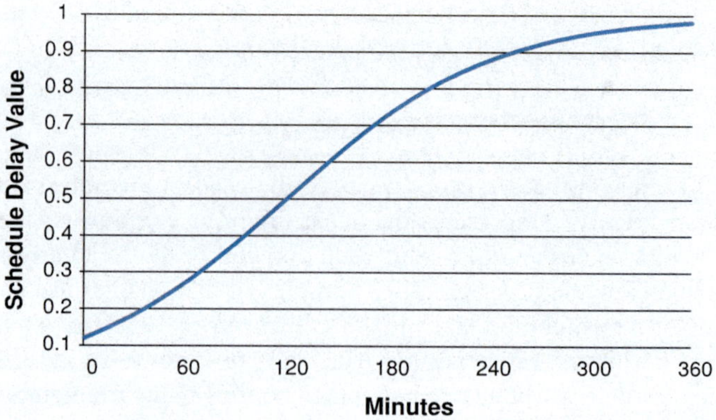

Fig. 2.5 Impact of logit function schedule delay on itinerary value *Source*: Reprinted from Koppelman et al. (2008) with permission of Elsevier

where

DT_i is the departure time for itinerary i
TP_j is the start of each 15-min time period from 5:30 a.m. to 10:30 p.m.
$\text{sincosValue}_{TP_j}$ is the sum of the added terms for time period j.

The formulation based on schedule delay is found to fit the data better than the model based on continuous time-of-day preferences. Additional results that capture differnces in day-of-week departure time preferences as well as early and late departure (or arrival) delays by outbound and inbound itineraries are reported in Koppelman et al. (2008).

2.4.7 Summary

This section focused on describing two major types of market share models found in scheduling models: those based on the QSI methodology and those based on discrete choice methods. An emphasis was placed on identifying concepts that in the authors' experiences are commonly misunderstood in practice. This includes the definition of nested logit models and schedule delay functions.

Based on our interviews with industry experts, we learned that many airlines currently using logit-based methodologies are contemplating reintroducing QSI methodologies due to the perceived complexity of logit models and difficulty in maintaining parameter estimates. However, in our opinion, we believe that the fundamental problems currently being observed are not due to the use of a logit model, but rather over-parameterized utility functions. One of the primary differences between the published MNL model of a major U.S. carrier (which was clearly seen to dominate their QSI model) and the logit models used in practice relates to the number of variables (and estimated β coefficients).

In the published MNL model, each regional entity has 36 parameter estimates in addition to estimates associated with each airline carrier. Further, 18 of these parameters, which are associated with dummy variables for time-of-day preferences, can be further reduced via incorporation of an appropriate continuous schedule delay function. This is in comparison with alternative logit models reported to have hundreds, *if not thousands*, of parameter estimates. However, a simple, yet well-specified MNL utility function can lead to superior predictive performance over a QSI model. Complexity should not be driven by the number of variables included in the model, but rather by the desire or need to obtain more accurate substitution patterns than those imposed by the MNL. Further, more flexible patterns can be incorporated via the use of more advanced GEV or mixed logit models discussed in the Customer Modeling chapter.

2.5 Airline Fleet Assignment Process and Schedule Development

2.5.1 Introduction and Scope

The fleet assignment process represents one of the most important and well studied applications of operations research in the airline industry. In many ways the schedule development and fleeting process embodies the complexities and computational difficulties characteristic of many aspects of the airline industry. To begin, many carriers use the fleet assignment process to help finalize market frequencies, flight times and enforce various operational requirements on the schedule. These may include operational needs such as station purity in which particular stations are limited to one or two types of fleet to meet maintenance and engineering capabilities, the incorporation of minimum revenue guarantees (MRG) in which municipalities contract for service to their airport, and the increase or reduction of available aircraft due to retirements and new deliveries.

Later in the schedule development process, the fleet assignment process and optimization tools are used to finalize the fleet assignments, distribute various subfleets within the network based on operational limitations such as range, and incorporate maintenance opportunities and crew considerations. For example, a carrier might fly several markets with a Boeing 737 but some of the markets may require a 737–800 rather than a 737–200 due to range limitations. Incorporating maintenance opportunities may involve having a specified number of aircraft of a specific type on the ground for 12 hours beginning between 1800 (6:00 p.m.) and 2000 (8:00 p.m.) in the evening to ensure that enough aircraft are available to launch operations the following morning. The carrier may also want their flight crews to stay with the same aircraft for as long as possible to minimize “crew connections” in which the crew leaves one aircraft upon arrival and flies another aircraft for their next scheduled flight. Having the crew stay with the aircraft saves time for both the crew and the airline and results in a more efficient operation and better utilization of the aircraft. It also facilitates a more effective line maintenance operation during the operating day due to the opportunity for maintenance personnel to discuss issues with the crew during aircraft turns when needed.

The efficient utilization of expensive resources is an objective of any profitable airline. One important aspect of this utilization process is fleet assignment. Fleet assignment involves the optimal allocation of a limited number of fleet types to flight legs in the schedule subject to various operational constraints. The most common form of the FAM makes simplifying assumptions about passenger demands, revenues and network flows to approximate the expected revenue for each flight leg in the schedule. These simplifying assumptions provide a point estimate of the expected revenue for each leg in the schedule given various capacity options. The following section presents the basic development of the most common form of the fleet assignment model. In addition, the following section will also present two potential enhancements to the typical fleet assignment model that incorporates the O&D passenger flows into the process. Following the development of the fleet assignment model and its enhancements, we compare and contrast two formulations and present example results using actual airline schedules.

2.5.2 Fleet Assignment Model Development

The fleet assignment problem is typically posed as a binary assignment model in which a specific aircraft type is assigned to each leg in the schedule. The basic fleet assignment model maximizes overall profit subject to three primary constraints: (1) plane count, (2) balance and (3) cover. The plane count constraint stipulates that, for each fleet type, the number of planes used to fly the schedule cannot exceed the total number of planes available. The balance constraint stipulates that the number of arrivals must equal the number of departures for each station, time event and fleet type. The cover constraint requires that a fleet type be assigned to each leg in the network. Most airlines include numerous additional operational constraints that help tailor the solution to the specific operational requirements of the airline.

Typically, fleet assignment models pose the problem as a multi-commodity flow problem. For the fleet assignment problem, arcs represent the arrival and departure of flights and aircraft on the ground. The nodes define the specific time and station where these activities take place. Figure 2.6 illustrates the basic timeline approach used by the fleet assignment model. Figure 2.6a portrays the actual timeline of flights arriving and departing a single station. Figure 2.6b presents the node/arc representation of the timeline for the same station. Figure 2.6c provides a detailed schematic of the decision variables that represent the selection of aircraft type i assigned to flight leg j and ground arcs flowing into a single node at time t and station s within the timeline.

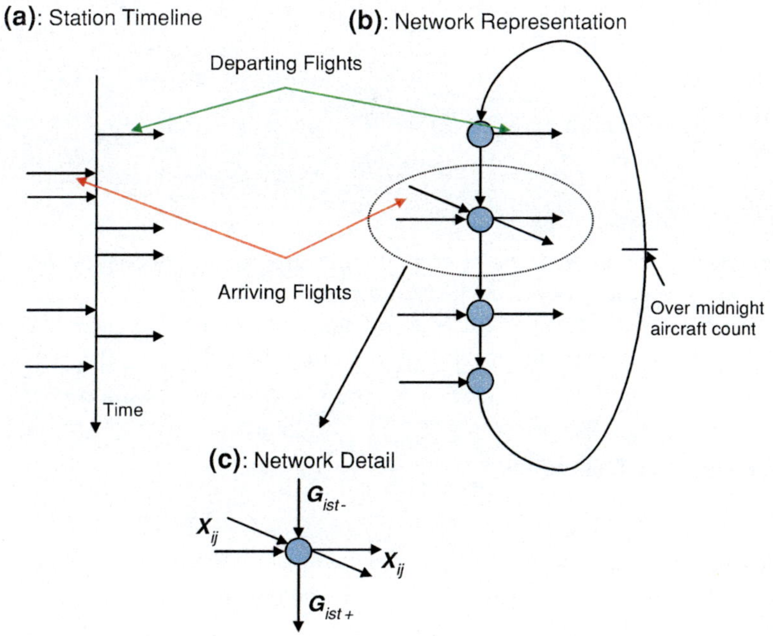

Fig. 2.6 Network representation of a typical FAM formulation. **a** Station timeline. **b** Network representation. **c** Network detail

Using this network representation, we formally develop the basic FAM proposed by Hane et al. (1995) using notation similar to that used by Lohatepanont (2001) and Smith and Johnson (2006).

2.5.2.1 Definition of Sets

S: set of stations or airports indexed by *s*.
J: set of flight legs indexed by *j*.
F: set of fleet types (e.g. S80, 737) indexed by *i*.
T: set of all departure and arrival events indexed by *t*.
Re(*i*): set of all flight legs for fleet type *i* crossing the counting line (e.g. midnight) indexed by *j*.
IN(*i,s,t*): set of flight legs inbound to $\{i,s,t\}$.
OUT(*i,s,t*): set of flight legs outbound from $\{i,s,t\}$.

2.5.2.2 Decision Variables

$$x_{ij} = \begin{cases} 1 & \text{if aircraft type } i \in \text{F is assigned to schedule leg } j \in J, \\ 0 & \text{otherwise} \end{cases}.$$

G_{ist^+} represents the number of aircraft on the ground for fleet type $i \in F$, at stations $s \in S$, on the ground arc just following time $t \in T$.
G_{ist^-} represents the number of aircraft on the ground for fleet type $i \in F$, at stations $s \in S$, on the ground arc just prior to time $t \in T$.

2.5.2.3 Parameter Definitions

R_{ij} represents the expected revenue associated with assigning aircraft type $i \in F$ to flight leg $j \in J$ and is a function of expected demand, spill and unit revenue per passenger.
C_{ij} represents the expected costs associated with assigning aircraft type $i \in F$ to flight leg $j \in J$ as a function of fixed, ownership and variable costs.
NP_i represents the number of available aircraft of type $i \in F$.

2.5.2.4 Conventional Leg-Based FAM Formulation

$$\max\ P = \sum_{j \in J} \sum_{i \in F} (R_{ij} - C_{ij})\, x_{ij} \quad \text{(Objective : Maximize Profit)} \tag{2.1}$$

subject to:

$$\sum_{j \in Re(i)} x_{ij} + \sum_{s \in S} G_{is0^-} \leq NP_i \quad \forall i \in F \quad \text{(Plane Count)} \tag{2.2}$$

$$G_{ist^-} - G_{ist^+} + \sum_{j \in IN(i,s,t)} x_{ij} - \sum_{j \in OUT(i,s,t)} x_{ij} = 0 \quad \forall i \in F\ , s \in S, t \in T \quad \text{(Balance)} \tag{2.3}$$

$$\sum_{i \in F} x_{ij} = 1 \quad \forall j \in J \quad \text{(Cover)} \tag{2.4}$$

$$x_{ij} \in \{0, 1\} \quad \forall i \in F, \quad \forall j \in J$$
$$G_{isj} \geq 0 \quad \forall i \in F \text{ , } s \in S, \ t \in T \tag{2.5}$$

Constraints (2.2) represent resource constraints and states that the number of planes of each fleet type i cannot exceed the total number of planes available, NP_i. Constraints (2.3) represent the balance constraints stating that, at any station and time, the arrival of an aircraft must be matched by the departure of the aircraft. Aircraft can arrive at a station from another station or from the same station in the previous time event, $t - 1$. The time events in this formulation represent an arrival or departure event or a combination of arrivals and departures at the station. Constraints (2.4) represent the cover constraints that stipulate each flight leg must be assigned a fleet type. Constraints (2.5) define the decision variable for assigning fleet type i to flight leg j as a binary variable and specify non-negativity for the ground arcs. Several references present overviews of the general leg-based FAM (see Abara 1989; Subramanian et al. 1994; Hane et al. 1995).

The conventional fleet assignment model described by Eqs. 2.1–2.5 is used for both long-term planning of the airline schedule and near-term finalization of the schedule fleet allocation. Depending on the carrier, this model can be used for long-term planning to fleet a typical daily schedule or a weekly schedule. Most U.S. based carriers tend to plan schedules based on a typical day while European and Asian carriers tend to focus on weekly schedules.

During the planning process, carriers will often need to reduce or expand their schedules to better match available capacity to demand during off seasons such as the fall and early winter or high demand seasons such as Christmas and New Year's holidays. In these cases, airlines can use the FAM to help select and fleet flights that best contribute to the overall profitability of the schedule while dropping other flights that do not contribute. To reduce the schedule, airlines can run the FAM in "reduction mode" in which we relax the cover constraint described by Eq. 2.4. Relaxing Eq. 2.4 as a less than or equal to constraint allows the model to drop flights that do not help maximize the overall profitability of the schedule.

Similarly, the relaxed version of the FAM can be used to expand the schedule to capture the need for more capacity during high demand seasons. To accomplish this, many airlines "overbuild" schedules to include more flights than they expect to operate. Using the overbuilt schedule as input, the airline then optimizes the schedule using the FAM in reduction mode to drop less desirable flights from the schedule.

In both cases mentioned above, the airline must add a number of operational constraints to the FAM to prevent undesirable results. Often these added constraints place bounds on the number of frequencies allowed in any market and limit the number of flights dropped overall. This prevents the model from dropping out of markets entirely or radically reducing the resulting schedule in the name of expected profitability. One caveat that should be kept in mind involves the revenues and costs used to drive the objective function when using reduction mode. The FAM described by Eqs. 2.1–2.5 requires accurate reflections of the revenues and costs associated with each potential fleet/flight combination considered by the model. These revenue and cost estimates are dependent on the original schedule used to forecast demand and passenger traffic. Relaxing the cover constraint (Eq. 2.4) compromises the accuracy of these forecasts and the overall estimates for revenue and costs. Many carriers try to limit this undesirable impact by iteratively updating the forecasted revenues and costs to ensure more accurate results. However, this problem extends beyond schedule reductions. In fact, the revenues, costs and fleeting results can be influenced by the overall mix of local and connecting passengers. As a result, several enhancements to the FAM have been proposed to incorporate the influence of connecting passengers and accurately reflect the change in revenues due to limited schedule reductions.

During the near-term planning process, the FAM can be used to finalize the overall fleeting (allocate sub-fleets), incorporate crew considerations into the final schedule and build transition schedules that bridge one seasonal schedule into the next. To build a transition schedule, an airline can formulate the FAM using the final fleet assignments from the two seasonal schedules as inputs and allow the model to optimize the fleet assignments to connect the two schedules. In addition, the FAM can be used to re-fleet portions of the schedule to better match overall demand to available capacity near the day of departure. We present an actual case study of this type of application near the end of the chapter.

As highlighted above, the major problem with the leg-FAM approach described by Eqs. 2.1–2.5 is that it does not accurately incorporate the O&D marketing effects and expected passenger flows throughout the network. The fleet assignment process should account for multiple markets utilizing each leg of the schedule, multiple classes within each market, and network interactions caused by the various markets competing for space.

Several approaches to incorporating RM aspects into FAM have been investigated over the past 10 years to develop an Origin–Destination Passenger-based Fleet Assignment Model (ODFAM). These approaches have dealt with the size and non linearity of ODFAM through various decomposition approaches.

Farkas (1996) demonstrates that RM has a significant impact on traffic volume and mix and by ignoring these effects FAM can yield sub-optimal solutions. His analysis illustrated the necessity of modeling the effects of both network flow and stochastic demand to improve FAM performance. He concludes that incorporating RM directly into FAM is not practical. He proposes three approaches to this problem:

- Column generation. Where each column represents a complete fleeting solution. The master evaluates traffic and revenue, and ensures that allocations do not exceed capacity. The columns are generated using a multi-commodity formulation. Although no computational results are published Farkas states that the subproblem is relatively slow to solve (40 min) and is impractical for operational use.
- Leg Class revenue management FAM. Since many airlines do not have full network control in their RM systems, Farkas investigates the impact of leg class revenue management control on FAM. He shows that for a typical airline fare

structure, the revenue function could be non-concave. This non-concavity makes this formulation unattractive in terms of computational efficiency.
- Decomposing the flight schedule into subnetworks between which there are limited or no leg-interactions. Fleeting solutions for each sub-network are generated, the traffic and revenue for each sub-network is evaluated with a Monte Carlo simulation. In the FAM formulation, each of the assignments for a subnetwork is represented by one meta-variable. By starting with a feasible leg-FAM solution, this approach should always produce improving solutions. No computational results are available.

Knicker (1998) and Barnhart et al. (2002) investigate the interactions between RM and FAM. In this work, the authors develop a Passenger Mix Model (PMM) that gives a schedule with known flight capacities and a set of passenger demands with known fare, and determines optimal traffic and revenue. PMM includes aspects of customer choice modeling and includes recapture (the probability that a customer who is spilled from one flight leg books on another of the same airline). PMM assumes that demand is deterministic and that the airline has complete knowledge and control of which passengers they accept. PMM could be formulated as a multi-commodity flow problem but due to the large number of passenger types and potential paths this approach is impractical. Kniker reduces the problem by using key-paths, the originally desired itinerary for each passenger. Alternate itineraries are necessary only when passengers are spilled from their preferred itinerary. The problem is solved using column generation, with each column representing passengers spilled from one itinerary and recaptured on another. Kniker formulates the stochastic version but does not present results.

Kniker combines PMM and FAM. The integrated problem, IFAM, is solvable but suffers from increased fractionality versus leg-FAM in which aggregate leg revenues and costs are used to reflect profitability of different fleets on a flight leg. He improves performance through coefficient reduction and additional cuts, but the MIP is still much more difficult to solve than the corresponding leg-FAM MIP. Kniker compares performance of various approaches using a Monte Carlo simulation model. By comparing models that capture the network effects assuming deterministic demand versus stochastic models that ignore network effects, he shows that if flow demand is at least 25% of the total demand, then capturing network effects is more important than capturing stochastic effects. Knicker does not formulate a version of FAM that addresses both stochastic demand and network effects.

Lohatepanont (2001) continues the analysis of IFAM. He investigates the sensitivity of IFAM to several of the simplifying assumptions in its formulation:

- *Demand uncertainty*. IFAM assumes that demand is fixed and known. The demands used in FAM are forecasts subject to random and systematic errors
- *Imperfect control*. PMM assumes that airlines have complete control over which passengers are accommodated
- *Recapture rate errors*. PMM assumes that recapture rate is known.

Table 2.1 Itinerary demand information

Market	Itinerary (sequence of flights)	Number of passengers	Average fare
ABQ-DFW	001	75	$220
DFW-BOS	002	120	$225
ABQ-BOS	001–002	80	$400

Through simulation analysis of IFAM and PMM Lohatepanont shows that while relaxing these assumptions, to make the models more realistic, reduces the benefit of IFAM versus leg-FAM, IFAM consistently outperforms FAM. Barnhart et al. (2002) provide an excellent recapitulation of Kniker and Lohatepanont's work and the relationship between capacity assignments and RM passenger allocations in a deterministic setting. We present the model proposed by Barnhart et al. (2002) in the next section.

Erdmann et al. (1997) proposes a sequential approach to the itinerary FAM problem. They solve FAM and then the passenger mix problem. Kliewer (2000) proposes an approach that integrates FAM and RM using simulated annealing. Kliewer uses a neighborhood search strategy, starting with an initial feasible solution and looks for improving assignment swaps. He accepts or rejects new solutions based on a simulated annealing strategy. The revenue is evaluated with a deterministic passenger flow model.

The model proposed by Barnhart et al. 2002 improves the conventional leg-based FAM by explicitly incorporating the *network* and *recapture effects* into the fleeting process. To better understand this motivation, consider the following example.

Network and Recapture Effects: An Illustrative Example

Consider a small airline network with two flight legs: Flight 001 ABQ-DFW and Flight 002 DFW-BOS. Table 2.1 shows demand and fare data in three OD markets ABQ-DFW, DFW-BOS and ABQ-BOS (connecting through DFW). If 100-seat aircraft is assigned to both flight legs in the network, the optimal revenue for the network of $41,875 is obtained by accommodating 75, 75 and 25 passengers from ABQ-DFW, DFW-BOS, and ABQ-BOS markets respectively.

In conventional FAM, flight legs are assumed to be independent; consequently, fares of connecting passengers have to be allocated to corresponding flight legs in the itineraries. Knicker (1998) experiments with a number of fare allocation schemes and shows that no single allocation scheme, which is applicable to all networks, exists. In this example, we use a simple "equal-fare" allocation, in which the connecting fare is divided equally among the flight legs making up the itinerary. Thus, in our example, the ABQ-BOS fare of $400 is equally divided and allocated to ABQ-DFW and DFW-BOS flights ($200 each). With this allocation, the optimal leg-based revenue is obtained by maximizing the revenue on each flight independently. As a result, the optimal passenger mix is 75 and 25 passengers from ABQ-DFW and ABQ-BOS markets respectively for Flight 001, and 100 passengers from DFW-BOS market alone for Flight 002. The optimal revenue is $44,000. Notice that the resulting optimal mix of passenger is infeasible because none of the ABQ-BOS passengers get on Flight 002, and thus the revenue of $44,000 is inaccurate and unachievable.

Alternatively, one can view this as leg-based FAM's inability to calculate *spill* consistently in the network. When the total passenger demand for a flight leg exceeds the capacity of that flight leg, some passengers are not accommodated or are *spilled.* In this example, with leg-based FAM, 55 ABQ-BOS passengers are spilled from Flight 001, but 80 ABQ-BOS passengers are spilled from Flight 002. On the other hand, with the optimal passenger mix given at the beginning of this example, 55 ABQ-DFW passengers are consistently spilled from both legs.

Next, we introduce the concept of recapture. Normally, spilled passengers are either (1) lost to the airline (that is, they choose to travel on competing airlines or choose not to travel by air) or (2) *recaptured* on alternative flights in the network of the original airline. These recaptured passengers generate *recaptured revenue* for the airline on alternative flights. Most leg-based fleet assignment models ignore totally these recaptured revenues in their estimation of flight-leg revenues because inconsistent spills cannot possibly lead to accurate recapture estimates.

In conclusion, leg-based FAM cannot estimate flight-leg revenues accurately because it assumes flight-leg independency. Specifically, the inaccuracy is a result of (1) the inconsistent estimates of spills due to the *network effect* and, consequently, (2) the inaccurate estimates of (possibly significant) recapture revenues due to the *recapture effect.*

Further, this example demonstrates how decisions made independently for each flight leg are incorrect and suboptimal. To get an accurate estimate of passenger revenue, one needs to take a holistic look at the entire network.

Passenger Mix Model

As the example above shows, a leg-based view of the network does not accurately capture the network effects due to O&D passenger flows and recapture. We need a tool that can estimate the network revenue more accurately. Specifically, we need a tool that can estimate spill consistently throughout the network and allow recaptured revenue to be estimated systematically. Knicker (1998) proposes the PMM for this purpose. The objective of PMM is to find the optimal itinerary-based mix of passengers that maximizes the total revenue (including recaptured revenue) or, equivalently, minimizes the total *spill cost*, the revenue loss due to spilled passengers. PMM is formulated as follows:

$$\text{Min} \sum_{p \in P} \sum_{r \in P} \left(\text{fare}_p - b_p^r \text{fare}_r \right) \cdot t_p^r \tag{2.6}$$

$$\text{Subject to}: \sum_{p \in P} \sum_{r \in P} \delta_i^p t_p^r - \sum_{r \in P} \sum_{p \in P} \delta_i^p b_r^p t_r^p \geq Q_i - CAP_i \quad \forall i \in L \tag{2.7}$$

$$\sum_{r \in P} t_p^r \leq D_p \quad \forall p \in P \tag{2.8}$$

$$t_p^r \geq 0 \quad \forall p, r \in P. \tag{2.9}$$

This formulation utilizes a special set of variables (*keypath* variables t_p^r), first proposed by Barnhart et al. (1995), to enhance model solution. Specifically, the model a priori assigns passengers to their desired itineraries; next, if the capacities on some flights are insufficient, the model finds an optimal way to spill passengers off from these flights such that the total *spill cost* (the revenue loss due to spilled passengers) is minimized. This model incorporates recaptures using a set of Quantitative Service Index (QSI) based parameters called recapture rates, b_p^r, which is defined as the recapture rate from itinerary p to itinerary r or the fraction of passengers spilled from itinerary p that the airline succeeds in redirecting to itinerary r.

Let P be the set of all itineraries and L be the set of all flight legs. The decision variable, t_p^r, is the number of passengers who are redirected from their desired itinerary p to an alternative itinerary r. The parameter fare_p denotes the averaged fare for itinerary p. The objective function (Eq. 2.6) minimizes the total spill cost $\left(\sum_{p \in P} \sum_{r \in P} \text{fare}_p \cdot t_p^r \right)$ less the recaptured revenue $\left(\sum_{p \in P} \sum_{r \in P} b_p^r \text{fare}_r \cdot t_p^r \right)$. Constraints (2.7) ensure that the total number of passengers for each flight leg i (which equals to the original passengers desiring this flight, Q_i, less the total passengers spilled from this flight, $\sum_{p \in P} \sum_{r \in P} \delta_i^p t_p^r$, plus the total passengers recaptured from other flights, $\sum_{r \in P} \sum_{p \in P} \delta_i^p b_r^p t_r^p$) does not exceed the capacity of that flight, CAP_i. δ_i^p equals 1 if itinerary p utilizes flight leg i, 0 otherwise. Constraints (2.8) and (2.9) ensure the number of spilled passengers for each itinerary does not exceed the demand for that itinerary and is not less than zero.

PMM is a large-scale LP model that requires specialized solution algorithm. Knicker (1998) proposes a column and row generation-based algorithm for the model. Specifically, only a small set of variables is included in the original master problem and, as the algorithm progresses, more columns (variables redirecting passengers to alternative itineraries) are generated as necessary. Notice also that in the optimal solution most of Constraints (2.8) are not binding because most of the passengers are traveling on their desired (keypath) itineraries. Thus, Constraints (2.8) can be initially omitted and subsequently generated back in as necessary.

Itinerary-Based Fleet Assignment Model

The Itinerary-Based Fleet Assignment Model (IFAM) is the integration of the PMM and the leg-based FAM. The IFAM formulation is:

$$\text{Min} \sum_{k \in K} \sum_{i \in L} c_{k,i} f_{k,i} + \sum_{p \in P} \sum_{r \in P} \left(\text{fare}_p - b_p^r \text{fare}_r\right) \cdot t_p^r \tag{2.10}$$

$$\text{Subject to}: \sum_{k \in K} f_{k,i} = 1 \quad \forall i \in L \tag{2.11}$$

$$y_{k,o,t^-} + \sum_{i \in IN(k,o,t)} f_{k,i} - y_{k,o,t^+} + \sum_{i \in OUT(k,o,t)} f_{k,i} = 0 \quad \forall k, o, t \in N \tag{2.12}$$

$$\sum_{o \in A} y_{k,o,t_m} + \sum_{i \in CL(k)} f_{k,i} \leq N_k \quad \forall k \in K \tag{2.13}$$

$$\sum_{k \in K} \text{SEAT}_k \cdot f_{k,i} + \sum_{p \in P} \sum_{r \in P} \delta_i^p t_p^r - \sum_{r \in P} \sum_{p \in P} \delta_i^p b_r^p t_r^p \geq Q_i \quad \forall i \in L \tag{2.14}$$

$$\sum_{r \in P} t_p^r \leq D_p \quad \forall p \in P \tag{2.15}$$

$$f_{k,i} \in \{0,1\} \quad \forall k \in K, \forall i \in L \tag{2.16}$$

$$y_{k,o,t} \geq 0 \quad \forall k, o, t \in N \tag{2.17}$$

$$t_p^r \geq 0 \quad \forall p, r \in P. \tag{2.18}$$

The objective function (Eq. 2.10) minimizes the total operating cost ($\sum_{k \in K} \sum_{i \in L} c_{k,i} f_{k,i}$) and the total spill (less recapture) cost $\left(\sum_{p \in P} \sum_{r \in P} \left(\text{fare}_p - b_p^r \text{fare}_r\right) \cdot t_p^r\right)$. Constraints (2.11)–(2.13) are original FAM constraints—coverage, balance and count constraints, respectively. Constraints (2.14) are capacity constraints, which dictate that for a given flight leg i the capacity of the chosen assignment ($\sum_{k \in K} \text{SEAT}_k \cdot f_{k,i}$) must exceed the total traffic $\left(Q_i - \sum_{p \in P} \sum_{r \in P} \delta_i^p t_p^r + \sum_{r \in P} \sum_{p \in P} \delta_i^p b_r^p t_r^p\right)$. Constraints (2.15) guarantee that no spills exceed demands. Constraints (2.17) ensure binary selectivity. And finally Constraints (2.17)–(2.18) ensure non-negativity.

Because IFAM is an integration of two large-scale models, Barnhart et al. (2002) propose a column and row generation-based solution algorithm for solving IFAM. Specifically, the column and row generations are applied to the PMM part of the model, that is, to the traffic variables (t_p^r) and demand constraints (Constraints (2.8)).

Barnhart et al. (2002) test the model and algorithm on actual large-scale data set of a major U.S. airline. Their results indicate significant savings from the optimal assignment of aircraft types to flight legs, taking into account network and recapture effects. Further, they present experiments to validate IFAM's three key assumptions/parameters, namely, (1) deterministic demand, (2) recapture rate and (3) optimal control of passenger mix through PMM. Their findings indicate that (1) in a simulation test using stochastic demand generator, IFAM fleeting decisions consistently outperform FAM fleeting decisions, (2) IFAM fleeting decisions are not particularly sensitive to a reasonable range of recapture rates and (3) in a simulation test where suboptimal control of passenger mix is simulated, IFAM fleeting decisions again consistently outperform FAM decisions. For detailed information and discussion, readers are referred to Barnhart et al. (2002).

Another approach for incorporating both network effects and the stochastic nature of demand first proposed by Jacobs et al. (1999, 2000, 2008) uses revenue management controls to drive the fleeting process. This model uses a Benders decomposition approach to integrate the FAM model with a stochastic O&D Revenue Management model. We refer to this approach as O&D FAM. The revenue associated with any FAM solution depends on the capacity assignment for all flight legs. Given an assignment solution, the O&D revenue is estimated by the O&D RM sub-problem. The revenue function for the entire network is approximated in the master problem (FAM) using a series of Benders cuts. Each cut improves the accuracy of the revenue approximation in the master FAM problem. Once a specified accuracy is achieved in the relaxed master problem, the assignment variables are changed to integer variables and the MIP is solved.

This approach is appealing because it addresses both passenger flows within the network and demand uncertainty. It also provides a method of incorporating the passenger mix optimization model used for revenue management directly into the fleet assignment process.

Typically, airlines estimate the expected revenue for each fleet-flight combination using a proportional spill model (Swan 1983) and an average fare per passenger. This process applies a spill model to the total demand for each leg in the schedule individually. As a result, the leg-based revenue and profit estimates cannot capture the effects of network flow on the traffic of individual legs. Using this formulation, the total revenue for each leg in the schedule reflects an independent point estimate of the revenue function. This leads to errors in estimating the expected traffic and revenue for each leg in the schedule.

In reality, the revenue function accounts for the cumulative effect of all market classes flowing over leg j as a function of its capacity and incorporates the interaction between all the legs in the schedule. The revenue function is actually a concave function with respect to leg capacity, CAP_j, resulting from O&D network flow. Figure 2.7 presents an illustrative example of the revenue function for a single leg in the network.

Network flow or O&D yield management (O&D YM) solutions yield a set of bidprices for each leg which represents the dual value of the O&D YM capacity constraint and equals the slope of the revenue function at a given leg capacity,

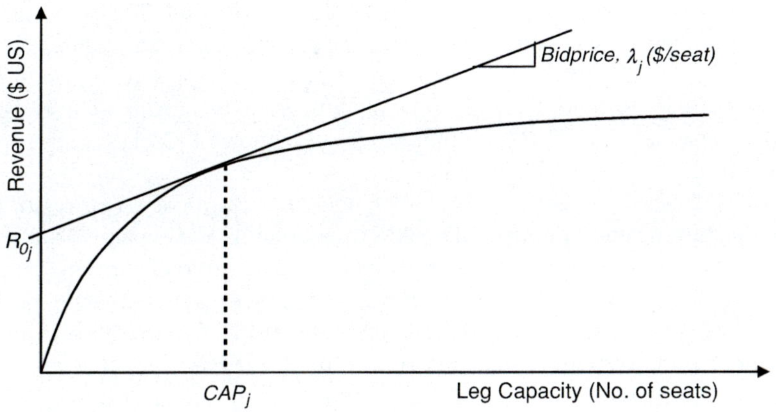

Fig. 2.7 Expected revenue as a function of leg capacity

CAP_j (Fig. 2.3). This slope can be used to define a linear approximation and upper bound to the revenue function. Mathematically, this upper bound is expressed as:

$$R_{0_j} + \lambda_j \, \mathrm{CAP}_j \geq R \, (\mathrm{CAP}_j) \tag{2.19}$$

where R_{0_j} represents the right-hand side of the linear approximation to the revenue function and $R\,(\mathrm{CAP}_j)$ equals the total revenue as a function of the capacity of leg j, CAP_j. λ_j defines the marginal value of an extra seat on leg j (the bidprice) resulting from the O&D yield management. Mathematically, the bidprice for leg j is defined by:

$$\lambda_j = \frac{\partial R(\mathrm{CAP_j})}{\partial \mathrm{CAP}_j} \tag{2.20}$$

In practical terms, the bidprice represents the minimum acceptable price of a seat. More importantly, the bidprice represents the change in the *total system revenue* due to a unit change in the capacity of leg j. Therefore, the bidprice captures the cumulative effects of the market classes flowing over leg j and the interactions between leg j and the other legs in the network. Please see Appendix A for a complete formulation and review of the O&D Yield Management (O&D YM) model formulation.

For any solution of FAM and a corresponding set of bidprices, the total revenue for the schedule is the sum of the revenues realized on each leg. Summing over all the legs in the network yields the following upper bound on the total revenue (R_{Total}):

$$\sum_{j \in J} R_{0_{jv}} + \sum_{j \in J} \lambda_{jv} \mathrm{CAP}_{jv} \geq R_{\mathrm{Total}} \quad \forall v \in V \tag{2.21}$$

where v is the index in the set V for a specific FAM and YM solution. This relationship represents a Bender's cut (Parker and Rardin 1988; Nemhauser and Wolsey 1988; Bradley et al. 1977) and defines an overall upper bound on the total revenue for the schedule as a function of network flow results. Therefore, the overall revenue used in FAM is limited by a function of O&D passenger flow resulting from the O&D YM process.

Constraints (2.21) relate the FAM to the network flow model. This relationship allows decomposition of the O&D FAM model into two separate but related problems: (1) a linear fleet assignment model and (2) a nonlinear network flow model. Separately, each of these models can be solved using conventional IP or NLP methods. Using Constraints (2.21), the general FAM is modified to include O&D effects. The resulting *linear* FAM used by O&D FAM is defined as:

Linear FAM Formulation

$$\max\ P = R_{\text{Total}} - C_{\text{Total}} \quad \text{(Objective : Maximize Profit)} \tag{2.22}$$

subject to:

$$\sum_{j \in \text{Re}(i)} x_{ij} + \sum_{s \in S} G_{is0^-} \leq NP_i \quad \forall i \in F \quad \text{(Plane Count)} \tag{2.2}$$

$$G_{ist^-} - G_{ist^+} + \sum_{j \in IN(i,s,t)} x_{ij} - \sum_{j \in OUT(i,s,t)} x_{ij} = 0 \quad \forall i \in F, s \in S, t \in T \quad \text{(Balance)} \tag{2.3}$$

$$\sum_{i \in F} x_{ij} = 1 \quad \forall j \in J \quad \text{(Cover)} \tag{2.4}$$

$$\sum_{j \in J} R_{0_{jv}} + \sum_{j \in J} \lambda_{jv} \left(\sum_{i \in F} \text{CAP}_{ij} x_{ij} \right) - R_{\text{Total}} \geq 0 \quad \forall v \in V \quad \text{(Revenue)} \tag{2.23}$$

$$C_{\text{Total}} - \sum_{j \in J} \sum_{i \in F} C_{ij} x_{ij} = 0 \quad \text{(Cost)} \tag{2.24}$$

$$\begin{aligned} & x_{ij} \in \{0, 1\} \quad \forall i \in F, \quad \forall j \in J \\ & G_{isj} \geq 0 \quad \forall i \in F\,,\ s \in S,\ t \in T \end{aligned} \tag{2.5}$$

For this formulation, the objective function is modified and two new constraints are added. Constraints (2.23) represent a variation of Constraints (2.21) and allows for the incorporation of the original binary decision variable, x_{ij}. Constraint (2.24) simply redefines the total cost of the fleet assignment as a constraint. O&D FAM explicitly incorporates network effects by utilizing the bidprices provided by solving an O&D network flow model to estimate the revenue function of FAM. The network passenger flow model represents the O&D yield management or O&D revenue management process.

Conceptually, O&D FAM is very different than leg-based FAM. The revenue estimates for leg-based FAM are made on a leg-by-leg basis. Therefore, the revenue estimates and subsequent FAM formulation do not capture the network effects due to O&D yield management. Even using a nonlinear approximation to the revenue function for each leg in the network, leg-based FAM cannot accurately approximate the impact of up-line and down-line capacity constraints. On the other hand, O&D FAM incorporates the network effects directly into FAM through Constraints (2.23). Constraints (2.23) represent an upper bound on the *total system revenue* for the network and provide a link between fleet assignment and network flow as a function of the bidprices for all the legs within the network. As a result, the approximation is also a function of the O&D revenue management effects throughout the network.

For O&D FAM formulation presented above, the O&D revenue Management or O&D Yield Management (O&D YM) process is modeled using a nonlinear network flow model that maximizes overall system expected revenue subject to capacity constraints for each flight in the network. The decision variable represents the number of seats allocated to each O&D fare class itinerary. The overall expected revenue is based on the expected traffic for each O&D itinerary and is a function of the seats allocated to the itineraries competing for space over each flight leg in the network. The nonlinear O&D YM model is typically solved using a sub-gradient algorithm. Appendix A presents a complete formulation and review of the O&D revenue management problem.

To accurately implement O&D FAM, several network flow solutions corresponding to feasible fleet assignments are needed. One possible iterative approach for solving this problem is shown in Fig. 2.8. To begin the iterative algorithm, initial fleet assignments or bidprices for each leg in the network are needed. These can be obtained by: (1) assuming arbitrary initial capacities for each leg in the schedule and solving the network flow model or (2) assuming some initial bidprices and solving O&D FAM or (3) using a standard leg-FAM solution to define the initial capacities. Either option is acceptable and will not affect the final solution of the algorithm. For the work presented here, we assume initial capacities for each leg in the schedule and start the algorithm by solving the network flow model. For each iteration of the algorithm shown in Fig. 2.8, the linear FAM model is subjected to a subset of the revenue Constraints (2.23). To define the parameters needed for Constraints (2.23), the total O&D revenue ($R_{\text{Total }YM}$) is estimated using the Network Flow (O&D revenue management) model.

By incorporating O&D passenger flow aspects into the model, O&D FAM avoids many of the problems inherent to Leg-FAM. Using the results of the network flow model, O&D FAM provides a more realistic estimation of the *total system revenue* function for the network schedule. In addition, O&D FAM incorporates the impact of traffic flow in a manner that is consistent with revenue management practices. This approach uses a series of linear approximations based

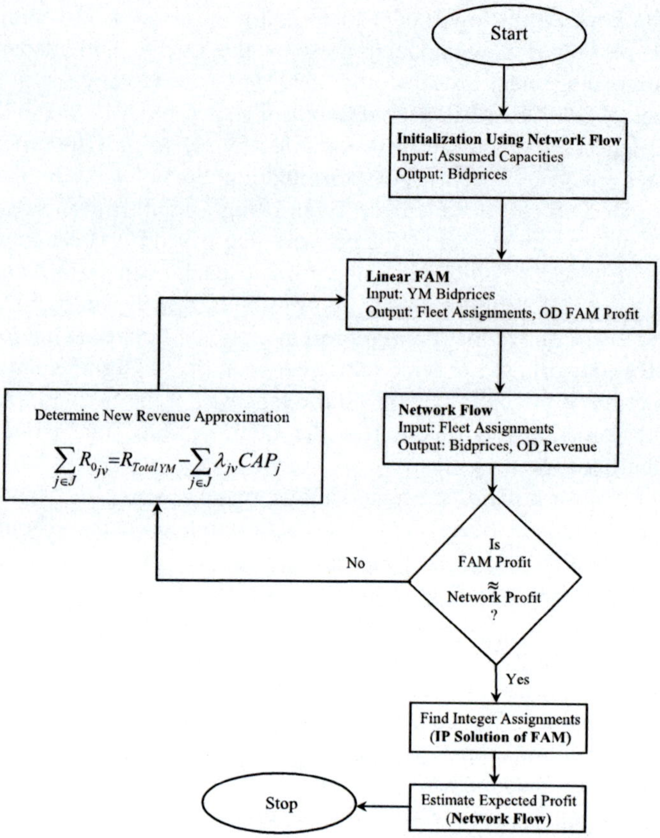

Fig. 2.8 Basic O&D FAM algorithm

on the expected network traffic to determine a linear upper bound of the total revenue function instead of point estimates of the total expected revenue for each leg. This approach prevents the propagation of errors and possible inferior fleet assignments due to inaccurate estimates of the revenues based on leg-level estimates of demand instead of the O&D traffic throughout the network. Although the decision variable is defined as a binary variable, it is computationally beneficial to relax this condition throughout the iterative portion of the solution process. This allows fractional solutions to be used in approximating the revenue function.

Although the iterative approach presented in Fig. 2.8 provides a more accurate estimation of the revenue and can result in a more efficient fleet assignment, it does require multiple solutions to the relaxed fleet assignment model. This problem is compounded by the fact that accurately modeling the revenue function may require several linear approximations of the revenue function. This can become extremely inefficient and computationally cumbersome. To avoid this problem, we can use convex combinations to generate additional bidprices and linear approximations of the true revenue function without resolving the complete FAM formulation. Jacobs et al. (1999, 2008) provides a detailed description of how to improve the computational efficiency of this formulation using convex combinations.

Now that we have developed a conventional FAM and developed two enhancements to that model to incorporate the network effects, we turn our focus to applying these models to the fleeting of actual airline schedules. For illustration, we will consider the use of these models for general fleeting as would used in long-term schedule planning.

To illustrate the conventional Leg-FAM and O&D FAM approaches, a ten-city, 48 leg example consisting of 534 O&D service markets and two fleets is presented and compared to the results obtained using two different revenue estimation schemes for the leg-based FAM formulation. The first leg-based formulation uses a prorating scheme to approximate the expected revenues from each leg in the schedule. The second formulation simply uses the total revenue for each O&D market class to approximate the revenue for each leg. Results indicate that O&D FAM outperforms both leg-based FAM formulations. Fleet assignments from O&D FAM show a 2.8% improvement in expected profit over the results of the leg-based FAM formulations.

To compare the conventional leg-based FAM formulation and O&D FAM, we benchmark the benefits and practicality of using an O&D approach to fleet assignment. We present a case study comparing the Leg-FAM and O&D FAM approaches using actual airline schedules with more than 4,000 daily operations. For these benchmarks, we present results of a general fleet assignment process in which all the scheduled flights must be assigned an aircraft type and a schedule reduction run in which non-profitable flights can be canceled.

2.5.3 Illustrative Example

To illustrate and compare the typical FAM and O&D FAM approaches, we use the simple ten-city example shown in Fig. 2.9. This network consists of 48 flight legs, 534 O&D market classes and two fleet types. In addition to direct flights to and from DFW, the network includes six non-stop flights. The market class prices are representative of actual coach fares for each O&D reported in SABRE. Typical market class demands are assumed. For this illustrative example, the model was formulated and solved using AMPL, the commercially available LP/IP solver CPLEX and the O&D passenger flow model presented in Appendix A.

The network flow model determines the optimal network traffic and bidprices using an approach similar to the O&D revenue management model used by many airlines. Standard deviations for determining traffic and seat allocation for each market class were calculated using a constant coefficient of variation (CV). Traffic and demand were modeled using a Gamma distribution. Table 2.2 presents a summary of the data used to formulate and solve the model.

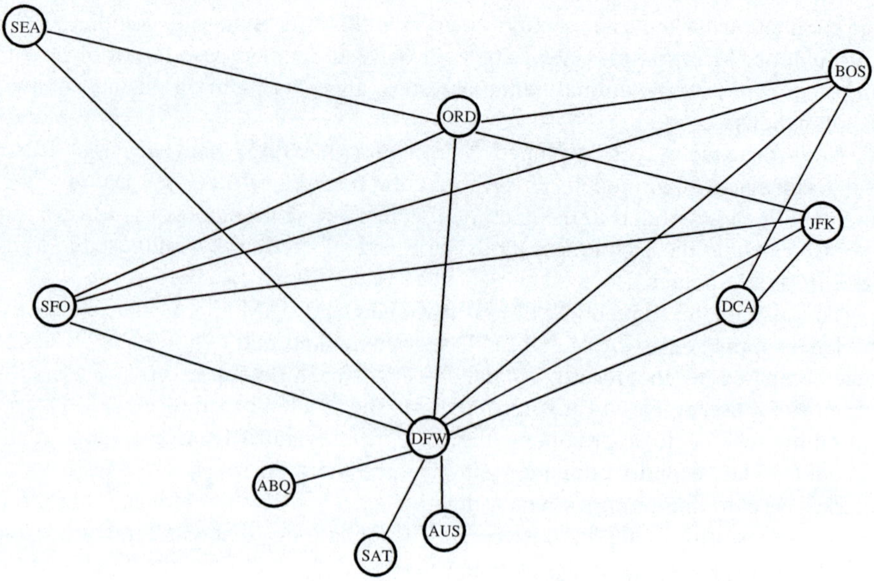

Fig. 2.9 Network for illustrative example (Jacobs, Ratliff and Smith, 2000)

Table 2.2 Summary of data for illustrative example

Model parameter	Value
Number of stations (cities)	10
Number of flight legs	48
O&D market classes	534
Number of hub (DFW) complexes	2
Fleet types:	2
Boeing 757	
Capacity	200
Cost ($/min)	$79.00
Available aircraft	7
MD Super 80	
Capacity(No. of passengers)	150
Cost ($/min)	$66.45
Available aircraft	8
Average O&D demand	
Y Class	5
M Class	15
Q Class	25
Coefficient of variation (CV)	0.35

To compare with the O&D FAM solution, Leg-FAM must be solved using average or estimated leg-based revenues and costs. For this illustrative example,

Table 2.3 Results for prorated revenue case

	Leg-FAM		O&D FAM	
(1)	Obj. function value (2)	Expected value (3)	Obj. function value (4)	Expected value (5)
Continuous solution	$55,202	$66,080	$69,886	–
Integer solution	$55,202	$66,080	$69,519	$67,954

two common and simple approaches were used to estimate the leg revenues. In the first case, the expected revenue is estimated by prorating the average fare paid by a customer proportional to the distance of each flight leg in the itinerary. In the second case, the revenues are estimated by averaging the total revenue paid by a customer. This second approach double counts the revenue and represents an extreme attempt to capture the O&D revenue. A complete description of these approaches is presented in Appendix B.

2.5.3.1 Results and Comparison with the Prorated Revenue Case

Table 2.3 compares the results from solving the ten-city example using O&D FAM with typical Leg-FAM results for the same test scenario. The results presented in Table 2.3 represent expected profit. For the results presented here, the Leg-FAM formulation used prorated leg revenues. The prorated revenues used in this case were proportional to the block times for each leg in the itinerary.

Column 2 of Table 2.3 represents the expected profit from Leg-FAM. For this example, the initial Leg-FAM solution was integer and no branch and bound was necessary. To estimate the total expected profit on an O&D level, the Leg-FAM solution is used as input to the network flow model (see Appendix A). The network flow model estimates the optimal traffic and revenue by intelligently allocating space to competing O&D fare classes. Column 3 represents the total expected O&D profit using the fleet assignments from Leg-FAM.

Column 4 presents the objective function value for the O&D FAM solution. The first entry in this column represents the continuous solution for O&D FAM. The solution to this illustrative problem required 5 iterations of the main algorithm and 2 iterations of the convex combination algorithm. The second entry represents the expected profit when the O&D FAM model is forced to an integer solution subject to the Benders approximations to the revenue function. As expected when the problem is further constrained, the total revenue decreases. This example required 19 branch and bound nodes to determine the optimal integer solution. Because this solution is constrained by the linear approximations to the total revenue, it is an over estimate of the true profit. Column 5 presents the expected profit for the integer O&D FAM solution. For this example, the O&D FAM model required eight iterations to converge to a 1% difference in the linear FAM Profit

Table 2.4 Solution comparison for flight leg JFKDFW1 total leg demand = 179.5

(1)	Leg-FAM Inputs S80 (2)	757 (3)	Network solution (O&D FAM) S80 (4)	Network solution (Leg-FAM) 757 (5)
Traffic	138.9	165.4	122.6	153
Revenue per passenger	$152.55	$152.55	$193.83	$166.99
Load factor	0.93	0.83	0.82	0.77
E(revenue)	$21,186	$25,232	$23,764	$25,550
Leg cost	$11,961	$14,220	$11,961	$14,220
E(profit)	$9,226	$11,012	$11,803	$11,330

and the Network Profit. This performance varies based on the specific attributes of the problem and the convergence tolerance. The figure in column 5 represents the expected profit found by solving the network flow model for the final integer assignments from O&D FAM.

In the scenario presented here, O&D FAM shows a 2.8% improvement in total expected profit over the Leg-FAM solution. For this example, the iterative O&D FAM algorithm required approximately 20 s of real time to converge. The solution time for Leg-FAM was nearly instantaneous.

Appendix C presents the final fleet assignments and expected traffic for the O&D FAM and Leg-FAM solutions. The average overall network load factor for these results equals 66.9% for the Leg-FAM solution and 66.2% for the O&D FAM solution.

The results presented in Table 2.3 illustrate that O&D FAM can produce superior fleet assignments. However, to clearly understand the fundamental differences between the Leg-FAM and O&D FAM solutions, it is necessary to look at the solution at the leg level. Table 2.4 presents specific details a leg in which the optimal aircraft assignment differs for the two methods. In addition, this table provides details concerning the expected revenue used as input for the Leg-FAM formulation.

Table 2.4 compares the results for the JFKDFW1 flight leg in the network. For this example, Leg-FAM assigned a Boeing 757 and O&D FAM assigned a S80 to the flight leg. This fight leg serves 27 individual market classes and has a total average leg demand of 179.5 passengers. Columns 2 and 3 of Table 2.4 present the expected traffic, revenues and profit figures for use as input to Leg-FAM. The traffic values presented in columns 2 and 3 are determined using a normally distributed spill model with a coefficient of variation (CV) of 0.35. The revenue per passenger represents a weighted average based on demand of the prorated leg revenues for each market class itinerary traveling on the leg. The leg cost is the cost of flying the leg and is based on the block times for the individual flight leg. Column 4 presents the expected traffic, revenues and profits for the final O&D FAM solution. Column 5 presents the expected traffic, revenues and profit figures for the final Leg-FAM solution. For columns 4 and 5, the average revenue per

Table 2.5 Solution comparison for flight leg ORDDFW2 total leg demand = 168

(1)	Leg-FAM Inputs S80 (2)	Leg-FAM Inputs 757 (3)	Network solution (O&D FAM) S80 (4)	Network solution (Leg-FAM) 757 (5)
Traffic	135.7	158.3	130.3	155.3
Revenue per passenger)	$77.23	$77.23	$83.97	$78.05
Load factor	0.90	0.79	0.87	0.78
E(revenue)	$10,479	$12,222	$10,945	$12,123
Leg cost	$9,303	$11,060	$9,303	$11,060
E(profit)	$1,176	$1,162	$1,642	$1,063

passenger represents the quotient of the total expected revenue and the expected traffic.

The results presented in Table 2.4 illustrate two consistent differences between O&D FAM and Leg-FAM solutions. First, Leg-FAM consistently *underestimates* the average revenue per passenger. Second, Leg-FAM *overestimates* the total expected traffic resulting in a load factor that is too high. These differences are due to the fact that Leg-FAM uses weighted average revenue based on total leg demand to estimate the total revenues for each leg. The spill model used to estimate traffic for Leg-FAM considers only the total demand for the leg. No consideration is given to the allocation of seats to different O&D market fare classes. In essence, using total demand to estimate the traffic assumes that all the passengers are alike and have the same chance of obtaining a seat. In reality, seats are allocated based on price, expected demand and overall revenue. The network flow model tends to reduce leg traffic and increase the average revenue per passenger by discriminating between the individual O&D market classes bidding for space on the flight leg. The total revenue received for a network schedule is a function of the individual fares and actual traffic for each O&D market class throughout the network.

The concept of leg-based revenue is misleading. For example, the traffic for the JFKDFW1 flight leg presented in Table 2.5 consists of 27 individual market classes. Only three market classes using this flight leg were local with itineraries consisting of a single flight leg. The remaining traffic traveling on this leg had itineraries consisting of two legs. These inconsistencies between the network flow solution and capacity planning lead to less profitable fleet assignments. Table 2.5 presents results for a second leg in the example network.

For Leg-FAM to yield accurate and meaningful results, the input assumptions must reflect the expected results. The results presented in Tables 2.4 and 2.5 show that the inputs to Leg-FAM differ significantly from the results with respect to revenue per passenger, load factor and total revenue.

Table 2.6 Results for total revenue case

	Leg-FAM		O&D FAM	
(1)	Obj. function value (2)	Expected value (3)	Obj. function value (4)	Expected value (5)
Continuous solution	$543,841	$66,080	$69,886	–
Integer solution	$543,841	$66,080	$69,519	$67,954

2.5.3.2 Results of Total Revenue Case

To fully appreciate the benefits of O&D FAM, the results are compared to a Leg-FAM solution in which total O&D revenue is used to estimate the leg revenue. Table 2.6 compares the results from solving the ten-city example using O&D FAM with results obtained using Leg-FAM. As before, the results presented represent profits. For the results presented here, the Leg-FAM formulation used average total itinerary revenues to estimate the expected revenue for each leg in the network. The average itinerary revenues used in this case were weighted proportional to the itinerary demand. The expected revenue for each leg represents the product of the expected traffic and the average total itinerary fare paid by an individual traveling on the leg. As in the prorated case, the initial solution to Leg-FAM was integer.

In this case, the Leg-FAM solution using total revenue yields the same fleet assignment as the solution using prorated revenues. For this example network, the solution indicates that there is no benefit from using prorated revenues over total revenues in the objective function of Leg-FAM. However, the results presented in Table 2.6 do show that using total revenues to formulate Leg-FAM results in a substantial overestimate of the total expected revenue and profit. In this case, the Leg-FAM solution overestimates the *total system revenue* by 73%. This results in a Leg-FAM profit that is eight times the actual expected profit.

Since the total revenue example uses an average of the total revenue for itineraries traveling on any given leg in the network and double counts revenue for connecting passengers, it is difficult to compare individual leg assignments and draw any conclusions about the expected revenue per passenger. This type of solution only allows network-wide comparisons of the overall profit.

The illustrative examples presented above illustrate the use of both Leg-FAM and O&D FAM for fleeting airline schedules. In the ten-station, 48 flight leg example presented above, O&D FAM outperformed Leg-FAM by 2.8% in profit. This scenario consisted of 534 O&D market classes with three fare classes per O&D and two equipment types. Fleet assignments from O&D FAM show significant improvement in expected overall profit over Leg-FAM. The algorithm is computationally efficient and an integer solution for this example can be found by solving relatively few branch and bound sub-problems. Although this example is relatively small, it provides an initial benchmark for assessing the effectiveness of the O&D FAM algorithm. This formulation of O&D FAM provides a solid platform to develop and test more sophisticated models. For this formulation of O&D FAM, all nonlinear aspects of the problem are isolated in network flow model. The O&D market effects captured by the network flow model are transferred to FAM in the form of linear approximations to the total revenue function. The fact that the nonlinearities associated with the O&D passenger flow within O&D FAM are isolated in the network flow model presents opportunities for enhancing the model to include additional aspects important to the fleeting and schedule planning process. These include: price optimization, passenger recapture and added crew and station fixed and manning costs in the fleet assignment process.

Next we need to apply these same models to actual airline schedules to illustrate their effectiveness, scalability and practicality. In the next section, we present a long-term planning case in which Leg-FAM and O&D FAM are used to fleet an actual airline schedule.

2.5.4 Application in Practice: General Fleet Assignment for Schedule Planning

To illustrate the benefits associated with the O&D FAM approach presented earlier, we consider the assignment of fleets to a commercial airline planning schedule and compare the expected profitability to the fleet assignments using the typical leg-FAM approach. For this illustration, we consider two types of fleeting scenarios. The first scenario represents a switching run in which all the flight legs in the schedule must be assigned a fleet. This type of scenario represents the typical resource allocation process used by many airlines to finalize their fleet assignments during the scheduling cycle. The second scenario represents a reduction run in which we relax the cover Constraints (2.4) to allow cancelation of unprofitable flying. This represents the process used by many airlines to prune a draft schedule during the long-range planning cycle.

The passenger revenue, traffic and costs used for this application were forecasted by American Airlines' proprietary Integrated Forecasting System (IFS). IFS forecasts market share, passenger demand, spill, traffic, revenue and cost (both variable and fixed) at either the O&D itinerary or leg level based on the host airline schedule and OA competitive schedules. In addition, IFS takes into account numerous attributes such as passenger recapture, fuel burn and cost, station and crew manning costs, minimum revenue guarantees (MRG), frequent flyer program (FFP) revenue allocations, code-share and proviso agreements and revenue accounting allocations more representative of the actual business process. For Leg-FAM, IFS allocates passenger revenue to each leg based on local and connecting passenger demand and spill, the up-line and down-line contribution of connecting passengers to the network. For O&D FAM, IFS forecasts passenger demand and revenue at the O&D itinerary level. For this application of Leg-FAM and O&D

Table 2.7 Comparison of O&D FAM versus Leg-FAM for switching scenarios

Scenario	Daily operations	Change in passenger traffic (%)	Change in revenue (%)	Change in cost (%)	Change in profit (%)
Fall	4,034	(3.65)	(1.80)	(2.40)	1.21
Spring	4,434	0.10	0.71	(1.20)	7.20

Table 2.8 Comparison of Leg-FAM and O&D FAM schedule reduction results

Fleeting scenario	Daily operations	Change in passenger traffic (%)	Change in revenue (%)	Change in cost (%)	Change in profit (%)
Base	4,930	–	–	–	–
Leg-FAM	4,569	(1.9)	(2.5)	(5.2)	11.6
O&D FAM	4,281	(13.1)	(9.5)	(14.9)	18.1

FAM, IFS was used to forecast the inputs to both models and evaluate the resulting schedules produced by each of the models

In the first case, we present the results of a switching run on two different planning schedules. Table 2.7 presents the results of the switching scenario.

The switching results presented in Table 2.7 illustrate the economic improvements associated with using an O&D FAM approach over the traditional Leg-FAM approach. For the fall schedule we see that the O&D FAM approach results in a reduction in system traffic and revenue along with an even bigger reduction in overall system costs. This resulted in an overall system profit increase of 1.21% (0.2% of revenue gain). For the spring schedule, the O&D FAM approach increased the overall system traffic and revenue while decreasing the overall system costs. This resulted in an overall profit increase of 7.20% (1.1% of revenue gain). For the switching scenarios presented in Table 2.7, O&D FAM required 10–12 iterations of the Benders algorithm to converge.

In the second case, we present the results of a reduction run which allows the cancelation of unprofitable flying. Table 2.8 presents the results of the reduction scenario.

The results presented in Table 2.8 clearly show an improvement in overall profitability for the schedule reduction scenario. Traditional leg-FAM reduced the scheduled operations by nearly 2% while O&D FAM reduced the scheduled operations by approximately 13%. In this case the overall profitability increase resulted from the model's ability to decrease costs by dropping unprofitable flying and reducing aircraft gauge significantly more than revenue reduction suffered by reducing the overall O&D services available.

In both the switching and reduction runs presented above, O&D FAM tends to reduce the overall system costs significantly more than the traditional leg-FAM approach. This is due to the fact that O&D FAM explicitly models the O&D passenger flows within the assignment process. As a result, O&D FAM properly spills low value O&D connecting traffic in favor of cost savings while leg-FAM assigns aircraft based on a leg-centric view of profitability in which the traffic and revenue on each leg is assumed to be independent of the capacity of other flights within the schedule. This means that leg-FAM assumes the traffic and revenue for a flight leg remains constant even if the flights used by connecting passengers are down-graded or canceled. However, O&D FAM improves on the traditional leg-based methodology by incorporating the O&D passenger flow into the assignment process. As a result, passenger spill happens at the O&D level and the corresponding traffic and revenue automatically adjust as capacity assignments change or flights are canceled.

In the spring case where demand is high, O&D FAM repositions capacity such that more profitable passengers are allocated capacity while still reducing overall system costs. In these cases, O&D FAM properly balances the revenue benefits of local and connecting traffic with the potential reduction in costs. This yields a significant improvement in overall system profit.

The benchmark results presented here further illustrate the benefits associated with using an O&D approach to the fleet assignment problem. In these cases, both the switching and reduction scenarios showed a significant improvement in overall system profitability. In the switching cases, O&D FAM found solutions that significantly cut overall system costs while either marginally decreasing revenue by spilling low value connecting passengers or increasing overall system revenue by properly allocating capacity to accommodate profitable passengers.

2.5.5 Section Summary and Conclusion

This section has presented a brief overview of the scheduling and fleet assignment process. In addition, this section has provided an introduction to some of the optimization models commonly used to develop and finalize an airline schedule prior to publication and shown how these same processes and models can be used in the near-term to improve schedules to better match overall demand to allocated capacity. In the next section, we introduce some of the concepts models associated with routing a final schedule to meet crew and maintenance and engineering requirements.

2.6 Aircraft Rotations and Maintenance Routing Planning

Once the fleet assignment process has been completed, schedule planners focus on evaluating potential rotations and routing solutions for the aircraft used to fly the schedule. Aircraft rotations or turns link arriving flights to departing flights within a station while routings consider the complete path of an aircraft over the day, week or month. The aircraft rotation solution is governed by local operational requirements, operational efficiency and aircraft count. The aircraft routing process

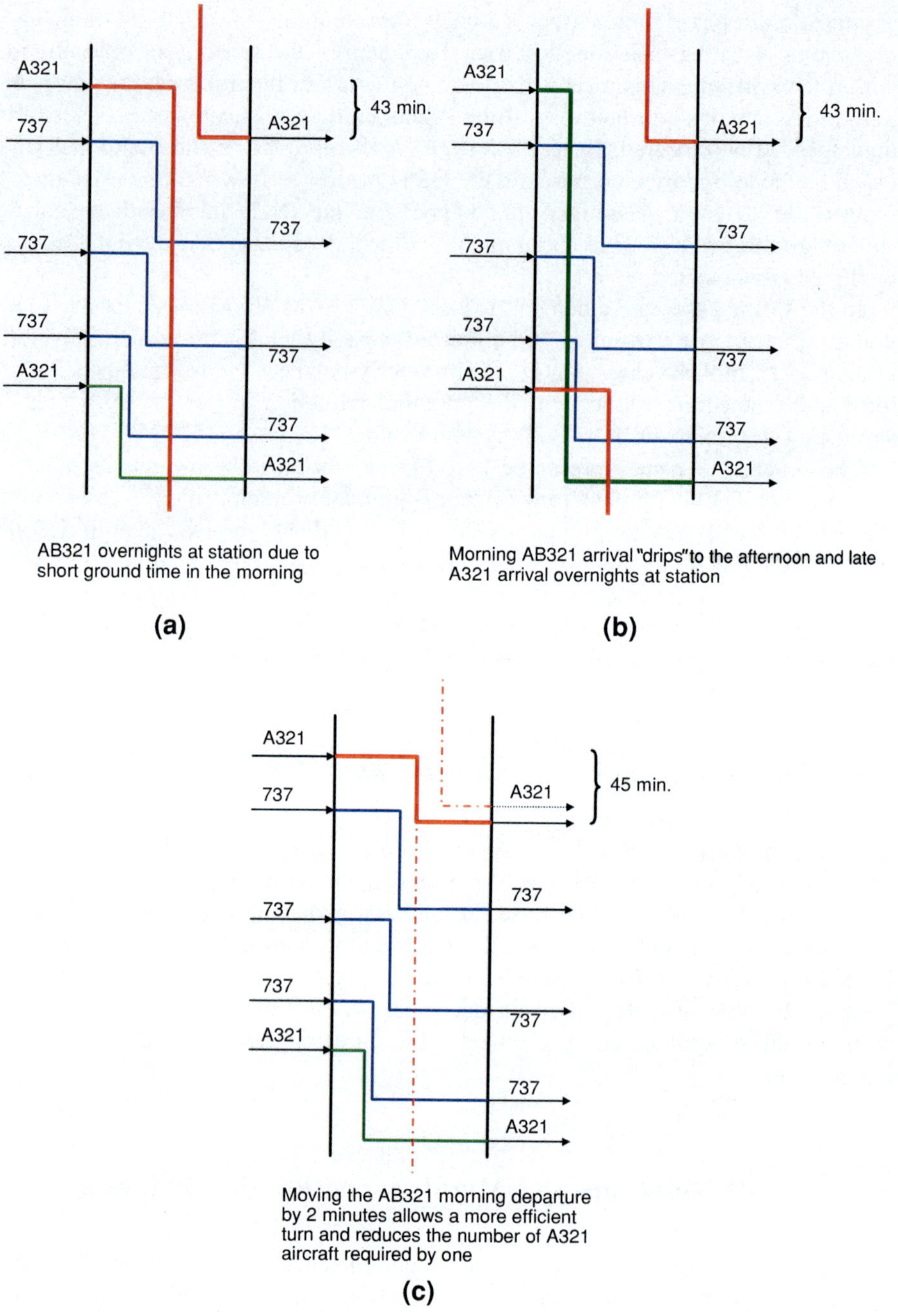

Fig. 2.10 Impact of aircraft turns. **a** AB321 overnights at station due to short ground time in the morning43. **b** Morning AB321 arrival "drips"to the afternoon and late A321 arrival overnights at station. **c** Moving the AB321 morning departure by 2 minutes allows a more efficient turn and reduces the number of A321 aircraft required by one

involves multiple levels of detail and sophistication and is usually governed by maintenance requirements. In addition, aircraft routings typically provide maintenance opportunities rather than actual maintenance scheduling for specific aircraft. The following sections provide an overview of some of the ways airlines determine rotations and evaluate routings with the schedule.

2.6.1 Aircraft Rotations

Generally, the FAM and the fleeting process are driven by overall system profit or cost and do not determine aircraft turns within each station in the network. As a result, the schedule planners must determine the best aircraft rotations by "connecting" or "turning" the schedule in an efficient manner. During this process, schedule planners will attempt to maximize the efficiency of the station by effectively managing ground time and removing any excessively long ground times known as "drips". The set of rotations presented in Fig. 2.10a and b show a typical set of aircraft rotations resulting from a fleet assignment solution and applying a minimum ground time of 45 min at the station. In Fig. 2.10a, the early morning A321 arrival cannot turn to the morning A321 departure because the available ground time of 43 min violates the minimum ground time for the station. As a result, the aircraft sits at the station until the next morning when it supports the A321 morning departure. An alternative rotation pattern is shown in Fig. 2.10b. In this case, the morning A321 arrival connects to the last departure of the day. Again, one aircraft sits on the ground all day and the last arrival of the day sits overnight to support the morning departure. Both of these patterns represent inefficient rotations and increase the number of aircraft required to fly the schedule without any revenue benefit.

Moving the morning departure by 2 min improves the rotation pattern and reduces the number of aircraft required to fly the schedule by one (Fig. 2.10c). In cases where the aircraft count constraints are not tight, the FAM can produce fleet assignments that maximize overall profit but yield inefficient operations at the station level. This example only represents one of many possible scenarios uncovered by the schedule planners following the fleet assignment process.

Unfortunately many commercial carriers use simple last-in/first-out (LIFO) and first-in/first-out (FIFO) strategies to rotate aircraft. These strategies typically require a great deal of manual intervention by the schedule planners to remove inefficiencies and insure the final schedule meets all operational constraints and adheres to aircraft count limitations.

The process of determining the most efficient aircraft rotations can also be formulated as an optimization model in which the objective represents the minimization of excess or unnecessary ground time within the schedule. The decision variables reflect the aircraft connections or rotations for each station and the final timing of the flights.

Excess ground time represents the extra time an aircraft spends on the ground between flights compared to minimum ground time (MGT) needed to turn or rotate the aircraft for the next flight. When it occurs during the day, it signifies a lower utilization of the aircraft. In the early schedule planning phase, excessive ground time means extra aircraft must be used to cover the scheduled flights. Later in the scheduling process when additional operational considerations such as crew with equipment and updated gating availability limitations become a priority, the model can be used to effectively manage ground time to improve operations.

Conceptually, effectively managing aircraft rotations and ground time within the schedule requires a model that explicitly represents connections between arriving and departing aircraft and Terminator to Originator (T to O) connections that connect some arrivals to departures the following day. In addition, the model must allow limited flight retiming to facilitate improved rotations that reduce excess ground time while constrained by aircraft count.

One possible model formulation to determine efficient aircraft rotations while managing ground time within the schedule is presented below. This formulation uses the following notation:

Decision Variables

F_{jk} binary variable for flight j, retime candidate k

$C^{s}_{jk_1,lk_2}$ binary variable that indicates flight j candidate k_1 is connected to flight l candidate k_2 at station s

T_{sjk} binary variable, is 1 if flight j candidate k is selected as a terminator, overnight at station s.

O_{sjk} binary variable, is 1 if flight j candidate k is selected as an originator, departing from station s

TO_{shf} integer variable, signifies overnight arc at station s for station h and assigned fleet f

Sets and Parameters

a^{in}_{sj} mapping parameter is 1 if flight j is an inbound flight for station s

a^{out}_{sl} mapping parameter is 1 if flight l is an outbound flight for station s

b_{sjk_1} mapping parameter is 1 if flight j candidate k_1 can become a terminator at station s

b_{slk_2} mapping parameter is 1 if flight l candidate k_2 can become an originator at station s

h_{sjhf} mapping parameter, is 1 if flight j is O/T at station s but is confined to hub-isolation hub h, and assigned fleet f

$p_{jk_1,lk}$ penalty associated for connecting flight j candidate k_1 and flight l candidate k_2 (=square of GT)

p_f incentive associated with freeing one extra plane from fleet f

p_k penalty associated with flight j candidate k

p_{sjk} penalty associated with originator/terminators on station s and flight j

FL set of all flights considered

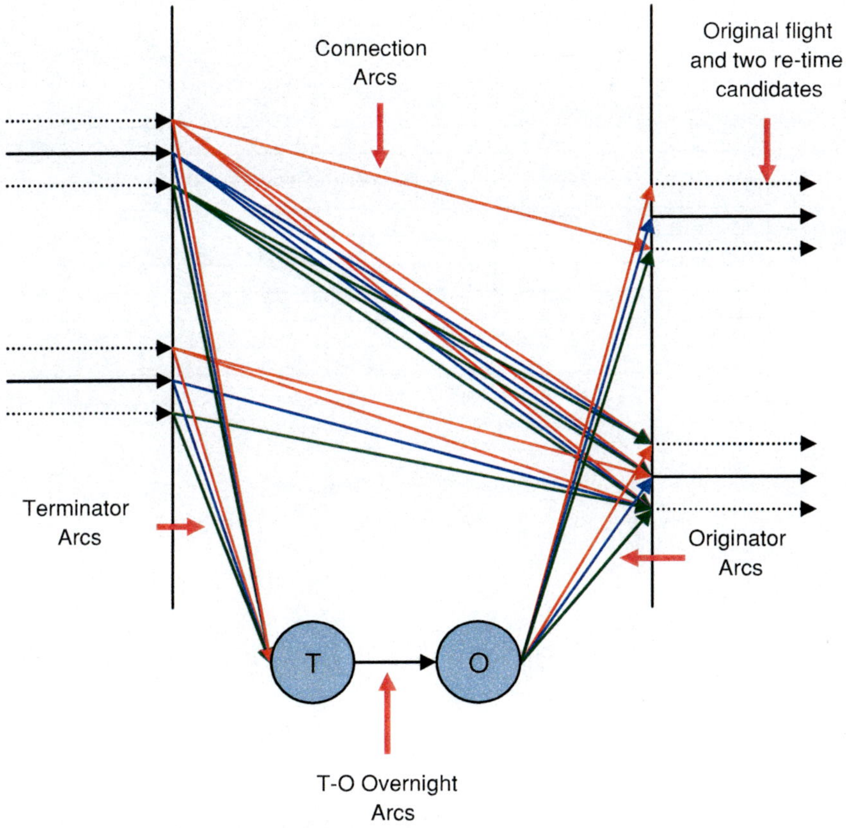

Fig. 2.11 Ground time management model arc formulation

S set of all stations considered
H set of all hubs considered for hub isolation (should include free-hub + 1 hub iso, like ORD)
K(*j*) set of retime candidate for flight *j*
$E(j,k_1)$ set of outbound flights which is allowed to connect to flight *j* candidate k_1
$E(l,k_2)$ set of inbound flights which is allowed to connect to flight *l* candidate k_2
F set of fleet

Using these definitions, the problem of determining efficient aircraft rotations while managing excess ground time and aircraft count is formulated as:

$$\text{Minimize} \sum_{l \in E(j,k_1)} p_{jk_1,lk} C^{s}_{jk_1,lk_2} - \sum_{f \in F} p_f s_f + \sum_{j \in F} \sum_{k \in K(j)} p_k F_{jk} + \sum p_{sjk} O_{sjk_1} + \sum p_{sjk} T_{sjk_1} \tag{2.25}$$

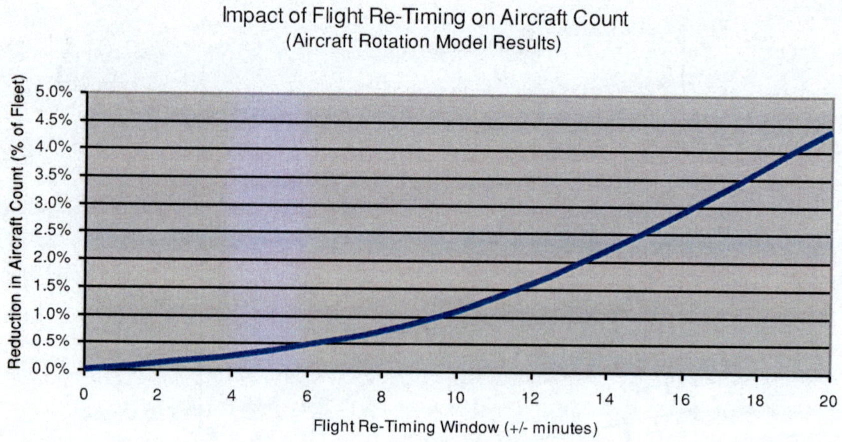

Fig. 2.12 Impact of various re-timing windows on aircraft count

Subject to:

$$\sum_{k \in K(j)} F_{jk} = 1, \quad \forall j \in FL \tag{2.26}$$

$$a_{sj}^{in} F_{jk_1} - \sum_{l \in E(j,k_1)} C_{jk_1,lk_2}^{s} - b_{sjk_1} O_{sjk_1} = 0, \quad \forall k_1 \in K(j), \quad \forall j \in FL \tag{2.27}$$

$$a_{sl}^{out} F_{lk_2} - \sum_{j \in E(l,k_2)} C_{jk_1,lk_2}^{s} - b_{slk_2} T_{slk_2} = 0, \quad \forall k_2 \in K(l), \quad \forall l \in FL \tag{2.28}$$

$$TO_{shf} - \sum_{j} \sum_{k_1 \in K(j)} h_{sjhf} O_{sjk_1} = 0, \quad \forall s \in S, \quad \forall h \in H, \quad \forall f \in F \tag{2.29}$$

$$TO_{shf} - \sum_{j} \sum_{k \in K(l)} h_{slhf} T_{slk_2} = 0, \quad \forall s \in S, \quad \forall h \in H, \quad \forall f \in F \tag{2.30}$$

$$\sum_{h \in H} \sum_{s \in S} TO_{shf} + s_f = PL_f, \quad \forall f \in F \tag{2.31}$$

Solutions to this formulation determine efficient aircraft rotations that minimize the excess ground time and releases as many aircraft as possible from the schedule. The objective function (Eq. 2.25) minimizes the penalty associated with each aircraft rotation. For this formulation, the penalty for connecting an arrival to a viable departure increases as a function of the elapsed ground time between the arrival and departure. Second term in the objective function provides an incentive for the model to release or un-assign aircraft from the schedule. The remaining

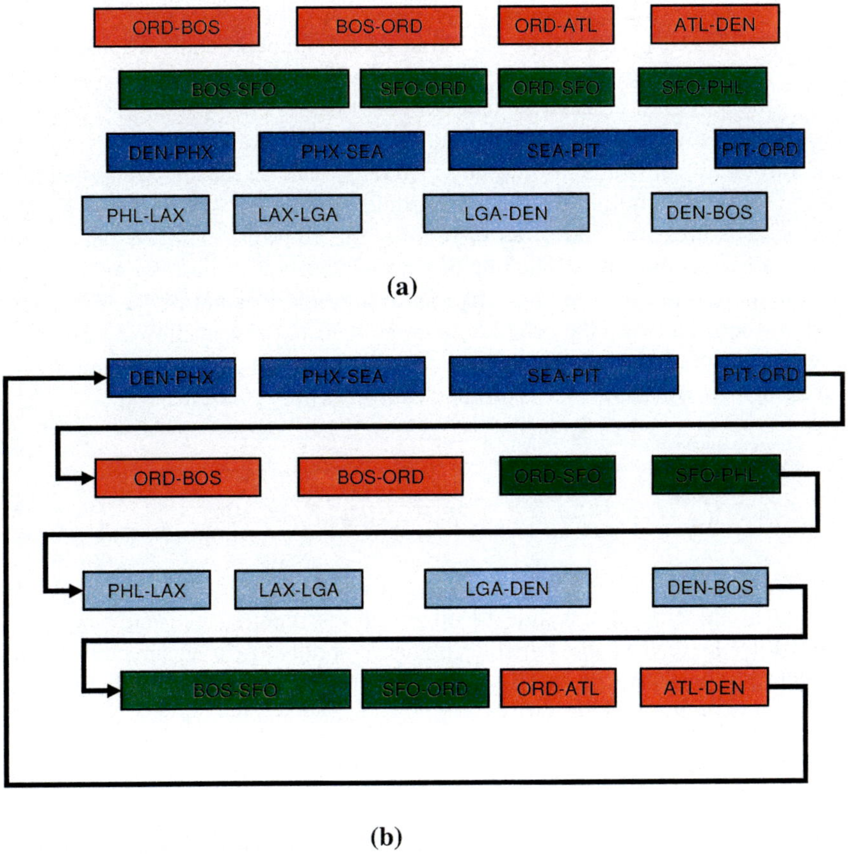

Fig. 2.13 Creating an Euler tour for aircraft routings

penalties help to limit the number of re-timed flights, terminators and originators within the finished schedule.

The model constraints, Constraints 2.2 through 2.6, describe various details of flow balance within the schedule. The first constraint, Constraints 2.2, insures that only one retime candidate be selected for each flight within the schedule. Constraints 2.3 represent the flow balance constraint for inbound flights and potential connections to departing flights. Similarly, Constraints 2.28 represents the balance constraint for the outbound flights and each potential connection to a viable inbound flight. In addition to connecting to a departing flight on the same day, a flight can also stay overnight at the station. In this case, the inbound flight connects to a terminator arc. Similarly, an originating flight will connect to an outbound flight the following day. The terminators and originators are connected via the integer T–O arc. Figure 2.11 presents an illustration of the connections for a single station assuming a 1 day schedule. Constraints 2.29 and 2.30 link the terminator arcs with the T–O overnight arc and the T–O overnight arc to the originator arcs, respectively. Finally, Constraint 2.31 limits the number of aircraft used in the schedule to be less than or equal to the aircraft available. Because this formulation only considers aircraft connection and not fleet assignments, the aircraft count can be modeled by simply adding up the number of T–O aircraft. The slack variable in the aircraft count constraint allows the schedule planner to incentivize reducing the number of aircraft required to fly the schedule.

Figure 2.12 presents the results of running the aircraft rotation model for various retiming windows for a schedule of approximately 225 aircraft. As the results show, for small retiming windows, the model solution does not reduce the number of aircraft required to fly the schedule a great deal. However, allowing the model to use a maximum retiming window of ±20 min can reduce the number of required aircraft by nearly 4.5% or about ten aircraft.

2.6.2 Aircraft Maintenance Opportunity Routing

The initial aircraft maintenance routing process focuses on evaluating routing feasibility of the fleeted schedule to meet various maintenance and operational requirements. In reality, the schedule planners simply try to insure that aircraft have ample opportunity to receive maintenance within specified time intervals. The routing of individual aircraft within the schedule happens much closer to the day of departure and changes frequently due to operational events such as weather delays throughout the network.

FAA regulations require periodic maintenance checks for commercial aircraft based on metrics such as flown hours, the number of take-offs and landings and elapsed time in service since the last maintenance check. These safety and maintenance requirements are very strict and aircraft not meeting the minimum requirements will be grounded.

In addition to meeting regulatory requirements, many carriers try to uniformly add wear and tear to the aircraft. For example, over a period of one month it is better to add approximately 300 flying hours to all aircraft of a given fleet rather than add 400 hours to half of the fleet and only 200 hours to the remaining fleet. To accomplish this, many carriers attempt to develop routings that represent an Euler Tour. For the maintenance routing problem, the nodes represent stations or airports and the arcs represent lines of flying over the day. An Euler Tour in a directed graph represents a closed tour such that each arc or line of flying is traversed exactly once even though nodes or airports may be traversed multiple times. Once an Euler Tour exists, all aircraft will repeatedly experience the same sequence of dial routes though in any given day, each aircraft is assigned a different daily route.

Figure 2.13 presents a simple example of potential daily aircraft routings for a fleet of four aircraft. These lines of flying can be determined using a number of techniques such as LIFO (Last In-First Out) or FIFO (First In-First Out) routings.

For this example, assume Chicago (ORD) has a qualified maintenance base for the airline and that routine maintenance checks (an A-Check) must be performed once every 4 days. Figure 2.13a illustrates the original daily routings for the four aircraft.

If we link the original lines of flying by matching up the final airport in any line of flying to a common originating airport, we end up with two separate tours. The first tour links the first line of flying to the third and the second tour links the second line of flying to the fourth. This results in one tour that provides overnight access at ORD and one tour that does not because the aircraft overnights at either BOS or PHL which are not maintenance bases. These types of tours are often referred to as "locked rotations" due to the fact that the aircraft flying these rotations have no opportunity for overnight maintenance. Left alone, the two aircraft flying these locked rotations would eventually violate the mandated maintenance requirements and be grounded. In such cases, the carrier would be required to ferry the aircraft to a maintenance base for the required maintenance before resuming commercial service.

However, for our four-aircraft example, we can remove the locked rotation by making a swap when two aircraft are both on the ground at the same station. To build an Euler Tour given the lines of flying shown in Fig. 2.13a, we must break two or more of the original lines of flying. For our example, this can be done by breaking and swapping the first and second lines of flying of when both aircraft are on the ground in ORD. Making this swap allows us to link the daily routings as shown in Fig. 2.13b. As a result, the locked rotation has been removed and each of the four aircraft will spend the night at the maintenance base in ORD once every 4 days. In addition, each of the aircraft will follow the same path over the 4 day maintenance interval. By routing the aircraft this way (Fig. 2.13b), the four aircraft will all accrue roughly the same number of flying hours, take-offs and landings over time. This will help the airline control the wear and tear for the fleet over time, better plan parts inventories and allow for better long-term fleet planning.

Although the two-step algorithm outlined above only considers feasibility via maintenance opportunities and ignores cost and capacity issues, it represents a common and popular approach used by many airlines and freight carriers. For the most part, it provides the schedule planners with a degree of confidence that enough maintenance opportunities exist within the schedule to meet any regulatory needs. However, additional constraints can complicate matters. For one, many airlines have consolidated the number of maintenance bases to reduce costs. In addition, some carriers perform a variety of checks (such as A, B and heavy C checks) at a single maintenance base. As a result, many carriers currently have capacity limitations associated the number and combination of checks that can be performed on any overnight shift.

The literature is replete with heuristics and formal optimization approaches to determine feasible routing schemes for a carrier. Qi et al. (2004) present a nice overview of many of the approaches used to develop feasible aircraft maintenance routings during the schedule planning phase. Clarke et al. (1997) model the aircraft routing problem as a network flow model with side constraints. This formulation contains both flight arcs and "waiting" arcs while the nodes represent flight departures and arrivals. In this formulation, maintenance opportunities are included within the waiting arcs with enough idle time to cover the maintenance check at a specified maintenance base.

In addition, Goplan and Talluri (1998) present a polynomial time approach for solving the aircraft routing problem when maintenance must be performed at least every 3 days. In this case, they develop an interative two-phase algorithm that builds one-day routings in the first phase and determines a tour satisfying the maintenance requirements in the second phase. If no cycle exists, a new iteration begins by modifying the one-day routes from the first phase using a heuristic that includes changing aircraft type for specific flights. Talluri (1998) shows that expanding this work to consider the four-day or more maintenance routing problem results in an NP-complete problem in the second phase. Barnhart et al. (1998) explore the use of string models for aircraft assignment and routing. In this work, they present a daily version of the string model that contains additional connectivity constraints to ensure that the strings can be concatenated to form a feasible maintenance tour or cycle. Lacasse-Guay et al. (2010) compares and contrasts several of the available aircraft routing models available in the literature. More recently, Zhe et al. (2011) present a new compact network representation of the maintenance routing problem and develop a new mixed-integer linear programing formulation to solve the problem. The proposed method can be integrated with other aspects of the airline planning process such as fleet assignment and crew pairing. In addition, Gronkvist (2005) has proposed a column generation—constraint programing hybrid approach for scheduling/routing aircraft while making tail assignments.

To develop planned maintenance routings, many carriers have implemented commercially available software applications such as Sabre's ARM application to help the schedule planner make sure that the fleeted schedule provides adequate maintenance opportunities every 3–4 days. Most of these applications use the two-phase approach outlines above and use a heuristic search and swap algorithms to find break daily lines of flying to develop feasible aircraft maintenance tours.

The reader should keep in mind that aircraft maintenance routing during the schedule planning process focuses on routing feasibility. No tail assignments are done at this time and often, the actual routings are planned much closer to the day of departure. In addition, many of the final tail assignments and routings are changed during the day of departure. In fact, at large carriers only about 80–85% of the expected terminators at 1,100 in the morning actually end up terminating at the planned stations that evening. Often, these changes are due to schedule disruptions due to weather events, crew misconnects or short-term maintenance issues that delay an aircraft and force a substitution or move-up (when an available aircraft is moved up to cover a delayed aircraft) in the schedule.

2.7 More Recent Developments in Schedule Planning and Development

In addition to the typical tasks associated with schedule planning and development, there have been a number of new developments that improve the schedule efficiency and profitability. These include the integration of crew and scheduling processes to develop more efficient flying schedules, the implementation of demand driven dispatch where fleet assignment changes are made between crew compatible aircraft close to the day of departure to better match passenger demand and the addition of probabilistic evaluation techniques to better evaluate the likely performance of specific markets.

2.7.1 Integration of Crew and Scheduling Processes

Originally, scheduling, crew planning, aircraft routing and revenue management processes were all optimized within separate functional silos. This was predominately due to the fact that each of these processes was done by hand. Later, when computers were available and mathematical models were developed to optimize aspects of the business, computational power was limited and airlines could not solve the more integrated and complex problems. However, many researchers have recently begun exploring integration ideas throughout the airline business. In addition to the integration of fleeting and revenue management presented earlier, Sandhu and Klabjan (2007) propose the integration of schedule fleeting, aircraft routing and crew pairing. Their approach combines these three aspects of the airline planning process to a single solution framework to develop a simultaneous solution to all three planning problems. In this work they present two different decomposition strategies to solve the integration problem: the first uses a Lagrangian relaxation and delayed column generation while the second uses a Benders decomposition strategy. Gao et al. (2009) investigate crew and fleet assignment integration using station purity. In their work, Gao et al. focus on three challenges that included the influence of fleet assignment on crew scheduling, addressing crew scheduling in a tractable way within the integrated model and to produce a robust schedule. Their work proposes a new approach that integrates crew connections within the FAM and imposes station purity by limiting the number of fleet types and crew bases allowed at each station.

More recently, Ruther (2010) has proposed a multi-commodity flow model for integrating aircraft routing, crew scheduling and tail assignment. In this work, Ruther assumes crews are initially only told when they will work. This allows the generation of a schedule much closer to the planning horizon. However, for many U.S.-based carriers, this assumption would violate existing collective bargaining agreements with crew rank and file. Ruther presents solutions for very small sample cases but acknowledges that solving larger more realistic problems would be very difficult and would require other techniques to decompose the problem.

Recently, the integration of crew scheduling aspects has yielded significant cost savings associated with reduced crew connections where crews move from one aircraft to another following their arrival at a station and better feedback between the schedule development process and crew scheduling process. Integrating the scheduling and crew planning processes provides opportunities for having the crew and aircraft stay together as much as possible during the day. This saves connection time for the crew members during aircraft rotations and increases their efficiency for the airline. Some U.S.-based carriers have started the integration of crew and scheduling processes by refining the fleet assignments between sub-fleets to match crew pairings.

One of the easiest ways to accomplish this is to incentivize the FAM to consider potential crew trips within the fleeting process. This can be done using forced turns or soft-forced turns in which a desired aircraft turn corresponds to a crew turn within a pairing or trip. Soft-forced turns represent a relaxation of the typical forced turn and allow a crew connection to be made at a cost to determine a feasible scheduling solution. In this case, the penalties or costs associated with creating a crew connection are included in the objective function. In reality, this represents an iterative process between the fleeting model and the crew pairing optimizer.

American Airlines applied this strategy in 2006 and was able to reduce the number of flight deck crew connections for the Super 80 fleet from about 330 to 35 per day. In this case, the collective bargaining agreement gave pilots an extra 10 min to close out the aircraft paper work when connecting. By minimizing the number of crew connects, the airline was able to save approximately 100 hours of pilot and first officer labor per day of operation for the Super 80 fleet. Other carriers like Southwest Airlines have also had success with more closely integrating their crew routings with the aircraft to keep the crew and equipment together throughout the day.

2.7.2 Demand Driven Dispatch

Another area of great interest today focuses on the re-allocation of capacity close to the day of departure. Often referred to as Demand Driven Dispatch or D^3 and first proposed by Berge and Hopperstad (1993) this process attempts to increase overall profitability by making strategic aircraft swaps between crew compatible aircraft near the day of departure driven by updated O&D passenger demand forecasts. Three primary factors driving the increased profitability include (1) daily forecast variability, (2) forecast error and (3) inconsistencies between the schedule fleeting and revenue management methodologies used to produce the schedule and manage the seat inventory.

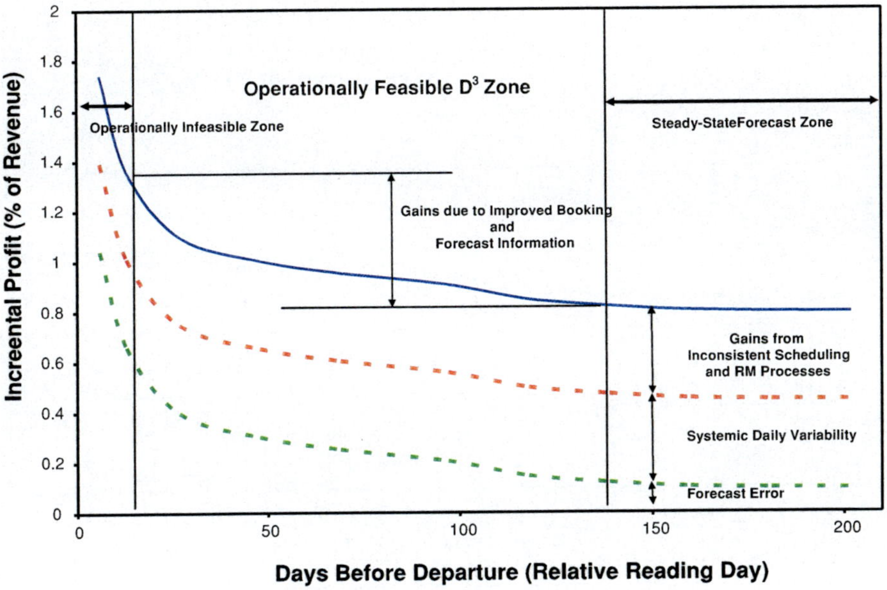

Fig. 2.14 Incremental benefit of D^3 as a function of time before departure

Table 2.9 Initial D^3 benchmark results

Measure	Input schedule	D^3 solution
Incremental profit gain (% of total revenue)		0.64
Switched flights		114
Segments flown		
RJ3	230	198
RJ4	336	368
Utilization		
RJ3	10:31	9:37
RJ4	10:02	10:14

Table 2.10 Trade-off between revenue gain and swap limit

Swap limit	Daily profit increase (% of revenue)	Cumulative percent of total (%)
25	0.25	39
50	0.35	56
75	0.50	78
100	0.60	94
114	0.64	100

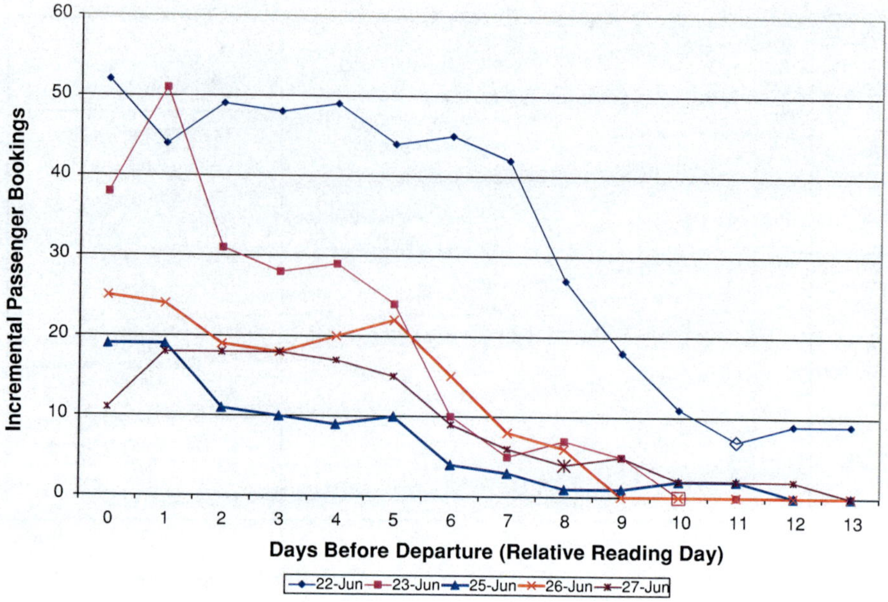

Fig. 2.15 Incremental passenger bookings after D^3 equipment swaps

Most airlines develop schedules based on typical day or typical day-of-week forecast data. However, in reality, passenger demand can vary substantially by day or day-of-week throughout the schedule's effective life span. D^3 capitalizes on opportunities created by the systemic variation of O&D passenger demand flowing through the network by identifying strategic and feasible capacity realignments that increase total system profit. D^3 exploits these opportunities by allocating additional seat inventory in markets with increased demand and decreasing seat availability in markets where expected demand has not materialized.

D^3 also improves schedule profitability by using improved forecast data nearer the day of departure. Most airlines develop schedules far in advance of their operation based on average demand forecasts for the schedule period. Time lines for developing schedules vary by airline but typically range from 6 to 18 months prior to departure. The errors associated with the demand forecasts used by the revenue management systems decrease closer to the day of departure as bookings materialize.

Lastly, many airlines fleet the schedule using leg-based methods while managing the seat inventory using O&D-based strategies. This leads to an inconsistent matching of supply (aircraft capacity) and passenger demand. Using the O&D FAM methodology presented earlier, D^3 can help compensate for the inconsistencies inherent in the schedule development and revenue management processes by better matching O&D passenger demand and capacity.

Table 2.11 Summary of traffic changes for June 22nd D^3 swaps

Flight	Original fleet assignment	D^3 fleet assignment	Expected traffic (passengers)	Actual traffic (passengers)
1	EM3	EM4	21	21
2	EM3	EM4	48	44
3	EM3	EM4	27	32
4	EM3	EM4	47	33
5	EM3	EM4	43	38
6	EM3	EM4	36	25
7	EM3	EM4	23	27
8	EM3	EM4	48	50
9	EM3	EM4	37	34
10	EM3	EM4	46	42
11	EM3	EM4	45	50
12	EM3	EM$	48	43
13	EM3	EM4	44	26
14	EM3	EM4	48	33
15	EM3	EM4	38	44
16	EM3	EM4	21	6
17	EM4	EM3	30	30
18	EM4	EM3	32	37
19	EM4	EM3	29	Canceled
20	EM4	EM3	36	Canceled
21	EM4	EM3	26	31
22	EM4	EM3	34	37
23	EM4	EM3	36	30
24	EM4	EM3	33	26

Table 2.12 Yearly summary of D^3 incremental traffic gains

	Incremental passenger accommodated	Number of days D^3 implemented during the month	Average number of incremental passengers per day
January	56	13	4.31
February	61	22	2.77
March	276	28	9.86
April	124	29	4.28
May	239	31	7.71
June	263	28	9.39
July	189	22	8.59
August	315	31	10.16
September	138	20	6.9
October	427	31	13.77
November	327	30	10.90
December	394	20	19.70
Yearly Total	**2,809**	**305**	**9.21**

Each of these factors contributes to the potential profit opportunities of implementing a D^3 strategy within a traditional network carrier. To illustrate the benefits of D^3 scenario, we performed a series of benchmarks and pilot studies for American Eagle Airlines (a subsidiary of American Airlines). American Eagle operates approximately 2,200 daily flights within the United States, Mexico, Canada and the Caribbean Islands.

To illustrate the benefits of D^3, Jacobs et al. (2001, 2008) applied D^3 to American Eagles' Embraer 135 and 145 aircraft during the spring and summer of 2001. The pilot study included 5 days of departures at Chicago's O'Hare International Airport. The D^3 swaps were identified and implemented using Sabre's commercial O&D FAM (Jacobs et al. 2000) implementation 14 days before departure. This timeframe was selected to maximize the potential profit benefits without impacting maintenance planning, operations and scheduling. Figure 2.14 presents the incremental benefit of D^3 as a function of the time before departure. Fourteen days before departure presented a convenient time to make D^3 swaps due to 14-day advanced purchase requirements. In addition, inside 7–10 days before departure, equipment swaps can disrupt maintenance planning.

Figure 2.14 also illustrates the individual components (daily variation in demand, improved forecasting and an O&D FAM) that make up the incremental benefit of D^3 swaps as a function of time.

Table 2.9 presents the overall results of the benchmark study using a daily schedule. The initial benchmark study did not include any restrictions on the number of equipment swaps made by the D^3 process. As a result, this benchmark result represents an upper bound on the profit benefits associated with this specific data. Overall, the unrestricted profit benefits associated with this demand and revenue data amount to a 0.64% margin profit gain (as a percent of total system revenue) and required 114 equipment swaps. To evaluate the sensitivity of D^3 to equipment swap limits, we ran a series of benchmark scenarios with various swap limits in O&D FAM. Table 2.10 presents the trade-off between the swap limits and the expected daily benefit.

To confirm the benchmark results, we performed a pilot study using 5 days of departures at Chicago's O'Hare International Airport (ORD). The 5 days included high, medium and low demand days. The pilot study allowed up to 25 equipment swaps and was driven by O&D revenue management data (itinerary fares, passenger demand and existing bookings). Table 2.9 presents the average daily results for the pilot study. Overall, the pilot study accommodated 33 additional passengers per day with an incremental profit of 0.27% with respect to total revenue. Figure 2.15 presents the cumulative number of additional passengers booked with respect to days before departure. Table 2.11 summarizes the change in traffic for D^3 swapped flights for June 22nd departures.

Following the benchmark and pilot studies, American Eagle implemented the D^3 process using O&D FAM for operations in its Chicago (ORD) and Dallas–Fort Worth (DFW) hubs. Table 2.12 summarizes the results for a full year of D^3 operations in Chicago (ORD) using Embraer 140 and Embraer 145 aircraft with a six-seat difference and a swap limit of 25. A swap limit of 25 generally constrains

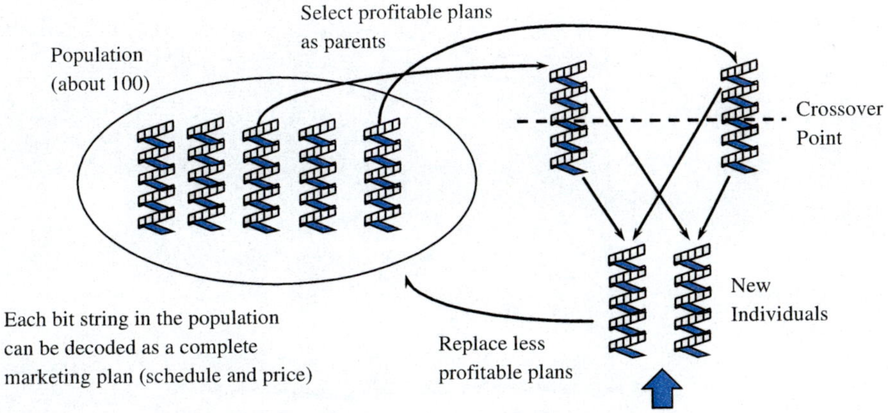

Fig. 2.16 Illustration of the genetic algorithm proposed by Walker (2001)

the number of upgraded flights to 12 per day. In addition, most of the Embraer flying out of ORD and DFW represents "out and back" round trips which mean only six upgraded flights drive the D^3 swaps.

The first column lists the month of operation, the second column represents the number of incremental passengers above the initial aircraft capacity gained by the D^3 process, the third column indicated the total number of days D^3 was used for each month and the fourth column represents the average number of passengers gained by the D^3 process per day for each month.

For this application of D^3, results show that adjusting fleeting decisions closer to the day of departure clearly identifies capacity changes that increase the O&D passenger revenue and profit in a manner consistent with revenue management controls. The D^3 swaps provided substantial incremental revenue without hampering operations at either hub.

2.7.3 *Operational and Market Performance Evaluation*

Another recent focus area involves the probabilistic evaluation of schedule performance. Several researchers and technology companies have developed ways to evaluate the profitability and operational performance of a proposed schedule. Rosenberger et al. (2002) and Lee et al. (2003) have developed and tested an operational performance simulation routine called SimAir to measure airline performance. This approach uses a modular set of simulation routines that reflect various aspects of the airline operations. This model also includes the ability to simulate the impact of disruptions to the airline due to inclement weather events or other unplanned delays. Green (2002) applied SimAir to evaluate the impact of various disruption scenarios on the overall operational performance of an

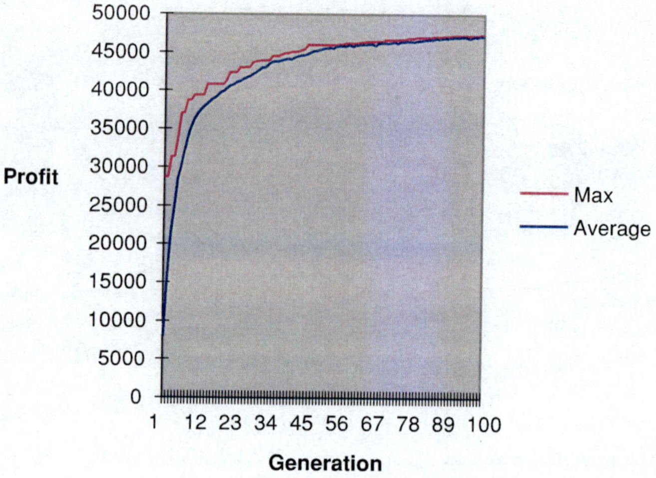

Fig. 2.17 GA performance with respect to network profit

American Airlines schedule. The output of SimAir consists of estimates of many of the operational metrics used by most airlines and the U.S. Department of Transportation (DOT). These include D0, A+14, completion factor and block performance.

Sabre and a number of airlines have combined efforts to develop a revenue simulation routine that can be used to estimate the impact of various revenue management strategies on overall network revenue (Ratliff 2008). This simulation routine provides participating airlines with a flexible mechanism to plug in and test various revenue management and pricing ideas and measure the impact to overall system revenue. The simulation modules use Bayesian updating scheme to develop booking scenarios that serve as inputs to a Monte Carlo simulation engine.

Wong and Jacobs (2009) and Jacobs (2005) present a risk-based methodology for quickly evaluating likely market performance of individual markets using estimates of revenue and cost factors. This methodology was encouraged by the fact that many carriers use simple estimates of revenue per available seat mile and cost per available seat mile to estimate if a market will yield profitable results without accounting for the potential variability associated with factors like market changes, operational disruptions and demand fluctuations. This approach uses the first and second moments of the revenue and cost factors to estimate the probability that the overall revenue will exceed the overall cost within the market. This methodology provides an attractive and easy mechanism for incorporating chance constraints into optimization models such as the FAM to help guide scheduling decisions based on probabilistic considerations. Parker (2008) has enhanced this approach to associate a confidence level with market share estimations and passenger behavior.

2.7.4 Automated Schedule Creation Methodologies

Ideally, schedulers would like to have a method for automatically developing a "clean-sheet" schedule that includes pricing points and accurately incorporates the revenue management, operational and cost factors into the process. Walker (2001) presents an interesting Genetic Algorithm (GA) approach for developing a "clean-sheet" schedule or incrementally improving an existing schedule. In this work, the GA represents a complete market plan (schedule and price) as a bit string. These bit strings or market plans are then combined and mutated to produce a series of new market plans that replace existing plans that represent inferior solutions for the market. This process is repeated for a set number of iterations to develop an improved overall schedule that includes both local and flow markets. Figure 2.16 illustrates this basic concept.

For the GA approach, Walker uses a simple passenger choice model to determine the attractiveness of a service within the market plan. Using this, the expected traffic is determined using the same revenue management model used in the development of the O&D FAM formulation presented earlier (See Appendix for a detailed description of the model). Costs were estimated as a function of the flights based on equipment, distance and the appropriate unit costs. Figure 2.17 presents a graphical representation of the GA performance with respect to network profitability for a small illustrative schedule design problem.

Appendix A: Network Flow and the O&D Yield Management Formulation

The O&D yield management (O&D YM) problem is a nonlinear network flow problem. The formal decision variable represents the number of seats allocated to a specific O&D market class. The objective is to maximize the overall revenue of the nonlinear objective function. The model is subject to one structural constraint that limits the total number of seats allocated to all market classes across any flight leg in the schedule to be less than the capacity of the aircraft assigned to the leg. Formally, the O&D YM model is defined by:

$$\max R_{\text{Total}_{YM}} = \sum_{s \in S} R_s \left[\int_0^{\hat{a}_s} y_s f(y_s) dy_s + \hat{a}_s \int_{\hat{a}_s}^{\infty} f(y_s)\, dy_s \right] \tag{A.1}$$

subject to:

$$\sum_{s \in S(j)} \hat{a}_s \leq \text{CAP}_j \quad \forall j \in J \tag{A.2}$$

$$\hat{a}_s \geq 0 \quad \forall s \in S \tag{A.3}$$

where $\hat{a}_s$ is the decision variable which represents the number of seats allocated to service s contained in the overall set of services S. $S(j)$ represents the set of services that involve flight j and is a subset of the total services S. A service represents a path specific O&D market fare class and can involve one or more flight legs. For example, all Y class passengers flying from Seattle to Boston through DFW on given flights are considered a single service. CAP_j represents the capacity of the fleet type assigned to leg j. $y_s$represents the stochastic demand for service s and $f(y_s)$ is the probability density function of the service demand. R_s equals the revenue per passenger resulting from the sale of a seat for service s. The probability density function for this formulation is usually modeled using a Gamma function. The O&D YM formulation is solved using Lagrangian relaxation (Reeves 1993) and the sub-gradient algorithm.

The solution of this model results in a set of seat allocations for each O&D service on each flight leg in the schedule. More importantly, the dual variables associated with each flight leg represent the marginal value of having one additional seat for the flight leg. This value is commonly referred to as the bidprice or hurdle rate and represents the slope of the revenue function at that leg capacity. In addition, the bidprice can be used as a metric to determine if it is profitable to sell additional tickets for a specific service.

For example, if the bidprice for traveling from Seattle to DFW is \$123 and the cost of a discount ticket for that service is \$90.00, it would not be profitable to sell the seat to a discount passenger. On the other hand, if the full fare coach ticket costs \$130, it would be profitable to sell additional full fare coach tickets.

Appendix B: Estimating Leg Revenue in FAM

One of the most common ways to formulate a FAM is using average or estimated leg-based revenues and costs. The leg-based revenues represent point estimates of the expected revenues over each leg in the schedule. Typically, there are two ways of estimating the expected revenues for each leg.

The first approach defines the expected revenues for each leg as the product of the expected traffic and the average fare paid by an individual traveling on the leg. There are two major problems with this approach: double counting of the through passenger traffic and the averaging of the revenue per passenger. This method of estimating revenues results in total expected revenues much higher than the revenues actually realized by the schedule. This is due to the fact that this approach double counts revenue by using O&D fares for connecting passengers to estimate the expected revenues for each leg in the itinerary. In addition, this approach assumes that all the demand has a per passenger revenue equal to the mean revenue for the leg. Taken together, these shortcomings can add a significant amount of error to the expected revenues for each leg in the schedule. Further more, this approach tends to bias leg-FAM solutions to favor the assignment and use of larger aircraft. As a result, the fleeted schedule tends to under perform with respect to expected profit.

The second approach improves upon the first by prorating the revenue estimates of through traffic with respect to factors such as the length of the individual legs within the itinerary. This approach provides a more accurate and reasonable estimate of the revenue for each leg. However, this approach still assumes that the total demand on each leg has a leg revenue per passenger equal to the mean revenue for the leg.

The first approach significantly over emphasizes the value of connecting traffic on each leg within the network. This can result in the false perception that a large capacity aircraft is needed for a moderately loaded leg *fleeting differences* in the network. The second approach over emphasizes the value of the local traffic on the network legs. This approach also may result in the false perception about the size of aircraft that needed to meet the traffic needs of the leg. Reality is somewhere in between these approaches and depends on the overall network load. In cases when the average leg revenue is consistent with the O&D revenue expected on the schedule legs, Leg-FAM provides a good O&D solution. However, this is rarely the case and the errors in the revenue estimates can lead to inferior fleet assignments.

Appendix C: Final Assignments for Leg-FAM and O&D FAM (Prorated Revenue Case) (Shaded rows represent differences between Leg-FAM and O&D FAM)

	Leg-based	FAM solution	O&D FAM	Solution
Flight Leg	Equipment	Expected traffic	Equipment	Expected traffic
ABQDFW1	757	165	757	165
ABQDFW2	757	170	757	171
AUSDFW1	757	163	757	163
AUSDFW2	757	166	757	167
BOSDCA1	S80	38	S80	38
BOSDFW1	S80	124	S80	123

(continued)

(continued)

	Leg-based	FAM solution	O&D FAM	Solution
BOSDFW2	757	153	757	147
BOSORD1	757	34	757	34
BOSSFO1	S80	34	S80	34
DCABOS1	S80	38	S80	38
DCADFW1	757	158	757	157
DCADFW2	757	161	757	159
DCAJFK1	S80	34	757	34
DFWABQ1	757	170	757	171
DFWABQ2	757	148	757	149
DFWAUS1	757	166	757	170
DFWAUS2	757	145	757	144
DFWBOS1	757	162	757	158
DFWBOS2	S80	97	S80	97
DFWDCA1	757	169	757	168
DFWDCA2	757	119	757	117
DFWJFK1	757	160	757	160
DFWJFK2	757	137	S80	112
DFWORD1	757	167	S80	131
DFWORD2	S80	124	S80	124
DFWSAT1	757	169	757	170
DFWSAT2	757	149	757	149
DFWSEA1	S80	136	S80	136
DFWSEA2	757	153	757	153
DFWSFO1	S80	131	S80	131
DFWSFO2	757	145	757	144
JFKDCA1	S80	38	757	38
JFKDFW1	757	153	S80	122
JFKDFW2	757	158	757	158
JFKSFO1	S80	34	S80	34
ORDBOS1	757	63	757	63
ORDDFW1	757	131	757	131
ORDDFW2	757	155	S80	130
ORDSEA1	S80	39	S80	39
SATDFW1	757	170	757	170
SATDFW2	757	153	757	154
SEADFW1	757	183	757	183
SEADFW2	S80	24	S80	24
SEAORD1	S80	63	757	63
SFOBOS1	S80	34	S80	34
SFODFW1	757	178	757	176
SFODFW2	S80	53	S80	53
SFOJFK1	S80	39	S80	39

References

Abara J (1989) Applying integer linear programming to the fleet assignment problem. Interfaces 19(4):20–28

Algers S, Beser M (2001) Modelling choice of flight and booking class—a study using stated preference and reveled preference data. Int J Serv Technol Manag 2:28–45

Barnhart C, Boland NL, Clarke LW, Johnson EL, Nemhauser GL, Shenoi R (1998) Flight string models for aircraft fleeting and routing. Transp Sci 32(3):208–220

Barnhart C, Kniker T, Lohatepanont M (2002) Itinerary-based fleet assignment. Transp Sci 36(2):199–217

Berge ME, Hopperstad CA (1993) Demand driven dispatchdispatch: a method for dynamic aircraft capacity assignment, models and algorithms. Oper Res 41(1):153–168

Bradley SP, Hax AC, Magnanti TL (1977) Applied mathematical programming. Addison-Wesley, Menlo Park

Carrier E (2008) Modeling the choice of an airline itinerary and fare product using booking and seat availability data. Ph.D. Dissertation, Department of civil and environmental engineering, Massachusetts Institute of Technology, Cambridge, MA

Civil Aeronautics Board (1970) Effect on total market traffic of changes in quality of service (QSI). Docket 21136 (box 1196). Exhibit BOR-R-300. In Exhibit Series *Rebuttal Exhibits of the Bureau of Operating Rights* by J.F. Adley and C.J. Caridi

Clarke L, Johnson E, Nemhauser G, Zhu Z (1997) The aircraft rotation problem. Ann Oper Res 69:33–46

Coldren, GM (2005) Modeling the competitive dynamic among air-travel itineraries with generalized extreme value models, Ph.D. Dissertation, Department of Civil and Environmental Engineering, Northwestern University, Evanston

Coldren GM, Koppelman FS (2005a) Modeling the competition among air-travel itinerary shares: GEV model development. Transp Res Part A 39(4):345–365

Coldren GM, Koppelman FS (2005b) Modeling the proximate covariance property of air travel itineraries along the time-of-day dimension. Transp Res Rec 1915:112–123

Coldren GM, Koppelman FS, Kasturirangan K, Mukherjee A (2003) Modeling aggregate air-travel itinerary shares: logit model development at a major U.S. airline. J Air Transp Manag 9(6):361–369

Data Base Products Inc. (2010) Superset, Dallas, TX, USA

Erdmann A, Kiahaschemi M, Noltemeier A, Schrader R (1997) Fleet assignment with respect to itineraries. Presentation at mathematical programming conference, Lausanne

Ervin F, Lonsdale R, Parker R, Zhang J (2005) The effect of alliances in scheduling and strategic planning. Paper presented at the AGIFORS scheduling and strategic planning conference, Seattle

Farkas A (1996) The Influence of Network Effects and Yield Management on Airline Assignment Decisions, Ph.D. Dissertation, Department of Civil and Environmental Engineering, Massachusetts Institute of Technology, Cambridge, MA

Gao G, Johnson E, Smith B (2009) Integrated airline fleet and crew robust planning. Transp Sci 43(1):2–16

Garrow LA (2010) Discrete choice modelling and air travel demand: theory and applications. Ashgate, Aldershot

Goplan R, Talluri K (1998) The aircraft maintenance routing problem. Oper Res 46(2):260–271

Gramming J, Hujer R, Scheidler M (2005) Discrete choice modeling in airline network management. J Appl Econom 20:467–486

Green T (2002) Evaluating airline schedules for improved operational dependability. In: Proceedings of the AGIFORS annual symposium. Honolulu, Hawaii

Gronkvist M (2005) The tail assignment problem. Thesis submitted to the Department of Computer Science and Engineering, Chalmers University of Technology and Goteborg University, Goteborg, Sweden

Hane C, Barnhart C, Johnson EL, Marsten R, Nemhauser GL, Sigismendi G (1995) The fleet assignment problem: solving a large-scale integer program. Math Program 70:211–232
Jacobs TL (2005) Quantifying market performance risks: a probabilistic approach. In: Proceedings of the AGIFORS scheduling and strategic planning group, Seattle, Washington
Jacobs TL, Johnson EL, Smith BC (1999) O&D FAM: incorporating passenger flows into the fleeting process. In: AGIFORS symposium proceedings 39, New Orleans
Jacobs TL, Hunt E, Korol M (2001) Scheduling and revenue management process integration: benefits and hurdles. In: Proceedings of the AGIFORS revenue management and distribution study group. Bangkok, Thailand
Jacobs TL, Ratliff RM, Smith BC (2000) Soaring with synchronized systems: coordinated airling scheduling, yield management and pricing decisions can make airline revenue rake off. ORMS Today 27:4
Jacobs TL, Smith BC, Johnson EL (2008) Incorporating network flow effects into the fleet assignment process. Transp Sci 42(4):514–529
Jacobs TL, Smith BC, Johnson E (2000) System and method for incorporating origination and destination effects into a vehicle assignment process. US patent no. 6,076,067, United States Patent Office, Washington D.C. (Commercially known as O&D FAM™)
Kliewer G (2000) Cooperative approaches for market modeling and fleet assignment. Presentation to ISMP 2000. University of Paderborn, Germany
Knicker TS (1998) Itinerary-based fleet assignment, Ph.D. Dissertation, Massachusetts Institute of Technology
Koppelman FS, Coldren GC, Parker RA (2008) Schedule delay impacts on air-travel itinerary demand. Transp Res Part B 42(3):263–273
Lacasse-Guay E, Desaulniers G, Soumis F (2010) Aircraft routing under different business processes. J Air Transp Manag 16(5):258–263
Lee LH, Huang HC, Lee C, Chew EP, Jaruphongsa W, Yong YY, Liang Z, Leong CH, Tan YP, Namburi K, Johnson EL, Banks J (2003) Simulation of airports/aviation systems: discrete event simulation model for airline operations: SIMAIR. In: Proceedings of the 35th conference on Winter simulation: driving innovation, New Orleans, Louisiana
Lohatepanont M (2001) Airline fleet assignment and schedule design: integrated models and algorithms, Ph.D. Dissertation, Massachusetts Institute of Technology
McFadden D (1974) Conditional logit analysis of qualitative choice behavior. In: Zarembka P (ed) Frontiers in econometrics. Academic Press, New York, pp 105–142
Nako SM (1992) Frequent Flyer programs and business travellers: an empirical investigation. Logist Transp Rev 28(4):395–414
Nemhauser GL, Wolsey LA (1988) Integer and combinatorial optimization. Wiley, New York
OAGOAG Worldwide Limited (2010) Official Airline Guide, Bedfordshire, LU5 4HB, UK
Parker R (2008) The connectedness of the world's airline network: assessing the error introduced when isolating sub-networks of the world's airline networks. In: Proceedings of the 48th AGIFORS annual symposium, Montreal, Canada
Parker RG, Rardin RL (1988) Discrete optimization. Academic Press, Boston
Parker R, Lonsdale R, Glans T, Zhang Z (2005) The na universal market simulator. Presentation to the advanced market techniques forum of the american marketing association, Coeur d' Alene, Idaho
Proussaloglou K, Koppelman FS (1999) The choice of air carrier, flight and fare class. J Air Transp Manag 5(4):193–201
Qi X, Yang J, Yu G (2004) Chapter 50—Scheduling problems in the airline industry. In: Leung JY-T (ed) Handbook of scheduling algorithms, models and performance analysis. Chapman and Hall/CRC, New York
Ratliff R (2008) Multi-flight demand untruncation: simulation results. In: Proceedings of the AGIFORS revenue management and distribution study group meeting. Tahiti, French Polynesia
Reeves CR (1993) Modern heuristic techniques for combinatorial problems. Wiley, New York

Rosenberger JM, Schaefer AJ, Goldsman D, Johnson EL, Kleywegt AJ, Nemhauser GL (2002) A stochastic model of airline operations. Transp Sci 36:357–377

Ruther S (2010) A multi-commodity flow formulation for the integrated aircraft routing, crew pairing, and tail assignment problem. In: Proceedings of the 45th annual conference of the ORSNZ, November 2010

Sandhu R, Klabjan D (2007) Integrated airline fleeting and crew pairing decisions. Oper Res 55:430–438

Smith BC, Jacobs TL (1997) Airline planning and marketing in a changing environment. J Infrastruct Syst 3(1):1–3

Smith BC, Johnson EL (2006) Robust airline fleet assignment: imposing station purity using station decomposition. Transp Sci 40(4):497–516

Subramanian R, Scheff RP Jr, Quillinan JD, Wiper DS, Marsten RE (1994) Coldstart: fleet assignment at delta air lines. Interfaces 24(1):104–120

Suzuki Y, Tyworth JE, Novack RA (2001) Airline market share and customer service quality: a reference-dependent model. Transp Res Part A 35(9):773–788

Swan WM (1983) Traffic losses at high load factors. In: AGIFORS symposium proceedings 23, Olive Branch, Mississippi

Talluri K (1998) The four-day aircraft maintenance routing problem. Transp Sci 32(1):43–53

Transportation Research Board (2002) Transportation research E-circular E-C040, aviation demand forecasting: a survey of methodologies. Transportation Research Board, Washington D.C

Walker A (2001) Macro-level scheduling and pricing using a genetic algorithm. In: Proceedings of the 41st AGIFORS annual symposium. Sydney, Australia

Wong D, Jacobs TL (2009) A probabilistic approach to quantifying network & market performance risk. In: Proceedings of the 49th AGIFORS annual symposium. Atlanta, Georgia

Zeid MA, Rossi TF, Gardner B (2006) Modeling time of day choice in the context of tour and activity based models. Paper presented at the 2006 annual meeting of the transportation research board, Washington D.C

Zhe L, Chaovalitwongse WA, Huang HC, Johnson EL (2011) Network model for aircraft routing problem. Transp Sci 45(1):109–120

Chapter 3
Revenue Management

Darius Walczak, E. Andrew Boyd, and Roxy Cramer

3.1 Introduction

Revenue management is arguably the most celebrated application of mathematical modeling used in the travel industry. Originating in the airline industry around the time of deregulation in the late 1970s, revenue management has consistently been credited with increasing airline revenues by over six percent. In a frequently cited reference, Thomas Cook, former CEO of SABRE Decision Technologies, claimed incremental revenue improvements at American Airlines of almost $1 billion annually (Cook 1998).

The basic concept is simple. Airlines sell seats by creating different fare classes carrying different restrictions and prices. For example, on a flight from Seattle to Chicago, a carrier might offer an unrestricted Y class fare of $500 and a $200 Q class fare requiring a Saturday night stay and carrying change penalties. If an airline chose to, it could make both fare classes available until the flight's departure or until the plane ran out of seats. Airlines were quick to realize, however, that there was no need to keep the shelves fully stocked. With 10 seats left on a plane and a predicted demand of more than 10 passengers expected to purchase the more expensive Y class ticket, it does not make sense to sell any Q

D. Walczak (✉) · R. Cramer
Senior Scientist and Director, PROS Revenue Management, 3100 Main Street, Houston, TX 77002, USA
e-mail: dwalczak@prosrm.com

R. Cramer
e-mail: rdc_766@yahoo.com

E. Andrew Boyd
Chief Scientist and Senior Vice President, PROS Revenue Management, 3100 Main Street, University of Houston, Houston, TX 77002, USA

C. Barnhart and B. C. Smith, LLC (eds.), *Quantitative Problem Solving Methods in the Airline Industry*, International Series in Operations Research & Management Science 169,
DOI: 10.1007/978-1-4614-1608-1_3, © Springer Science+Business Media, LLC 2012

class tickets. Revenue management is the practice of determining what fare classes should be left on the shelf at any given time.

Of course, the process of actually determining what should be left on the shelf can be quite sophisticated, involving overbooking, forecasting, optimization, distribution channel, price, and a host of other factors (see Weatherford 1991). Effective revenue management incorporates the planning and control of these many dimensions.

3.1.1 History

The existence of fare class products traces its roots to the regulatory era. The Civil Aeronautics Board ensured that every airline charged the same price for a flight from city A to city B, but it recognized that there were instances where different prices for certain types of service were acceptable. Thus, tickets for children or tickets for late night flights from the west coast to the east coast—so-called "red-eye" flights—could be priced differently from regular fares, but would still be the same from carrier to carrier.

Recognizing the need to accommodate bookings by fare class, computerized airline reservations systems were built with fare classes in mind (see Chap. 4 "Airline Distribution"). For each flight, a reservations system maintained the number of seats available for sale in each fare class. If a seat was available it could be sold and the reservations system would automatically decrease the number of seats available in that class and on the plane. Passengers had tremendous freedom to cancel reservations at will, without penalty. Thus, airlines found it necessary to overbook on popular flights—to sell more seats than the physical capacity of a plane with the expectation that some booked passengers would not show up. The earliest revenue management efforts focused on how to overbook. Rothstein (1985) has a good account of this colorful period in the history of revenue management. Smith et al. (1992) describe early overbooking as practiced at American Airlines.

With the arrival of deregulation, airlines were freed to schedule, price, and manage inventory however they saw fit, and the question of how many tickets to make available in each fare class became a competitive necessity. It was an incredible change for the industry. Fare products proliferated as airlines sought to offer various forms of discounted fares. In the competitive new world, revenue management became a survival skill, as attested to by the many airlines that did not embrace it and subsequently went out of business (see, for example, Cross 1997, p. 125).

At first, simply managing the many new fare classes was a challenge. But over time, increasingly sophisticated scientific analysis emerged. Reservations systems kept record of vast numbers of historical bookings that in turn could be used to predict what would happen in the future. Computerized revenue management systems evolved that used this historical information to forecast future bookings

Table 3.1 An example of nested inventory levels for a 100-seat aircraft

Fare class	Price ($)	Protection levels	Booking limits
Y	500	25	100
M	300	55	75
Q	200	100	45

and cancellations, establish overbooking levels, determine the optimal number of seats to set aside in each fare class on each flight, and periodically load the optimal fare class inventory levels into an airline's reservations system. Reservations systems then managed the sale of tickets, recording bookings and cancellations based on the numbers provided by the revenue management system.

Today's revenue management systems are tools that connect revenue management analysts with an airline's most important asset: its seat inventory. While these systems typically embody sophisticated mathematics, analysts are required to manage holidays, special events, respond to competitive situations, and review results that fall outside of pre-established bounds. Good revenue management systems take care of the majority of inventory setting decisions automatically, freeing analysts to focus their time on the most important flights or markets.

3.1.2 Inventory Control

To better comprehend the mechanics of inventory control and lay the foundation for understanding the underlying mathematical models, consider an aircraft flying from Seattle to Houston with 100 coach seats and three fare classes, a $500 Y class fare, a $300 M class fare, and a $200 Q class fare. Assuming for simplicity that everyone who purchases a ticket actually shows up to fly (there are no cancellations or day of departure no-shows), an airline might partition inventory, setting aside 25 seats for Y class, 30 seats for M class, and 45 for Q class. The problem with partitioned inventory is that it has the potential for a higher value fare class such as Y to sell out while there still remains lower value M or Q class inventory. If this occurred, passengers wanting to pay $500 would be turned away or forced to purchase a less expensive M or Q class ticket.

To avoid this problem, airlines typically use nested inventory levels. With nested inventory levels, any seats available for lower priced classes are also available for higher priced classes. The overlapping seat availabilities can be represented in a number of different ways. Protection levels refer to the number of seats protected for sale only to a given class and all the classes with higher fare value. Booking limits refer to the number of seats available for sale to a given class and all classes of lower fare value. Observe that if protection levels are known then so are booking limits, and vice versa. In our example, a reservations system would maintain information similar to that found in Table 3.1.

While nesting is particularly important in environments where reoptimization is performed less frequently, it is not quite correct to say that with frequent reoptimization nested controls are not needed. Many provably optimal controls are of the nesting type and this is no coincidence since the partitioned allocations solve a problem that is different from the real one where classes higher in a hierarchy are indeed allowed to use allocations of classes lower in a hierarchy. In particular, nested controls have better pooling properties: classes with fares that are close to each other will be treated similarly. This is in contrast to partitioned controls which tend to produce inferior results when class demand variance is significant and there are many such classes. Two popular methods that utilize partitioned controls are EMSRa (single leg) and Probabilistic Nonlinear Program ("PNLP ") on a network. Talluri and van Ryzin (2004), have illustrative examples of such behavior on pages 46 and 96–97, for the EMSRa and the PNLP cases, respectively; see also references therein to simulations studies that support their theoretical considerations.

When nested inventory is used, the process of updating booking limits as seats are sold is slightly more complicated. For example, if no adjustments are made to the number of available Y and M class seats, an airline selling 45 Q class seats would show 100 Y class seats and 75 M class seats still available, even though there are only 30 seats left on the plane [see Vinod (2006) for a detailed discussion of the ways in which nested inventory can be decremented]. For business reasons, not all airlines use purely nested inventory control. For example, seat inventory is often partitioned to separate space allocated to group requests or block sales (sales to wholesalers) because they are controlled separately from individual bookings. But most of the mathematical analysis has focused on nested inventory control, and in practice variants of this form of control remain predominant.

When airlines sell tickets on an itinerary consisting of multiple flight legs, the ticket is issued in a single class: different classes are not issued on different flight legs. The reason for this practice stems from the fact that airlines view their products as a fare class from the city of origin to the city of destination, and products are priced in this way. There is a Y fare and there is a Q fare for traveling from Seattle to Miami, but there is not a "Y/Q" fare. Nothing stops a passenger from booking two separate flight legs in two separate fare classes, but the price would normally be higher than the lowest available fare class priced from Seattle to Miami. The limitation of booking in a single fare class is likely to become less important as airline distribution continues to evolve, but as of now it remains a significant consideration.

Since itineraries with multiple flight legs are booked in one fare class, airline reservations systems normally check all flight legs to determine class availability. For example, if an itinerary consisted of two flight legs and there were Q class seats available on each, then Q class would be available. If Q class was not available on either or both legs, it would not be available for the itinerary consisting of both flight legs.

The problem with controlling inventory in this way is easy to see, as illustrated in Fig. 3.1. Imagine that our Seattle to Miami itinerary connects through Atlanta, and

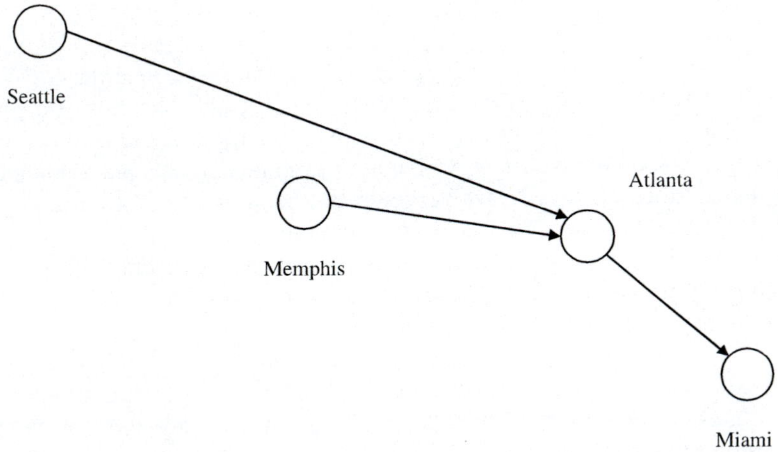

Fig. 3.1 Q class on the Seattle–Atlanta–Miami itinerary sells for $600, while Q class on the Memphis–Atlanta–Miami itinerary sells for $300. Both itineraries draw from the same Q class inventory on the Atlanta–Miami flight leg

that the Q class fare is $600. At the same time, suppose the price of a Q class fare from Memphis to Miami connecting through Atlanta is only $300. Both of these itineraries are looking at the same Q class inventory on the Atlanta–Miami flight leg, yet one ticket is worth far more than the other. The problem is multiplied many times over when all the possible itineraries to Miami connecting through Atlanta are considered. Fare class control is fine for airlines with no connecting traffic. When an airline's route structure consists of a network of connecting itineraries, however, fare class control is inadequate. The fact that airline products are defined and priced at one level and inventory is controlled at another represents one of the single largest flaws in modern day airline reservations systems and poses many practical challenges for the application of sophisticated mathematical models.

Setting aside the fact that the airline inventory control environment has fundamental flaws, airline revenue management must nonetheless operate within that environment. Historically, two conceptual approaches to revenue management have dominated both industry applications and the academic literature.

Leg revenue management starts from the perspective that the flight network consists of a collection of individual flight legs. Demand forecasts are made by leg and fare class, and fare class booking limits are calculated using an optimization algorithm that looks only at single flight legs (see Sects. 3.2.2, 3.3.2, and 3.3.3.1). Leg revenue management not only simplifies forecasting and optimization, but there is a natural tendency on the part of airline revenue management analysts to think in terms of "seats on flights," not "seats on itineraries." As a result, leg-based revenue management systems remain predominant in practice.

A significant difficulty with leg revenue management is the choice of fare for each fare class since, as discussed above, tickets with many different fares draw from the same leg/fare class inventory. Fares are important inputs to inventory allocation optimization algorithms. When using leg revenue management, the choice of fare is accomplished by choosing an appropriate "average" fare for each class, where many different definitions of "average" have been proposed and used. One class of average fares, called pseudo-fares, work reasonably well. Pseudo-fares are discussed in greater detail in Section "Mathematical Programming Models".

Network revenue management, which more commonly goes under the name of origin and destination or O&D revenue management, addresses the underlying network optimization problem by dealing directly with the flow of passengers across multiple flight legs—and not through the use of average fares as leg methods do. Within O&D, two primary methods of control dominate both practice and the literature: virtual nesting and bid price control.

The most sophisticated forms of virtual nesting start by defining virtual buckets on each leg. For example, virtual fare bucket 1 on an Atlanta–Miami flight might be for "fares" in the range \$700 and up, virtual bucket 2 for fares in the range \$600–700, virtual bucket 3 for fares in the range \$500–600, and so on. Itinerary/class fares are then prorated in some way to the individual flight legs in the itinerary, and each leg component of the itinerary fare is assigned to the appropriate virtual bucket. For example, if a \$600 Seattle–Atlanta–Miami Q class fare was prorated so that \$50 was assigned to the Seattle–Atlanta leg and \$550 was assigned to the Atlanta–Miami leg, then on the Atlanta–Miami leg it would be assigned to virtual bucket 3 (and on the Seattle–Atlanta leg to the appropriate fare bucket). If seat inventory is available in the assigned virtual buckets for this itinerary/fare class, Q class is available on each leg and therefore available on the Seattle–Atlanta–Miami itinerary. Virtual nesting is similar in spirit to leg revenue management—forecasting and optimization are typically done in the same way as for leg revenue management but using virtual buckets instead of fare classes. However, virtual nesting does not suffer from the need to use an average fare for each bucket (the buckets are defined tightly enough to allow the midpoint fare to be a good point estimate). The buckets are designed to hold a particular range of values, and itineraries are assigned to buckets based on their value.

The most common forms of bid price control begin by calculating a bid price for each of the flight legs in the network. An itinerary/fare class is available for sale if its price is at least as large as the sum of the bid prices on the itinerary. For example, if the Seattle–Atlanta flight leg has bid price of \$200 and the Atlanta–Miami flight leg has a \$350 bid price, then a \$600 Seattle–Atlanta–Miami Q class ticket is available for sale, since $\$600 \geq \$200 + \$350$. Bid price control is discussed more extensively in Section "Mathematical Programming Models". Additional details on both virtual nesting bid price control can be found in Phillips (2005, Chap. 8).

3.1.3 Displacement Costs

One of the most important concepts in all of revenue management is that of displacement cost of the inventory left on hand. A \$600 Seattle–Atlanta–Miami Q class fare seems to be more valuable than a $300 Memphis–Atlanta–Miami Q class fare, but it depends upon what passengers are being displaced on the Seattle–Atlanta and Memphis–Atlanta flight legs. If the Memphis–Atlanta flight is expected to fly with empty seats, then selling an additional seat on this flight doesn't displace any potential passengers. The displacement cost is 0. If on the other hand, the Seattle–Atlanta flight can be filled with passengers willing to pay the $700 Y fare to Atlanta, then the displacement cost of selling a seat on the Seattle–Atlanta flight leg is (at least) $700. The $300 Memphis–Atlanta–Miami Q class fare turns out to be more valuable than the $600 Seattle–Atlanta–Miami Q class fare once the cost of displacing other passengers is taken into account.

Displacement costs go under a variety of names—marginal costs, marginal revenues, opportunity costs, bid prices (Simpson 1989) and are not surprisingly related, if not identical, to dual variables from a constrained optimization problem. In most cases the revenue management optimization models assume risk-neutrality and maximization of the expected revenue. This makes the displacement costs essential since it is necessary to compare revenue from the immediate decision (the immediate reward from the fare in hand) with the impact that this decision will have on the future revenue stream. This logic applies regardless of whether the optimization approach is based on a heuristic such as EMSRb or exact solution such as dynamic programming, in a single flight leg or in a network setting. In all cases it is necessary to get a handle on the displacement cost, be it by estimating it as expected marginal seat revenue from the top nest of classes coming on a first-come-first-serve basis (EMSRb) or as an increment of the value function when inventory drops by one unit. In the network context this is further complicated since the displacement cost for a booking request using multiple legs is not necessarily equal to the sum of the displacement costs across the legs involved (although it is often approximated as such). See Bertsimas and Popescu (2003) for a good discussion and Section "Approximate Dynamic Programming" on approximate dynamic programming for some insights.

3.1.4 Revenue Management and Pricing

The relationship between revenue management and pricing departments within an airline is often confusing. Airline pricing departments create and price fare class products in much the same way dishwasher manufacturers create and price different models of dishwasher: how do the different products serve different market segments, how do product features and prices promote up-sell, and perhaps most importantly, what products and prices are being offered by competitors. The notion of what products are actually available for sale at any given time—the central focus of revenue management departments—is not a primary concern for pricing departments. Pricing is a strategic activity; revenue management, tactical. Pricing departments establish the rules of competitive engagement; revenue management departments manage seat inventory subject to these rules. With increased fare transparency—both to booking customers as well as to competitors—the two levels of decision making will have to converge.

Nonetheless, although revenue management departments do not change prices, they are often viewed as performing operational pricing since opening and closing fare classes lead to changes in the lowest price the customer sees. If an airline suddenly stops selling $200 Q class tickets and the customer only sees $500 Y class tickets, many customers will conclude that "the price has gone up by $300" although no prices have actually changed.

This example highlights a subtle and confusing aspect of revenue management. Are fare classes viewed by consumers as different products or simply different prices for the same product? The more general question of demand modeling is discussed at length in what follows.

3.1.5 The Changing Business Environment

Led in large part by academic research, revenue management forecasting and optimization have evolved to a highly advanced level. While the late 1990s and early 2000s saw a push for more and more sophisticated algorithms, the focus has since shifted to the underlying mathematical models. This shift was sparked in large part by two important business considerations: the growth of internet ticket sales and competition from a new generation of low fare carriers.

Touting a simplified fare structure with few or no ticketing restrictions, low fare carriers forced legacy carriers to part with such profitable activities as requiring passengers to stay over a Saturday night to get a reduced fare. Legacy carriers worry about the pricing power wielded by their low fare adversaries, and in many cases are reacting by emulating their practices.

The internet has contributed to shifting priorities by making competitive fares transparent. With the click of a mouse, revenue management analysts can instantly find the market fares that a passenger can purchase. Not just the fares associated with different fare classes, but what fare classes are actually available for sale. The same technology that has made it easy for us as passengers to search for the best ticket price has provided ideal tools for revenue management analysts to window shop the competition's prices. For many carriers, this newfound visibility into what the competition is offering has promoted a tendency for analysts to call into question the results of traditional revenue management algorithms.

The immediate result of these changes has been to bring existing revenue management practices under scrutiny, and to shift the scientific focus to models that address the new business environment. In response to carriers' sensitivity about what the competition is doing, one modeling theme that has emerged is the incorporation of competitive fare information. While competitive fare information has the potential to provide value, models that are reactive in nature—forms of fare matching, for example—pose potential dangers, and it remains to be seen what practical conclusions carriers will ultimately reach. It is ironic that a reactive response to competitive fares is what revenue management was in part designed to overcome. Bob Crandall's pioneering revenue management efforts at American Airlines to combat the low fare upstart People Express were an effort to avoid straight price matching. Using the tools of revenue management, Crandall could nominally offer competitive fares, yet hold back inventory to sell at a higher price after People Express sold out at a low price. He ultimately drove People Express out of business (Cross 1997) with better management of seat inventory and better service.

3.1.6 About What Follows

In the remainder of this chapter we look at the two important mathematical components of revenue management—forecasting and optimization. Both have received extensive attention in the context of revenue management, and it is not our intention to attempt a complete summarization of this work. Nor is it our intention to delve too deeply into the intricacies of the mathematical models, many of which are quite sophisticated. Instead, our goal is to provide insight into the way forecasting and optimization is actually practiced at many airlines around the world. Although the details vary from carrier to carrier, and while the discipline is constantly evolving, there are terms and concepts that have developed over the years that are common not only to airline revenue management but to revenue management in general. It is our hope that readers unfamiliar with revenue management will gain a basic understanding of these fundamentals as a background for further reading.

3.2 Forecasting

3.2.1 The Role of Forecasting

The most important role of forecasting in revenue management applications is to provide required inputs to an optimization problem that determines what fare classes are available on what flights. How a specific forecasting model is designed depends on how the forecasts will be used. Leg revenue management algorithms use forecasts at the flight leg/fare class level to produce nested inventory levels. The forecasting task is therefore to estimate the demand in each fare class on each flight leg. O&D revenue management algorithms require a much finer level of detail involving customer itineraries.

Forecasting requirements have grown more complex as optimization models have become more sophisticated. In the earliest models, only net demand (defined as expected gross bookings less expected cancellations) estimates were required. Refinements to the optimization and additional business cases were made by considering the individual components of the demand process: gross bookings, cancellations, show-ups, and no-shows. To help establish a context for our forecasting discussion, we next briefly review the different components and characteristics of air travel demand.

3.2.1.1 Bookings, Cancellations, Show-Ups, No-Shows, and Overbooking

Demand accumulates during the reservation period, or booking cycle—the time prior to a flight's departure during which reservations are accepted. Reservation periods begin close to a year in advance of the departure date. On any day prior to departure, new bookings can arrive and existing bookings can cancel, and they may do so up to a few days or hours prior to departure. On the day of departure, new bookings are typically called walk-ups or go-shows, if they have other bookings, and cancellations are called no-shows. (In hotel and car rental revenue management the terminology is slightly different: the customers who show up right before departure without a reservation are called walk-ins and walk-ups, respectively.) No-shows also include reserved passengers who simply do not show up without notifying the airline. The actual number of occupied seats after the plane departs, sometimes called materialized demand is the figure used in an airline's load factor calculations (Figs. 3.2, 3.3).

If airlines only sold as many seats as were available in an aircraft, then due to cancellations and no-shows planes could fly with empty seats, a problem referred to as spoilage. To avoid this situation, airlines overbook flights, selling more seats than physically available with the expectation that some booked passengers will not materialize. When more than expected do show up for the flight—as is bound to happen with overbooking—some passengers will be denied a seat. Overbooking algorithms balance the cost of denying a seat to a passenger (typically, they must be offered a seat on an alternative flight, sometimes hotel vouchers or coupons for future travel) against the opportunity cost of flying with empty seats. At any time during the booking cycle, overbooking limits will depend on bookings currently in hand, expected new bookings to come, expected cancellations, and expected no-shows and show-ups.

3.2.1.2 Constrained Data

Because of fixed capacity and inventory control, throughout a flight's life cycle some demand is almost always turned away, leading to bookings on hand that are almost always biased low. Bookings on hand can overestimate true demand if enough passengers double-book flights, an activity that can cause what the airlines know as churn in reservations systems. For the most part in RM systems, attempts to account for churn reside in the observation building process—the process of summarizing PNR data to the level of the forecasting models. It is the low bias induced by constrained inventory that every RM model must directly account for, one way or another.

Not accounting for the constrained nature of the available bookings data will underestimate true demand and will lead to suboptimal solutions. In particular, if we ignore the fact that data is constrained in our forecasting models and go about the business of setting controls, we will at each step capture less potential revenue, an effect known as spiral-down. Queenan et al. (2007) illustrate the spiral-down effect on revenues when no attempt to unconstrain bookings is made in the EMSR-b heuristic for setting seat protections. In Sect. 3.2.1.2 we discuss some common approaches to unconstraining. Unconstraining, and the product segmentation discussed next are two facets of a more general issue: the model misspecification. Cooper et al. (2006a, b) were the first to rigorously investigate the spiral-down phenomenon within the context of revenue management. In particular, they analyzed the impact of incorrect forecasting and optimization assumptions and their interplay in the repeated forecast and optimize cycles.

Revenue management models frequently assume that fare classes are different products and that demand arrives independently for each product. Even if fare classes are considered as different products by passengers, the independence assumption is clearly a simplification. Demand is correlated because of many external factors, including availability, seasonality, economic trends, and holiday or special events that influence overall demand. There are many ways to model these effects in forecasting so that most of the dependence can be accounted for, even while modeling demand for each class separately. The problem becomes far more difficult when the fare class restrictions or fences are not adequate to create different products in the minds of consumers, and they purchase based on price. We touch upon this broad topic of customer choice in Sect. 3.2.4 and see Chap. 1 "Customer Modeling" for a much more comprehensive analysis.

3.2.2 Leg Forecasting

3.2.2.1 Motivation

One of the most well known and widely used methods for determining nested inventory levels is the expected marginal seat revenue method, version b, more commonly known as EMSRb (see Sect. 3.3.2.1). Two distinguishing characteristics of EMSRb and other early methods are (a) an assumption that demand arrives in search of a specific fare class (passengers view fare classes as products), and (b) only net demand-to-come is required for calculations (there are no random processes describing the order in which passengers arrive). As a result, early forecasting methods focused on estimating the mean and variance of the demand-to-come in each fare class by combining forecasts of bookings-to-come, cancellations, show-ups, and no-shows. These components were not always forecast separately, but nonetheless made up the conceptual elements of what went into

Table 3.2 Number of bookings in Y class for flight 100, departed on thursday, 9/21

Days prior to departure											
180	90	40	30	20	10	8	6	4	2	0	Departure date
0	3	4	5	10	11	13	18	19	23	30	9/21

There are 10 data collection points

estimates of the demand-to-come. EMSRb and most other early methods for determining nested inventory levels also assumed that demand was normally distributed within each leg/fare class.

More recently, employed dynamic programming algorithms relax both of the assumptions found in EMSRb . Dynamic programming allows much richer demand behavior to be incorporated than passengers arriving in search of a specific fare class (the product model of demand), and the stochastic order in which passengers arrive is modeled explicitly. As a result, dynamic programming algorithms typically require forecasts of passenger bookings and cancellation rates at different points during the booking cycle, and these forecasts may not be limited to individual fare classes.

We present much of the following discussion in the context of forecasting leg/fare class bookings. The basic concepts, however, are applicable for forecasting cancellations and other components of demand, which we return to later in this chapter.

3.2.2.2 Modeling Bookings

One of the more interesting features of airline revenue management forecasting has to do with the way in which data is structured for forecasting purposes. The structure has evolved relatively consistently throughout the industry, largely through cross-fertilization brought about by conferences and people moving from one airline to another. Understanding how the data is structured is fundamental to understanding how revenue management forecasting is performed.

Consider flight 100 departing from Houston at 9:00 a.m. on Thursday, 9/21, bound for Orlando, and suppose bookings can be made up to 360 days in advance of departure. Airlines typically break up this 360 days booking cycle into discrete data collection points, or DCPs. Each data collection point is defined by a number of days prior to the flight's departure. The number of people who have booked in a particular fare class by each data collection point is then recorded as shown in Table 3.2. The data collection points and where they are located in the booking cycle define a discrete booking curve. Data collection points go under a variety of different names, including reading points.

Airlines typically forecast not just the expected number of passengers in a given fare class due on the day of departure, but the expected number of passengers at different points in the booking cycle. To forecast Y class demand in DCP 10, for flight 100 from Houston to Orlando leaving at 9:00 a.m. on Thursday 12/7, for example, we would logically use historical Y-class bookings in DCP 10 for Thursday, 9:00 a.m. flights to Orlando originating in Houston. Such a model is

Table 3.3 Historical bookings in Y class for flight 100 on different departure dates as recorded on 10/16

Days before departure											
180	90	40	30	20	10	8	6	4	2	0	Departure index: date
0	3	4	5	10	11	13	18	19	23	30	1: 9/21
2	3	4	4	5	11	16	20	22	27	27	2: 9/28
2	5	9	9	11	18	18	18	23	28	28	3: 10/5
1	1	1	3	7	14	20	22	24	24	27	4: 10/12
2	3	4	3	14	21	28	28	32	xx	xx	5: 10/19
1	2	6	9	15	22	xx	xx	xx	xx	xx	6: 10/26
1	2	5	6	6	xx	xx	xx	xx	xx	xx	7: 11/2
2	3	6	9	xx	xx	xx	xx	xx	xx	xx	8: 11/9
0	0	0	xx	xx	xx	xx	xx	xx	xx	xx	9: 11/16
0	1	4	xx	xx	xx	xx	xx	xx	xx	xx	10: 11/23
0	0	xx	xx	xx	xx	xx	xx	xx	xx	xx	11: 11/30
0	4	xx	xx	xx	xx	xx	xx	xx	xx	xx	12: 12/7

"xx" denotes that no observation has yet been made because the date of 10/16 has not passed

commonly called a "day of week" (or DOW) model and will work reasonably well when schedules and demand patterns are similar by day of week. Other constructions of historical data are possible, say if adjacent days were strongly correlated, as might occur in a market undergoing strong trends or high drift.

Historical demand can be represented in a natural tabular format, with each row corresponding to the departure date and each column to the number of days prior to departure, as shown in Table 3.3. The data in this table on a given calendar date is referred to as a snapshot.

The shape of the historical observations in Table 3.3 arises from the fact that the booking cycles for the different departure dates of the flight 100 overlap. This is best seen by observing the calendar date on which an observation is made, as depicted in Table 3.4. As time passes, observations fill in the table down and to the right along the ragged diagonal. Due to the shape of the table and its "downward movement," it is often referred to as the data guillotine, although Nevada table, the wedge, and other similarly descriptive names are also used.

For algorithms such as EMSRb , the forecasting task is to estimate the bookings-to-come from the last observed data collection point to the day of departure; in pictorial terms, the time period in each row corresponding to the non-shaded cells. Dynamic programming algorithms require forecasts of the arrival rates between each data collection point—effectively, one forecast for each empty cell in the data guillotine. For notational simplicity, we limit our discussion to estimates of total bookings-to-come.

3.2.2.3 Utilizing Information from Flights Not Yet Departed

The implications the data guillotine has on forecasting model design and configuration are interesting and complex. For example, as discussed in L'Heureux (1986),

Table 3.4 Calendar date on which an observation of demand is made

Calendar date of data collection point											Departure index: date
3/25	6/23	8/12	8/22	9/1	9/11	9/13	9/15	9/17	9/19	9/21	1: 9/21
4/1	6/30	8/19	8/29	9/8	9/18	9/20	9/22	9/24	9/26	9/28	2: 9/28
4/8	7/7	8/26	9/5	9/15	9/25	9/27	9/29	10/1	10/3	10/5	3: 10/5
4/15	7/14	9/2	9/12	9/22	10/2	10/4	10/6	10/8	10/10	10/12	4: 10/12
4/22	7/21	9/9	9/19	9/29	10/9	10/11	10/13	10/15	10/17	10/19	5: 10/19
4/29	7/28	9/16	9/26	10/6	10/16	10/18	10/20	10/22	10/24	10/26	6: 10/26
5/6	8/4	9/23	10/3	10/13	10/23	10/25	10/27	10/29	10/31	11/2	7: 11/2
5/13	8/11	9/30	10/10	10/20	10/30	11/1	11/3	11/5	11/7	11/9	8: 11/9
5/20	8/18	10/7	10/17	10/27	11/6	11/8	11/10	11/12	11/14	11/16	9: 11/16
5/27	8/25	10/14	10/24	11/3	11/13	11/15	11/17	11/19	11/21	11/23	10: 11/23
6/3	9/1	10/21	10/31	11/10	11/20	11/22	11/24	11/26	11/28	11/30	11: 11/30
6/10	9/8	10/28	11/7	11/17	11/27	11/29	12/1	12/3	12/5	12/7	12: 12/7

A shaded square indicates a date that occurred on or before 10/16 forecast methods which place more weight on recent observations and use data from not yet departed flights (incomplete rows) are less sensitive to abnormal departure dates but are more sensitive to unusual booking dates. That is, forecasts tend to overreact to changes that occur along the guillotine edge. As L'Heureux points out, this is an important consideration since airline seat sales and promotions often impact all the observations along the guillotine edge, rather than just a single row or departure date.

3.2.2.4 Unconstraining

As mentioned earlier, because of capacity constraints and fare class closures, historical bookings are always less than or equal to the actual demand. The common way of dealing with this is to unconstrain, or detruncate, bookings. All unconstraining methods are some form of data imputation—replacing constrained or missing data with an estimate of the data that might have been obtained without any constraints in place. The unconstrained data is then used as the basis for forecasting.

There are many methods for unconstraining. Which particular method is employed depends upon the forecasting model being used and the preferences of the modeler. One established method for unconstraining bookings observations and arriving at demand-to-come forecasts is the expectation maximization (EM) method.

The EM method (Dempster et al. 1977) is named for the two steps that comprise one iteration. Let $A_{d,s}$ represent the true demand for a given fare class from the beginning of the booking cycle to d days before departure for the flight departing on date index s. In the context of unconstraining bookings, it is assumed that $A_{d,s}$ for a fixed d and all date indices s are defined by independent, identically distributed random variables X_s, but that actual historical observations have been truncated at a value c_s (possibly ∞) because of reaching a limit on the number of seats that could be sold. The expectation, or E-step, is to find the conditional expectation $\mathrm{E}[X_s|X_s \geq C_s]$, where the distribution of the X_s is defined by an initial set of estimates of its defining parameters—usually the mean and variance of a normal distribution. The constrained value is then replaced with the conditional expectation found in the E-Step, and the revised data is in turn used to find the maximum likelihood estimates (MLEs) of the distributional parameters. This is the M-Step, and it gives rise to a new set of estimates for the parameters defining the distribution of the X_s. The E-Step and M-Step are repeated until the MLEs converge. The essence of the EM method is that under very general conditions the MLEs converge to the parameters of the previously unknown distribution of the X_s. Once the MLEs converge, the truncated observations are replaced with their conditional expectations and are considered unconstrained. For an in-depth treatment of the EM method and other computational algorithms, see Tanner (1993). For a very readable review of the EM and other methods as well as numerical results, see Zeni (2001).

To apply any of these methods requires that we know if an observation is constrained or not. Airlines may track when a fare class was open or closed for sale, typically on a daily basis. This raw information must be summarized appropriately to the level at which the unconstraining is performed—for example, from the daily level to the DCP level. While conceptually not too difficult, summarizing open/closed data adds a great deal of data crunching to the forecasting task.

3.2.2.5 Pick-up Models and Regression

To illustrate how the data guillotine can be used for forecasting purposes, we consider a collection of straightforward models that have been employed at various times in the history of revenue management. For simplicity, we assume that all of the bookings displayed in Table 3.3 represent unconstrained bookings $A_{d,s}$. We also define $B_{d,s}$, the bookings-to-come from d days before departure for flight date index s, by $B_{d,s} = A_{0,s} - A_{d,s}$, where d varies, usually, according to a preselection of DCPs. From Table 3.3 we have $B_{20,2} = A_{0,2} - A_{20,2} = 27 - 5 = 22$, while $B_{4,5}$, $B_{10,6}$, and $B_{20,7}$ all require forecasts.

One method of forecasting the expected bookings-to-come is using a simple historical average. For example, an estimate of the number of Y class bookings for the flight departing on 10/26 can be calculated from the numbers in Table 3.3 as $(30 + 27 + 28 + 27)/4 = 28$. The expected bookings-to-come $B_{10,6}$ are then estimated as $6 = 28 - 22$, the latter number being the bookings on hand $A_{10,6}$.

This seems to be a reasonable estimate, but almost immediately we see how it might be improved. One clear disadvantage of this method is that it fails to take into account the fact that the 22 bookings for the 10/26 departure is considerably larger than the average number of bookings at 10 days before departure, which is $(11 + 11 + 18 + 14 + 21 + 22)/6 = 16.2$. Thus, it is reasonable to hypothesize that the number of bookings on the day of departure would also be larger than average. We may therefore want to ask about the growth, or *pick-up,* we can expect in bookings between now and the day of departure.

Pick-up models usually assume that the bookings-to-come $B_{d,s}$ are defined by a normal random variable with mean μ_d and variance σ_d. One advantage is that the normal distribution is characterized fully by its first two moments so that there are no other parameters to estimate or fit. The normality assumption also has a theoretical justification in the Central Limit Theorem, which states that the sample mean of independent observations tends toward normality under certain regularity conditions. The independence assumption in the standard Central Limit Theorem is usually violated by time series data, but there are versions of the theorem that cover dependent data (see Fuller 1996). Using historical observations from the data in Table 3.3 an estimate of the bookings-to-come from ten days before departure is

$$((30 - 11) + (27 - 11) + (28 - 18) + (27 - 14))/4 = 14.5.$$

Simple linear regression is another method used to estimate bookings-to-come. The most commonly used model is $A_{0,s} = a + b \cdot A_{d,s} + \varepsilon$, with estimates of a and b derived from historical observations $(A_{d,s}, A_{0,s})$. Using simple linear regression on the known observations from $d = 10$ days before departure—(11, 30), (11, 27), (18, 28), and (14, 27)—yields least squares estimates $a = 29.2$, $b = -0.1$ and a resultant estimate of 27 bookings for the flight departing on 10/26. The forecast of bookings-to-come is then calculated as $5 = 27 - 22$.

3.2.2.6 Linear Model Extensions

Although generating perfectly reasonable estimates from a statistical perspective, the methods discussed in the previous section weigh each observation equally and therefore do not capture trend or level changes that make the recent departure dates more relevant. For this reason some type of moving average is almost always used. A very popular model is the exponentially weighted moving average (EWMA). For example, in the pick-up or demand-to-come framework we may use

$$\begin{aligned} \hat{B}_{d,s} &= \alpha B_{d,s-1} + (1-\alpha)\hat{B}_{d,s-1} \\ &= \alpha(A_{0,s-1} - A_{d,s-1}) + (1-\alpha)\hat{B}_{d,s-1} \end{aligned} \tag{3.1}$$

where $\hat{B}_{d,s}$ is the estimator of $B_{d,s}$ and $0 < \alpha < 1$ is the smoothing parameter. Other, more advanced techniques have been employed, including Kalman filtering (Kalman 1960; Kalman and Bucy 1961) and Bayesian hierarchical forecasting (West and Harrison 1997; Gelman et al. 1995).

To illustrate the concept in Table 3.3 above we have 10 data series, indexed by DCP and by departure date. We can model the demand process for each DCP over time rather than the demand to come process directly. Define $X_{d.t}$ to be the demand observed in DCP indexed by d (there is no loss in generality if d is an index instead of the number of days prior). Similar to Eq. 3.1, an EWMA model for $X_{d.t}$ looks like

$$\begin{aligned} \hat{x}_{d,s+1} &= \alpha x_{d,s} + (1-\alpha)\hat{x}_{d,s} \\ &= \hat{x}_{d,s} + \alpha(x_{d,s} - \hat{x}_{d,s}) \end{aligned} \tag{3.2}$$

Note that EWMA does not truncate historical data to a fixed window size although the weight on any particular observation asymptotes over time to 0, slower or faster depending on the choice of the smoothing parameter α. For example, $\alpha = 0.25$ weighs the most recent observation 25% and the prior mean estimate 75%, so that at time s, $x_{d,1}$ receives a weight of $0.25(0.75)^{s-2}$ which decays rather quickly. A variety of calibration and fitting exercises can help fine-tune a choice of α.

Note that Eq. 3.2 uses only the most recent observation and the most recent forecast. This means we only have to store two values (and the parameter) each time, reducing storage costs and minimizing read–writes to the database. For applications the size of a typical airline RM system, such computational advantages can be critically important.

Moving average models are very effective when trend or level changes (drift) exist in the demand process, and are therefore most effective for single period ahead forecasting. As is, they will smooth over periodic effects, such as day of week and seasonality. Both effects are prominent in airline demand and are essential to obtaining longer horizon forecasts. Day of week effects have traditionally been dealt with by pooling similar days of week together: the guillotine structure contains data for a particular day of week, the so-called DOW models we mentioned above. The pooling concept has also been used to separate demand by season, for example, using only data for summer months to forecast demand for summer months. More traditional seasonality methods have been employed as well, whereby a factor such as a monthly ratio is used to deseasonalize historical observations, forecasts are made, and then the deseasonalized forecasts are reseasonalized. Spikes in demand due to holidays or special events can be handled similar to seasonality, but user intervention is normally required to help define when holidays and special events occur.

An extension of the EWMA model known as the Holt–Winters method (Holt 1957; Winters 1960) accounts for linear trend and seasonal effects directly, as follows. First, assume that our bookings process is

$$X_{d,s} = (A_{d,s} + T_{d,s})S_{d,s-L} \tag{3.3}$$

where T is a level change due to trend, and S is a seasonal multiplier. The question is how to model these separate components. After observing new data at time s, each component is updated in turn:

$$\begin{aligned} A_{d,s} &= \alpha x_{d,s} + (1-\alpha)(\hat{x}_{d,s} + T_{d,s}) \\ T_{d,s} &= \beta(\hat{x}_{d,s} - \hat{x}_{d,s-1}) + (1-\beta)T_{d,s-1} \\ S_{d,s} &= \gamma\left(\frac{x_{d,s}}{S_{d,s}}\right) + (1-\gamma)S_{d,s-L} \end{aligned} \tag{3.4}$$

where α, β, γ are smoothing parameters and the index L is the seasonal lag parameter. For example, if there is annual seasonality, L is a lag of one year. The forecast for time $s + 1$ is given by

$$\hat{x}_{d,s+1} = (A_{d,s} + T_{d,s})S_{d,s+1-L} \tag{3.5}$$

(For obvious reasons, the Holt–Winters approach is also called *triple exponential smoothing*.) The EWMA and Holt–Winters approaches are easy to implement, since we only need to set smoothing parameters and store only the latest state each update. They do not assume any structure on the data generation process nor do they provide any estimates for variance. More advanced methods address these limitations as we will discuss in Sect. 3.2.2.7.

A critical assumption in using any time series is that the demand process is stable, that the future will look a lot like the past, and that models should be designed to capture the effects as they evolve over time. Time series can fail when there are structural shifts in the demand process, to the extent that history and future do not match up very well. Such disruptions are common in airline demand due to schedule and route changes.

We can address the flux in schedules at least partly by not relying on flight numbers as product dimensions, nor on precise departure times as a product attribute, instead treating departure time in blocks such as early, mid, late morning, afternoon, and evening. Customers do not see a great difference between a 7:30 p.m. and an 8:05 p.m. departures, at least not to the extent that the demand would fundamentally change if the departure time shifts from 7:30 p.m. to 35 min later. Time blocks such as these help distinguish different demand behavior and smooth over effects of schedule change. Other product attributes that might also be leveraged in helping to predict demand behavior are the number of stops, total travel time, frequent flier miles, and more. Assessing the strength of any one factor on demand and exploiting that effect in predicting demand intersects with the realm of Customer Choice Modeling, to which we return in Sect. 3.3.3.3.

3.2.2.7 More Advanced Time Series Methods

In this section we will review two other methods that are popular for airline demand forecasting, namely the Kalman filter (Kalman 1960; Kalman and Bucy 1961) and the Bayesian hierarchical methods (West and Harrison 1997; Gelman et al. 1995).

While EWMA is generally considered a model-free approach, the Kalman filter applies to a structural or state-space model. State space models are characterized by observation process and a related—but unobservable—mean state process. To illustrate, for a quantity of interest $X_{d.t}$ a Kalman filter arises from solving the following system of equations:

$$\begin{aligned} \mu_{d,t} &= A\mu_{d,t-1} + w_{t-1} \\ x_{d,t} &= H\mu_{d,t} + v_t \end{aligned} \tag{3.6}$$

where w_t, v_t are error processes and A, H are (scalar) parameters that may or may not depend on time (we assume here they do not change over time). The observable process is $x_{d,t}$ and the state or system process is $\mu_{d,t}$.

When w_t, v_t, are individually uncorrelated with finite constant variances W, V, respectively, the following recursion holds

$$\begin{aligned} \hat{\mu}^-_{d,t} &= A\hat{\mu}_{d,t-1} \\ \hat{\mu}_{d,t} &= \hat{\mu}^-_{d,t} + K_t(x_{d,t} - H\hat{\mu}^-_{d,t}) \end{aligned} \tag{3.7}$$

where $\hat{\mu}^-_{d,t+1} = E[\mu_{d,t+1}|x_{d,s}, s = 0, 1, \ldots, t]$ is the conditional expectation of the mean process given observations up to time $t-1$ and the parameter K_t is the gain or

Kalman gain factor. If in addition the error processes are normally distributed and uncorrelated processes with mean 0, then the following recursion holds for K_t :

$$\hat{\sigma}_t^{2-} = A^2 \hat{\sigma}_{t-1}^2 + W$$
$$K_t = \frac{H\hat{\sigma}_t^{2-}}{H^2(\hat{\sigma}_t^{2-} + V)} \tag{3.8}$$

where $\hat{\sigma}_t^2$ is an estimate of the error variance $\sigma_t^2 = E[(\mu_t - \hat{\mu}_t)^2]$ and where $\hat{\sigma}_t^{2-}$ is an estimate of the error variance $\sigma_t^{2-} = E[(\mu_t - \hat{\mu}_t^-)^2]$ given the data up to time $t-1$.

Equations 3.6 and 3.7 together suggest a stepwise process for obtaining a forecast for $X_{d,t+1}$: from Eq. 3.7 the first step updates our estimate of the process mean for time t with data up to time $t-1$; the second step adjusts our estimate of the process mean with the newly observed $x_{d,t}$ Then, following Eq. 3.6, a new forecast for $X_{d,t+1}$ is produced as $H\hat{\mu}_{d,t}$. Welch and Bishop (2006) provide a more complete explanation of the Kalman filter, including several numerical examples. Like EWMA, the Kalman filter only needs the new observation and its most recent parameter state, but in addition it also provides a means to model the uncertainty in predictions. Further, it has optimal prediction properties under certain conditions [among many, see Gijbels et al. (1999) for a discussion and comparison between Kalman filter and EWMA].

The Bayesian Hierarchical method applies the fundamental Bayesian concept of a prior distribution, a data distribution, and a posterior distribution. The data distribution governs the observations that we see, for example, the actual bookings (which, for the moment, we assume are adequately unconstrained). The data distribution depends on parameters—typically a mean and variance—which are unknown but follow a distribution of their own, called a prior distribution. After we observe new bookings, we will update the prior distribution, obtaining the posterior distribution. An important mathematical result known as Bayes' Theorem provides the mechanism. Suppose we have a data distribution $f(x|\theta)$ for a random variable X which depends on an unknown parameter, θ having a distribution $g(\theta)$. Then, the posterior distribution is found by

$$g(\theta|X = x) = \frac{g(\theta)f(x|\theta)}{\int g(\theta)f(x|\theta)d\theta}, \tag{3.9}$$

The denominator in Eq. 3.9 is known as the marginal data distribution, sometimes denoted by $m(x)$. The integration is over the range of θ, weighted by the likelihood $f(x|\theta)$ as determined by the prior distribution $g(\theta)$.

In the time series context, the posterior then serves as the new prior distribution before we observe a new data point, i.e., before time advances and we gather more evidence about the data distribution parameters. For some selections of the data and prior distributions (namely Gaussian for the data and prior mean distribution, conditional on variance) the Bayesian update formulas are equivalent to a Kalman filter. Rao (2001) makes the relationship precise.

Like the Kalman filter, the Bayesian method requires more work in specifying the model than simpler smoothing methods such as EWMA (which only has a smoothing parameter). In particular, we have to specify a distributional form of the data and that of its parameters. These together via Bayes' Theorem determine the form of the posterior distribution. The literature and research is vast on the topic of priors and selecting them and various sampling techniques can be employed when the posterior is not of a closed form. Fortunately, for our purposes, there are pairs of distributions that have a nice property known as conjugacy. Conjugate pairings of prior and data distributions yield posterior distributions that have the same functional form as that of the prior, and there are such pairs for our favorite distributions in modeling airline demand.

For example, the gamma prior and Poisson data distribution form such a conjugate pair: if we have a Poisson data distribution with arrival rate, λ, and if λ follows a gamma distribution, then given a new observation x, the posterior will again be gamma. If we have a Gaussian data distribution with a mean parameter μ and known variance, σ^2, and μ follows a normal distribution, then the posterior will again be normal. Other useful conjugate pairs are the beta-binomial, for modeling rates, and Dirichlet-multinomial, for modeling multinomial responses (such as proportions across a fixed number of mutually exclusive categories).

To illustrate the Bayesian approach, let us assume that demand by DCP $X_{d.t}$ follows a Poisson distribution with arrival rate λ, which in turn has a gamma prior with parameters $\alpha_{d,s}$ and $\beta_{d,s}$. Specifically

$$f(x_{d,s}|\lambda) = \frac{\lambda^{x_{d,s}} e^{-\lambda x_{d,s}}}{x_{d,s}!} \tag{3.10}$$

and

$$g(\lambda_{d,s}|\alpha_{d,s}, \beta_{d,s}) = \frac{\beta_{d,s}^{\alpha_{d,s}}}{\Gamma(\alpha_{d,s})} \lambda_{d,s}^{\alpha_{d,s}-1} e^{-\beta_{d,s}\lambda_{d,s}} \tag{3.11}$$

(Technically, in Eq. 3.11 all the parameters should have "hats" because we are estimating them at each time index. We have omitted the hats to keep the notation a little less crowded.)

By applying Bayes' rule it can be shown that the posterior after observing $x_{d.s+1}$ is gamma with parameters $\alpha_{d,s+1} = \alpha_{d,s} + x_{d,s}$ and $\beta_{d,s+1} = \beta_{d,s} + 1$. To obtain a forecast for time $s + 2$ we use the expected value of λ under the posterior distribution just obtained. That is,

$$\begin{aligned} \hat{x}_{d,s+2} &= E[X_{d,s+2}|\lambda_{d,s+2}] \\ &= \lambda_{d,s+2} \\ &= E[\lambda_{d,s+2}|\alpha_{d,s+1}, \beta_{d,s+1}] \\ &= \frac{\alpha_{d,s+1}}{\beta_{d,s+1}} \end{aligned} \tag{3.12}$$

Table 3.5 Simulation of gamma-poisson model

Time(s)	alpha	beta	Expected value for lambda	Variance of lambda	X (*d,s*)
0	1	0.5	2.0	4.0	45
1	46	1.5	30.7	20.4	29
2	75	2.5	30.0	12.0	30
3	105	3.5	30.0	8.6	35
4	140	4.5	31.1	6.9	21
5	161	5.5	29.3	5.3	13
6	174	6.5	26.8	4.1	11
7	185	7.5	24.7	3.3	32
8	217	8.5	25.5	3.0	50
9	267	9.5	28.1	3.0	42

In Table 3.5 we illustrate the Bayesian updating process for a gamma-Poisson model. For simplicity, the process is assumed to be stable (no evolving changes in the parameter distributions). At time 0, our prior opinion regarding λ is represented by the values of $\alpha = 1$ and $\beta = 0.5$.

Graphically, we can see how the effect of our initial prior washes out over time and how our uncertainty (variance) decreases as we obtain more information.

The method here can be extended—and generally is in forecasting applications—by adding drift in the distribution for λ, to account for gradual changes over time. In this way, Bayesian models are made dynamic. A further extension is

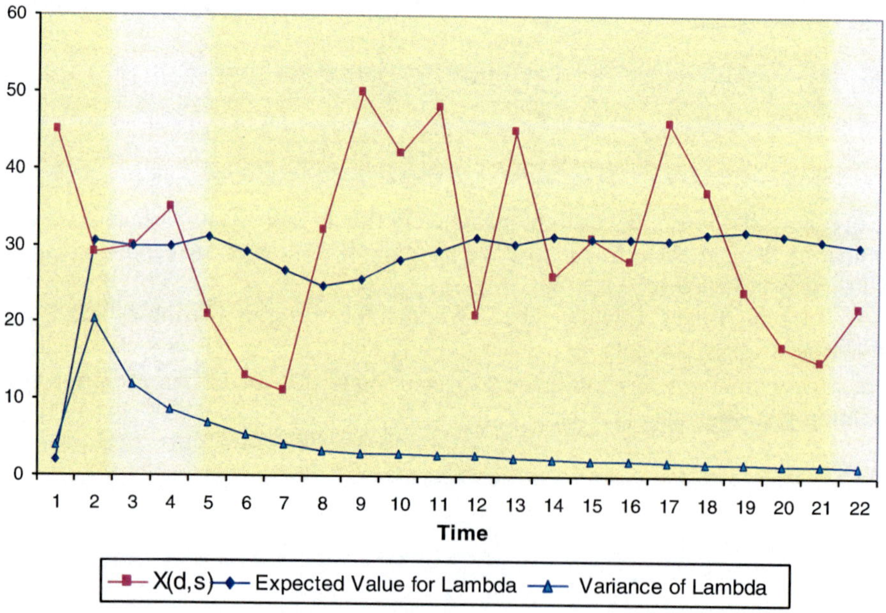

Fig. 3.2 Simulation of gamma -poisson model

achieved by placing priors on the priors (called hyper-priors), thus building a hierarchical model. The hierarchy is a way to model different processes related by a common underlying distribution. Through a hierarchical model, data gathered on one process can be leveraged in estimating the other processes, an appealing concept in sparse data situations like O&D forecasting, For example, a hierarchy allows us to model the overall demand for the O&D at one level and demand for different fare classes at the lower level, and then at an even lower level, we can model demand by DCP.

The Bayesian framework—with prior distributions and hierarchies—encompasses a nearly limitless choice of different models. In fact using a Gaussian prior on the starting state, the Kalman filter has a nice Bayesian interpretation see Rao (2001)]. The task of finding the best model is thus a daunting one.

3.2.2.8 Measuring Forecast Performance

As we have already noted, low bias in RM forecasts (due to constrained data) causes revenue loss by contributing to spiral-down. In that case getting better forecasts meant getting better inputs, independent of what model or method was used. But the importance of having good forecasts invariably leads us to ask, which is the better method, of all the methods we could implement? Clements and Hendry (1998) remind us that predictability is a quality of the underlying data process, and not of the forecasting method. The forecasting method is just the procedure for attempting to describe the underlying data process. Thus there is no definitive model or method that is best in all situations. But although there is no definitive answer for all cases, a model can always emerge as best in any given situation. In this section we highlight some of the main concepts and challenges to measuring the forecast performance.

Assuming we have collected a history of forecasts (f) and the (adequately unconstrained) actuals (a), the most common error metrics for evaluating forecast performance are:

Mean error (ME)

$$\text{ME} = \frac{1}{T}\sum_{t=1}^{T}(f_t - a_t)$$

Mean squared error (MSE)

$$\text{MSE} = \frac{1}{T}\sum_{t=1}^{T}(f_t - a_t)^2$$

Mean percentage error (MPE)

$$\text{MPE} = \frac{1}{T}\sum_{t=1}^{T}\frac{(f_t - a_t)}{a_t}$$

Mean absolute percentage error (MAPE)

$$\text{MAPE} = \frac{1}{T}\sum_{t=1}^{T}\left|\frac{f_t - a_t}{a_t}\right|$$

Mean absolute error (MAE, or sometimes, mean absolute deviation MAD)

$$\text{MAE} = \frac{1}{T}\sum_{t=1}^{T}|f_t - a_t|$$

The mean percentage error is a widely used metric, perhaps because people find percentages fairly easy to understand and interpret. But the MPE can be misleading, particularly for small actual values because the percentage error grows unbounded, and MPE is undefined if any of the actuals are 0. As we noted above, a systematic low bias (negative ME) will result in spiral-down, and unbiasedness (ME = 0) is usually desirable. But ME by itself is insufficient, as a 0 ME will be true whether every error is 0 or half of the forecast errors are +10000 and half are −10000. The ME should always be reported along with measures of spread, like MSE and MAE. MSE is susceptible to outliers, because large errors are being squared. Hence, MAE is also often reported. While no single error measure can tell the whole story, MSE is the most important loss function in theory and in practice, in the sense that most forecasting models are judged by how well they improve (i.e., lessen) the MSE. One reason is because MSE = VAR + BIAS^2, so minimizing MSE in a sense manages the tradeoff between variability and bias.

Not only is it a good idea to evaluate more than one error metric, but also we should evaluate the errors produced by more than one model on the same data. Error values really have no valid interpretation by themselves. If a forecasting model yields a 10% mean percentage error, should we be satisfied with it? Maybe a different model on the same data will do better. We will have no way of knowing whether 10% is reasonable performance without another baseline or competing measure. Baseline models are frequently the simplest, such as averages or moving averages, and then each alternative model can be made incrementally more complex. Furthermore, forecast performance depends greatly on how predictable the underlying data is. Suppose we are forecasting heads or tails in a fair flip of a fair coin. We will get either a 100% error or a 0% error on each flip and on average we will hit around 50%. How can we improve upon that? Each time there is a 50–50 chance of getting a head and of our forecast being right. Unless we have additional information that helps us predict the outcome, we have done the best we can do. On average, we achieved the bound dictated by the inherent variability (or noise) in the process. Again, predictability is a characteristic of the data generation process, and not of the forecasting model. This is why—although people attempt it all the time—we cannot preordain a level of forecast accuracy and make it a business requirement for a forecasting system.

We may achieve incremental improvements by comparing models on the same data processes, an exercise worth doing on a routine basis. But any model will suffer from bad inputs and dramatic structural breaks in the demand generation process. In a live system, preventing forecast failure has a lot more to do with the feedback and visibility expert users have into the system than whether it is a Kalman filter or simple moving average model. Checks and balances should be built-in. The system should detect data outliers and provide a straightforward way for expert users to investigate and censor outliers out of model updates. The system should detect extreme forecast errors and enable expert users to redefine models as needed, with little or no disruption to the RM process. In short, successful forecasting in the RM environment entails much more than the models.

3.2.2.9 Cancellations, Show-Ups, and No-Shows

The same data guillotine used for bookings can be used to record and estimate cancellations. Cancellations and no-shows are typically modeled as rates—the probability that a booking will choose to cancel or not show up. Thus, the predicted number of cancellations and no-shows depends not only upon the forecast rate but on the actual number of bookings. Cancellation rates tend to be higher for business travelers—who generally buy less restricted fares. Rates also vary during the booking cycle, with higher rates closer to departure. Thus cancellations, similar to bookings, are observed and modeled within the data guillotine structure. Show-ups and no-shows, being associated with day of departure demand and not the entire booking cycle do not require the data guillotine.

Modeling cancellation rates throughout the booking cycle often involves modeling a rate for each DCP, whereas simpler models are weighted averages of the ratios of booked to cancels without a good deal of attention paid to variations due to calendar effects. Is it reasonable to expect that a passenger booking for a Wednesday departure is more or less likely to cancel? A more compelling conjecture is that cancellation behavior is related to the type of customer and the product they are buying. For example, a leisure customer likely books well in advance, when lower prices and inventories are more available. Furthermore, because of penalties in cancelling hotel or resort reservations, they are less likely to cancel, especially later in the booking cycle. Business travelers as a rule will have higher cancellation rates and exhibit this behavior relatively late in the booking cycle.

Forecasting cancellation or no-show rates on a more detailed level (and sometimes even on O&D level for small markets) runs into the small numbers problem as the usual O&D forecasting for original-destination pairs that have very few observations. For an example of the problem consider an itinerary for which we have recorded only one booking and which was subsequently cancelled. Clearly, a cancellation rate forecast of 1 would be very suspicious. In practice the rates are either forecasted at an appropriate aggregate level or the rates are blended from rate estimates at lower—more detailed levels while accounting for high variance of rates coming from sparse entities.

Relatively recently, airlines have begun to experiment with information about individual customers to improve no-show forecasts. If members of the frequent flyer program tend to no-show at a higher rate than passengers who are not members, then knowing how many of the booked passengers are frequent flyers can improve the forecasts. For example, in Lawrence et al. (2003), the authors compare no-show rate forecasts using conventional ratio estimates with models that use passenger-level information such as frequent flyer, the time and length of the segments, whether or not the passenger made a special meal request, and more. Generally they show revenue improvements on the order of 0.3–3.5% from leveraging additional passenger information in predicting no-show rates.

3.2.3 O&D Forecasting

3.2.3.1 Motivation

The mathematical reasons for O&D forecasting arise from the requirements of network optimization models. Almost by definition, network optimization models require forecasts at the itinerary/fare class level. No longer is the Q class passenger flying from Seattle through Houston to Miami a "Q class passenger on the Seattle to Houston flight" and a "Q class passenger on the Houston to Miami flight." Instead, she's a "Q class passenger on the Seattle-Houston-Miami itinerary." We return to discuss the many advantages for optimization in Sect. 3.3, focusing here on the additional challenges posed by O&D forecasting.

In actuality, "O&D forecasting" is a misleading term, since it typically refers to the process of generating forecasts at the itinerary/fare class level or lower, such as itinerary/fare class/point of sale, and not the more aggregate origin/destination/fare class level. We adopt this convention in the discussion that follows. To provide some perspective, the number of origin/destination/itinerary/fare class/POS combinations can reach into millions for a typical US carrier while the origin/destination/fare class is on the order of hundreds of thousands.

As a general rule, optimization models prefer forecasts at very low levels of detail while the process of forecasting is less difficult at more aggregate levels. The different approaches to revenue management are in part a reflection of how airlines have chosen to balance these two conflicting objectives.

3.2.3.2 Challenges

The low level of detail required of O&D forecasts presents new challenges for forecasting methods and the software systems that support them. Perhaps the most obvious challenge is that the number of forecast entities grows dramatically. Where a flight network with 200 daily flights and 10 fare classes would maintain 2000 leg/fare class forecasts for each active departure date, the number of O&D

forecasts would easily run into hundreds of thousands. Another challenge is the size of the numbers being forecasted. While a leg/fare class forecast might be a number on the order of a few dozen, the average demand at the itinerary/class level is generally much less than one.

To some extent, the small numbers problem can be addressed by directly forecasting only the most significant O&Ds while treating others at a more aggregate level. Modeling tools are also available to help mitigate data sparseness, such as transformations designed to better approximate normality and model hierarchies, which correlate the demand of similar products while still maintaining separate forecasts.

Another challenge has to do with how data is maintained. One of the ways in which airlines store data is through inventory records on each flight; that is, for each flight leg the computerized reservations system keeps a record of how many seats have been sold in each fare class, and a record of how many seats can be sold in each fare class (generated from a revenue management system). A list of who these passengers are, including information about when they booked, the flights in the flight itinerary, references to the price paid for the ticket, number of people in the party, any history of changes made to the reservation, frequent flyer information, special requests, and more, are maintained in passenger name record or PNR data. PNR data warehouses can be extremely large, easily running into terabytes of data depending upon how many historical transactions an airline wishes to keep.

The preferred method of generating O&D forecasts is to build historical transactions suitable for forecasting purposes from the PNR data. While the process is simple in concept, actual transaction data at the PNR level can be quite complicated and the process of building good forecasting observations is a significant undertaking. Processed PNR's provide a wealth of historic data that can be mined not only for O&D forecasting but also for inferring customer-centric behavior, such as the likelihood of a cancellation or price sensitivity.

Once historical observations are built, the O&D forecasting process is similar to that of leg forecasting. Data is stored in a data guillotine and forecasts are generated for different days of departure on a rolling basis. Because data is often sparse, various techniques are used to reduce computer memory requirements, but these techniques are more technical than conceptual in nature (see, for example, Davis 2006).

Unconstraining ODF demand is more difficult than in the leg-based system. Primarily, an ODF can be constrained by capacity on one or more of the legs it overlaps while still being open on the O&D level according to bid price control. Thus for O&D unconstraining it is more important to know the degree of constraint in unconstraining algorithms; this introduces further practical challenges, since all the conditions including leg-level constraints across the network at any given time must be tracked. Otherwise, the system will have imperfect knowledge of when observed bookings are constrained.

Other methods for generating O&D forecasts have been proposed and used. One technique involves continuing to forecast at the leg/fare class level, but using itinerary/fare class data from airlines' revenue accounting systems to reconstruct O&D forecasts. As the name implies, revenue accounting systems are used by airlines to track payments, and as such the data may arrive weeks to months after a flight is booked. Thus, the itinerary/fare class forecasts constructed from this data are typically done on a much less frequent basis than their leg counterparts, and are used to decompose leg/fare class forecasts. The approach contains a number of approximations that many people are uncomfortable with, but it represents a reasonable engineering solution, balancing the tradeoff between the many complexities that arise from direct itinerary/fare class forecasting and the simplicity of leg/fare class forecasting.

One other method of O&D forecasting really is not an O&D forecasting method at all, but nonetheless supports network optimization algorithms based on virtual nesting. In fact, proponents of virtual nesting point to its lack of dependence on O&D forecasts as one of its advantages over other network optimization methods.

Instead of O&D forecasts, virtual nesting requires forecasts at the level of leg/virtual bucket, also known as leg/virtual class. In this method, some numbers of virtual buckets are defined for each flight leg, say 10, with bucket 1 corresponding to the "highest value" itinerary/fare classes and bucket 10 corresponding to the "lowest value" fare classes. The way in which each itinerary/fare class is assigned to a virtual bucket is an important part of the algorithm and is described in Sect. 3.1. In the dual index version of virtual nesting, different flight legs in an itinerary/fare class can be assigned to different virtual buckets.

With itinerary/fare classes assigned to virtual buckets, the forecasting problem becomes one of estimating how many passengers are expected in each leg/virtual bucket. From an operational standpoint, this is identical to leg/fare class forecasting and offers the advantages of forecasting at this level. The added complexity comes in assigning each itinerary/fare class to a virtual bucket. In practice, the methods that generate the most revenue require that the assignment be based on information from a static network optimization problem such as LP, which in turn requires O&D forecasts as input. Thus, the O&D forecasting question is not circumvented, but is relegated to the background. O&D forecasts for this background network optimization problem are often generated using approximations from data contained in revenue accounting systems.

Virtual nesting methods have the further complication that itinerary/fare classes must be periodically remapped to different virtual buckets and, at least theoretically, this should be done on a very frequent basis. As this remapping occurs, the history of demand in each virtual bucket changes as well, significantly complicating the forecasting process. There are many ways the details of virtual nesting algorithms can be resolved, representing tradeoffs between complexity and the amount of incremental revenue the approach will generate. Network optimization based on virtual nesting remains popular and is used at some of the major US carriers such as United Airlines and Delta.

3.2.4 Customer Choice Modeling

Forecasting methods have historically focused on estimating either leg/fare class or itinerary/fare class demand not only because of the needs of optimization algorithms, but also because this is what airlines observe. Week after week, airlines look at how many passengers showed up and in what fare class they booked. It therefore seems natural to forecast demand based on these historical observations. At least as a first approximation, if you want to know how many Y class passengers will show up for flight 100 next Wednesday, it makes sense to look at how many passengers showed up for flight 100 on Wednesdays in the past.

The limitations of this independent demand model have long been recognized, but it was not until recent years that the problems with this approach were fully appreciated. The earliest observation that "what you see isn't what you get" came from recognizing the need for unconstraining. There are far more subtle reasons, however, why observed demand is not necessarily a good indicator of what to expect in the future. The independent demand model assumes that customers have a preference for a Y or a Q class fare product in the same way that customers looking for a dishwasher have a preference for a sophisticated unit with dozens of different settings or a simple unit with nothing more than an "on" button. The models assume that customers arrive with a desire to purchase a particular fare class and walk away if it is not available. But what if customers perceive fare classes as nothing more than different prices for the same product—a seat on an airplane—and purchase the lowest fare class available? In this case, passengers are not revealing their actual willingness-to-pay through what they buy, but a lower bound on that value. Thus, using the independent demand model, forecasting algorithms will underestimate passengers' true willingness-to-pay. The consequences for airline revenues can be devastating (Boyd et al. 2001; Cooper et al. 2006a, b) and we discuss it more in Sect. 3.3.3.3. When passengers purchase based on price rather than product, revenue management is identical to pricing in the classical sense: a (discrete) demand curve is estimated and a price point is chosen that maximizes revenue. Using the same historical data, the two different models of how people behave lead to very different estimates of what they are willing to pay.

Rightly or wrongly, most of the academic literature has historically focused on the product model as have most of the revenue management models used in practice. (A notable exception in the academic domain is the 1994 and 1997 work of Gallego and van Ryzin, who explicitly address a pure pricing model in a continuous time setting, for a single-leg and the multiple-leg problems, respectively.)

In cases where airlines offer only one type of ticket but vary the price—as is the case with many of the recent generation of low cost carriers—the price model is appropriate. In other cases the situation is not so clear. Historical data about carriers that use restriction-based fare classes shows that there are many instances where customers purchase tickets in fare classes that are not the lowest available fare.

One recent approach is to assume that both product and price oriented customers exist, forecasting each, and developing optimization models that account for both types of demand (Boyd and Kallesen 2004).

An alternative approach is to assume that customers care not only about product and price, but attributes such as departure time, arrival time, trip duration, and aircraft type, among others. Discrete choice models, discussed at length in Chap. 1 "Customer Modeling", seek to determine the relative importance of these attributes in the passengers' decision-making process. The use of discrete choice models in revenue management applications was raised and formalized by Talluri and van Ryzin (2004). Ratliff (2006) has also pointed out some specific practical advantages of choice models such as possibility of accounting for effects of upsell, downsell, and recapture rates in the presence of competitive alternatives. The upsell and downsell refer, correspondingly, to customers buying a more profitable (for the airline) or a less profitable fare product if their original choice cannot be satisfied due to capacity restrictions; recapture happens when customers buy seats at another flight of the same airline if the flight of their first choice becomes full. While discrete choice models can be used at the flight leg or itinerary level, they are typically proposed for forecasting demand on all itineraries from origin city A to destination city B simultaneously, as discussed in Sect. 3.3.3.3. After determining the probability that a customer will choose each particular itinerary/fare class combination based on its different attributes, the itinerary/fare class forecasts required by optimization algorithms are determined by multiplying a forecast of the total demand between city A and city B by the associated probability.

Discrete choice models are conceptually very appealing. In practice, they face significant obstacles when employed in a revenue management setting. While we understand from personal experience that we are influenced by many different factors when purchasing plane tickets, personal experience also reminds us that we would have difficulty explaining exactly how we weigh the many factors to arrive at our decisions. Like all models of the way the world works, discrete choice models make certain assumptions. But the underlying assumptions behind discrete choice models are different from those behind, for example, the calculation of a simple average. And as a result, discrete choice models can give very different results than a simple average.

Much to the detriment of forecasting, most revenue management analysts evaluate a forecast by looking at what happened in the past. If 20 people purchased Y class tickets on flight 100 for the last 30 weeks, a user expects a forecast of 20. A discrete choice model could return with a forecast of 10. In fact, this is not a problem faced only by discrete choice models, but any forecasting method that looks beyond how many people purchased Y class tickets on flight 100 in the past. When forecasting methods attempt to untangle the many factors and interrelationships affecting demand, the results can be counterintuitive and therefore unacceptable to analysts.

Unlike many of the planning functions where discrete choice models are typically used, revenue management forecasts are scrutinized on a daily basis by individuals who are not formally trained statisticians. Analyst involvement is necessary because events occur—snowstorms, floods, terrorist attacks—that affect the normal operation of revenue management forecasting. As a very practical matter, revenue management forecasting must strike a balance between mathematics and what analysts will accept.

Between the technical challenge of finding demand models that better reflect consumer behavior than the independent demand model and the human challenge of keeping the process comprehensible to analysts, the area of demand modeling remains a centerpiece of revenue management research and practice.

3.3 Optimization

In the context of airline revenue management, optimization refers to any algorithm used to choose the highest value mix of passengers to fill the plane as efficiently as possible. Overbooking, upgrading passengers from one cabin to another, deciding whether groups should be accepted, and determining how much inventory should be made available through different distribution channels are all problems that fall under the purview of revenue management optimization. All of these problems are important, and each could be discussed at length. However, we here continue to focus most of our attention on the fare mix problem—the problem of determining what fare classes should be available at any given time—both for the sake of brevity and because it plays a central role in both the history and practice of revenue management. We begin, however, with a digression motivated by the historical progression of revenue management.

3.3.1 Overbooking

Overbooking attempts to balance spoilage (flying with empty seats) with the costs of denying boarding to an overbooked customer. The typical business measures of interest when optimizing overbooking levels are therefore expected profit contribution, probability of show-ups exceeding capacity or the expected number of denied boardings. The models usually stochastically optimize one of the measures or use a mathematical program to optimize one of them subject to constraints on others.

Overbooking, upgrades, comfort seating, and, to some degree, group requests, as practiced at most airlines are all parts of the capacity management process that is fairly static in nature and takes place whenever reoptimization of a flight takes place. It is usually done by flight leg and so many of the models proposed in the sixties and seventies as well as the assumptions they make are still relevant. The analysis usually starts with the very common assumption that customers cancel or do not show up independently from each other and from everything else and with the same probability (which could be time-varying). This implies that the number of passengers showing up at departure has a binomial distribution and for larger booking numbers this can be approximated by the normal distribution.

Shlifer and Vardi (1975) propose a simple model where the number of customers showing up is normally distributed with mean equal to the product of the current reservation level N and some parameter a, and with standard deviation equal to the product of N and another parameter b; the physical capacity of the plane being M. With $\Phi(\cdot)$ denoting the standard normal cumulative distribution function the probability $\eta(M)$ that the number of booked passengers who show-up at departure does not exceed the capacity can be expressed as

$$\eta(M) = \Phi\left(\frac{Na - M}{\sqrt{N}b}\right). \tag{3.13}$$

From this relationship it is possible to determine the maximum number of reservations guaranteeing that this probability will not exceed a prescribed level.

Including cancellations, no-shows and overbooking in a single-leg dynamic programming models is fairly straightforward. Subramanian et al. (1999) have done so, with a suitably rewritten state space if necessary (for instance, it may be desirable to keep track of the number of seats booked in each category if the cancellations or no-shows differ by category). The authors present a fairly comprehensive model and provide some useful insights. Perhaps this was the reason why cancellations and overbooking did not enjoy greater popularity in the literature as a standalone research topic (although they are cited in almost every revenue management research paper as future research extensions for whatever the authors investigate). A good recent review of overbooking issues can be found in Klophaus and Poelt (2007).

In practice, it turns out that forecasting cancellation rates at an arbitrary level of detail is not an easy task. As we discussed in the section on forecasting for sparse origin/destination/fare class combinations one often runs into the small numbers problem. The theoretical models are very sensitive both to cancellation rates as well as to expected overbooking costs (other than refunds). The latter are not easy to quantify in practice since they involve less tangible quantities such as loss of goodwill in addition to a simple compensation. Simon (1968, 1972) provides an interesting reading about the evolution of compensation mechanisms, a subject that never fails to stir interest among the public, cf. WSJ Opinion (2010).

Properly calibrated cancellation rates and overbooking costs are particularly important to dynamic models that make explicit use of those quantities. But even when proper inputs to the model are available, the behavior of the bid prices (the displacement or opportunity costs) may be different from what practitioners are familiar with from years of working with EMSR and simple DP models: the bid price does not have to be monotone in time to departure (cf. Sect. 3.1 in Subramanian et al. 1999, and Chap. 5 in Walczak 2007).

Somewhat easier to handle are methods that do not focus on expected revenue but instead look at the expected number of denied boardings or the probability of denying boarding, and thus do not require estimates of the overbooking cost. The above example from Shlifer and Vardi (1975) falls into that category.

Given all these issues a class of popular approaches in the industry involves working with net demand; that is, an expected demand judiciously adjusted to account for cancellations along with no-show behavior. In its simplest rendering it amounts to multiplying expected class demand by the cancellation rate and/or the show-up rate. This also seems to be the industry best practice at present, especially when the simpler optimization models are used where cancellations do not appear explicitly. For this reason we omit the overbooking aspect from most of the optimization models that follow.

Overbooking is doing quite well for a technique that started over half a century ago. Klophaus and Poelt (2007) report an estimated profit contribution of 105 million Euros in 2005 due to optimized handling of close to 5 million passengers who failed to show up. Klophaus and Poelt also point out one other important aspect of overbooking, namely, that passengers benefit as well from greater seat availability and reduced overall costs due to more efficient airline operations.

3.3.2 Single Leg Inventory Control

Early optimization models made the following three major assumptions:

1. Passengers arrive in search of a specific fare class and walk away if that fare class is not available.
2. The demand for each fare class is independent and continuous.
3. Lower fare demand arrives before higher fare demand; that is, if there are J fare classes with fares $f_1 > f_2 > f_3 > \cdots > f_J$, and a passenger arrives wanting to purchase a ticket in fare class j, then no additional passengers will arrive seeking to purchase tickets in classes $k = j + 1, \ldots, J$.

All of these assumptions are restrictive, and more recent modeling efforts have sought to relax them. However, for a variety of reasons, the optimization models these assumptions gave rise to are still in frequent use today.

The first analytical approach to controlling the availability of inventory classes was conceived by Kenneth Littlewood (1972) at British Airways . The essence of Littlewood's method was captured in the now famous Littlewood's rule. Devised for the case of two fare classes, the rule provides an explicit means for determining when low fare ticket requests should be rejected. Letting D_1 be the random variable describing the high fare demand, and letting f_1 and f_2 represent the fares of the high and the low fare classes respectively, Littlewood's rule can be stated as follows.

Littlewood's Rule: Protect n_1 seats for high fare demand, where n_1 is determined by the equation

$$f_2 = f_1 \Pr(D_1 \geq n_1). \tag{3.14}$$

The interpretation of this equation is straightforward. The term $\Pr(D_1 \geq n_1)$ is the probability that an n_1th customer will arrive, so that $f_1 \Pr(D_1 \geq n_1)$ represents the expected marginal revenue associated with setting aside seat n_1 for the high fare demand. Littlewood's rule states that when the expected marginal revenue drops below the lower fare f_2, no more seats should be protected for the higher fare demand. It is interesting to note that the optimal value n_1 determined by Littlewood's rule for nested inventory is different from the optimal value n_1 for partitioned inventory, which seeks to balance the expected marginal revenue between the two classes:

$$f_2 \Pr(D_2 \geq n_2) = f_1 \Pr(D_1 \geq n_1). \tag{3.15}$$

Here, D_2 is the random variable describing the low fare demand, n_2 is the number of seats set aside for the low value fare class, and these two values are chosen so that their sum equals the total number of available seats.

Littlewood's seminal work was recognized in 2004 when he was awarded the first INFORMS Revenue Management and Pricing Section historical prize. In his acceptance letter, Littlewood indicated that his idea initially met considerable resistance, and it took many years before it was actually put into practice. It is interesting to observe that the more recent generation of network-based optimization models, dynamic programming models, and discrete choice models have met similar resistance decades later.

3.3.2.1 Single Leg Heuristics

One of the limitations of Littlewood's rule is that it addresses the case of only two fare classes. The Expected Marginal Seat Revenue (EMSR) heuristics put forth by Belobaba extended the ideas found in Littlewood's rule to the case of J fare classes with corresponding fares $f_1 > f_2 > \cdots > f_J$ and demands described by random variables $D_1, \ldots, D_J$ (Belobaba 1987, 1989; Belobaba and Weatherford 1996). The original EMSR heuristic—which came to be known as EMSRa after the introduction of a second EMSR heuristic deemed, appropriately, EMSRb— calculates the protection level for fare classes $1, \ldots, j$ by applying Littlewood's rule $O(j^2)$ distinct times.

As an example of EMSRa, consider the case of three fare classes. The protection level n_2 for fare classes 1 and 2 would be calculated as $n_2 = m_1 + m_2$ where m_1 and m_2 are defined by

$$f_3 = f_1 \Pr(D_1 \geq m_1) \tag{3.16}$$

and

$$f_3 = f_2 \Pr(D_1 \geq m_2) \tag{3.17}$$

while the protection level $n_1 = m_1$ would be calculated from

$$f_2 = f_1 \Pr(D_1 \geq m_1). \tag{3.18}$$

In general, the protection level for fare classes 1,..., j is calculated as

$$n_j = m_1 + m_2 + \cdots + m_j$$

where

$$f_{j+1} = f_i \Pr(D_i \geq m_i) \quad i = 1,\ldots,j-1.$$

Observe that m_i changes from the calculation of one protection level to the next.

EMSRb also builds upon Littlewood's rule, but in a different way. Again using a three fare class example, EMSRb applies Littlewood's rule to calculate the protection level for fare classes 1 and 2, but uses the sum of the demands of fare classes 1 and 2. Letting F_2 be the demand weighted average fare of fare classes 1 and 2, the protection level n_2 is calculated from the intuitively reasonable expression

$$f_3 = F_2 \Pr(D_1 + D_2 \geq n_2).$$

In general, EMSRb calculates the protection level n_j for the first j fare classes as

$$f_{j+1} = F_j \Pr\left(D_1 + D_2 + \cdots + D_j \geq n_j\right) \tag{3.19}$$

where

$$F_j = \left(f_1 E[D_1] + f_2 E[D_2] + \cdots + f_j E\left[D_j\right]\right) / \left(E[D_1] + E[D_2] + \cdots + E\left[D_j\right]\right). \tag{3.20}$$

Under the assumption that each D_i is normally distributed, as is usually done, the distribution of their sum is also normally distributed, greatly simplifying the EMSRb calculations.

The EMSR heuristics are easy to understand, familiar to most practitioners, have proven to be surprisingly robust, and require only forecasts of total demand-to-come (the mean and variance of the D_i when they are assumed to be normally distributed), and fares. The run-time requirements are low, with the most complex calculations involving standard normal cumulative distribution functions. While early simulations based on simple data examples reported in Brumelle and McGill (1993), and Wollmer (1992) show an optimality gap on the order of 0.5–1%, EMSR suffers from the same problem as most heuristics, namely, in certain situations they can yield results that are far from optimal. This had already been noted in Robinson (1995).

The most significant modeling limitations of the EMSR heuristics, however, are twofold: the independent demand assumption (along with the predetermined order of arrivals) and the fact that EMSR does not take into account the dynamic nature of the booking process; specifically, the stochastic nature of passenger arrivals over time and future booking decisions.

The early focus on single-leg inventory control using forecasts of the total demand-to-come was easy to understand. In the mid 1980s revenue management was not an established practice, and each step led researchers into new territory (Boyd 2007). Further, as discussed in Sect. 3.1, airline reservations systems were (and for the most part still are) designed around nested fare class inventory control. Early revenue management consisted of someone reviewing bookings on a flight, looking at how many people were booked in each class, estimating how many people were likely to arrive in each class, and setting protection levels accordingly. This business process, supported by the reservations systems and the global distribution network they gave rise to, was extremely influential in establishing the framework for early research in revenue management, and even today serves as one of the most significant impediments to the adoption of more sophisticated approaches to revenue management. Starting from scratch with the knowledge garnered over many decades, reservations systems, global distribution systems, and the surrounding business processes would certainly be designed differently. But with millions of dollars already invested in a particular way of doing business, change is slow and difficult.

3.3.2.2 Single Leg Optimization Algorithms

A number of authors developed conditions for calculating optimal protection levels under the same set of assumptions as the EMSR heuristics, among them Brumelle and McGill (1993), Curry (1990), Robinson (1995) and Wollmer (1992). With the exception of Robinson, all of these published references were preceded by presentations or technical reports (Brumelle and McGill 1988; Curry 1988, Wollmer 1988). Each paper provides its own perspective on the problem. One of the more readable presentations can be found in Talluri and van Ryzin (2004).

We do not dwell upon these results here since they were never broadly adopted and put into practice, being surpassed by the models discussed in Sect. 3.3.3. Instead, we briefly describe conditions under which optimality was demonstrated: Brumelle and McGill (1993) assume Low-Before-High ("LBH") order of arrivals and continuous demand distributions; Robinson does away with LBH but still requires that fare class arrivals are not interspersed; Curry (1990) assumes LBH and continuous demands to calculate revenue functions for each single-leg sub-problem and then combines those to formulate a network problem using mathematical programming and concavity properties of revenues thus calculated on individual legs; Wollmer (1992) assumes LBH and discrete demand distributions. All the papers share the assumption of independent demand by fare class.

We do mention one interesting result by Brumelle and McGill (1993) that characterizes conditions for a set of protection levels to be optimal. Specifically, they show that in a continuous demand setting, the protection levels $n_1,\ldots, n_{J-1}$ are optimal if

$$\begin{aligned} f_2 &= f_1 \Pr(D_1 \geq n_1), \\ f_3 &= f_1 \Pr(D_1 \geq n_1 \text{ and } D_1 + D_2 \geq n_2), \\ &\ldots \\ f_{j+1} &= f_1 \Pr\big(D_1 \geq n_1 \text{ and } D_1 + D_2 \geq n_2 \text{ and } D_1 + D_2 + \cdots D_j \geq n_j\big). \end{aligned} \tag{3.21}$$

As pointed out by the authors, these expressions show that the probability a fare class nest will sell out (under the optimal policy) is proportional to the ratio of the fares. For example, the probability that the fare class nest consisting of classes 1 and 2 will sell out is equal to Pr ($D_1 \geq n_1$ and $D_1 + D_2 \geq n_2$). By the expression given above, if the protection levels n_1 and n_2 are set optimally, then the proportion of time classes 1 and 2 both sell out relative to the total number of flight departures should be approximately f_3/f_1; similarly, the proportion of time fare class 1 will sell out should be approximately f_2/f_1. While the test is not used in practice—the assumptions underlying the model are too far from reality—it remains an important result for at least two reasons. First, it represents an effort to provide a measure that could be used in practice to quickly and simply check an airline's revenue management practices (if the assumptions underlying the model were closer to reality). Tests of this type are highly valued by practitioners. Second, the result provides a quantitative measure of how often an airline should expect to sell out the seats on a flight if it is properly managing its inventory. As the result points out, it is a healthy sign when all of the seats are not always sold, something that practitioners frequently forget.

3.3.3 Three Important Mathematical Themes

As revenue management advanced, at least three major themes evolved. The first was the introduction of time dynamics, replacing fare class demand-to-come estimates with stochastic processes describing fare class arrivals. Beyond providing greater realism, this change brought to light an alternative to nested protection levels for controlling the sale of inventory; a method discussed in more detail in the remaining sections.

A second theme was that of capturing network effects. As airlines developed route structures that required passengers to travel on more than one flight, the limitations of leg-level models became increasingly clear to practitioners and mathematical modelers alike. For practitioners, the limitations manifested themselves in questions such as "what is the Y class fare on a leg when there are 100 different itineraries drawing upon Y class inventory, each with its own fare?" Mathematical modelers not only saw the need for network models, but devoted considerable effort to the problem while developing important theoretical insights along the way.

Just as important were business considerations. For airlines, which often operate under razor-thin profit margins, the quest for fraction-of-a-percentage point improvements in revenue made good sense. Theoretical results as well as simulations were promising since every new class of models was more comprehensive and better reflective of actual customer behavior. While not every airline jumps on the bandwagon of any new theoretical development, models that are dynamic in nature and/or capture network effects explicitly have become accepted worldwide since they deliver promised improvements. Implementation of those run-time consuming methods was in no small measure made possible by the incredible increase in computing power available at little cost, and to better numerical approaches.

A third theme was the development of better demand models, as was touched upon in Sect. 3.2.4. In the remainder of this section we address each of these topics in more detail. In contrast to the first two themes, the development of the demand models was mainly driven by airlines suffering losses due to shifts in passenger buying behavior caused by several factors. We will have more to say about this phenomenon in Sect. 3.3.3.3.

3.3.3.1 Fare Class Arrivals as Stochastic Processes

Unlike the early models that assumed only estimates of total demand to come D_i for each fare class, the next generation of models assumed that demand for each fare class arrived according to a well-defined stochastic process—almost always a Poisson process, including its time-nonhomogeneous variant where the arrival rates λ_i take on different values during different time intervals in the booking cycle. Among the many benefits of this approach, the assumption that passengers arrive in a low-before-high order could be relaxed while still yielding a well-defined optimization problem.

The assumption of Poisson arrivals leads naturally to the use of dynamic programming algorithms. For a single-flight leg, the dynamic programming state space consists of the number of remaining seats on a flight while the stage space is defined by the time to departure. State transitions correspond to bookings, and the decision in each stage/state pair is whether to accept an arrival in a given fare class. Formally, if p_i represents the probability of an arrival in fare class i, then the expected value $V_t(x)$ of having x bookings with t time units remaining satisfies

$$V_t(x) = \sum_{j=1}^{J} p_j \max\{f_j + V_{t-1}(x-1), V_{t-1}(x)\} + p_0 V_{t-1}(x) \tag{3.22}$$

where $V_0(x) = 0$ for all x, and J is the number of fare classes. Observe that in the summation there is a lowest fare f_i such that for f_i and every higher fare the decision is to accept the booking, whereas for every fare lower than f_i the decision is to reject.

The formulation is actually a discrete approximation of a continuous time problem where, to first order, the probability of an arrival from fare class i is $\lambda_i\, dt$ and the probability of no arrival is $1-\, (\lambda_1\, dt + \lambda_2\, dt + \cdots + \lambda_J\, dt)$. The probabilities p_i used in the discrete formulation are then approximated for a small but finite time interval Δt as $p_i = \lambda_i\, \Delta t$ and

$$p_0 = 1 - (\lambda_1 \Delta t + \lambda_2 \Delta t + \cdots + \lambda_J \Delta t).$$

There are several excellent papers on leg dynamic programming for revenue management applications as indicated by McGill and van Ryzin (1999), Stone and Diamond (1992), Sun (1992), Lee and Hersh (1993); Shaykevich (1994); Young and van Slyke (1994), Brumelle and Walczak (1997), Zhao and Zheng (1998a, b), Lautenbacher and Stidham (1999), Subramanian et al. (1999) and Zhao (1999). The settings in these works range from the typical, yieldable demand assumption with unit requests, through batch or group requests, to priceable demand models (with a limited number of classes), and may include overbooking and cancellations as well as more general arrival processes. In aggregate these works cover all relevant structural properties.

Several Asian and European airlines as well as some major American airlines are currently known to use DP-based solutions. Of particular interest is that while dynamic programming provides the expected monetary value $V_t(x)$ of having x seats available at time point t in the booking cycle, it does not directly generate nested protection levels for booking classes. The dynamic programming recursive equation, however, suggests a natural alternative to protection levels: given x seats with t time remaining, accept a passenger in fare class i if

$$f_i \geq V_{t-1}(x) - V_{t-1}(x-1) \tag{3.23}$$

That is, accept the passenger if the fare exceeds the expected value of giving up a seat. In this context, the value $v = V_{t-1}(x) - V_{t-1}(x-1)$ is typically referred to as the bid price, and the process of using a monetary value v whereby a fare class is closed if its fare is less than v is called bid price control. Because most reservations systems were not originally built to support bid price control, modifications are usually required. And since reservations systems must communicate within the global distribution environment—which still uses fare classes as the primary means of sharing information on what's available—bid price controls are often converted to "number of seats available in a given fare class" at some point. This is especially straightforward when fare classes are ordered by fare. As will be discussed in the following sections, bid price control extends naturally to flight networks.

The process of converting bid prices into nested protection levels follows the same general mechanism as in EMSRb except that the "EMSR's" are now produced by a dynamic program: for each fare class we determine the first inventory level where the bid price exceeds the fare; the capacity less that inventory level gives us the protection level for the nest of higher fare classes. An example with five fare classes and a cabin capacity of 20 is shown in Tables 3.6 and 3.7. First note that the lowest bid price is greater than class fare for classes 3, 4, and 5 and therefore their protection levels are zero. For fare class 2 the bid price at inventory level 15 exceeds its fare of $200, and therefore the protection level for the fare nest consisting of fare classes 1 and 2 is 20 and the protection level for fare class 1 is 15.

The dynamic programming formulation presented here is extremely simple, but is easily modified to account for cancellations, arrivals of parties consisting of more than one person, no-shows, and other considerations. We show how to introduce some of those features. Let p_c be the probability of a single cancellation at time, so that $p_0 = 1 - p_c - \sum_{j=1}^{J} p_j$ is the probability of having neither one of booking requests nor a cancellation. For $x \leq c - 1$ the Bellman equation is (c is the capacity that now includes an overbooking pad, the maximum number of allowable overbookings)

$$V_t(x) = \sum_{j=1}^{J} p_j \max\{f_j + V_{t-1}(x-1), V_{t-1}(x)\} + p_c V_{t-1}(x+1) + p_0 V_{t-1}(x), \tag{3.24}$$

while for $x = 0$ we have [$V_t(0)$ is no longer zero since a cancellation can now happen]

$$V_t(0) = p_c V_{t-1}(1) + (1 - p_c) V_{t-1}(0)$$

and for $x = c$ we have (since no cancellation can happen now)

$$V_t(c) = \sum_{j=1}^{J} p_j \max\{f_j + V_{t-1}(c-1), V_{t-1}(c)\} + p_0 V_{t-1}(c),$$

with an initial condition

$$V_0(x) = y(x).$$

Note that the cancellation probability (rate) will in most cases depend on the number of booked customers $c - x$, i.e. $p_c = p_c(x, c)$, it can also depend on time. The terminal reward $y(x)$ is the negative of the expected value of the overbooking cost when there are x seats left at departure. Its absolute value is typically assumed to be increasing and convex in x to reflect an increasing cost per unit as more customers are denied boarding. As Subramanian et al. (1999) note, the overbooking and cancellation refunds are allowed to be fare class-dependent when included in the expectation of the class fare (so that f_j becomes an expected value of fare less applicable refunds).

Perhaps the ease with which these additional considerations can be modeled is the reason why they have not received the same attention in the literature as they

Table 3.6

Fare class	Demand	Fare($)
1	10	250
2	15	200
3	20	150
4	35	125
5	30	100

Table 3.7

Capacity	Bid price	Nest 1	Nest 1,2
1	250	1	1,2
2	249.95	1	1,2
3	249.75	1	1,2
4	249.09	1	1,2
5	247.51	1	1,2
6	244.51	1	1,2
7	239.81	1	1,2
8	233.56	1	1,2
9	226.37	1	1,2
10	219.12	1	1,2
11	212.65	1	1,2
12	207.52	1	1,2
13	203.9	1	1,2
14	201.65	1	1,2
15	200.43	1	1,2
16	199.87		1,2
17	199.27		1,2
18	198.21		1,2
19	196.56		1,2
20	194.22		1,2

did before the introduction of dynamic programming. Overbooking levels, for example, were one of the earliest topics ever addressed in the revenue management literature. Before determining how many seats to protect for each fare class, airlines first calculated overbooking levels using expected bookings, cancellations, and no-shows. These levels were then used to establish the number of seats an airline would allow itself to sell, and an algorithm such as one of the EMSR heuristics would be used to calculate nested booking limits on this expanded number of seats. Yet, at least in theory, the overbooking problem essentially disappears when using dynamic programming, being captured in the bid prices when cancellations, no-shows, and overbooking costs are included in the model. Chapters 2 and 3 of Talluri and van Ryzin (2004) have theoretical results on provably optimal controls. In practice, revenue lifts ranging from 0.5 to more than 5% have been reported compared to EMSR-type of controls.

3.3.3.2 Network Effects

Many different network optimization models have been proposed and evaluated in the research literature. For a complete yet concise mathematical overview, the reader is directed to Chap. 3 of Talluri and van Ryzin (2004). Here we provide an introduction to methods that have received focused attention in the industry.

Mathematical Programming Models

Some of the earliest and most successful network revenue management models were based on mathematical programming formulations. Perhaps the most well known is the demand-to-come linear program which has arisen in many different contexts, including but not limited to Glover et al. (1982), Alstrup et al. (1986), and Wollmer (1986). The formulation looks at the expected demand-to-come on all itineraries and in all fare classes and chooses the collection of passengers that yields the maximum network revenue while not exceeding the network capacity.

$$\begin{aligned} \text{Max} \quad & fy \\ \text{s.t.} \quad & Ay \leq c(\lambda) \\ & y \leq d(\omega) \\ & y \geq 0 \end{aligned} \tag{3.25}$$

Here, A is a matrix with one row for each flight leg in the network, with column A_{ij} being the incidence vector for itinerary i, fare class j; that is, column A_{ij} has a 1 in entry k if the flight leg k is part of the itinerary, and a 0 otherwise (observe that the matrix a has identical columns for different fare classes j on the same itinerary i). The vector f represents the fares of the different itinerary/fare classes, while d is a vector such that the d_{ij} is an estimate of the demand for itinerary i, fare class j. The vector c contains the remaining seat capacities for each flight leg; the variables λ and ω are the dual variables of the corresponding constraints; y is the class allocation vector (the primal solution of the above LP).

The demand-to-come linear program has a number of apparent weaknesses. Unlike leg formulations, demand variability is not incorporated in the model. Further, the solution y provides the recommended number of seats in each itinerary/fare class, so it is a type of partitioned control. Even a small network can give rise to tens of thousands of such values, most of which are small fractions. The itinerary/fare class levels can be converted into leg/fare class levels by summing the availabilities for all itineraries on a given leg, but since fare class nesting is not explicitly incorporated in the optimization model, the aggregated values do not have any direct link with nested protection levels. As a result, in practice y is abandoned in favor of the dual solution.

The standard mathematical programming interpretation of the dual variables λ_ℓ is the marginal value of a seat on the flight leg ℓ. Thus, the sum of the λ_ℓ on all flight legs in an itinerary can be interpreted as the marginal revenue v that is being displaced if a seat is sold on that itinerary. This suggests a form of network bid price control. Accept a passenger in fare class j on the flight itinerary i consisting of flight legs 1,..., M if

$$f_{ij} \geq \lambda_1 + \lambda_2 + \cdots + \lambda_M. \tag{3.26}$$

This form of network bid price control has proven to work surprisingly well and has been used in practice (Talluri and van Ryzin 2004). More often, however, the demand-to-come linear program is used as part of a multi-step procedure for generating inventory controls. For example, it is common to use the demand-to-come dual variables to prorate itinerary/fare class fares to flight legs. Here, we understand proration as any mechanism that determines what portion of an itinerary fare to assign to individual flight legs when the network is broken into single-leg subproblems. We include commonly used forms of shadow price-based displacement adjustment in this definition. The terms prorated fares, pseudo-fares, and displacement adjusted fares are all used depending on the context.

We note that prorated product fares need not sum up to the original product fare across all the legs that the product uses. One simple example is to take a two-leg network and a product with a fare of \$100 that uses each of those legs. Assume that total expected demands across all products sums to something just below each leg's capacity, so that the shadow prices are zero for each leg. In its raw form, the displacement adjustment using the shadow prices of zero will results in prorated fares for this product to be \$100 on each of the legs (so they add up to \$200). In practice, more sophisticated ways of adjustments are used that handle these types of borderline cases.

For example, take a \$600 Seattle–Atlanta–Miami Q class fare and attribute \$250 to the Seattle–Atlanta leg and \$325 to the Atlanta–Miami leg (as already mentioned, the prorated fares may, but are not required, to total \$600). With fares apportioned in this way, a forecasted unit of Seattle–Atlanta–Miami demand can be treated as a \$250 unit of demand on the Seattle–Atlanta flight leg and a \$325 unit of demand on the Atlanta–Miami flight leg, and the problem is effectively decomposed into a collection of individual flight leg problems. The quality of the inventory controls generated by these flight leg problems is, of course, highly dependent on the method of fare proration, and methods that use information from a network optimization problem have not surprisingly proven to perform best among the various proration methods that have been proposed. In some formulations, the fare prorations and leg-level problems arise from a formal mathematical decomposition of the original problem (Boyd and Bilegan 2003). At this point we should mention that the dual variable-based bid prices are sometimes used directly while in other, more sophisticated approaches they are used to (essentially) decompose the network. This is how they are used in the dynamic programming decomposition that we describe next.

Dynamic programming (DP) decomposition (Talluri and van Ryzin 2004) solves the individual flight leg problems as dynamic programs similar to those described in Sect. 3.3.3.1. The breakdown of the network into leg subproblems is achieved by using pseudo-fares for each leg as described below.

The bid prices arising from these leg-level dynamic programs are then used instead of the dual variables from the mathematical program. This method, sometimes called DP Decomposition falls into the category of approximate DP discussed below. Virtual nesting, described in Sect. 3.1.1, uses leg prorated fares to form virtual buckets of like-valued demand on each flight leg, after which leg-level solution procedures are employed.

Leg prorated fares can also be used to help carriers establish a coherent set of prices for their different fare classes across their flight networks in a process known as fare class rationalization (Fellner et al. 2006). Each single-leg flight in a network can be part of many customer itineraries bought at different points of sales and for varying amounts of money. For example, a Y class ticket may have seen many actual fares that were low, for, say, reward travel or companion tickets, as well as some very high fares for expensive long-haul itineraries. When forecasting what a class fare on the leg in question is, it can happen that classes lower in a hierarchy have higher fare forecasts than a class higher in a hierarchy (a phenomenon called fare inversion). Fare class rationality uses network information to prevent fare inversions and to ensure smaller fare spread within each class. See Fellner et al. (2006) for an overview of the basic ideas.

There are many ways to prorate fares using the demand-to-come dual variables. One of the historically most popular methods is to take the original fare f_{ij} for an itinerary/fare class consisting of legs 1,..., M and calculate the displacement adjusted fare π_k for leg k as

$$\pi_k = f_{ij} - \lambda_1 - \lambda_2 - \cdots - \lambda_{k-1} - \lambda_{k+1} - \cdots - \lambda_M. \tag{3.27}$$

The reasoning behind this calculation is that the fare attributable to leg k should be penalized to account for the fact that there is a cost to reducing the seat inventory on other legs in the itinerary. The intuition is reasonable and it can also be mathematically justified when the value function in the network DP problem is approximated as a sum of affine functions corresponding to each leg (see van Ryzin and Liu 2004; Adelman 2007). The resultant prorated fares from this calculation are frequently referred to as pseudo-fares.

It should be noted that mathematical programs other than the demand-to-come linear program can be formulated. Williamson (1992) investigates a number of such formulations. Among the most well known of the alternative formulations is the probabilistic linear programming formulation (see Talluri and van Ryzin 2004). Additionally, the demand-to-come linear programs, such as the one introduced in the beginning of this section, can be shown to provide an upper bound on the true optimal expected revenue from the network. It also possesses the reassuring asymptotic property that the bound is tight in a well-defined sense (Cooper 2002).

Network Dynamic Programming

Conceptually, the network optimization problem can be modeled as a dynamic program almost as easily as the flight leg optimization problem. The network formulation is multidimensional in that the state space no longer consists of the number of seats on a particular flight leg, but the numbers of seats on each separate flight leg in the network. State transitions correspond to bookings on itineraries, and the decision in each stage/state pair is whether to accept an arrival in a given fare class.

To express the model formally, let A and f be as defined in Section "Mathematical Programming Models", and let p be a vector such that p_{ij} is the probability that the next arrival will be for itinerary i, fare class j. As with the leg formulation, these probabilities are typically derived from the arrival rates of presumed Poisson processes for each itinerary/fare class, with p_0 (not included in the vector p) representing the probability of no arrival on any itinerary during the prescribed time interval. Letting x be commensurate with the rows of A, with x_i corresponding to the number of seats on flight leg i, the network dynamic program can be written as

$$V_t(x) = \sum_i \sum_{j=1}^{J} p_{ij} \max\{f_{ij} + V_{t-1}(x - A_{ij}), V_{t-1}(x)\} + p_0 V_{t-1}(x) \tag{3.28}$$

where $V_0(x) = 0$ for all x. As with leg-level dynamic programming, a natural decision mechanism given the state x of the of the network with t time remaining is to accept a booking on itinerary i in fare class j if

$$f_{ij} \geq V_{t-1}(x) - V_{t-1}(x - A_{ij}). \tag{3.29}$$

Of course, the state space grows far too quickly for the model to be of any practical use. With m an approximation of the number of seats on each aircraft and n the number of flight legs in the network, the state space at each stage is m^n. Networks consisting of as few as three flight legs can become computationally prohibitive when interwoven within a complete revenue management system.

Another difficult aspect is the nonlinearity of the value function and the displacement cost as a function of capacity and time. These facts have prompted a number of different approaches to solving approximations to the problem. Chief among them is bid price control, which in essence linearizes the value function and reduces the number of states to a polynomial level.

Network dynamic programming has become something of a Holy Grail in revenue management and continues to generate research interest both in academia and in the industry. This is because of three reasons:

- It is the provably optimal solution under typical assumptions and as such carries the promise of additional revenue improvements
- Many features of the real-life process can be added to the model formulation in a fairly natural way
- It provides explanatory power.

Approximate Dynamic Programming

In what follows, we interpret approximate dynamic programming very broadly to include a variety of methods attempting to solve simpler versions of the full network dynamic program, including value function approximation methods.

One avenue researched in the literature is based on the fact that a network dynamic program can be reformulated as a multi-stage stochastic linear program with binary decision variables as done in Cooper et al. (2006a, 2006b). This equivalence leads to a number of approximate solutions that can be found in de Boer et al. (2002), Chen (2006), and Higle (2007).

Another class of approximate solutions uses simulation to approximate the value function $V_t(x)$ (Bertsimas and de Boer 2005; Cooper et al. (2006a, 2006b), who combine stochastic linear program with simulation).

Quite recently, substantial work has been done on approximating the network dynamic program in the state or policy spaces, and also on approximating the value function (Bertsimas and Popescu 2003; de Farias and Van Roy 2003; Walczak 2004; Farias and Van Roy 2006). The value function approximation seems very promising and provides theoretical basis for already existing heuristics.

For example, the DP Decomposition described earlier can be considered an approximate DP where the value function is decomposed into leg-level affine basis functions that are time and capacity independent. We can replace this family of functions with a richer one to obtain better displacements costs (and so better controls). For example, Adelman (2007) allows time dependent leg-level functions. This has been further extended to allow inventory-dependence (with degrees of aggregation) by Meissner and Strauss (2008) as well as customer choice (Zhang and Adelman 2007; Meissner and Strauss 2008).

An interesting question arises in this context: at which point should the more "precise" shadow prices from the LP be used directly as bid prices rather than used to decompose the network via displacement adjustments and then have the bid prices produced by single-leg DPs.

3.3.3.3 Choice Modeling and Optimization

Whether forecasts are made of total demand-to-come or of arrival rates λ, all of the optimization algorithms discussed thus far assume passengers arrive looking for a particular fare class. If that fare class is available, a purchase is made. If not, the passenger walks away. The inadequacies of this assumption were recognized early in the history of revenue management, and various efforts were made to account for them through modeling of *buy-up* and *buy-down* (passengers purchasing a more expensive or less expensive fare product when their desired fare product is not available) and the closely related *spill* (losing passengers when their desired flight and fare class are unavailable) and *recapture* (booking a spilled passenger on another flight and/or fare class). The historical literature on these topics lacks the coherency of many other revenue management topics, perhaps reflecting the fact that practitioners were seeking to come to terms with the right modeling abstraction. Two recent models that have helped frame the discussion are those of Boyd and Kallesen (2004) and Talluri and van Ryzin (2004), both of which were discussed in Sect. 3.2.4.

The change in demand models leads to corresponding changes in optimization models, though in the case of dynamic programming solution methods the changes are quite natural. The model proposed by Boyd and Kallesen postulates two types of demand which they refer to as *priceable* and *yieldable*. Priceable demand comes in search of the lowest available fare, while yieldable demand seeks a particular fare class. Models that seek to account for both types of demand have come to be known as *hybrid* demand models.

From the standpoint of dynamic programming, both types of demand are simply different arrival streams and can be incorporated quite simply. Isler (2004) and Walczak and Oosten (2003, 2005) demonstrate how, for computational purposes, priceable and yieldable demand can be converted from one into the other through a data transformation, which we discuss in Section "Equivalent Optimization Models". Other types of demand have been suggested, including *market priceable* demand, consisting of customers who seek the lowest available fare in the whole market and thus are not dedicated to one carrier.

The model proposed by Talluri and van Ryzin, which is discussed in the context of flight leg optimization, uses discrete choice modeling to determine the probability that an arriving passenger, faced with some subset S of fare classes, will choose a particular fare class $j \in S$ or will walk away. The probabilities can be based on any available information about the fare classes, including characteristics such as class restrictions in addition to price. These sets can be used in the standard leg-level dynamic programming formulation, where the decision at each stage (time before departure) in each state (number of seats remaining) is which subset S, if offered, will yield the highest expected revenue. As a practical matter, with n fare classes there are 2^n potential subsets S to choose from, but for several popular choice models the authors describe the structure of subsets S with the potential for being optimal in terms of an efficient frontier of solutions. In the single-leg case the class of subsets used in the dynamic program can then be limited to these efficient sets, and their existence makes the model computationally more tractable for most practical problems. Even in the network case efficient sets are useful, thanks to their asymptotic optimality and also whenever the optimal sets can be shown to be efficient, e.g., as in the parallel flights model of Zhang and Cooper (2005).

Conceptually, the model can easily be expanded to an entire network or to a subset of a network. This is particularly advantageous when an airline flies multiple itineraries from city A to city B, in which case departure time, flight duration, number of stops, and other characteristics can be included in the choice models. Unfortunately, the rapid growth of the state space is accompanied by growth in the number of efficient sets S, which now represent what fare classes to leave open on all of the different flight legs under consideration. This makes direct application of the model impractical for networks consisting of more than a handful of flight legs.

It is worth noting that the demand-to-come linear program can be extended to accommodate discrete choice models of demand. The idea was first raised in Iyengar et al. (2004) in the context of introducing more complex products than fare classes, and elaborated upon by van Ryzin and Liu (2004). For completeness, we present this model below after introducing some notation. Let S be a set of products offered to the customer (the *offer set*) and $D_j(S)$ the amount of product j sold when S is offered; S is a subset of a set of all available products N. We define the revenue $R(S) = \sum_{j \varepsilon S} F_j D_j(S)$ and the resource consumption $Q(S) = A\ D(S)$, where A is the incidence matrix, D is the product demand vector, and F_j is the product price.

$$
\begin{aligned}
& V(c) = \max \sum_{S} R(S)t(S) \\
& \text{s.t.} \\
& \sum_{S \subseteq N} Q(S)t(S) \leq c \\
& \sum_{S \subseteq N} t(S) \leq 1 \\
& t(S) \geq 0, \text{for all } S \text{ in } N.
\end{aligned}
\tag{3.30}
$$

The decision variables $t(S)$ now have the interpretation of the proportion of time that set S is offered. Note that we can have only one set offered at a time and that this deterministic program does not determine the order in which the sets have to be offered.

Like the dynamic programming formulation, the linear program has a size that is dependent on the 2^n potential subsets S of the number of fare classes, although the dual variables corresponding to flight leg capacity constraints, which are relatively small in number, are typically of the most interest for revenue management purposes. Unfortunately, the concept of an efficient frontier of sets S is somewhat less useful in the dynamic network problem since not all optimal offer sets have to be efficient. However, in the above linear programming formulations the optimal offer sets are efficient and column generation techniques can be used for the problem's solution.

One popular assumption that makes the problem easier and can be combined with the column generation techniques is to assume that there exist customer segments that have *disjoint consideration sets.* Such an assumption makes the problem separate by consideration sets into smaller linear programs. Once these are solved, then the dual solution to the choice linear program can be used to obtain the pseudo-fares and thus decompose the network. Van Ryzin and Liu (2004) propose this close analog to the traditional (based on yieldable demand) DP decomposition and DAVN, and offer comparisons and review of issues surrounding choice modeling on the network.

As the search for better demand models continues, airlines are faced with several questions of practical importance, given the possibility of changes:

- Are new forecasting methods required for the new models?
- Are new optimization routines required for the new models?
- Are new control mechanisms (reservations systems) needed to implement the optimal recommendations generated from the new models?

The manner in which we address these questions will impact the future development of airline RM, so we will sketch out some of those possible future directions in Sect. 3.4. In general, while the answer to the first question is yes, the answers to the last two depend on what new models are proposed. For certain classes of choice models, such as the hybrid or the multinomial logit (MNL) models, it is possible to reuse the existing optimization routines and capabilities of the reservations system by means of a relatively simple mathematical device that allows their conversion into the simpler, traditional yieldable model.

This data transformation also provides insights into dilution effects as well as into efficiency of the pricing points; its other byproduct is the run-time benefits resulting from converting into a model with simpler decision making (accept/deny). On the theoretical side, the transformed yieldable model is also mathematically simpler, and demonstrating structural properties is more easily accomplished, a fact recognized by Kincaid and Darling (1963).

In the following two sections we use simple examples taken from Walczak et al. (2008) to illustrate the concept of efficiency and its usefulness in converting a priceable demand problem into yieldable problem.

Efficient Fares

Assume that we have a pricing problem with the following three price points and expected demands:

Price ($)	Cumulative demand
300	3
200	4
100	10

It can be shown that some prices will never be used no matter how many units of a product we have to sell. In this example the $200 price point will never be used. Formally, only the price points that are part of the efficient frontier should be used in optimization. This is a well-known fact and more about the concept as well as an algorithm for determining the efficient frontier can be found in Talluri and van Ryzin (2004). In the airline context it means that the availability of an inefficient class will be the same as that of the closest lower class that is efficient.

Remaining capacity (*x*)	Optimal price (*p*) ($)	Demand (*d*)	Revenue *p**min (*x*, *d*) ($)
1	300	3	300
2	300	3	600
3	300	3	900
4	300	3	900
5	300	3	900
6	300	3	900
7	300	3	900
8	300	3	900
9	100	10	900
10	100	10	1000

Equivalent Optimization Models

The equivalence between a more general choice-based and the yield models holds only for efficient fare classes; otherwise it cannot be ascertained. (But one can always imbed a yieldable model in a more complex choice model such as the priceable one. In order to do this it is enough to sort the yieldable products by fare and then form the usual fare nests. By construction all the nests will be efficient.) The efficient fare classes constitute the domain of the data transformation and therefore determining the efficient frontier is the first step in the conversion process.

The equivalence between optimization models means here that the expected revenues (value functions) are identical for both models and that the optimal solutions can be mapped from one to another. The conversion is achieved by a relatively simple transformation of prices (fares) and expected demands. We refer the reader to the works cited below for mathematical details and will limit the presentation here to just one example (adapted from Walczak et al. 2008).

We focus here on a typical airline situation where a fare class can be identified with a nest of fares where it is the lowest-priced class. Let j index the set of efficient fare classes (or efficient nests), ordered by increasing demand. The expected (priceable) demand at price point j is D_j and the fare (or expected fare) is F_j where we assume that $F_j < F_{j-1}$ and $D_j > D_{j-1}$ with $D_0 = 0$.

The essence of the transformation is to calculate for each two consecutive (and efficient) nests the increment in expected revenue and divide it by the increment in the expected demand. This gives us the marginal expected revenue which can be directly compared to the marginal opportunity cost without going through the more complex maximization.

Let d_j be the transformed yieldable demand corresponding to price point j and f_j be the transformed fare. We define the demand transformation as

$$d_j = D_j - D_{j-1}. \tag{3.31}$$

The transformed demand d_i represents incremental demand if the next efficient offer set is open (the same as one additional class opened if all fare classes are efficient). The fare transformation is defined as

$$f_j = \frac{F_j D_j - F_{j-1} D_{j-1}}{D_j - D_{j-1}}. \tag{3.32}$$

We refer to Walczak and Oosten (2008) for a proof and related results, but encourage the reader to go back to our example above with three price points (and with the middle price point inefficient) to check that the transformed fares are \$300 and (after rounding) \$14 with the corresponding demands equal to 3 and 7. What it means is that opening the bottom two classes in addition to the top one provides seven units of additional demand but at a much lower fare of \$14 after accounting for expected losses due to customers buying down.

The equivalence between yieldable and priceable problems dates back at least to the seminal paper by Kincaid and Darling (1963), and some of its concepts seem to have been rediscovered in the last decade or so. For example Zhao (2000) converts a dynamic program with yieldable demand with two classes into a dynamic program with priceable demand; Chatwin (2002) describes a framework that includes the two types of demand by means of a transaction function, and discusses several notions of equivalence; Talluri and van Ryzin (2004) focus on the "increase in expected revenue per unit increase in expected demand," which is a concept tightly linked to the transformation. Ternoey (1997) looks at converting the priceable problem into the yieldable one, and points to the same practical reasons that we do here. The transformation itself seems to have been reinvented in the 2000s, primarily in the industry; see Isler (2004, 2005), Petersen and Fiig (2005), Walczak (2003), and Oosten and Walczak (2005).

The very advantage of the transformation is that it is implemented in practice as fare adjustments, and thus makes it easier to reuse the existing reservations systems.

Let us consider a simple example (Walczak et al. 2008) with hybrid demand, that is, with priceable and yieldable demand components. The table columns provide the following information: the second column has yieldable (independent) demand for each fare class, third column totals yieldable demands for each nest (starting at the top), the fourth has cumulative priceable demand, e.g., we expect 20 units of demand if FC2 is the lowest class opened (in addition to all the

Table 3.8 Demand, fare, and revenue for each class

Class	Yield Dmd	Cum Dmd	CumPrice Dmd	Total Dmd	Fare ($)	Revenues from yield passengers ($)	Revenue from priceable Passengers ($)	Total revenues ($)
FC1	5	5	10	15	200	1,000	2,000	3,000
FC2	7	12	20	32	150	2,050	3,000	5,050
FC3	2	14	30	44	90	2,230	2,700	4,930
FC4	10	24	50	74	75	2,980	3,750	6,730

yieldable demand), and the fifth column sums up the cumulative yieldable and the cumulative demand (so it is the total demand for the nest). Next to the column with fare classes we calculate the expected revenue for each nest, e.g., $2,980 is the expected revenue from the yieldable demand when all the fare classes are open (a nest with FC4 being the lowest available fare class); this is column 7. Column 8 has the expected revenue from the priceable demand and the last column has the total, i.e. the sum of the two previous columns.

Looking at the efficient frontier (the concave envelope) below we notice that fare class FC3 (or the nest {FC1, FC2, FC3}) is not efficient.

Then we apply our formulas to the three remaining fare classes. The table below has the transformation results. Columns 2, 4, and 5 correspond to columns 5, 9, and 6 in Table 3.8; column 3 has the transformed demand and column 6 has the transformed fare. For example, the transformed demand for FC4 (which

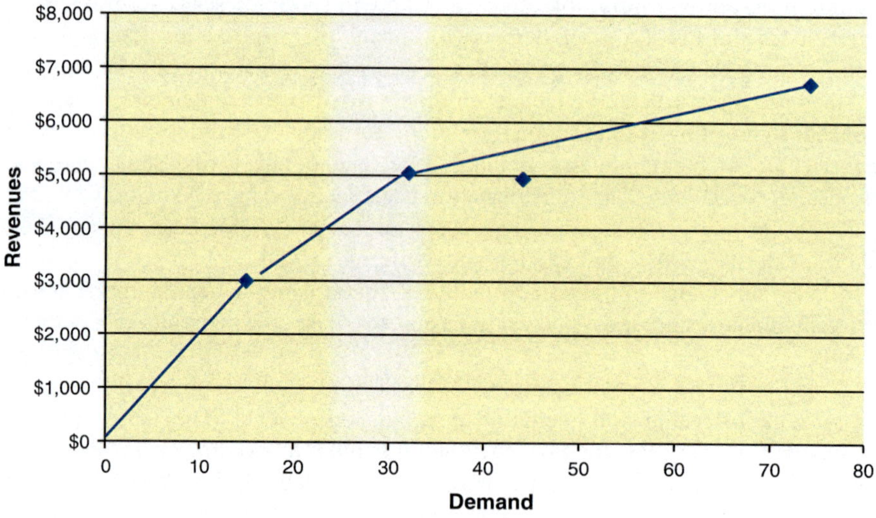

Fig. 3.3 The efficient frontier (the concave envelope)

Table 3.9 Transformed fare class demand

Class	Total demand	Transformed demand	Total revenues ($)	Fare ($)	Transformed fare ($)
FC1	15	15	3,000	200	200
FC2	32	17	5,050	150	121
FC3	44	0	4,930	90	40
FC4	74	42	6,730	75	40

Table 3.10 Transformed class fares

Class	Fare ($)	Fare adjustment ($)	Transformed fare ($)
FC1	200	0	200
FC2	150	−29	121
FC3	90	−50	40
FC4	75	−25	40

corresponds to all fare classes being open) is 42 and the transformed fare is 40, meaning that by opening all four the incremental increase in demand compared to opening only the two top classes is 42 but the effective fare is only $40 which is less than $75 (FC4 original fare) after accounting for buy-down from higher classes (Tables 3.9, 3.10).

Notice the following:

- While the transformed fare for FC1 stays the same, for FC2 we have transformed fare f_2 reduced to ($5050 − $3000)/(32 − 15) = $121 which is less than the original fare; the same is done for FC4.
- FC3 is not on the efficient frontier so the incremental demand (transformed demand) is set to 0 since FC3 should not be available for sale unless FC4 is available for sale.
- FC3 is open only when FC4 is open. Thus, by convention, these two classes have the same availability and the same transformed fares.

Without going into mathematical details let us gather some useful facts about the transformation:

- The transformed fares for efficient classes are never greater than their original fares; this represents the revenue loss due to dilution (buy-down). One can think of transformed fares as the original fare less a dilution adjustment.
- The fare adjustments (and so the transformed fares) can be expressed in terms of the price elasticity of demand; see Walczak and Oosten (2008) and Fiig et al. (2005).
- We obtain one transformed fare for each efficient nest (or for each efficient offer set in general).

Fare adjustments are becoming popular in the industry since they bring the optimization and control execution back into the "yieldable format"; they also provide insight into the impact of buy-down (see Isler 2005, Petersen and Fiig 2005; Gallego et al. 2007). Extending this approach is straightforward to network problems where buy-down happens within itineraries. Fortunately, typical airline networks are either already of this type or can be brought into that format and therefore the DP Decomposition or other Approximate DP approaches can be used. Also, a heuristic combining fare adjustments with EMSR within DAVN methodology is reported in the works just cited. Another important direction where the methodology has recently been applied are so-called fare families, cf. Fiig et al. (2010). Fare families are a way to model bundling of travel service (airline seats) with other services and restrictions. Different families come with different subsets of services or restriction levels and the products within each differ only by price. So buy-down is assumed to take place mainly within each family, but with some customers occasionally switching from one family to another.

With or without the data transformation the transition to different demand models is a significant shift for an industry whose habits, practices, and hardware developed around the yieldable demand assumption. While the need is clear and solutions to address that need exist, many experts working with the airline industry report a distinct preference for ad hoc approaches based on business rules or various flavors of forecast modification. So it is even more encouraging when the early adopters (such as those cited above) report successful results from systems based on sounder theoretical models. See Fiig et al. (2005) for SAS and Swiss experience with simulation and initial limited implementation, as well as Petersen and Fiig (2005) for positive results from implementation and a discussion of user experience at SAS new demand models; very recently Lufthansa (LH) announced (PROS 2011) implementation of the hybrid model in their revenue management system.

3.4 Conclusions and Future Directions

We tried to present the main ideas and developments in the airline revenue management in the four decades or so since its inception. Many of those ideas, both older, such as EMSR or the newer bid price control have become well established, integral parts of revenue management systems that are used everyday by the airlines. But the story of the airline revenue management has not yet been all told. The changing business environment, quest for new revenue sources, and improvements to the existing algorithms to wring out the incremental revenue and make demand-side management more efficient will keep the field alive for the foreseeable future.

Some of the changes are driven by technological improvements such as greater computer power that enables, on one hand implementation of more sophisticated algorithms and on the other bigger and easily accessible corporate-wide data warehouses will spur closer cooperation between airline business units. For example, carriers are expanding the online presence of their web sites and their capabilities, effectively taking over more functions of a typical travel agency and allowing customers to book components of a trip experience beyond the airline seat. These as well as ancillary revenue in the form of various upgrades such as additional leg room, upgraded services or certain fees can now be tracked, often in conjunction with frequent flyer information, and analyzed to present the booking customer with an optimal product bundle. The wealth of data will drive better product design as well as better, more dynamic integration with fleet assignment, scheduling, and passenger re-accommodation. Signs of approaching repositioning of airlines, GDS's and travel agencies are already emerging and have been reported in the business press, cf. Esterl (2011).

On the market side, carrier consolidation will lead, for the booking customer, to route expansion and, for the carrier, to route rationalization. Fewer but stronger alliances will seek real-time exchange of information between marketing and operating carriers, such as sharing capacity information and bid prices, cf. Poelt (2011). The lurking question of what constitutes a fair revenue sharing agreement based on what information can be shared *(prorate agreements)* is already coming to the forefront at some of the better integrated alliances.

A general research question that urgently needs good answers is that of demand modeling that at the same time encompasses competitive information, as discussed in Sects. 3.1.5, 3.2.4, and 3.3.3.3. A decade ago it was assumed that fare class booking history would serve as the primary input data for making forecasts, and the most important questions were what forecasting methods best estimate fare class demand and whether forecasts should be made at the leg/fare class or itinerary/fare class level. Today, it is increasingly understood that using past fare class bookings to forecast future bookings, even as a first-level engineering approximation, can lead to undesirable outcomes. The reality is that demand for a $600 Y class ticket is dependent upon many factors, and there are consequences for ignoring these factors. The challenge for academicians is to understand the strengths and weaknesses of various models of demand. The challenge for practitioners is to weigh the theoretical aspects of the models, how effectively model parameters can be estimated, the ability of users to comprehend and accept the models, and the consequences for the existing distribution environment.

A general consensus on how to model demand has yet to emerge, but early signs within the airline community point to evolutionary changes rather than adopting entirely new approaches to revenue management. Specifically, introducing types of customer behavior into traditional demand models is more common than a complete transition to customer choice models. It is very much in this spirit that the hybrid demand presented in Sect. 3.2.4 was developed.

Within the academic community a number of interesting research directions have surfaced. One important direction is a better understanding of the dynamics of forecasting and optimization when used in repeated cycles. Forecasting and optimization are intimately related, since the inventory controls that result from optimization impact the demand observations, which in turn impact the optimization. Cooper, Homem-de-Mello, and Kleywegt (2006a) provide one of the first theoretical efforts to understand the dynamics of revenue management and the spiral down phenomenon, and have done early work on the dynamics of revenue management when competitive prices are ignored (2006b). The dynamics of the forecasting/optimization cycle have long been understood as posing a potentially serious problem for revenue management, and, together with the closely related issue of demand modeling, will play a dominant role in shaping the future of revenue management.

Another research direction has to do with the question of what is the rational price for an airline seat given forecasts about market demand and a degree of information about competitor behavior. One way to answer it is to model airline competition as a dynamic game on a single flight. Talluri (2003), Isler and Imhof (2007), and Walczak (2007) show a number of properties of such models and numerically demonstrate the possibility of spiral-down type of phenomena (as well as assumptions to prevent them) even when both competitors have correct market information and use equilibrium prices. Isler and Imhof (2007) argue that airline competition might be better modeled as a repeated game rather than a dynamic one, and discuss the role of the revenue management analyst in the changing environment.

While modeling customer choice in the presence of competition is crucial, the thin profit margin environment in which most airlines operate will also drive improvements to existing network algorithms, and to overbooking and cancellation modeling, some of which we have described here. On the other hand, airline business metrics are not necessarily limited to profit or revenue: market share, load factors, long-term customer value, and various measures of risk are often as important. We have already seen it in our brief discussion of overbooking methods that metrics such as expected denied boarding cost have to be balanced against the expected revenue and its maximization. This concept of including other metrics and objectives into consideration applies more generally and has been relatively recently explored by Barz (2006, 2007), Feng and Xiao (2008), Levin et al. (2008), Walczak (2010), and Lieberman (2011). Going beyond revenue by implementing some of these ideas into practice will help revenue management and pricing optimization become an even more important and flexible tool to achieve strategic company goals.

3.4.1 Beyond Airlines

Revenue management is a remarkably simple idea with an easily communicated business case: adjust price based on available inventory and the demand that is expected to show up. Through a process of research, practice, discussion, and debate, practitioners and academics have made revenue management one of the most advanced applications of management science today—an application that has not only served the airline industry and the travel and transportation industry in general, but is serving as a model for all industries as they review their pricing practices.

As companies strive to emulate what airlines have achieved, they will undoubtedly go through many of the same struggles airlines did over 30 years ago when revenue management was in its infancy. It will take time to adapt the existing information technology infrastructures, and to transform organizations so that they think scientifically about their pricing. But with the example set by airlines, as well as the many lessons learned over the years, other industries should expect that the transformation will happen more quickly than it did for their forebears—even if not overnight.

This is not to say that airlines cannot benefit from other industries as well. Until recently, because of the reliance on the independent demand assumption and its implied demand segmentation, the revenue management community has done relatively little in the area of airline market research. There are definite opportunities to exploit from testing customer preferences at the market level, a practice that has matured in many industries outside airlines. The same is true for quantitative approaches to airline marketing, such as promotion optimization, that can be focused on customer segments and utilize airlines' own capacity information as well as competitors' prices.

References

Adelman D (2007) Dynamic bid prices in revenue management. Oper Res 55:647–661

Barz C, Waldmann K-H (2007) Risk-sensitive capacity control in revenue management. Math Methods Oper Res 65:565–579

Barz C, Uhrig-Homburg M, Waldmann K.-H (2006) Yield management: beyond expected revenue. Dreier T, Studer R, Weinhardt C (eds), Information management and market engineering, Universitätsverlag, Karlsruhe (TH), 71–84

Belobaba PP (1987) Air travel demand and airline seat inventory management. Ph.D. thesis, Flight Transportation Laboratory, MIT, Cambridge

Belobaba PP (1989) Application of a probabilistic decision model to airline seat inventory control. Oper Res 37(2):183–197

Belobaba PP (2004) Research highlights: pricing, distribution, and revenue management. Presentation to industry advisory board. See web.mit.edu/airlines/www/board-meetings/meeting-nov-2004/Pricing&RM-Belobaba.pdf. Accessed 1 May, 2007

Belobaba PP, Weatherford LR (1996) Comparing decision rules that incorporate customer diversion in perishable assett revenue management situations. Decis Sci 27(2):343–363

Bertsimas D, de Boer S (2005) Simulation-based booking limits for airline revenue management. Oper Res 53(1):90–106

Bertsimas D, Popescu I (2003) Revenue management in a dynamic network environment. Transp Sci 37(3):257–277

Boyd EA (2007) The future of pricing: how airline ticket pricing has inspired a revolution. Palgrave Macmillan, New York

Boyd A, Bilegan I (2003) Revenue management and e-commerce. Manag Sci 49(10):1363–1386

Boyd EA, Kallesen R (2004) The science of revenue management when passengers purchase the lowest available fare. J Rev Pric Manag 3(2):171–177

Boyd EA, Kambour E, Koushik D, Tama J (2001) The impact of buy down on sell up, unconstraining, and spiral down. Presented at the first annual INFORMS revenue management and pricing section conference. Columbia University, New York

Brumelle SL, McGill JI (1993) Airline seat allocation with multiple nested fare classes. Oper Res 41(1):127–137

Brumelle SL McGill JI (1988) Airline seat allocation with multiple nested fare classes. Presented at the INFORMS annual conference. Also presented at a University of British Columbia seminar, Denver, 1987

Brumelle S, Walczak D (1997) Dynamic allocation of airline seat inventory with batch arrivals. Air transport research group of the WCTR society proceedings. 3. Faculty of Commerce and Business Administration, University of British Columbia, Vancouver

Chatwin R (2002) On the equivalence of pricing and discount allocation problems in revenue management. Presented at INFORMS, San Jose

Chen L, Homem-de-Mello T (2006) Re-solving stochastic programming models for airline revenue management. Working paper, Department of Industrial, Welding and Systems Engineering, The Ohio State University, Columbus

Clements, Michael P, David F. Hendry (1998) Forecasting economic time series. Cambridge University Press

Cook TS (1998) Sabre soars. OR/MS Today 25(3):26–31

Cooper WL (2002) Asymptotic behavior of an allocation policy for revenue management. Oper Res 50:968–987

Cooper WL, Homem-de-Mello T, Kleywegt AJ (2006a) Models of the spiral-down effect in revenue management. Oper Res 54(5):968–987

Cooper WL, Homem-de-Mello T, Kleywegt AJ (2006b) Pricing dynamics of competitors whose models ignore competition. Presented at the 6th annual INFORMS revenue management and pricing section conference. Columbia University, New York

Cooper WL, Homem-de-Mello T. Some decomposition methods for revenue management. Transp Sci (to appear)

Cross RG (1997) Revenue management: hard-core tactics for market domination. Broadway Books, New York

Curry RE (1988) Optimum seat allocation with fare classes nested on segments and legs. Technical note 88-1. Aeronomics, Atlanta

Curry RE (1990) Optimal airline seat allocation with fare classes nested by origin and destinations. Transp Sci 24:193–204

Davis TA (2006) Direct methods for sparse linear systems. Society for Industrial and Applied Mathematics, Philadelphia

de Boer SV, Freling R, Piersma N (2002) Mathematical programming for network revenue management revisited. Eur J Oper Res 137:72–92

de Farias DP, Van Roy B (2003) The linear programming approach to approximate dynamic programming. Oper Res 51(6):850–865

Dempster AP, Laird NM, Rubin DB (1977) Maximum likelihood from incomplete data via the EM algorithm. J Roy Stat Soc B39:1–38

Esterl M (2011) Dogfight erupts in plane ticket sales. Wall Str J, Bus. January 6, 2011

Farias V, Van Roy B (2006) An approximate dynamic programming approach to network revenue management. Working paper, Electrical Engineering, Stanford University, California

Fellner Karolin, Kallesen R, Ruggiero A, Yuen B (2006) Improving revenue through fare rationalization and a new business process between revenue management and sales. J Rev Pric Manag 5(2):118–127

Feng Y, Xiao B (2008) A risk-sensitive model for managing perishable products. Oper Res 56(5):1305–1311

Fiig T, Isler K, Hopperstad C, Cleaz-Savoyen R (2005) DAVN-MR, a unified theory of O&D optimization in a mixed network with restricted and unrestricted fare products. Presented at AGIFORS reservations and yield management

Fiig T, Isler K, Hopperstad C, Olsen S (2010) Forecasting and optimization of fare families in network RM. Presented at AGIFORS revenue management and cargo

Fuller WA (1996) Introduction to statistical time series, 2nd edn. Wiley, New York

Gallego G, van Ryzin G (1994) Optimal dynamic pricing of inventories with stochastic demand over finite horizons. Manag Sci 40(8):999–1020

Gallego G, van Ryzin G (1997) A multiproduct dynamic pricing problem and its applications to network yield management. Oper Res 45:24–41

Gallego G, Lin L, Ratliff R (2007) Revenue management with customer choice models. Presented at AGIFORS, South Korea

Garrow L (2010) 50 years of our OR contributions. Presented at AGIFORS revenue management and cargo

Gelman A, Carlin JB, Stern HS, Rubin DB (1995) Bayesian data analysis. Chapman and Hall, London

Gijbels I, Pope A, Wand MP (1999) Understanding exponential smoothing via kernel regression. J R Stat Soc B 61(1):39–50

Higle JL (2007) Bid-price control with origin-destination demand: a stochastic programming approach. J Rev Pric Manag 5(4):291–304

Holt CC (1957) Forecasting trends and seasonals by exponenetially weighted moving averages, O.N.R. Memorandum 52, Carnegie Institute of Technology, Pittsburgh, PA

Isler K (2004) On the equivalence of airline revenue management and pricing problems. Operations research technical report, Swiss International Air Lines Ltd

Isler K (2005) From seamless availability to seamless quote. Presented at PROS RM conference, Houston

Isler K, Imhof H (2007) Revenue management with fenceless fares and competition. J Rev Pric Manag (to appear)

Iyengar G, Phillips R, Gallego G, Dubey A (2004) Managing flexible products on a network. CORC technical report TR-2004-01, Department of Industrial Engineering and Operations Research. Columbia University, New York

Kalman RE (1960) A new approach to linear filtering and prediction problems. J Basic Eng 92:35–45

Kalman RE, Bucy RS (1961) New results in linear filtering and prediction theory. J Basic Eng 83:95–108

Karaesmen I, van Ryzin G (2004) Overbooking with substitutable inventory classes. Oper Res 52(1):83–104

Kincaid WM, Darling DA (1963) An inventory pricing problem. J Math Anal Appl 7:183–208

Klophaus R, Poelt S (2007) Airline overbooking with dynamic spoilage costs. J Rev Pric Manag 6(1):9–18

L'Heureux E (1986) A new twist in forecasting short-term passenger pickup. AGIFORS Symposium Proc. **26**. Bowness-On-Windermere, England

Lautenbacher CJ, Stidham SJ (1999) The underlying markov decision process in the single-leg airline yield management problem. Transp Sci 33:136–146

Lawrence R, Se June Hong, Cherrier J (2003) Passenger based predictive modeling of airline no-show rates. SIGKDD. Aug 24–27

Lee TC, Hersh M (1993) A model for dynamic airline seat inventory control with multiple seat bookings. Transp Sci 27:252–265

Levin Y, McGill J, Nediak M (2008) Risk in revenue management and dynamic pricing. Oper Res 56(2):326–343

Lieberman W (2011) From yield management to price optimization: lessons learned. J Rev Pric Manag 10(1):40–43, 10th Anniversary Edition

Littlewood K (1972) Forecasting, control of passenger bookings. reprinted in 2005. J Rev Pric Manag 4:111–123

McGill JI, van Ryzin GJ (1999) Revenue management: research overview and prospects. Transp Sci 33:233–256

Meissner J, Strauss AK (2008) Network revenue management with inventory-sensitive bid prices and customer choice (Working paper). Department of Management Science, Lancaster University Management School, Lancaster

Oosten M, Walczak D (2005) Transforming pricing problems. Presented at the 5th annual INFORMS revenue management and pricing section conference. MIT, Cambridge

Petersen JP, Fiig T (2005) SAS Experiences with a network optimization in a mixed environment of restricted and unrestricted fare structures (OPTIMIX). Presented at PROS RM conference Houston

Poelt S (2011) The rise and fall of RM. J Rev Pric Manag 10(1):23–25, 10th Anniversary Edition

PROS (2011) Lufthansa extends PROS portfolio with implementation of O&D network optimizer. Press release, PROS corporate communications, Houston, TX

Queenan CC, Ferguson M, Higbie J, Kapoor R (2007) A comparison of unconstraining methods to improve revenue management systems. Prod Oper Manag 16(6):729–746

Rao CR (2001) A note on the Kalman filter. Proc Natl Acad Sci U S A 98(19):10557–10559

Ratliff R (2006) Multi-flight demand untruncation with recapture Presented at the 2006 AGIFORS revenue management and distribution study group conference, Cancun, Mexico

Robinson LW (1995) Optimal and approximate control policies for airline booking with sequential nonmonotonic fare classes. Oper Res 43:252–263

Rothstein M (1985) OR and the airline overbooking problem. Oper Res 33:237–248

Rothstein M, Stone AW (1967) Passenger booking levels. Proceedings of the seventh AGIFORS symposium, 392–435

Samu K (2010) WSJ opinion. Auctions for overbooking (a better idea than airline bumping). Wall St J, Rev Outlook, June 8, 2010

Shaykevich A (1994) Airline yield management: dynamic programming approach. Master's thesis, Department of Operations Research, University of North Carolina, Chapel Hill

Simon J (1968) An almost practical solution to the airline overbooking. J Trans Econ Policy 2:201–202

Simon J (1972) Airline overbooking: the state of the art—the reply. J Trans Econ Policy 6:254–256

Simpson RW (1989) Using network flow techniques to find shadow prices for market and seat inventory control technical report, MIT Flight Transportation Laboratory Memorandum M89-1 MIT, Cambridge

Smith BC, Penn CW (1988) Analysis of alternative origin-destination control strategies. AGIFORS Annual symposium proceedings

Smith BC, Leimkuhler JF, Darrow RM (1992) Yield management at American airlines. Interfaces 22:8–31

Stone R, Diamond M (1992) Optimal inventory control for a single flight leg. Working paper, Northwest Airlines, Operations Research Division, Minneapolis

Subramanian J, Lautenbacher CJ, Stidham SJ (1999) Yield management with overbooking, cancellations and no shows. Transp Sci 33:147–167

Talluri K (2003) on equilibria in duopolies with finite strategy spaces. Working paper. Universitat Pompeu Fabra, Barcelona

Talluri K, van Ryzin G (2004) Theory and practice of revenue management. Kluwer, Dordrecht

Tanner MA (1993) Tools for statistical inference: methods for the exploration of posterior distributions and likelihood functions. Springer, New York

Ternoey C (1997) Multi-period state contingent yield management. Presented at INFORMS Dallas 1997, Decision Focus/Aeronomics

van Ryzin G, Liu Q (2004) Decision, risk, and operations. Working papers series DRO-2004-04, Columbia University, New York

van Ryzin G, McGill J (2000) Revenue management without forecasting or optimization: an adaptive algorithm for choosing airline seat protection levels. Manage Sci 46(6):760–775

Vinod B (1989) A set partitioning algorithm for virtual nesting indexing using dynamic programming internal technical report, SABRE decision technologies

Vinod B (2006) Advances in inventory control. J Rev Pric Manag 4(4):367–381

Vinod B (1995) Origin-and-destination yield management. In: Jenkins D (ed). The handbook of airline economics. McGraw-Hill, New York, pp 459–468

Vulcano G, van Ryzin G, Ratliff R (2008) Estimating primary demand for substitutable products from sales transaction data. Working paper. Presented at the 8th INFORMS RM and pricing conference, Montreal

Walczak D (2003) Dynamic programs for revenue management and dynamic pricing. Presented at annual conference of Canadian OR Society (CORS), Vancouver
Walczak D (2004) An efficient approximate solution to revenue management pricing networks. Presented at the 4th INFORMS RM and pricing conference, MIT, Cambridge
Walczak D (2007) A pricing dynamic programming game. Presented at 2007 INFORMS RM and pricing conference. Barcelona, Spain
Walczak D (2010) A note on balancing expected profit, resource utilization. J Rev Pric Manag (submitted). Presented at INFORMS RM and pricing conference, Cornell University, Ithaca
Walczak D, Brumelle S (2007) Semi-Markov information model for revenue management and dynamic pricing. OR Spectr (Springer) 29(1):61–83
Walczak D, Oosten M (2008) Customer choice models and marginal revenue transformation in revenue management and pricing. (submitted)
Walczak D, Mardan S, Kallesen R (2008) Customer choice, fare adjustments and the marginal expected revenue data transformation. Presented at 2008 INFORMS RM and pricing conference, Montreal, Canada. J Rev Pric Manag 9(1/2):94–109
Weatherford LR (1991) Perishable asset revenue management in general business situations Doctoral thesis Darden Graduate School of Business Administration, University of Virginia, Charlottesville
Wei YJ (1997) Airline O-D control using network displacement concepts. Masters thesis, MIT, Cambridge
Welch G, Bishop G (2006) Introduction to the Kalman filter. UNC-Chapel Hill, TR 95-041
West M, Harrison J (1997) Bayesian forecasting and dynamic models, 2nd edn. Springer, New York
Williamson EL (1992) Airline network seat inventory control: methodologies and revenue impacts. Ph.D. thesis, MIT, Cambridge
Winters PR (1960) Forecasting sales by exponentially weighted moving averages. Manag Sci 6:324–342
Wollmer RD (1988) An airline seat management model for a single leg route when lower fare classes book first. Fall ORSA/TIMS Conference, Denver
Wollmer RD (1992) An airline seat management model for a single leg route when lower fare classes book first. Oper Res 40:26–37
Young Y, van Slyke R (1994) Stochastic knapsack models of yield management. Technical Report 94–76. Polytechnic University, New York
Zeni RH (2001) Improved forecast accuracy in airline revenue management by unconstraining demand estimates from censored data. Ph.D. thesis, Rutgers University, Newark
Zhang D, Adelman D (2007) An approximate dynamic programming approach to network revenue management with customer choice. Working paper. http://faculty.chicagogsp.edu/daniel.adelman/research/network.pdf
Zhang D, Cooper WL (2005) Revenue management for parallel flights with customer-choice behavior. Oper Res 53(3):415–431
Zhao W (1999) Dynamic and static yield management models. Ph.D. thesis, The Wharton School, Operations and Information Management Department, University of Pennsylvania, Philadelphia
Zhao W, Zheng YS (1998a) General dynamic models for airline seat allocation. Working paper, The Wharton School, University of Pennsylvania, Philadelphia
Zhao W, Zheng YS (1998b) Optimal policy for multi-class seat allocation with nonstationary demand. Working paper, The Wharton School, University of Pennsylvania, Philadelphia

Chapter 4
Airline Distribution

Dirk Gunther, Richard Ratliff, and Abdoul Sylla

4.1 Introduction

Air travel distribution is concerned with all tasks and processes surrounding the sale of airline seats to travelers. Today, international air travel distribution is a US $500 billion per year industry around a complex network of suppliers, travelers and intermediaries; industry passenger sales are growing at over 7% per year (IATA 2008). On a business level, airlines try to identify those paths in the network that give them the lowest cost access to the highest yielding customer base. On the other hand, customers try to identify the paths that allow them to purchase the air travel products (i.e. the airline seat and other related sales such as in-flight shopping) that best meet their needs at the lowest prices. Operations research models are used in many phases of the air travel process to optimize distribution decisions for airlines, customers and other air travel intermediaries.

Each supplier–traveler path in the distribution network can be divided into six general phases based on the product life cycle:

1. Demand stimulation
2. Shopping
3. Selling
4. Confirming/fulfilling
5. Travel experience and
6. Post-travel

D. Gunther (✉) · R. Ratliff · A. Sylla
LLC, Willowgate, Dallas, TX, USA
e-mail: dirk.gunther@sabre.com

C. Barnhart and B. C. Smith, LLC (eds.), *Quantitative Problem Solving Methods in the Airline Industry*, International Series in Operations Research & Management Science 169,
DOI: 10.1007/978-1-4614-1608-1_4,

These different phases are known collectively as the travel value chain (Gilliland 2000).

The first four travel value chain phases are the primary focus of distribution-related decision making. *Demand stimulation* involves the advertisement of schedules, fares and promotions; in addition, airlines partner with sales and marketing intermediaries to increase their reach and visibility. During the *shopping* phase customers (or indirectly via agents) conduct their planning and research for future trips prior to the actual purchase. In the *selling* phase trip plans are finalized, and the product is actually sold. The *confirming/fulfilling* step consists of issuing the airline ticket for the trip (usually done electronically) as well as payment and financial processing of the ticket. Customers may use an airline's own platform (i.e. websites, reservation offices, and call centers) or an airline's sales partner (i.e. travel agency or travel broker) during shop, sales and confirm/fulfill activities.

The *travel experience* phase consists of the numerous activities pertaining to the actual travel itself including accessing flight departure information, flight check-in, airport security, flight boarding, in-flight processes, baggage retrieval, car rental, hotel stay, and destination activities (cruises, museums, amusement parks, etc.). Suppliers play the primary role in this phase (although travel agencies may also provide customer support). The last step in the travel value chain, *post-travel* activity, includes accounting actions, post-travel marketing messages and merchandising. Both suppliers and travel agencies are involved in post-travel activities.

Efficient automation in the travel value chain has always been a strategic priority for airlines, and the airline industry was a pioneer in establishing a global electronic network to facilitate these core business processes for travel distribution. Global Distribution Systems (GDS's) are a key part of this electronic network; they provide a capability for travel agencies to view schedules and fares and make travel bookings. The remainder of this chapter shows the evolution of airline distribution from manual-based to highly automated business processes.

The global airline electronic network provided a strong foundation for the development of new decision support applications based on OR modeling to provide airlines with competitive advantage. In Sect. 4.2, important background material is provided regarding the major constituents involved in air travel distribution including how they interact together and how the associated business processes and electronic distribution infrastructure evolved over time.

In subsequent sections, applications for OR Modeling are described. Section 4.3 describes many of the 'traditional' OR models that are well known and broadly adopted across the airline industry. Topics discussed include:

4.3.1 Customer choice and display management (including methods for estimating expected revenue from shopping results)
4.3.2 Availability caching
4.3.3 Low-fare search
4.3.4 E-mail campaign management

4.3.5 Opaque air pricing
4.3.6 Merchandizing and customer relationship management (CRM)
4.3.7 Agency supply management

Section 4.4 deals with 'emerging' new OR models for other important business processes that are showing some early successes in application. Topics include the following:

4.4.1 Content management (product and pricing decisions by selling channel)
4.4.2 Airline channel management (decisions on the selling channels to participate)
4.4.3 Travel agency valuation

Section 4.5 involves 'future' modeling opportunities that either do not yet exist or with limited deployment across the industry. This section includes:

4.5.1 Dynamic pricing
4.5.2 Risk inventory models
4.5.3 Auction-based distribution
4.5.4 Price unbundling and product bundling
4.5.5 Customer-direct distribution and social media

In order to better understand the application of these OR models, the next section provides an overview of the current travel distribution environment and its evolution (including both business and electronic automation dimensions).

4.2 Background

Air travel distribution is all about the flow of information and money between the airline, marketing intermediaries (e.g. GDS's), travel agencies and the customer; opportunities to use management science to improve these processes exist at all these levels. As mentioned earlier, international air travel distribution is a US $500 billion per year industry. The evolution of electronic distribution from the original private channels through to modern, Internet-based sales channels is discussed in this section and should be of interest to modern OR practitioners. Online air travel sales in the U.S. reached an estimated US $58 billion (PhoCusWright 2008), making air travel one of the largest categories of product sales on the Internet.

4.2.1 Overview of Modern Airline Distribution

In today's distribution environment we have four main constituents: (1) airlines which supply the inventory, (2) GDS's which distribute airline inventory to

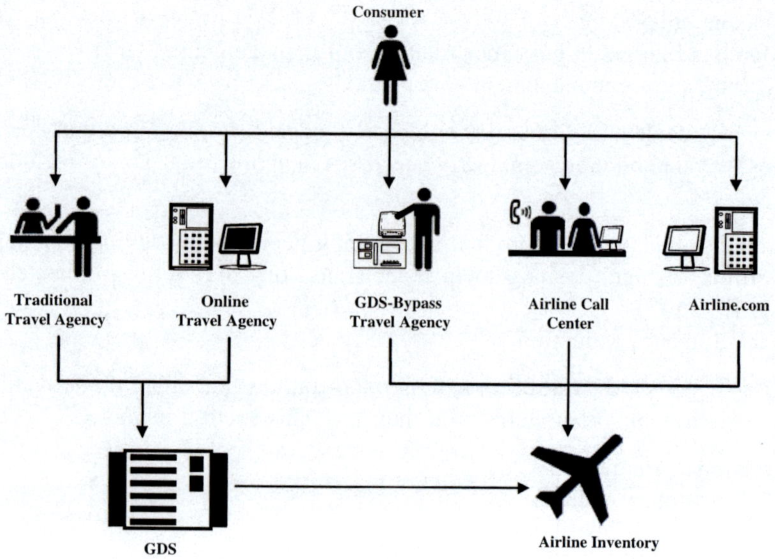

Fig. 4.1 Distribution outlets

subscribing travel agencies, (3) travel distribution providers or agencies which distribute travel content and (4) customers who purchase travel content and services.

We distinguish three principal distribution channels: GDS-based, GDS-bypass (or direct-connect) and supplier-direct. The GDS-based channel consists of agencies (both traditional offline and online) that use a GDS to connect to suppliers; the GDS-bypass channel consists of agencies that connect directly to suppliers. The supplier-direct channel consists of the airline's web site and sales offices (see Fig. 4.1). In most cases, agencies offer their products both online and offline—the predominant sales channel is used to classify the agency. The growth in the online channels has exceeded that of the offline ones, and indirect channels still comprise over half of all worldwide bookings (Vinod 2009).

Figure 4.2 shows the various distribution processes that link airlines with travel agencies. GDS's act as the central market place where schedule, fare and availability information from several carriers come together; this information is used to provide shopping and booking services to agencies and airlines. GDS's are compensated by booking fees collected from airlines and participation or usage fees from travel agencies.

Travel agencies use the information provided by GDS's (or directly by airlines) to purchase air fares on behalf of their customers. They are compensated by commissions from the airlines, GDS incentives and fees collected from the consumer.

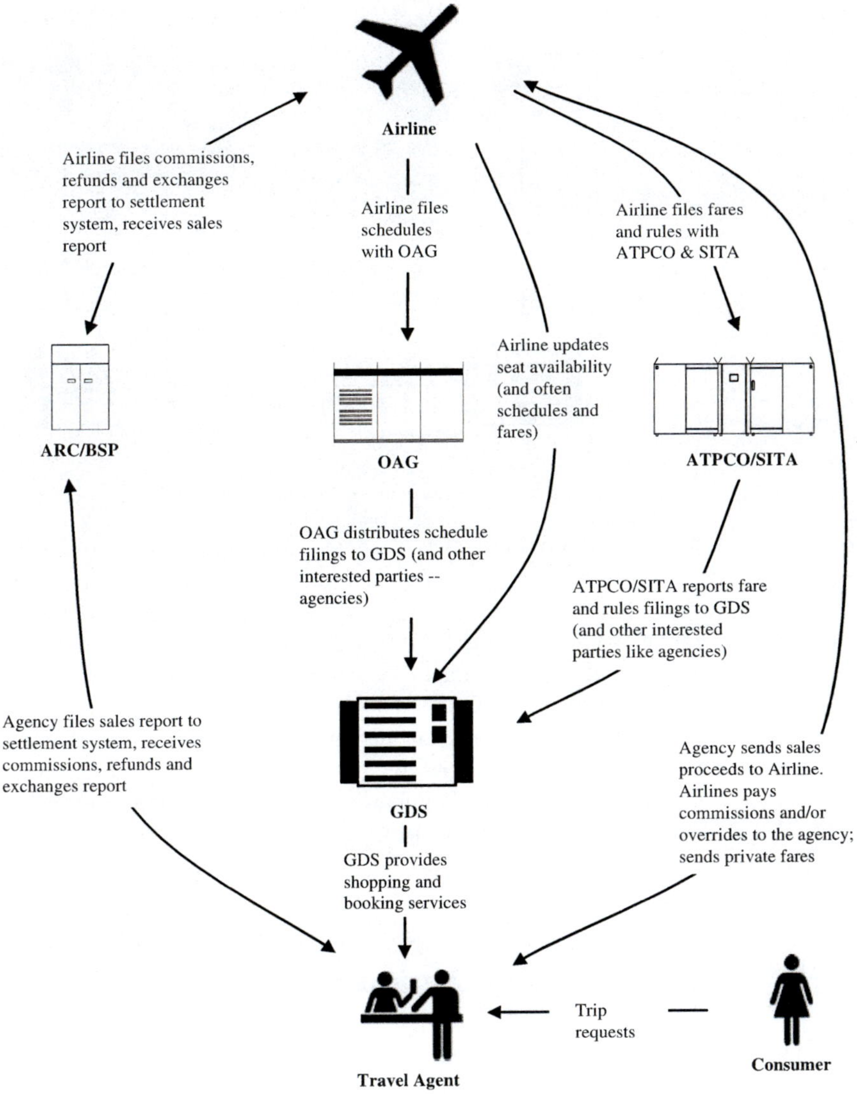

Fig. 4.2 Airline and agency distribution process flows

Other service intermediaries are also part of the picture. They combine information from all participating carriers and distribute it to all interested parties. OAG (schedule), ATPCO and SITA (fares), ARC and BSP (settlement) act as additional intermediaries in the distribution process.

The underlying flow of payments among the airlines and intermediaries is depicted in Fig. 4.3. Airlines pay booking fees to the GDS's for processing and storing the transaction information pertaining to sales (i.e. the PNR or passenger

Fig. 4.3 Distribution fees

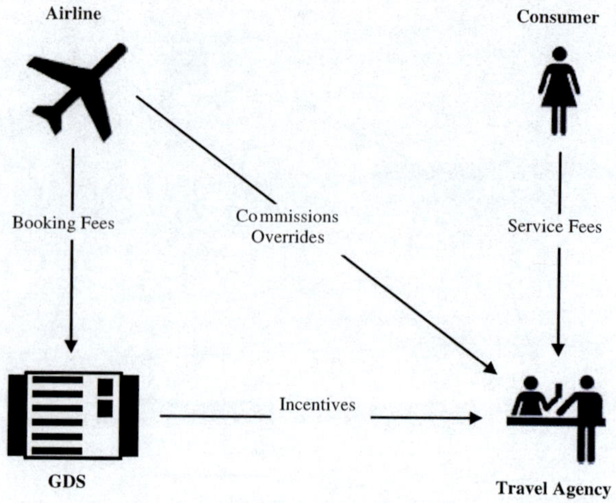

name record). Airlines also provide travel agencies commissions and override payments (e.g. when target sales volumes are met) to gain their favor. Also, in an effort to get travel agencies to use their booking systems, the GDS's provide travel agencies with incentive payments. Finally, travel agencies may also charge services fees when customers make bookings.

This complex arrangement of suppliers and multiple intermediaries linked together within a vast, modern electronic airline travel distribution infrastructure did not arise overnight. To provide the reader further insights into the circumstances and factors that led to the establishment of the current airline sales environment, the remainder of this section deals with the history of air travel distribution over the past century.

4.2.2 History of Airline Travel Distribution

When the first modern airlines were founded in the 1920s, air travel distribution was very simple (Airline 2009). Travelers and travel agents had to contact the airlines directly to make bookings and to obtain schedule and fare information. This practice changed in 1929 when Official Airline Guide Inc. published the first "Official Airline Guide" (OAG 2009) containing schedule information for 35 airlines and 300 flights. Travelers and travel agents, now had easy access to schedule information, but they still had to call the airlines to inquire about fare information and to make, change or cancel bookings.

In the USA, this rather inefficient environment did not change until 1938 when the US Congress created the Civil Aeronautics Board (United States Government role in civil aviation 2009). The goal of the CAB was to promote and regulate the development of air travel in the US. The CAB regulated air tariffs, routes and schedules. As a result of the CAB's work, the Air Transport Conference of America (ATC) was founded in 1945 by the Air Transport Association of America, an organization of 14 major airlines.

In its inaugural year, the ATC's tariff department started publishing fares (ATPCO 2009), and the ATC adopted the Sales Agency Agreement regulating travel agent accreditation (i.e. training and certification) and settlement of airline commission payments stemming from ticket sales. In recognition of their sales assistance, travel agents were paid commissions by the airlines and did not have to charge service fees to customers. These sales commissions were the main revenue source for travel agencies during this era. The ATC's tariff department was divested in 1965 and is known today as the Air Tariff Publishing Corporation (ATPCO). In 1964, the number of accredited US Travel Agents had reached 5,000. The Sales Agency Agreement was further developed to become the "Standard Ticket and Area Settlement Plan" (ASP). ASP facilitates and simplifies the accreditation, selling and remitting procedures and to improve the airline's financial control and cash flow. The ATC formed the Airline Reporting Corporation (ARC) to report on sales and the Standard Ticket and Area Settlement Plan (ARC 2009).

International regulatory bodies and regulations were developed in parallel to the US ones. In 1945, 57 airlines from 31 nations formed the International Air Transport Association (IATA). IATA has grown to become the most important international governing body in air traffic with more than 230 members (IATA 2009). In 1952, IATA signed into action the Billing and Settlement Plan (BSP) which acts as an international counterpart to the ASP. The BSP is administered by IATA.

The international counterpart to ATPCO is the Société Internationale de Télécommunications Aéronautiques (SITA). SITA was formed by ten European and one US airline (TWA) in 1949. Today, SITA provides communication and IT solutions for airlines, computer reservation systems, airports and other companies in over 200 countries (SITA 2009).

With OAG and the precursors of ARC, ATPCO, IATA and SITA, the marketplace had become much more efficient: travelers and agents had sources that combined schedule and tariff data across all airlines. Airlines had a trusted proxy that ensured payment of booked tickets. Though formed some forty years earlier, these organizations fulfilled a multi-airline payment role similar to that found today on the Internet by online payment service providers (Online Payment Services 2010) such as PayPal or RegOnline; these early precursors created a foundation from which modern electronic travel distribution evolved.

4.2.3 Computerized Reservation Systems

As discussed above, schedule and fare distribution as well as guaranteed ticket payment greatly contributed to making the market efficient. By the 1940s, the main bottleneck had become the actual booking process; to place, change or cancel a reservation or booking, travel agents and travelers would call the corresponding airline's reservation office. Airline sales agents tracked reservations, sales and remaining inventory manually using chalkboards and reservation slips. Booking an interline itinerary, i.e., a trip that involved flights from different airlines, required the agent or traveler to make calls to multiple reservation offices and to manually construct the travel itinerary. As airline networks grew larger and more complex, manual tracking of reservations, bookings and remaining inventory became unmanageable for the airlines. Processing a booking took on average of 90 min to complete (ReserVec 2009).

In 1946, American Airlines debuted "ReserVisor", the first automated reservation system, which was a simple electromechanical computer based on telephone switching system technology. The ReserVisor and its later enhancement, the Magnetronic ReserVisor (which incorporated magnetic data storage drums), were ultimately able to store information on 2,000 flights per day up to a month in the future. In 1953, Trans Canada Airlines began work with the University of Toronto on a computer-based network reservation system. Trans Canada Airline's efforts resulted in the ReserVec system, which used remote punch card terminals to query, book and cancel seats on future flights. ReserVec (which stood for "Reservations Electronically Controlled") started operation in 1962 and became fully functional in January of 1963. Although both of these systems automated the tracking of bookings and open seats, they still required operators to take calls, look up flight availability and update the records (ReserVec 2009).

In 1957, American Airlines entered into an agreement with IBM to build a computerized automated reservation system. This agreement resulted in development of the Semi-Automatic Business Research Environment (SABRE) which first went online in 1960 and took over all booking functions in 1964 (Sabre 2009; IBM 2009). In addition to maintaining booking counts for future flights, SABRE also stored passenger booking information. Other airlines followed soon thereafter. In 1968 Delta Air Lines launched the Delta Automated Travel Account System (DATAS), in 1971 United Airlines launched its Apollo Reservation System, and Trans World Airlines launched IBM's Programmed Airline Reservations System (PARS) (Computer reservations system 2009). Although other hardware vendors had been involved early in airline CRS development (e.g. Burroughs at TWA and Univac at United Airlines), IBM's PARS platform running the Transaction Processing Facility (TPF) operating system soon became the industry standard owing to its scalability to handle large transaction volumes (Vinod 2009). These TPF-based reservations and departure control platforms supported all the major distribution and day-of-travel functions in the travel value chain including shopping, selling, confirming/fulfilling, and passenger check-in. By the early 1970s, these airline systems comprised the world's largest private, real-time data processing networks (Karcher 1996).

As CRS's became established, data standards were developed to enable the electronic exchange of flight information, availability, etc. across the major airlines. As no single airline could serve all markets, this information exchange was necessary in order for airline reservations offices to sell interline bookings involving multiple carriers. ATPCO and OAG started electronic processing of fare and schedule data, respectively. The ARC Standard Ticket and Area Settlement Plan was adopted to correctly reimburse airlines for these interline ticket sales.

4.2.4 Creation of Global Distribution Systems

As the airlines continued to grow, travel agencies began to experience the same problems with keeping track of reservations and making bookings that the airlines had faced previously. In 1974, an industry effort was initiated to create a single reservation system for all the airlines and travel agencies (called JICRS for "Joint Industry Computer Reservation System") (Copeland et al. 1995). The intent was to create a single electronic mechanism to display flight schedules, check availability and make bookings across multiple airlines. The collaborative effort broke down in 1976 when United Airlines announced plans to provide a proprietary version of the Apollo System for use by travel agents.

This move by United caused a flurry of activity by other airlines to quickly introduce versions of their own reservations system into travel agency locations. By the end of 1976, American Airlines had installed SABRE at 130 travel agency locations (Sabre 2009), and United Airlines was marketing their Apollo System to agents in both North America and Japan (Galileo 2009). These agency-installed versions of the airline CRS's proliferated and later became known as GDS's (Global Distribution Systems). Current GDS providers include six major brands: Abacus (in Singapore), Amadeus (located in Madrid with data center operations in Erding, Germany), Sabre (located in Dallas with data center operations in Tulsa), Apollo, Galileo and Worldspan (the latter three are jointly operated in Atlanta by Travelport). Most of these GDS providers had the same basic approach to marketing: they offered travel agencies free equipment and training in exchange for a subscriber fee based on the volume of bookings that the travel agency makes. In the pre-Internet era, the GDS's (along with the telephone) were the primary means for agencies to access up-to-date travel information, and agencies could afford to pay the subscriber fees since they represented only a fraction of the standard 10% sales commission (Vinod 2009).

A major reason for the urgency by the airlines to implement their own, proprietary CRS at travel agencies was the value of display bias (Smith et al. 2001). Display bias pertains to the order in which search results are displayed, and it has a powerful impact on the likelihood of an item being selected (i.e. the first few results displayed are much more likely to be chosen by the user than later ones). To better understand the impact of display bias, one has to understand how agents have traditionally used the CRS to book a flight:

1. Agent requests schedule and availability information for a certain origin, destination and travel date.

2. The CRS uses schedule information, CRS rules, and routing restrictions to construct itineraries that satisfy the agent's request. The number of feasible itineraries can be large.
3. The CRS sorts all feasible itineraries according to CRS rules and returns them in fixed size batches to the agent.
4. The agent books one of itineraries displayed or issues a request for more itineraries matching the request.

With display bias, the airlines show their own flights at the top of the search results provided to the agent (thereby making competitor offerings less attractive). (Copeland et al. 1995) noted that 70% of all bookings were made from among the first six options displayed on the screen. An internal study by American Airlines in 1981 showed that over half of travel agents select the first option displayed (Sabre 2009). Since agents tend to select from among the first few options (rather than those located further down), CRS owners purposely positioned information on-screen to skew the purchase behavior towards their airline (i.e. the displays were not neutral) (Granados 2004). The impact of display bias on booking share became widely known as the "halo effect". In Sects. 4.3.1 and 4.3.5, we describe how customer choice models can be used to help quantify the halo effect (including an application known as "CRS-Sim" that was originally used by American Airlines). Even today, the halo effect is still an important consideration for most major airlines; one of the major reasons for the practice of airline codesharing is that a desirable operating flight can be displayed several times on the screen across different marketing carriers (providing a type of "multiplier effect").

Soon after the first GDS's sets were installed at travel agencies, the U.S. airline industry underwent deregulation in 1978. Deregulation in the U.S. allowed airlines to make their own decisions about schedules and prices without the need for government approval. It spawned new airlines and led existing airlines to quickly expand their route network. Moreover, airlines created complex fares structures and rules trying to maximize their revenues by segmenting the market.

4.2.5 GDS Regulation in the U.S

The increased competition in the U.S. due to airline deregulation also significantly increased the impact of the halo effect. It became so strong that it threatened the survival of airlines without a GDS. Copeland et al. (1995) estimated the additional value of the halo effect to be worth $100 million per year for American Airlines alone (and even more considering the other CRS's). Among allegations that this display bias was not in the best interest of the general public, in 1984 the U.S. Department of Transportation regulated the GDS necessitating several new requirements including:

1. "Neutral" (unbiased) displays of airline flight information.
2. Published rules for the sorting algorithms used by GDS in displaying flight alternatives.
3. Standardized schedule of times when fare changes are published by airlines.
4. GDS sales of competitive booking data across airlines (to allow tracking and monitoring of sales by airline across agencies).
5. GDS to charge the same booking fee for all airlines (to eliminate fee-based screen preference).

Similar European legislation to regulate GDS behavior (known as the 'Code of Conduct') was passed in 1989.

While the fees that GDS's charged the airlines were regulated, the airlines paid varying overrides to travel agencies enticing them to sell their product over that of competitors. Override programs pay agencies additional sales commissions if they achieve some carrier-defined performance targets. The performance targets vary by carrier and are based on some measure of sales volume or share over a defined measuring period (usually a quarter). Override programs allow carriers to create attractive commission structures for preferred agencies that deliver incremental value without raising commissions for agencies overall. Similarly, GDS tried to get travel agents to subscribe to their system by sharing part of the booking fees with travel agencies. Ultimately, decision support tools emerged to help airlines and travel agencies manage these performance-based commissions; such models are discussed in Sects. 4.3.8 and 4.4.3.

4.2.6 Frequent Flyer Programs

Airline deregulation led not only to increased levels of competition but, more, importantly, to the commoditization of airlines. In an attempt to set themselves apart and to build loyalty among agencies and customers, several airlines created override and frequent flyer programs (Frequent Flyer Program 2009).

Frequent flyer programs have followed a similar approach to override programs but are targeted at travelers: travelers accrue credits each time they fly with the airline. The credits can be applied to future travel purchases. The early programs used punch cards to track credits. American Airlines created "AAdvantage" in 1981 (Frequent flyer program 2009) which was the first computerized frequent flyer program. AAdvantage is viewed by many as the origin of today's frequent flyer programs. Frequent flyer credits are typically expressed in miles or number of trips. Travelers can redeem these credits in exchange for upgrades, travel discounts or free flights. Travelers can also purchase miles. Over time frequent flyer programs have become a very large business by themselves. Airlines sell their frequent flyer credits to other companies, such as credit card companies, who give them to their customers. Also, many airlines have partnered with hotel chains and car rental companies allowing travelers to accrue and to apply credits when booking hotel rooms or rental cars. Today, there are over 70 frequent flyer programs worldwide with more than 100 million members. Airlines that support a frequent flyer program reserve, on average, 5% of their entire inventory for frequent flyer rewards (Frequent Flier 2009).

4.2.7 Low Fare Search

Prior to deregulation in the U.S., fare search was simple. There were only a few fares in each market, the fare amounts were regulated by the U.S. government, and prices were the same across all airlines. In that era, travel agents (and indirectly the customers) shopped for air travel based on schedules and service. Post U.S. deregulation, all of that changed. Airlines began competing strongly based on prices and to practice price discrimination across various customer segments by offering a wide array of different fare products (with different sets of restrictions such as advance purchase requirements, Saturday night stays, point of sale, etc.). The deregulated environment and, in particular, the vast number of fares and schedules and complex fare rules it had spawned found agents struggling to find the cheapest itinerary for their customers.

Consider, for example, a customer looking for a one-way ticket from New York City to Los Angeles. There are more than 50 non-stops flights on a typical day between New York City and Los Angeles. Each flight is associated with about 20 fares. Additionally, there are more than 1,000 attractive one-stop itineraries when considering Chicago, Atlanta, Dallas, Denver and Salt Lake City as connect points. Each of these itineraries is associated with about 400 fares. In total, there are more than 400,000 flight-fare combinations just for the travel from New York City to Los Angeles.

To address this problem, in 1984, Sabre released Bargain Finder Plus, the industry's first low fare search function. Other intermediaries also implemented low fare search functions, all of which were designed to find the cheapest itineraries for a given set of trip parameters, such as origin and destination station, mandatory intermediate stations, travel dates, maximum number of stops, etc. Low fare search capabilities gained great importance and further evolved when travel distribution was taken to the Internet (Computerworld Honors Program 2009). A discussion of low fare search algorithms is provided in Sect. 4.3.3.

4.2.8 Internet-Based Distribution

Prior to the Internet, airline communication technology was based on private message exchange networks; e.g. X.25 protocols (Cisco 2009) were widely used in the 1970s (and continue in some parts of the world even today). However, as

Internet-based technology proliferated, lower cost TCP/IP protocols (Parziale et al. 2006) became more widely adopted.

Air travel distribution took to the Internet in 1985 when CompuServe, the leading Internet provider at the time, offered customers direct access to an electronic version of OAG and Sabre's EAAsySabre. For the first time travelers were able to check schedules (via OAG) and book travel (via EAAsySabre and later Worldspan's TravelShopper) directly over the Internet (Lewis 1992). However, Internet booking volumes were rather low as private Internet access was very limited; even by 1990, less than 1 M US households had Internet access.

The Internet as a data exchange formed a public counterpart to the privately owned GDS's. It allowed airlines and other travel suppliers to directly exchange data with travel agencies, customers, and among each other without having GDS's in the middle. The Internet also offered travel distributors great reach via a single and cheap to maintain sales outlet, i.e., a website.

In the beginning, Internet-based air travel distribution started out as websites that emulated a very limited subset of the functionality that GDS's provided to traditional travel agents. The new websites were referred to as online travel agencies (OTA's). The OTA's allowed customers to find and purchase itineraries for selected markets and departure dates. Itineraries could be sorted by schedule or fare.

Over time, numerous different OTA's have arisen providing consistent improvements in shopping functionality, enhanced itinerary construction and low fare search functionalities, plus other new features and services that enhance a customer's travel experience. EAAsySabre was a precursor to Travelocity (which is now one of the four major online travel agency websites in the USA) (Travelocity 2009). Three other popular online agency sites include Expedia [originally founded by Microsoft in 1995 (Expedia Inc. 2009)], Orbitz [originally started in 2000 (Orbitz 2009) by a five-airline consortium] and Priceline (which began in 1998 focusing on opaque product offerings). In their backend data processing, each of these online agencies use GDS's to check availability and make bookings (although Orbitz also provides a direct connection to airline CRS's for certain types of transactions). All three of these websites provide comparison shopping and booking capabilities for most airlines across all major markets. This multi-supplier retail perspective is one of the key distinguishing features of online agencies compared to airline-owned websites.

Airline websites were first created in 1995 (Southwest Airlines 2009) and, similar to OTA's, have seen rapid growth. Initially, these websites allowed customers to view only the airline's own schedules, fares, and availability and to place bookings. In late 1995, Alaska Airlines added online ticketing capabilities (Shelton 1995). Over time, functionality was added to also include the airline partner flights in the shopping and booking process. Customers are now able to book interline itineraries on most airline websites. Airline websites also added a variety of features similar to the ones offered by OTA's. The reader may ask "Why did airlines begin creating their own websites and duplicating the capabilities of OTA's?" There are two major reasons: (1) to create brand loyalty by bringing online customers directly to their own electronic 'store' and (2) to save paying commissions and booking fees to the OTA's and GDS's.

From a customer's perspective the Internet offered more transparency. For the first time customers could see competing products side by side and easily compare prices across a variety of different OTA and airline websites. Before the Internet, only travel agents had this ability. Moreover, the Internet allowed customers to quickly check a variety of stores, i.e., OTA's and airline websites to find the offering that best matched their needs. Decision support tools for managing important aspects of this complex array of different selling channels are described in Sects. 4.4.1 and 4.4.2.

Customers' ability to shop for travel content themselves had a profound impact on "look-to-book" ratios, i.e., the average number of shopping requests issued for each booking made; the look-to-book ratio soared by a factor of 10 and more (Ratliff 2003). This increase in shopping request activity had a profound impact on the airline data processing systems. Within only a few years, availability transaction processing workloads had doubled or tripled due to increased shopping activity by online OTA's, and it greatly strained the airline reservation systems. Response times to requests began exceeding the generally accepted four second limits, necessitating rapid (and costly) expansion of the airline reservation systems computers and network infrastructure. To lessen the impact of availability requests against the airline reservation systems, both OTA's and GDS's began developing fare and availability caches to reduce the cost per booking associated with the increased shopping activity. Developing good caching strategies has now become an important element of selling air travel on the Internet (Ratliff 2003). Decision support tools aimed at predicting how likely customers are to book flights are discussed in Sect. 4.3.6 (Merchandising and CRM).

Aside from replicating business models that already existed in the GDS world, Internet-based distribution also created a variety of new business models. Some examples include opaque products, bundled travel packages, branded fares, ancillary fees, travel meta-search engines and competitive market intelligence collection.

4.2.9 Opaque and Package Products

Consumers' primary focus on price and the high degree of price transparency across websites led online travel distributors to design other distribution models and travel products. The idea is to enable selling distressed inventory at low prices without diluting sales of higher-priced products. Two popular types of alternative distribution products are opaque and package sales. Although package products had been around before Internet-based travel distribution, their sales volumes increased dramatically due to OTA's and new variations arose in the online channels (e.g. reverse auctions and dynamic packaging).

Opaque fares represent one class of products spawned by the Internet. When purchasing an opaque product the customer does not know all the product details until after it has been purchased. For example, when shopping for flights, the customer may only see the fare, travel dates, number of stops, and the general departure time window (morning, afternoon or evening). The customer does not learn the specific flight details (i.e. exact departure time, carrier, total flight duration, etc.) until after the itinerary has been purchased. Despite this uncertainty, opaque products appeal to some customers because they offer savings over comparable non-opaque products. In 1998, Priceline.com was the first OTA to sell opaque products using reverse auctions (Priceline.com 2009). Reverse auctions allow customers to name the maximum price they are willing to pay for air travel, and the site chooses the specific supplier and itinerary. Hotwire.com (founded in 2000 and owned by Expedia) is another example of an OTA focused primarily on opaque travel products. GermanWings, a European low-cost airline, offers a blind destination feature; the consumer decides on the travel date, and is notified of the destination (from a pre-determined list) after payment. Information on decision support models used for opaque air pricing can be found in Sect. 4.3.5.

A second type of alternative distribution is the travel package. Travel packages consist of several adjacent travel components, such as air and hotel, air and car, car and hotel, or air, hotel and car. The customer sees a single overall price for the entire package. Traditional packages sold in non-Internet channels are so called static packages where the components are (aside from some upgrade options) based on a specific combination of flight(s), hotel and/or car type.

The Internet offers customers a more flexible version known as dynamic packaging.

Dynamic packaging, one of today's most popular methods to shop for travel online, was introduced in 2002 by Expedia.com (Mullaney 2004). It allows customers to create their own tailored combination of air, hotel and/or rental cars. Dynamic packages are frequently built around bulk fares; bulks are special fares that can only be sold when packaged with other travel products (i.e. hotel or car). Airlines use bulks to sell distressed inventory. From the airline's perspective bulks represent a different market segment in that they don't compete outright with the airline's regular fares; therefore, the dilution risk of bulks is reduced. Moreover, bulk fares often come with restrictions that make them unattractive for business travelers. A good overview of dynamic packaging in travel can be found in Smith et al. (2007).

Southwest Airlines used customers' focus on price to launch in 2005, "Ding!", the industry's first direct link to a customer's PC (Southwest Airlines 2009). Customers who download the application get an audible alert when Southwest offers new low fares on routes of interest. OTA's, such as Travelocity and Expedia, offer comparable functionality (including messaging to mobile phones). They alert customers via e-mail if fares in markets of interest fall below certain thresholds. A general description of models used in airline e-mail marketing campaigns can be found in Sect. 4.3.4.

4.2.10 Branded Fares and Ancillary Fees

In an attempt to escape commoditization and create product differentiation, in 2006 Air Canada introduced "ala carte fare options" (Air Canada 2011) which are more commonly known as branded fares (RM Forum 2007); these products were initially offered only on Air Canada's website. The various branded fare names (e.g. "Discount", "Standard" or "Deluxe") are used to distinguish fares that include (or exclude) certain services and benefits (e.g. preferred seating, the right of free itinerary changes or cancelation, etc.); some branded fares may require the customer to purchase certain services separately (e.g. extra fees for checked-in luggage). Such separate charges for specific services are commonly referred to as ancillary fees and are becoming a growing trend in the travel industry (Peterson 2009). US airlines collected nearly US$ 8 billion in ancillary sales revenue in 2009 (Grossman 2010).

Both branded fares and the widespread growth in types of ancillary fees being levied have had very profound impact on air travel distribution because they do not fit into the standards established by the airline industry. GDS's were not initially equipped to support either distribution of branded fares nor many of the new types of ancillary fees being assessed by travel suppliers; development work by the GDS's in this area is ongoing. In Europe there are also discussions under way on regulating the way branded fares, in particular the total price for all services, need to be displayed to the customer.

4.2.11 Travel Shopping Data

Travel shopping data involves capturing and storing low fare search results that are displayed to customers and travel agents. It is important due to the use of such information by leading consumer advocate groups to compare and rank the major Internet-based travel distributors (e.g. Consumer Reports 2004). The dominant criterion used in establishing the rankings became a distributor's ability to find the lowest priced flight itinerary for a given shopping request. Other itinerary characteristics such as departure time, number of stops or total travel time were accorded only secondary importance. Due to the increased media attention by consumer advocate groups, both consumers and travel suppliers began to look more closely at online price competitiveness. As such, gathering information on low fare search results across the various travel websites became more important and fueled the growth of travel meta-search engines (consumer-focused) and competitive intelligence data brokers (supplier-focused).

Meta-Search Engines leverage the product transparency and the ease of data exchange and collection offered by the Internet; meta-search engines go out and shop several travel websites for customers and return the best options (Grossman 2006). Meta-search websites have seen impressive growth in traffic;

in 2006, 9% of all online shoppers used air meta-search (up from only 3% in 2005). These websites generally do not have a capability to make bookings but instead refer the customer directly to the airline websites; their business models are based on advertising revenue and referral fees. From a customer viewpoint, they offer a quick comparison shopping capability while avoiding any additional travel agency booking charges. Some modern examples of travel meta-search engines include websites such as Kayak, Mobissimo, and SideStep. Another example includes a special kind of meta-search site known as Farecast.com. In addition to scraping travel outlets for current prices, Farecast.com also predicts future prices for the requested markets and travel dates to advise the customer when to purchase (including offering its "Fareguard" fare change insurance product).

Price pressures and transparency created by the Internet also led to the creation of competitive intelligence data brokers in the late 1990's. These corporations collect pricing data from OTA's and supplier websites using online robotics referred to as "shopbots". The information they collect is sold to airlines and online travel agencies who use this information to ensure that they are price competitive in the marketplace (Wright et al. 2008). In effect, these competitive intelligence data brokers act as meta-search engines for travel suppliers rather than consumers.

Initially, competitive data collection was done entirely via robotics to scrape websites of interest. This action resulted in high costs for the website operators as they incurred shopping data processing costs without realizing revenues from sales; naturally they aggressively fought this practice. Recognizing the value of their pricing data to competitors and the high costs of being shopped by robotics, some airlines (Adler 2003) have shut out and sued companies that collect competitive shopping data from their website.

Today, most of these competitive intelligence data brokers have entered agreements with the travel websites that they shop. These agreements regulate website compensation as well as when and how much data can be collected. In some cases the data collection company may get, under the terms of the agreement, access to web services which allow for more efficient and reliable data access than traditional website scraping.

Although these competitive intelligence information brokers have provided at least a partial solution to the problem of online competitive fares, they do not provide insights into the actual customer shopping process itself. Because web-based customers typically shop numerous times prior to actually making a purchase decision (Moe and Fader 2004a), such data provides valuable information for several applications such as, for example forecasting future demand for travel. This aspect of shopping data collection still remains largely untapped at the present time.

4.2.12 GDS Deregulation in the U.S.

Internet-based air travel distribution outlets were the reason for the end of GDS regulation in the U.S. in July 2004 (DOT 2004). Online travel agencies had become very popular and widely used by consumers, but these websites did not fall under the purview of the GDS regulation. The Internet made it relatively easy to detect intentional display bias on the part of a travel website by allowing quick comparison to the results of other competing websites. Also, the widespread usage of search engine display optimization in many industries across the Internet made the GDS legislation seem outdated. One author described the situation as follows: "Unfortunately, it has never been clearly explained why display preference in computerized reservation systems is any different from display preference on grocery shelves and soft drink machines" (Kleit 2009).

Without any direct means for the government to control the problem of display bias across other industries on the Internet, it seemed inappropriate to maintain the legacy display restrictions on travel GDS's specifically. Furthermore, while all GDS's were initially formed and owned by airlines, they later became large businesses on their own and were ultimately spun off as separate companies. This divestment of GDS ownership by the airlines also mitigated the conflicts of interest associated with screen bias.

4.2.13 Air Travel Distribution Chronology

In summary, OTA's started out as websites that emulated a very limited subset of the functionality GDS's provide to traditional travel agents. OTA's allowed customers to find and purchase itineraries for selected market and departure date. Itineraries could be sorted by schedule or fare. Over time, numerous different OTA's have arisen providing consistent improvements in shopping functionality, enhanced itinerary construction and low fare search capabilities, plus other new features and services that enhance a customer's travel experience. Table 4.1 highlights several key milestones along this rich history of air travel distribution.

4.2.14 OR Plays an Important Role in Airline e-Commerce

The original manual-based airline sales infrastructure that arose in the first half of the twentieth century evolved into a more complex, electronically enabled infrastructure in the second half of the twentieth century. The modern twenty-first century, triple channel structure (i.e. GDS-based, GDS-bypass, and supplier-direct outlets) provides many diverse opportunities for effective airline sales and marketing. However, its inherent complexity poses managerial challenges in ensuring

Table 4.1 Major milestones in travel distribution technology

1920	First modern airlines are founded
1929	OAG distributes first airline schedules
1938	U.S. Congress forms Civil Aeronautics Board (CAB)
1945	Air Travel Association of America (ATA) forms Air Traffic Council (ATC)
	ATA publishes fares
	ATA adopts agency sales agreement
1946	American Airlines debuts first automated reservation machine (ReserVisor)
1949	SITA is founded by eleven airlines to develop an international telecommunications network for the air transport industry
1952	IATA starts BSP (Billing and Settlement Plan) and the Standard Agency Agreement to facilitate selling, payment and accreditation procedures among international travel agencies and airlines
1953	Trans Canada Airline begins work on computerized reservation engine
	American Airlines and IBM enter agreement to develop computerized reservation engine
1960	American Airlines and IBM release first computerized reservations system (SABRE)
1965	ATA divests of tariff publishing unit to form Air Tariff Publishing (ATP)
1968	The DATAS II reservations system is started at Delta Air Lines
1971	The PARS reservations system is built by Trans World Airlines
	United Airlines' Apollo reservations system is launched
1975	ATP is reorganized and renamed to Air Tariff Publishing Corporation (ATPCO)
1976	United's Apollo CRS becomes available to agencies, making it the first GDS in the world
1978	US airline industry is deregulated
1979	Texas International Airlines launches industry's first frequent flyer program
1981	Eastern Airlines develops System One Direct Access, a computer reservations system for travel agencies (was later incorporated into Amadeus)
1984	U.S. Congress regulates GDS's
1984	Sabre launches low fare search functionality (Bargain Finder Plus)
1985	The CompuServe online network takes travel distribution to the Internet via EAAsySabre
1995	Southwest Airlines launches the first airline web site
	The Alaska Airlines website offers online ticketing
1996	The online agency Travelocity.com is launched as a partnership of Sabre Interactive and Worldview
	Expedia.com is launched by Microsoft
1998	Priceline.com becomes industry's first reverse auction web site (with their "Name Your Own Price" slogan)
1999	ITA software launches their new low fare search engine (later integrated into Orbitz.com)
2000	Sidestep.com becomes first travel meta-search engine
2001	The airline-owned Orbitz.com is launched becoming the third major online travel agency
2002	Orbitz.com launches Supplier Link, the first airline reservations system direct booking capability
	American Express completes its TravelBahn communications network including the "SuperPNR" allowing greatly expanded content in the passenger name record
2004	US Congress deregulates GDS's
	"G2 Switchworks" starts and offers a new low cost GDS-bypass technology based on airline direct bookings
2005	Southwest releases "DING!", industry's first direct link to customer desktop
2006	FareCast launches industry's first fare forecasting site
2006	Air Canada introduces branded fares

the various channels operate both productively and cost effectively across all major constituents. In general, airline network optimization problems are very large and complex to solve. Distribution-related costs amounting to about 10% of total air ticket value, and air travel sales span a complex network with multiple players and channels (each with different revenue performance). As described in the next three sections of this chapter (traditional, emerging and future models), the reader will see that travel distribution management is a rich area for Operations Research efforts.

4.3 Traditional OR Modeling in Distribution

In this section we discuss several long-standing OR models that have been developed for travel distribution problem solving. We start with customer choice models that predict customer behavior and their application in the display of low fare search results. Next we provide details on how availability and low fare search algorithms work, followed by models to support merchandising and CRM related decision making.

4.3.1 Customer Choice and Display Management

Customer choice models (see Chap. 1 for a full description of the methodology) have a variety of uses in travel distribution decision support modeling. However, calibrating such models can be challenging. One of the best sources of information for calibrating travel customer choice models is based on shopping data, and it consists of three elements:

1. Request attributes: what did the customer ask for?
2. Choice set: which options were presented to the customer?
3. Selection: which option did the customer book, if any?

We distinguish two classes of choice models: micro-models that predict customer behavior in a sales session and macro-models that predict behavior across sessions (and even channels). In particular, micro-models address the questions:

1. Given a customer request and choice set what is the likelihood of a specific option in the choice set being selected given that the customer chooses an option?
2. Given a customer request and choice set, what is the likelihood that the customer chooses any of the options (i.e. makes a booking)?

Macro-models are concerned with the question of: "Given a set of product-channel combinations, what is the likelihood of a specific product-channel combination being chosen by a generic customer?"

These customer choice models are tremendously important in understanding the impact of GDS display bias (previously mentioned in Sect. 4.2). Screen display rules were originally intended to make the shopping process efficient for travel agents and consumers by identifying and showing the "best" options associated with the travel request. However, airlines who owned the original GDS's had co-opted this process with preferred product positioning, in effect turning screens into indirect advertising of their products. Any display process (whether biased or unbiased) is serving multiple constituents (all with slightly different objectives). This tension plays out differently in each channel (whether airline CRS's, online agencies or GDS's), and OR models are often used to help quantify the business effect of display rule changes.

4.3.1.1 CRS Simulation

One of the first incarnations of customer choice modeling functionality was CRSSim, a commercial simulation tool developed by Sabre that computed an airline's screen presence and predicted, in turn, the associated market share. CRSSim was used to identify profitable schedule adjustments, i.e. changes in departure times. CRSSim was also used to tailor GDS rules (within DOT regulations) in favor of the owners of the GDS's.

Once they became widely available to the travel agency community, GDS's quickly became the main channel of distribution. Several studies showed that a carrier's GDS screen presence had a large impact on its market share (Copeland et al. 1995). In this context GDS screen presence refers to the number and position of a carrier's itineraries returned by the GDS after the travel agent requests itineraries for a given date and city pair.

A carrier's screen presence on the GDS is a function of the GDS's display rules and characteristics of the carrier's and competitors' schedules. GDS rules can be divided into two groups: editing and ordering rules. Editing rules determine which itineraries are presented to the user while ordering rules determine the order in which the itineraries are presented.

The purpose of editing rules is to suppress itineraries that have no apparent value to the user. Consider, for example, a request for itineraries from Austin, TX, US to London, UK. The GDS may find among others the following two itineraries that have the same second leg and are operated by the same carrier:

1. 5 h layover in Dallas Fort Worth:
   a. Leg1: Austin to Dallas Fort Worth leaving at 11:30 a.m. arriving at 12:30 p.m.
   b. Leg2: Dallas Fort Worth to London Heathrow leaving at 5:30 p.m. arriving at 9 a.m. (next day)
2. 3 h layover in Dallas Fort Worth

a. Leg1: Austin to Dallas Fort Worth leaving at 1:30 p.m. arriving at 2:30 p.m.
b. Leg2: Dallas Fort Worth to London Heathrow leaving at 5:30 p.m. arriving at 9 a.m. (next day)

The first itinerary does not add any apparent value to the customer. A GDS's editing rules may suppress the first itinerary.

Ordering rules aim to sort itineraries in decreasing order of attractiveness. In most cases, itineraries are ordered first by their number of stops, i.e., non-stops before directly connecting, then by displacement and elapsed time. Displacement time is the absolute difference between the requested departure time and the itineraries start time. Elapsed time is an itinerary's total travel time from the departure of the first leg until the arrival of the last leg.

Besides GDS screen presence, carriers are also concerned with flight or itinerary-specific CRS screen presence. CRS screen presence is important for all direct sales outlets such as the carrier's call centers, reservation offices and the carrier's web site. CRS screen presence determines to a large degree which itineraries are sold through these outlets. Hence, a carrier is interested in adjusting CRS rules to prominently display the itineraries it is interested in selling in a given market.

Knowing the importance of GDS and CRS screen presence, carriers have shown a great interest in computing screen presence of a given schedule, adjusting schedules to improve screen presence, and adjust CRS rules to favorably display a given schedule. Computing screen presence in the face of GDS rules and a large set of competitors' schedules quickly becomes a large scale analytical task especially for carriers serving a large number of markets. CRS simulation tools were designed to tackle this task. Given a set of display rules, a carrier's schedule and a set of competitor schedules, a CRS simulation computes the carrier's screen presence and associated market share impact on the market, itinerary and flight level.

At the core of every CRS simulation is a Customer Choice Model that predicts a customer's choice, i.e., which option the customer will choose from the screen. The prediction is based on every option's characteristics including their screen position.

In this context a CRS simulation tool may evaluate screen position and probability of sale for one or more customer requests for each market. Different customer requests vary by assumed preferences or restrictions stated in the request. The assumed preferences and restrictions are representative of the requests observed in practice. The outcome for a given request is weighted by the request's observed relevance in practice. If only one customer request is used then, in most cases, default preferences are assumed and no restrictions are assumed.

CRS simulation tools have been used in three different areas:

- Schedule evaluation and improvement: Compute screen presence of a given schedule, evaluate the impact of schedule adjustments (i.e. departure time, block time or equipment changes) on screen presence/market share, and identify flights and itineraries with little or no screen presence.

- Code Share: Estimate the impact of entering a code share in certain markets on-screen presence and market share.
- CRS Rule Design: Identify CRS rule adjustments to favorably design a given schedule.

CRS simulation tools in their original form have lost some of their value with the arrival of the Internet and the shift to price driven changes. However, similar tools are being used today by online travel agencies and airlines alike to optimize the set of travel options offered to a customer.

4.3.1.2 Travel Agency Display Management

More recently, customer choice modeling has been used to determine which content to show to the customer in which order and form such that the expected profit is maximized:

Immediate profit from current visitor:

$$\text{ProfitNow} = \sum \text{p(sale)[Product]} \times \text{Margin[Product]}$$

Future profit from current visitors:

$$\text{ProfitFuture} = \text{p(future visit)} \times \sum \text{p(sale)[Product]} \times \text{Margin[Product]}$$

Future profit depends on the visitor's perception of the site's offerings. If the OTA puts too much weight on margin then the customer may perceive the site as unattractive and not include it in future searches. As a consequence most OTA's have gone over to the same "basic display algorithm". Price is the first sort criterion followed by service quality, departure time and carrier. All OTA's allow customers to re-sort the display according to their own preferences.

OTA's still use display management to:

1. Break ties in the basic display algorithm
2. Identify how to highlight content
3. Analyze the impact of "missing content"

Figure 4.4 outlines several business applications of display management methods.

The key to these applications is understanding the impact of content selection (what is shown), content placement (where is it shown on the screen) and content pricing (which mark up is applied) on customer decisions. Looking at a simple low fare search example demonstrates that content placement alone can be a difficult task. Consider the following example (see Table 4.2) based on low fare search results for a Dallas (DFW) to Atlanta (ATL) flight departing at 1:00 p.m.

Fig. 4.4 Business applications of display management

Online Travel Agencies • What is the long term impact of bad display management? • How powerful is the limited display management?	Brick-and-Mortar Travel Agencies • How much share can be moved by biasing against "default" ? • What is the impact of loyalty programs on the agency's ability to move share?
Airlines • Which content to show at which place? • How to treat code-share content on the carrier's website?	Corporations • How to bias the screen to maximize the savings from preferred supplier agreements? • How much share can be moved to a preferred carrier (i.e. what are achievable performance targets)?

Table 4.2 Example low fare search results from Dallas to Atlanta

Itinerary	Departure	Arrival	Airline	Travel time	Price ($)	Notes
1	1:03 p.m. DFW	4:19 p.m. ATL	American Airlines Flight 1952	Nonstop 2 h 16 min	230	Closest departure
2	1:31 p.m. DFW	4:38 p.m. ATL	AirTran Airways Flight 110	Nonstop 2 h 7 min	230	
3	12:11 p.m. DFW	3:19 p.m. ATL	American Airlines Flight 1750	Nonstop 2 h 8 min	240	
4	12:00 p.m. DFW	5:11 p.m. ATL	Flight 240/277	1 stop (CLT) 4 h 11 min	190	Lowest price
5	2:20 p.m. DFW	5:31 p.m. ATL	American Airlines Flight 2072	Nonstop 2 h 11 min	235	
6	3:55 p.m. DFW	6:50 p.m. ATL	AirTran Airways Flight 102	Nonstop 1 h 55 min	235	Fastest flight
7	10:00 a.m. DFW	1:00 p.m. ATL	Delta Flight 2160	Nonstop 2 h 0 min	250	

If viewed along a single criterion, three choices stand out (1, 4, and 6) involving three different carriers. When multiple criteria are considered simultaneously, there are other alternatives that would also be considered "good" (e.g. 3 and 5). So in this example, it appears there is no clear dominance among the various options. A customer choice model is very useful in this example because it helps to rapidly quantify the likelihood of selection among these alternatives.

Applying a basic multinomial logit (MNL) choice model to the example problem (see Chap. 1 for the MNL formulation), we use:

Table 4.3 Air choice parameters for multinomial logit model

Variable	Est. β coefficient	Description
Outbound displacement	– 0.004686	The absolute difference between the requested departure time and flight departure time from the origin city (in min)
Elapsed time	– 0.007984	The travel time of the complete forward itinerary (in min)
LN (fare)	– 11.3864	Natural logarithm of the fare. If it is a round-trip ticket, the fare is divided by 2
Origin point presence	+1.3771	The airline's capacity share percentage associated with all flights departing from the origin airport
Number of connections	– 1.5386	Number of connections in the outbound itinerary

$$\mathrm{p}(j|\ \mathrm{sale}) = \frac{e^{\beta_1 x_1^j + \beta_2 x_2^j + \cdots + \beta_n x_n^j}}{\sum_{k=1}^{m} \left(e^{\beta_1 x_1^k + \beta_2 x_2^k + \cdots + \beta_n x_n^k}\right)}$$

where:

1. $\mathrm{p}(j)$ is the probability of selection of alternative "j" given a sale occurs
2. "m" alternative choices
3. "n" different attributes
4. $x_1, x_2, \ldots, x_n$ are the attributes being considered
5. $\beta_1, \beta_2, \ldots, \beta_n$ are calibrated parameters for each attribute

Mishra et al. (2005) presented the following MNL model for solving problems such as our example. Their model used five different attributes to explain customer choice; Table 4.3 shows their MNL parameter values associated with each attribute.[1]

Based on the example, we already have information on four of the five model attributes considered. The following additional values were used for the "Origin Point Presence" (OPP) attribute: American = 69.73%, AirTran = 0.66%, Delta = 19.04%, and US Airways = 1.18%. OPP is defined as the percentage of total capacity that each airline provides at the originating airport (in this case DFW). As an explanatory variable, OPP is closely related to carrier brand preference. In practice it is more widely used as an explanatory variable; unlike carrier brand preference, OPP coefficients do not vary widely by market, yet they still

[1] This example was used because the MNL parameters were published. Unfortunately Mishra's model omits display position as an explanatory attribute; the authors would recommend including it in practice.

Table 4.4 Choice model estimated likelihood of selection by itinerary

Itinerary	Departure	Arrival	Airline	Travel time	Price	p (select)
1	1:03 p.m. DFW	4:19 p.m. ATL	American Airlines Flight 1952	Nonstop 2 h 16 min	$230	34.1%
2	1:31 p.m. DFW	4:38 p.m. ATL	AirTran Airways Flight 110	Nonstop 2 h 7 min	$230	12.4%
3	12:11 p.m. DFW	3:19 p.m. ATL	American Airlines Flight 1750	Nonstop 2 h 8 min	$240	18.0%
4	12:00 p.m. DFW	5:11 p.m. ATL	Flight 240/277	1 stop (CLT) 4 h 11 min	$190	7.6%
5	2:20 p.m. DFW	5:31 p.m. ATL	American Airlines Flight 2072	Nonstop 2 h 11 min	$235	19.3%
6	3:55 p.m. DFW	6:50 p.m. ATL	AirTran Airways Flight 102	Nonstop 1 h 55 min	$235	5.4%
7	10:00 a.m. DFW	1:00 p.m. ATL	Delta Flight 2160	Nonstop 2 h 0 min	$250	3.2%

capture the positive marketing impact of originating airport carrier presence arising from frequent flyer programs and increased local advertising.

Applying this model to the DFW-ATL example problem results in the following estimated selection probabilities among the seven different flight alternatives (see Table 4.4).

Based on the above results, it is readily apparent that American Airlines has the greatest likelihood of being selected (if something is purchased). All three of their itineraries were deemed likely to be selected based on the overall combination of attributes. The top-rated AirTran alternative (itinerary #2) was also attractive in most respects, but AirTran's origin point presence was very small compared to AA in this market. AirTran's other flight suffered from being too far away from the desired departure time by the customer. The US Airway's service, although lowest in price, was roughly twice the total flight time and hence deemed the least likely alternative. The sum total of AA's three itineraries is approximately 71% which should correspond to its "instantaneous" market share (i.e. until competitive conditions change).

Many other attributes could have been considered in constructing the MNL model. For example, one important additional variable could be the order of display. Based on Copeland and Mason's finding that 70% of travel agency bookings occur from among the first six options displayed (Copeland et al. 1995), one would expect display order to add additional predictive power. Computational experience by the authors confirms that display order is significant and is an important attribute to include if possible.

The estimated probability of selection is directly related to the expected sales share; as such, it is something that both the airlines and the agencies should monitor carefully. From an airline's perspective, it wants to ensure that its products are being displayed as well as possible. Also, monitoring shopping results provides a mechanism for airlines to determine if they are currently competitive or uncompetitive in a market. Agencies are also interested because many airlines will provide a bonus incentive payment if that agency sells at or above a target sales share amount. Both parties are interested in accurately estimating sales share from these shopping results.

4.3.1.3 Estimating Expected Revenue from Shopping Results

Another useful aspect of applying customer choice models to recent shopping data results is that, because we can compute a carrier's expected sales, we can similarly determine that carrier's expected revenue share (at different price points). This analysis is very helpful for carriers and agencies to determine the revenue impact of pricing discounts or premiums (on specific itineraries) considering current marketplace conditions. Using the previous methods discussed for finding an itinerary's probability of selection, the fare value associated with that itinerary can be used to estimate expected revenue.

$$E(\text{revenue}_j \,|\, \text{sale}) = p(j \,|\, \text{sale})^{*}\text{fare}_j$$

where:

- $E(\text{revenue}_j \mid \text{sale})$ is expected revenue of that itinerary given a sale occurs
- $p(j \mid \text{sale})$ is the probability that itinerary "*j*" is selected
- fare_j is the fare value associated with itinerary "*j*"

Applying the above formula to our DFW-ATL example problem, the following expected revenues can be obtained for each itinerary (see Table 4.5).

By summing up itineraries for the same carrier, carrier-level expected revenue for this specific market (and departure/arrival dates) were calculated, resulting in the following:

Carrier	E (revenue I sale) ($)
AirTran	41.16
American	166.87
Delta	8.10
US Airways	14.50
Total	230.63

Table 4.5 Expected revenue by itinerary

Itinerary	Departure	Arrival	Airline	Travel time	Price ($)	p (select) (%)	E (rev | sale) ($)
1	1:03 p.m. DFW	4:19 p.m. ATL	American Airlines Flight 1952	Nonstop 2 h 16 min	230	34.1	78.44
2	1:31 p.m. DFW	4:38 p.m. ATL	AirTran Airways Flight 110	Nonstop 2 h 7 min	230	12.4	28.42
3	12:11 p.m. DFW	3:19 p.m. ATL	American Airlines Flight 1750	Nonstop 2 h 8 min	240	18.0	43.12
4	12:00 p.m. DFW	5:11 p.m. ATL	Flight 240/277	1 stop (CLT) 4 h 11 min	190	7.6	14.50
5	2:20 p.m. DFW	5:31 p.m. ATL	American Airlines Flight 2072	Nonstop 2 h 11 min	235	19.3	45.30
6	3:55 p.m. DFW	6:50 p.m. ATL	AirTran Airways Flight 102	Nonstop 1 h 55 min	235	5.4	12.74
7	10:00 a.m. DFW	1:00 p.m. ATL	Delta Flight 2160	Nonstop 2 h 0 min	250	3.2	8.10

These results are an alternative way to view the market competitiveness by carrier. What makes the expected revenue view especially interesting is in determining the revenue impact of changes to the price levels (i.e. changing fares on selected itineraries when holding the others constant). In practice, if one carrier changes its fares (either by publishing a new fare value or simply opening/closing fare class availability), one or more of the competing carriers will likely respond to the change (e.g. by full or partial matching). However, because an analyst on the competing carrier will need to search to find that a change has been made, there is a time lag associated with that competitive response. That provides a window of opportunity for carriers who practice real-time (or near real-time) dynamic pricing. Variations of this technique are also used by airlines using historical trends instead of real-time data.

4.3.2 Availability Caching

4.3.2.1 Availability Caching History

Particularly in the context of low-fare search displays, having up-to-date information about an airline's seat availability is crucial in travel distribution. In the early days of GDS's communication between a GDS and an airline's reservation system was relatively slow and expensive. Therefore, GDS's developed availability caches to minimize communications with airline reservation systems.

Availability caches are local images of airline inventory systems. Initially availability caches only stored the status of an entire flight; the flight was either sold out or available for sale. Airlines broadcasted Availability Status (AVS) messages every time the status of a flight changed. Airline deregulation, increased complexity in pricing and inventory controls, and the advent of revenue management required GDS's to maintain more comprehensive availability caching systems. Consequently, availability caches were enhanced to store availability at the flight and fare class level. By the late 1980s, availability caches were further enhanced by the introduction of numeric availability status (NAVS, also referred to as AVN) messages. A NAVS message would contain not only the status (i.e. open or closed) of a fare class but the number of remaining seats in the class–up to a maximum number of seats such as 7 or 9.

The introduction of fare class-specific messages and NAVS caused a rapid increase in availability message traffic, to the point that availability messages far exceeded the number of actual requests. These traffic increases created a major bottleneck and degraded the communication network reliability (Taylor 2003).

When airlines began moving away from segment-based inventory systems to origin and destination (O&D) inventory systems, it became apparent that the concept of AVS/NAVS caching was no longer practical. O&D based systems define availability as a function of the customer's entire itinerary, not just the specific flight being considered, and, in some cases, even customer-specific information is used. Maintaining a complete local availability image that accounts for all these parameters is impractical with the AVS/AVN concept.

In the 1990s, GDS's deployed seamless availability; the availability of a flight would be queried in real-time during the shopping session. Seamless availability changed the concept of flight status from a "push" (from the airline to the GDS) to a "pull" (from the GDS to the airline) system. Inventory was queried only when needed, which limited the communication traffic volumes. Seamless availability worked well until online distribution rose in popularity. Online shopping dramatically increased "look-to-book" (LTB) ratios, causing more availability transactions for the same number of sales. In the traditional agency channels, look-to-book ratios were about 12:1 to 20:1. In the online channels look-to-book ratios range from 100:1 to well over 1,000:1 (Ratliff 2003).

The increased volume of availability requests became costly for airlines, GDS's and OTA's. Airlines and GDS's had to heavily invest in their infrastructures to support the increased shopping volumes. Also, online travel agencies had to pay GDS's per shop, so high LTB ratios were expensive for them. All parties involved, searched for solutions to the problem. Availability proxies and a new form of availability caching were two solutions that the industry quickly implemented.

Availability proxies reside locally (i.e. at the GDS or OTA) and mirror the process of an airline's reservation system. The airline sends updates of their Revenue Management controls and reservation holdings to the proxy's host. The proxy computes at the time of shopping the availability for a given request. Availability proxies have the advantage that network traffic and the airline's system load are independent of the shopping volume and LTB-ratios. Availability proxies have the disadvantage that they require the proxy host to build and maintain an exact replica of the airline's availability processor; this is very costly and carrier specific. It also requires an environment where the carrier openly shares its processes. The complex nature of availability proxies have led to few implementations.

On the other hand, seamless availability caching is being widely used by GDS's and OTA's. Rather than using availability status messages pushed by the airlines, this cache stores the responses of seamless availability calls. Due to size limitations caches usually store only a subset of all availability responses; itinerary-fare class combinations are stored for commonly requested markets and future departure dates. This type of cache can be easily implemented, does not rely on heavy status message traffic, covers more scenarios than traditional AVS caches, and does not require technology sharing between the airline's reservation system and the subscribers' systems. However, cache management is complex; data storage/retrieval costs, cache misses and cache data quality are all important factors. Storing (and retrieving) large amounts of data have obvious infrastructure costs, so volume/speed must be balanced against costs. Cache misses are also costly; availability transactions need to be polled from an airline's reservations system resulting in delayed response time for customers (and accompanying risk of customers abandoning their shopping activity). Cache data quality is typically the biggest customer concern. Over time, cached responses become invalid (i.e. "stale"), and false open/close errors affect consumer's satisfaction. For agencies, false open errors (i.e. the GDS indicates a flight-fare class is open when it is actually closed) increase the risk of being issued a debit memo from the airline. False closed errors are also a problem because using the (incorrect) cached response can result in missing some low fares (making the travel agent search less effective). To minimize the business impact of cache errors most OTA's and GDS's use availability caching only at the time of shopping; they always perform a live request to the airline's reservation system before ticketing.

The industry is constantly working on new caching strategies that maximize cache accuracy while minimizing the number of calls to the airline's reservation system. A typical caching strategy consists of determining which availability results to store, which cache entries to overwrite, when to "void" a cache entry, and when to refresh cache entries. The next subsections provide more details regarding the operation of a seamless availability cache.

4.3.2.2 Availability Request Processing

When an availability request is processed then the cache is first checked for a corresponding entry. If the cache does not have the required information then the airline reservation system is called. If the cache has a corresponding entry then the system needs to decide if the entry is still valid or if the airline's reservation system should be queried. This decision is, in general, based on the cache entry's "age" and the characteristics of the underlying itinerary. Clearly, the older a cache entry the lower the probability that the entry is still valid. The rate at which the expected accuracy of a cache entry deteriorates depends on how frequently the underlying itinerary is likely to be booked or cancelled. For example, a cache entry for an unrestricted fare class on an itinerary that departs 330 days in the future will be longer valid than a cache entry for a discount fare class on an itinerary that departs 21 days in the future. Unfortunately, this implies that the higher the rate of deterioration the greater the value of the entry and vice versa (since popular itineraries see more frequent changes and are more likely to be requested).

4.3.2.3 Availability Response Processing

When processing an availability response the system needs to decide if the availability response should be stored in the cache. If there are empty or voided cache entries then the entry will likely be stored – assuming that the cost of storing/overwriting a result is negligible. If the cache is full with valid entries then the system needs to find a cache entry that is replaced by the new availability response. The system is likely to look for the entry with the worst expected value. If this expected value is worse than the expected value of the availability response at hand then the existing cache entry should be replaced. Otherwise the new response should be discarded. The formula used to compute the "expected value" of a cache entry is, of course, system-specific and part of the caching strategy. In general, the expected value should reflect the expected number of (future) airline reservation system calls that can be avoided by the cache entry.

4.3.2.4 Cache Maintenance

Additional cache maintenance actions can be taken at periodic intervals to void or refresh cache entries. Performing maintenance actions at pre-defined intervals improves the cache response time when availability is requested and processed. In particular, the proactive refreshing of cache entries can improve response times notably since it can prevent cache misses. The frequency of the maintenance actions and the criteria that are used to void or refresh a cache entry are defined as part of the caching strategy. Caching strategies are considered a core competency by many large OTA's and GDS's and, hence, they have not published details of their strategies and findings.

While availability caches are the method of choice in today's distribution world their future looks questionable as airlines are investigating customer-specific availability concepts. These concepts require caches to store availability information at a more granular level. This reduces the availability space covered by a cache and, in turn, increases the number of cache misses. At the same time, the time between "like" availability requests will increase. For example, availability for a given flight might be requested 100 times per day. However, availability for the same flight and a given customer type might be requested only 50 times per day. Hence, the response cached for a customer-specific response is more likely to become obsolete by the time the next request for the same availability is made.

In the long-term airlines, GDS's and other airline subscribers will look at other solutions. Due to industry trends towards more detailed inventory controls, proxies may become a necessity as availability cache effectiveness decreases. Another option is an update of the communication infrastructure. Upgrading to today's technology would allow all participants to exchange availability information in a much more effective manner and eliminate (or at least significantly reduce) the reliance on availability caches (Taylor 2003).

4.3.3 Low Fare Search

Finding the cheapest travel itinerary that matches a customer's request is crucial to the success of travel distributors. Although major online Websites have access to the same airline-published fares distributed in the global distribution systems (GDS's), they use different algorithms to search for the lowest fares, often obtaining different results, which encourages consumers to comparison-shop. The process of finding the cheapest travel itinerary is referred to as low fare search. According to a study by PhoCusWright (Cannizzaro et al. 2004), more than 60% of online consumers buy the absolute lowest fare. Forrester (Harteveldt 2004) also found that 'price satisfaction' was the most important driver in Web-site loyalty further stressing the importance of good low fare search algorithms.

Searching for low fares may seem like a simple problem, but the number of possible fare combinations and the rules that govern how fares can be combined make it difficult. The basic unit of pricing, a fare, is defined as the price of one-way travel between two cities, regardless of the number of flights involved. Airlines update their lists of fares 10 times daily. With each fare comes a set of rules for its use. Low fares, for instance, often have requirements, such as two-week-advance purchase, a connection, or travel at an inconvenient time of day. Rules of another type restrict the way a fare can be combined with other fares. Fares are combined to priceable units. A priceable unit is a collection of fares and associated flights that have one of several possible geometries: a one-way trip or a round trip, for example, or an open-jaw trip, or a circle trip. Still other fare rules could require a round trip with a Saturday night stay or use in combination with an international flight.

Using the techniques of complexity theory, Sara Robinson showed that for a given fare, it can be NP-hard just to determine what existing flights satisfy restrictions; even if the flights are fixed, the problem of choosing fares to cover flights is NP-hard (Robinson 2002). This problem is further complicated by fare combinability. Consider a request for a round trip from Seattle (SEA) to Chicago

Fig. 4.5 Fare combinability example

(ORD) and back connecting in San Francisco (SFO). The cheapest fare for this itinerary could be that for a SEA-ORD round trip. Alternatively, it could be a SEA-SFO round trip plus a SFO-ORD round trip. Or it could be a SEA-ORD one-way fare plus an ORD-SFO one-way fare plus a SFO-SEA one-way fare. These combinations are illustrated in Fig. 4.5.

This last combination is known as an on-line triangle. Even though the itinerary is symmetrical, we could use one fare outbound and two fares for the return. If we substituted a second airline's flight and fare from SFO to SEA, we would have an interline triangle. All of these types of fare constructions are patterns the LFS process must consider to conduct a complete search. Because the combinations of airlines, itineraries, fares, and constructions are so extensive, brute-force enumeration is infeasible. A GDS's LFS process needs to respond to 500 requests per second for domestic markets.

4.3.3.1 Problem Definition

A customer specifies a *trip* and a set of side constraints. The aim is to find an itinerary that meets all constraints and ranks highest with respect to some customer-defined preferences. A customer's trip consists of one or more *segments*. Each segment has an origin and a destination station. In pricing, different types of trips are distinguished:

- *One-way trip*: Arbitrary sequence of segments.

One way trip

- *Simple round trip*: A trip of two segments that make the same connection in the opposite direction.

Round Trip

- *Circle trip*: A sequence of consecutive segments that starts and ends at the same station.

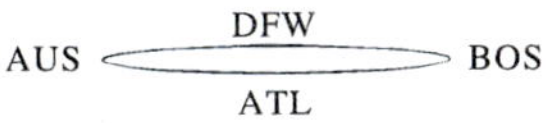

Circle Trip

- *Single open jaw*: A round trip where the return trip starts or ends at a station that is not part of the out-bound trip.

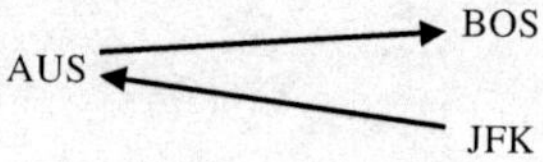

Single Open Jaw

- *Double open jaw*: A round trip where the return trip differs from the out-bound trip in origin and destination.

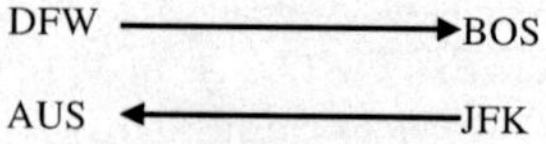

Double open jaw

The customer specifies a travel time window for each segment of the trip and a set of constraints. The algorithm needs to find a solution that satisfies all side constraints and allows the customer to travel within the specified time windows.

The customer defined side constraints can restrict some of the itinerary parameters, such as carrier, equipment type, number and location of intermediate stops, ground time or class of service. These constraints are usually easy to handle since they reduce the solution space and therefore speed up the search process.

Customer preferences are, in general, soft constraints. They are much more difficult to handle than hard constraints. Consider for example a cost driven customer, i.e., a customer who asks for the cheapest available fare. However, the customer might be willing to pay a *little* more than the cheapest fare, if this buys him a *much* more convenient itinerary, e.g., an itinerary with less intermediate stops or an itinerary that allows travel at more convenient days/times. It is hard to resolve conflicting soft constraints. Sometimes a customer is not even sure if a constraint is a hard or a soft constraint. Take for example preferred carrier vs. exclusive or undesired (prohibited) carrier. Customers impose these constraints to create a diverse solution that gives them a large number of interesting alternatives. We will discuss approaches to this dilemma toward the end of this document and assume for now that a customer is interested in the cheapest itinerary that satisfies all hard constraints.

The low fare search is also limited by *fare rules*, in particular *combinability rules*. Fare rules define conditions under which a fare can be applied, e.g., advance purchase requirements, Saturday night stay etc. Combinability rules are fare rules that limit the fares that that can be combined on an itinerary. Combinability rules make the problem particular hard, because they can be checked only after several fares have been selected/combined.

4.3.3.2 Algorithms

One approach to solving the low fare search problem is dynamic programing (Robinson 2002; de Marcken 2003). The dynamic programing algorithm uses priceable units as basic building blocks. A priceable unit is a collection of up to four fares that adhere to all fare rules and cover all flights contained in a one-way, simple round trip, single open jaw or circle trip.Given a set of flights, the dynamic programing algorithm finds a combination of priceable units that covers all flights and minimizes total price. When looking at a circle trip AUS-DFW-JFK-BOS, for example, the dynamic programing algorithm looks at the priceable unit for the circle trip as a whole as well as all combinations of smaller priceable units:

- Priceable Unit: Circle trip AUS-DFW-JFK-BOS
- Combination 1: One-way fares: AUS-BOS + BOS-AUS
- Partition 2: Round-trip fare: AUS-BOS-AUS
- Partition 3:
  - Open Jaw AUS-DFW-JFK-AUS
  - Priceable Units for DFW-BOS
  - Priceable Units for BOS-JFK, etc.

The algorithm was developed by the founders of ITA Software (www.itasoftware.com), the low fare search service provider for several airlines, corporate booking tools, and online travel agencies.

Another approach to solving the low fare search problem is to formulate it as a shortest path problem with side constraints on a *fare network*. The following sections discuss this approach in more detail.

4.3.3.3 Fare Network Model

A node of the fare network represents a station. An arc represents a fare that is filed for the associated O&D pair. A graphical illustration is provided in Fig. 4.6.

A cheapest path from a node A to a node B in the network gives a lower bound on the fare that one has to pay to get from A to B. Consider the set of paths that are feasible with respect to all customer-defined constraints and time windows as well as fare rules. The shortest path of this set gives the cheapest feasible fare combination for the customer request and hence a solution to the LFS problem. There are two difficulties with the approach:

1. The network of all filed fares is too large to be considered as a whole.
2. A path has to be constructed before its feasibility and in some cases (when the airline uses O&D availability controls) availability can be checked.

Therefore, fare aggregation and bounding techniques are used to overcome the first difficulty. A k-shortest path algorithm is used to overcome the second one, i.e.

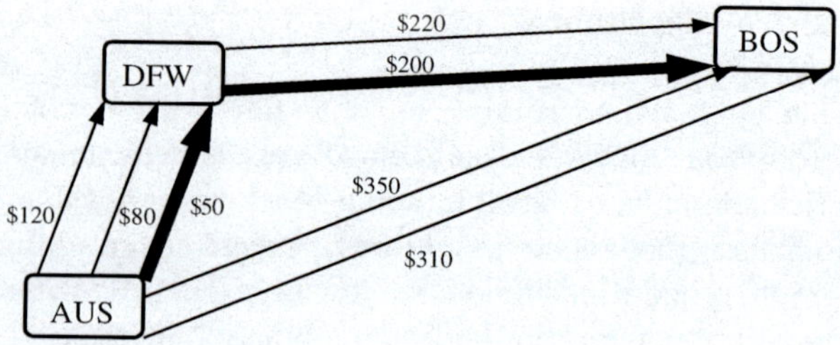

Fig. 4.6 Sample fare network (cheapest AUS-BOS-path along bold arcs)

we compute the shortest, second shortest and so on path, until we find a feasible path. That path is then the solution to our problem.

4.3.3.4 Data Aggregation and Bounding

Data aggregation is done offline. An *aggregated network* is constructed. A single arc replaces each bundle of parallel arcs in the fare network. A cost and a time label are attached to this arc. The cost label gives the cheapest available fare in the represented bundle. The time label is a lower bound on the traveling (flying) time on the flight leg that is associated with the arc bundle. Example illustrations are provided in Figs. 4.7 and 4.8.

Consider a *route* between two stations, i.e. a sequence of intermediate stations. The cost of the corresponding path in the aggregated network gives a lower bound on the total fare along the route. The lower bound is tight if all the cheapest fares along the route are available and combinable. Hence the cost of a path in the aggregated network represents a lower bound on the cost of any path in the detailed network that uses the same intermediate nodes. This property is used to speed up the shortest path algorithm:

1. Find a shortest path $P_{\mathbf{agg}}$ in the aggregated network.
2. Construct the part of the *detailed network* that contains all nodes that are on $P_{\mathbf{agg}}$; $P_{\mathbf{agg}}$ becomes an *expanded path.*
3. Compute the cheapest feasible path in the detailed network that uses only available capacity and store the detailed path in a priority queue.
4. If the cost of the cheapest detailed path is cheaper than the cost of the cheapest non-expanded path in the aggregated network, then it is the optimal solution.
5. Otherwise go to step 2 processing the next shortest path in the aggregated network.

The time labels of the arcs in the aggregated network can be used to perform feasibility checks during path construction. Again a lower bound property is exploited. The sum of the time labels along a path in the aggregated network gives a lower bound on the travel time of any itinerary that uses the same intermediate nodes. Hence, if the sum of the time labels along a path is greater than the customer-defined time window then the path can be eliminated without going to

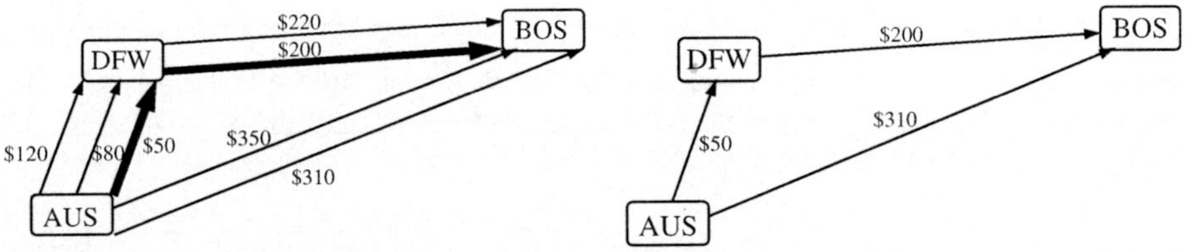

Fig. 4.7 Detailed and aggregated fare network

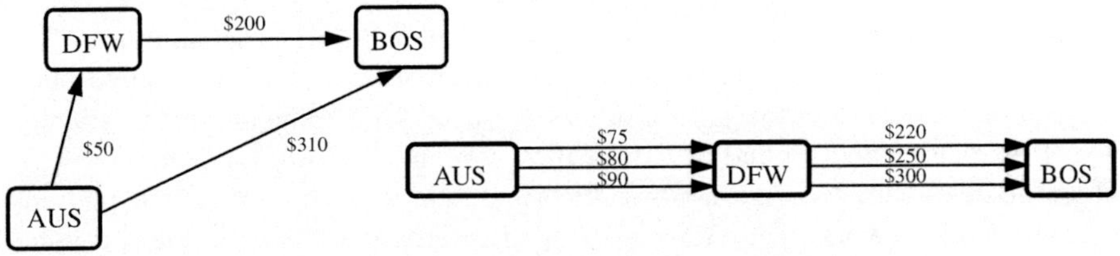

Fig. 4.8 Aggregated and detailed network for path AUS-DFW-BOS

the detailed network and sometimes even before the path is completed. A similar bounding can be performed using the great-circle distance between the cities, thus avoiding *circuitous connects*.

4.3.3.5 Constructing a Detailed Network

Fare and flight availability are taken into account when a detailed network is constructed. All flights that depart in the specified departure time window are pulled in order of increasing departure date/time. Fare availability is checked for all relevant flights. An arc is added to the detailed network for each fare that is applicable and available:

- A fare is applicable if all rules that can be checked at this point are satisfied. Some fare rules cannot be checked, since there is incomplete information about the itineraries that are generated for the other trip segments.
- A fare is available if it can be allocated to a flight that leaves in the requested departure time window.

Note that the lower bound on the fare amount that is stored in the aggregated network can be tightened when a detailed network is constructed. The lower bound can be replaced by the lowest fare that satisfies all trip based applicability and availability checks. However, the corrected lower bound is valid only for the current search.

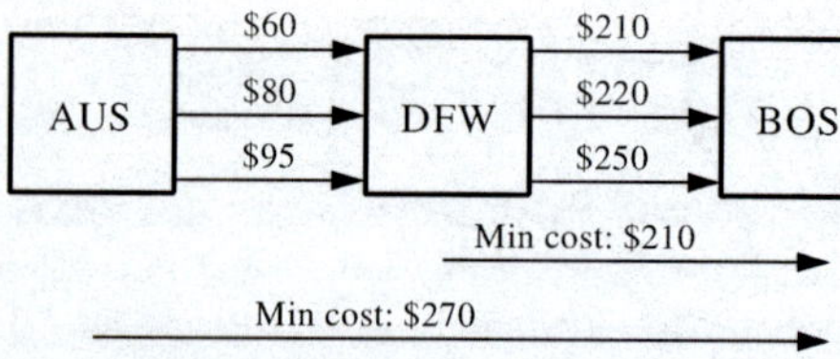

Fig. 4.9 Stage network with minimum cost arcs for non-terminal stages

4.3.3.6 Shortest Paths in Detailed Networks

Finding a shortest path in a detailed network is similar to the procedure used for the aggregated network. However additional information about the detailed network can be used to steer the branching and to speed up the algorithm. A detailed network is constructed around a path in the aggregated network. Each complete path in the detailed network consists of the same intermediate nodes as the path in the aggregated network. Therefore, a detailed fare network can be viewed as a stage network with each node representing a stage (see Fig. 4.9). The sum of the cheapest edges between stages i, $(i + 1),\ldots, n$ is a lower bound on the cost of getting from stage i to the final stage n.

The lower bound on the cost for the remaining stages and the actual cost of an incomplete (partial) path are added together. In each iteration the cheapest path with respect to this sum is expanded. All arcs out of the end node of the path are considered in increasing order of associated fare. A new path is generated for each outgoing arc until the cost of the path plus the fare of the next outgoing arc is higher than either

- the cost of any unexpanded path in the aggregated network
- the cost of any path in a detailed network.

If all outgoing arcs were considered before one of the stopping criteria is satisfied, then the old path is deleted. Otherwise, the cost label of this path is updated, and it is put back on the heap. A pointer to the first arc that leaves the last node of the path and that has not been used for path construction is stored with the path. When the path is read from the heap the next time, then no arc that has already been used to expand this path will be considered again.

4.3.3.7 Detailed and Aggregated Level Interaction: Example AUS-BOS

Consider a one way trip from Austin to Boston to illustrate the interaction between the aggregated and the detailed network level. The next few illustrations show the part of the aggregated network that is of interest for this example.

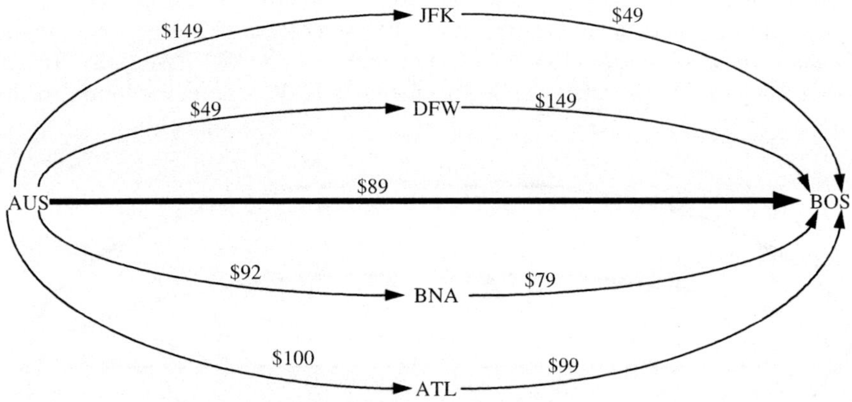

The shortest path in the network is the nonstop segment (see bold arc) with a cost of $89. A request for the lowest fare for the given date shows that the cheapest available fare is $140. A detailed network that contains all applicable and available nonstop fares is constructed:

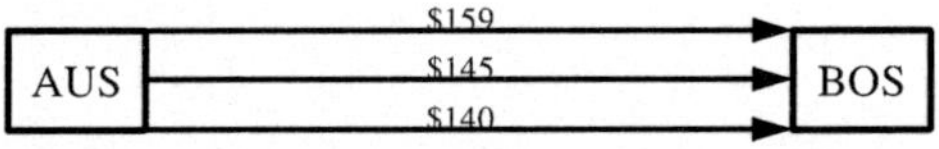

The cheapest path in the detailed network is priced at $140. This is less than the cost of the next cheapest path in the aggregated network (AUS-BNA-BOS):

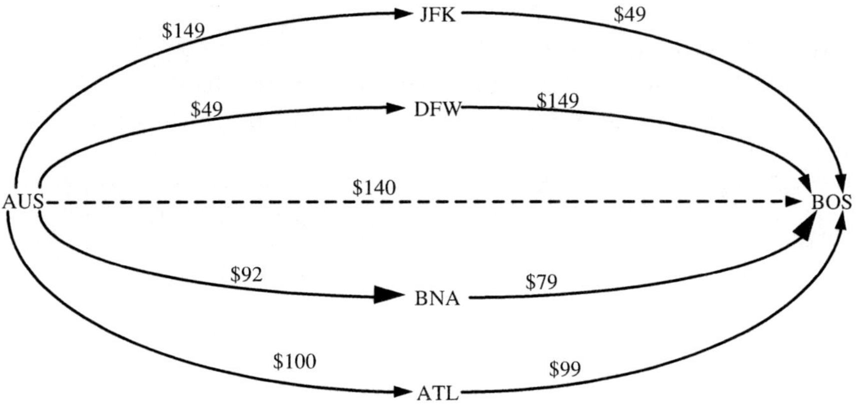

Therefore a new detailed network is constructed. This network contains all fares that are applicable and available for AUS-BNA and BNA-BOS:

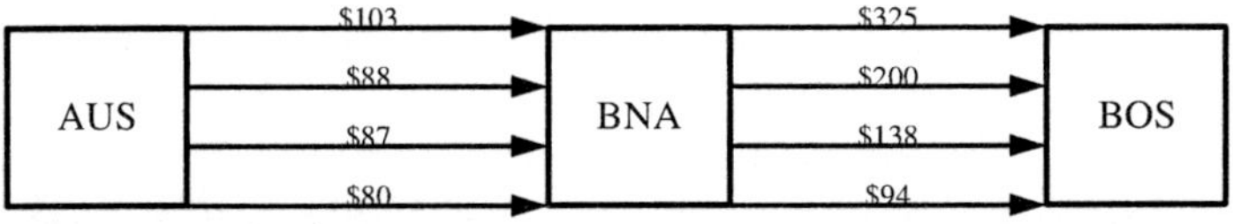

The shortest path in this network has cost $174. Therefore $140 is still the cheapest path on the detailed network level. However the next shortest path in the aggregated network costs $198 (AUS-JFK-BOS and AUS-DFW-BOS). Hence the shortest path in the detailed nonstop network is the optimal solution to the problem.

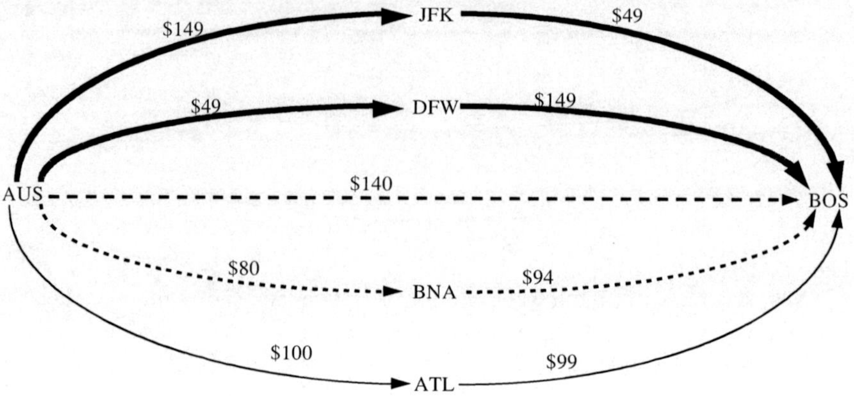

4.3.4 E-mail Campaign Management

This next section describes a technique for notifying customers of special deals on airfares in markets that they are interested in. Such information is usually of great interest to airline customers, but because airfares change frequently, there is a high risk of overloading customers with too many messages (i.e. 'spamming'). Unless the e-mail content is relevant and compelling, there is a risk that the customer will begin blocking messages from the airline.

The OR problem is to optimize the trade-off between notifying the customer of relevant airfare changes versus overloading him/her with too many e-mail messages. What comprises relevant content? This information typically is comprised of:

1. Customized messages containing content of interest to the specific customer being sent a message
2. The set of markets the customer is interested in (e.g. those markets in which the customer has previously purchased or has included in a notification list)
3. Price changes in target markets that now provide an exceptional value (compared to a reference level)

Travelocity has developed one possible solution to this problem as part of its 'Good Day to Buy' e-mail campaign management application. In practice, Travelocity has observed customer click-through rates to the 'Good Day to Buy' targeted e-mails that are up to ten times greater than unsolicited message rates, thus providing clear evidence of customer relevance. An excellent, detailed description of the Travelocity method can be found in Smith et al. (2007); their general approach is summarized in the remainder of this section.

The Travelocity approach is based on the observation that, if prices in a market drop below a certain threshold reference level, observed air and hotel booking volumes increase significantly. When prices in a market drop below this reference level, Travelocity generates a customized e-mail message notifying the customer of the good deals that are currently available.

Travelocity used a simple 30-day moving average to establish the reference fare value and calculates a relative price difference versus current market prices. A piecewise regression model was used to develop a demand model that relates the relative price difference to historical sales. Based on the observation that sales were significantly larger when current prices were below (rather than above) the reference, a two-piece regression curve was used. The two pieces intersected (i.e. the knot occurs) where the current market price equals the market reference price; an example of the resulting regression model is shown in Fig. 4.10. Travelocity estimated that use of these 'Good Day to Buy' targeted e-mail notification messages generated an additional $6.9 million sales revenue in 2006.

4.3.5 Opaque Air Pricing Model

In opaque pricing, the consumer is only aware of some attributes of the final trip during the shopping process; the destination, travel dates, carriers or flights could be hidden until a non-refundable payment has been made. Most consumers find this uncertainty unacceptable. Opaque products are cheaper than normal fares, however, and many consumers are willing to buy an opaque product if the price is right.

Opaque product segmentation provides airlines an alternative distribution outlet that enables incremental revenues by selling distressed inventory cheaply, but it poses the risk of diluting sales of non-opaque products. This dilution risk is higher for airlines that sell both non-opaque and opaque inventory on their own website; because only one product brand is applicable, the customer uncertainty is reduced, and the opaque product segmentation is weaker. As such, most airlines prefer to sell opaque products via third party online travel agencies because the presence of multiple airline brands result in higher opaque product uncertainty for the customer (providing stronger segmentation and less dilution risk). A related (but not identical) type of distribution outlet is found in "name your own price" channels (e.g. Priceline) as described by Wang et al. (2009).

The key decision in selling opaque air products is determining the appropriate markup amount. Retailer markup decisions involve setting a selling price at some positive amount above the acquisition cost of the goods being sold; the difference is referred to as the markup amount. Post (2010) discusses an interesting methodology to solve the opaque product pricing whereby the customer can vary the

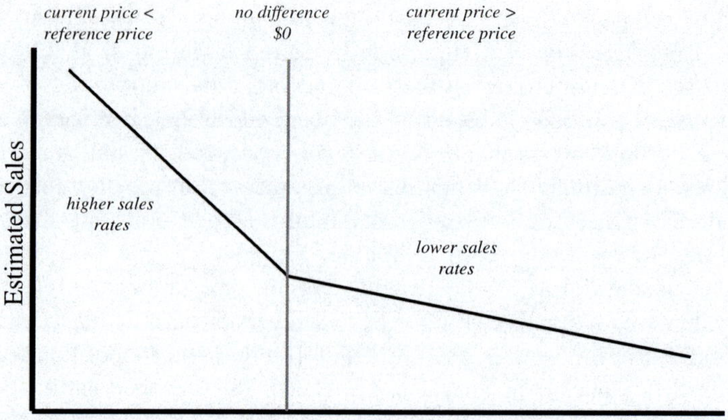

Fig. 4.10 Example: Two-piece demand regression model

level of "opacity", and the prices are automatically recalculated depending on how opaque the product really is. Rao and Zouaoui (2009) provide a detailed description of a different optimization approach. The remainder of this sub-section provides a summary of the Rao and Zouaoui method for optimizing dynamic markups of opaque air products.

Rao and Zouaoui base their optimization on a dynamic markup rather than the more widely used fixed or static markup approach. Because most online agencies are selling air products across numerous different airlines, they have access to a large, representative sample of current marketplace prices for air travel. Rao and Zouaoui used this current market information to find the minimum priced non-opaque fare available, denoted *l*, and use it to dynamically set the optimal markup amount, *m*, above the wholesale price, *w*. to obtain the optimal opaque air product selling price, *f*. Customer choice models utilizing market information for *l* along with values for *f* and *w* are used to predict the probabilities that a customer will: (1) buy an opaque itinerary, (2) buy a non-opaque itinerary, or (3) abandon the website. As such, the optimized opaque price, *f*, maximizes expected revenue considering the current customer alternatives. Figure 4.11 illustrates the major parameters considering in the dynamic markup decision. Rao and Zouaoui noted a 48% revenue improvement by using an optimal dynamic markup approach for a major online travel agency compared to a previous application based on fixed opaque markups.

The objective of the opaque air product dynamic markup problem is to set the opaque markup fee to maximize the expected profit from each shopping session that includes an opaque. The markup fee (m) can be expressed as a percentage of the difference between the lowest non-opaque fare (l) and the cost of the opaque fare (w). The expected profit per session is $p_0 \cdot m \cdot (l - w) + C \cdot p_n$, where C is the commission on each sale of non-opaque fares, p_o is the probability of selling the opaque fare, and p_n is the probability of selling a non-opaque fare. Note that there

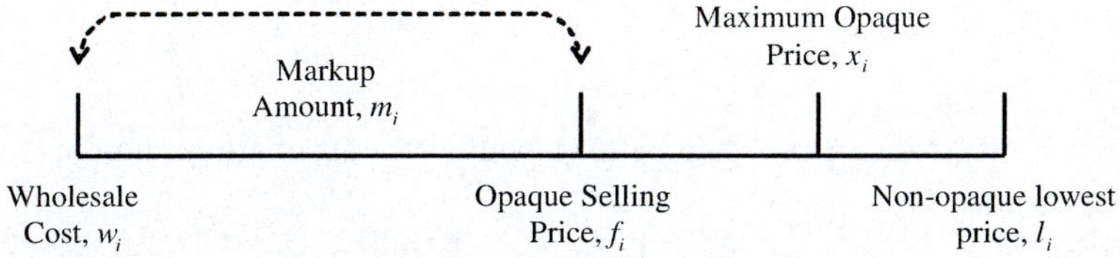

Fig. 4.11 Opaque dynamic markup decision parameters

is also the probability of not selling anything, p_a (i.e. abandonment), so the three possible cases are related according to the following equation: $p_o + p_n + p_a = 1$. Note that the probabilities of sale (p_o and p_n) are functions of *m*, *l*, and *w*.

Similar to the choice modeling approach described in Sect. 4.3.1, Rao and Zouaoui use a multinomial logit model incorporating several effects including: price, advance purchase, day of week and itinerary quality of service. The data used to calibrate the MNL model consists of customer shopping and selling sessions captured from website clickstream data to obtain information on what air travel options were presented to and purchased customer during their session (or not purchased in the case of abandonment without sales). The clickstream history data are used to calibrate the MNL models and estimate customer preferences for both opaque and non-opaque products. These models are subsequently used in actual sales sessions to dynamically optimize the opaque markup amounts. Customer responses from those sessions are captured to recalibrate the MNL models used in future sessions.

Using the standard notation for multinomial choice models, the equations for calculating p_o and p_n are as follows:

$$p_o = \frac{e^{\beta_o x_o}}{e^{\beta_a x_a} + e^{\beta_o x_o} + e^{\beta_n x_n}} \text{ and } p_n = \frac{e^{\beta_n x_n}}{e^{\beta_a x_a} + e^{\beta_o x_o} + e^{\beta_n x_n}}$$

where β is the vector of MNL regression model beta coefficients, and $\boldsymbol{x}$ is the vector of attributes considered.

The following optimization model is used to determine the optimal dynamic markup amount, m^*:

Maximize

$$p_0 \cdot m \cdot (l - w) + C \cdot p_n$$

s.t.

$$f = w + m \cdot (l - w)$$
$$(l - f) \geq k$$
$$m \geq 0$$

where (k) is defined as the maximum opaque selling price (based on the minimum separation that the retailer wants to keep between the opaque and lowest non-opaque items). Note that $k \leq l$.

Since the probabilities of selection depend on the markup amounts, the underlying optimization problem is highly non-linear. However, in practice it is usually adequate to specify the markup percentage to the nearest 1–2% value, so a simple grid linking markup percentage, m, to its associated p_o and p_n selection probabilities can be readily constructed. A grid search can then be performed to find m^* that satisfies the required marketing constraints. An example illustration of how the probability of a session sale and the expected profit typically vary as a function of the opaque markup percentage can be found in Fig. 4.12.

4.3.6 Merchandising and CRM

Previously we discussed the importance of screen presence in effective merchandising, but another crucial aspect involves the customer dimension. Customer conversion rate is the likelihood that a customer (or travel agent) will actually buy something during the shopping session. A similar metric conversion that is commonly used in travel is the "look-to-book" (LTB) ratio. The LTB ratio is inversely related to conversion rate but includes the additional "work" associated with multiple shopping requests made during the session. The equations below include both definitions:

Conversion rate = booked sessions/sessions without bookings
Look-to-book = shopping requests/bookings

Airlines can further improve their revenue effectiveness by targeting sessions with a high expected conversion rate. Forecasting conversion rates from shopping sessions can be very difficult, but there are at least four causal factors that can be used to help explain expected conversion rates. They are:

- Channel (typically online agency, online supplier, offline agency, offline supplier or e-mail campaign)
- Customer-history information
- Frequency of request (how many times in the past the same request has been made by this individual)
- Relative attractiveness of the items returned in the choice set (i.e. are they good deals or not)

Conversion rates vary greatly by channel. Ratliff (2003) reported that the productivity of professional travel agents (12-to-1 LTB) is much higher than a typical customer visiting a major online travel retail website (200-to-1 LTB). This difference occurs because travel agents can rely on their experience to find good travel deals more quickly. Interestingly, an article in The Economist (2005) showed the conversion rates associated with online supplier websites are much

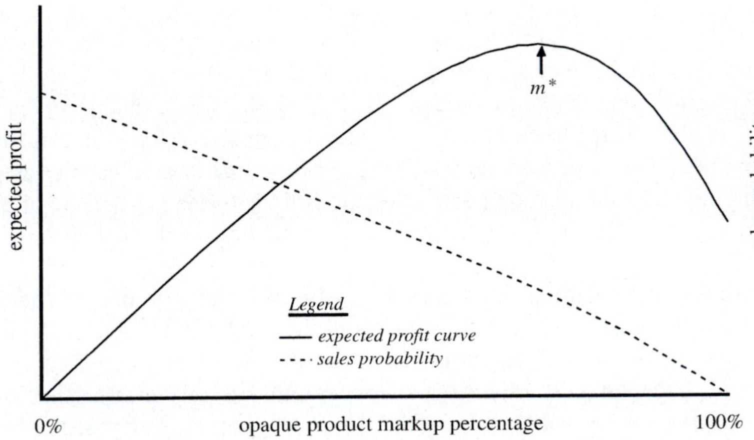

Fig. 4.12 Example expected profit and sales probability vs. opaque markup

higher than those of online travel retail sites. Airlines such as American, Delta and Southwest had conversion rates in the 10% range, whereas online travel websites including Expedia, Orbitz and Travelocity were less than half that amount (in the 4% range). The underlying reason is that most web-based shoppers appear to start their search using the online retail sites (where multiple carriers are displayed) but they often complete their booking directly on the airline website (e.g. to avoid booking fees assessed by the online agencies). Another contributing factor is web-based shopping robots because they will often check multiple online agencies to find the one which has the best price; since these robotic-based sessions generate many shops for each booking made, they tend to drive up online agency look-to-book ratios.

If the specific customer can be identified who is associated with the shopping session, his or her historical activity can be used to help determine the expected conversion rate. The customer could be identified directly (from login or other unique identification information obtained earlier in the session) or indirectly (by inspection of previous pages and requests made earlier in the session). For those customers known directly, the simplest approach involves keeping track of their historical conversion rate. For "anonymous" customers, it is still possible to gain some insights into their likes and dislikes by viewing their activities performed during the session. For example, if the customer has viewed both air and hotel web pages during the session, it is reasonable to conclude that he or she might be more likely to buy a package special (comprised of air and hotel combined). A good discussion of data mining techniques for determining these relationships (such as link and sequential analysis) can be found in Dunham (2002). Better understanding of a customer's likes and dislikes can help the airline or retailer to improve conversion rates by dynamically modifying their website content to match the customer's preferences. This dynamic content can come in the form of different banner ads as well as alternative flight results shown on the screen. One well-known retailer outside the travel industry is Amazon.com; its use of collaborative filtering helps improve their suggestion effectiveness (Linden et al. 2003).

Other recent research by Moe and Fader (2004b) shows that frequency of request is another factor in determining expected conversion rates. If a customer has made repeat visits to a website and asked for the same item in each case, Moe and Fader demonstrated a strong positive relationship between probability of a sale and the number of repeat visits. In the context of air travel, such behavior is commonplace.

A fourth factor in helping explain sales conversion rates is one of the most obvious, basically answering the question “Is this a good deal (or not)?” A simple (but effective) method of determining whether the items returned in low fare search results are generally attractive or unattractive is to compare the lowest priced item with the historical price in that air travel market. Travelocity uses an approach for its e-mail campaigns that they have called “Good Day to Buy” (Smith et al. 2007). The method works by establishing a historical price reference point for each market, and prices that fall at or below the reference price are deemed attractive. Other commonly used methods for determining overall attractiveness of items returned in shopping sessions involve the inclusive value of utility (Ben-Akiva 1985) considering all the items collectively in the choice set. Good discussions of the application of utility-based conversion rate forecasting can be found in Rao and Smith (2006), Moe (2006).

4.3.6.1 Multi-channel Considerations

In the previous section describing probability of selection of itinerary choices, expected revenue was expressed in conditional terms (i.e. given that a sale takes place). So the more general form of expected revenue considers the probability of a sale taking place (during that session). Sales conversion rates vary by customer and channel, so expected revenue also depends on the customer and channel being considered. When an airline or agency is making distribution optimization decisions, the distribution costs and expected revenue across the various (competing) channels need to be determined. Furthermore, the general mix of customers served by that channel should also be found. The graph shown in Fig. 4.13 shows the relative attractiveness of customer types considering the channel costs and revenue potential of the customer base.

4.3.7 Agency Supply Management

Airlines negotiate agreements with agencies they consider integral to their distribution objectives. Carriers try to motivate these agencies to favor their products by offering financial incentives, exclusive, merchant content, or a combination

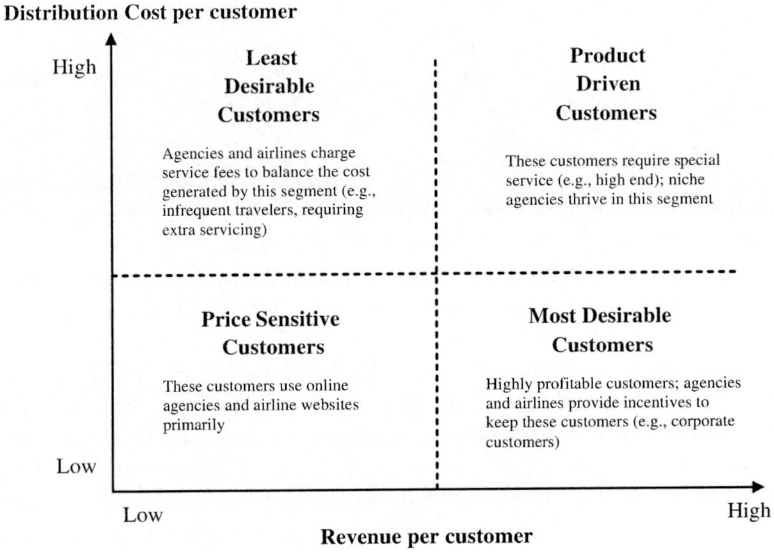

Fig. 4.13 Revenue and distribution cost per customer type

thereof. An agency-carrier agreement may or may not specify any of the following components:

- *Base commissions*: Commissions that are paid for each booking made by the agency. Base commissions may differ by market and ticket type. Base commissions can be a fixed dollar amount or a function of the ticket price.
- *Marketing Funds*: Funds the agency receives to promote the airline to its customers. Payment of marketing funds may or may not come with specific promotion requirements, such as online ads for example.
- *Exclusive Content Access*: Exclusive content is not available to the general agency population. The availability of exclusive content is controlled by the carrier.
- *Discounted Fare Access*: Discounted fares sell at a discount of published fares. The availability of a discounted fare is tied to the corresponding published fare.
- *Merchant Content:* Access to fares that the agency can mark up before selling them. The fare at which merchant content is made available, i.e. the net price, is defined in the contract with the agency. In most cases these fares are defined via a discount of comparable published fares, i.e. published fares with the same restrictions. The airline may restrict the markups the agency can apply.
- *Most Favored Nation (MFN) agreement*: Guarantees the agency that its access to the supplier's content will be at least as good as the access of any competitor in a specified competitor set.
- *Override Payment Schedule*: An override payment schedule guarantees the agency additional compensation, so called overrides, if it reaches certain

performance targets. Performance targets are specific to the deal and can be based on a variety of metrics such as sales, revenue or seat miles. The additional compensation can be defined as additional commissions, one-time payments, marketing funds, access to merchant or access to exclusive content. In general, an incentive deal contains multiple hierarchical performance targets. The higher the target the higher the perceived value of the associated compensation.

After the arrival of Internet-based distribution, the base commissions went down considerably and performance based compensation became more important. With the rise of new distribution models, such as dynamic packaging and opaque pricing, merchant content became more important. In this era all the big agencies, especially the Internet-based ones, had to ensure to get the same merchant content as their big competitors. Hence, they tried to negotiate MFN clauses covering as much content, including some types of merchant content, as possible. Lately, carriers have become more reluctant to grant access to merchant content as they have to lower regular fares to match low-cost competition and see merchant content as another driver of yield dilution. Since airlines started to reduce or completely eliminate commissions, the negotiation and management of performance-based compensation became more and more important to the financial success of agencies.

An agency has to consider several things when evaluating a supplier agreement. In particular, it has to compare potential performance targets with its estimated ability to shift demand from one carrier to another and to stimulate demand. An agency can vary service fees and use a variety merchandising tools such as promotions or e-mail campaigns to realize demand shift and stimulation. Online agencies can also use screen presence, i.e. what is shown to the customer and in which order, to influence demand.

The value of a deal also depends on the terms of the agency's other supplier deals. Therefore, an agency should evaluate a deal in the context of all deals that are already in place or on the horizon. This is particularly important for agencies that have one or more deals with performance-based components because:

1. Incentives are either paid in full, when the target is reached, at the end of the measuring period or, not at all. Hence, missing a target even by a small amount can result in large revenue reduction.
2. Supplier-agency deals with an incentive component have, in general, worse base economics than supplier deals with no incentive component. Hence, shifting demand to a carrier with an incentive agreement may result in lower total revenue than taking no action (see Fig. 4.14).

Aiming for one deal's performance target may reduce the likelihood of or even prevent achieving the performance targets of other deals.

Smith et al. (2007) developed a model to address the supply management problem for Travelocity. The model captures multiple planning periods and represents demand on the market-carrier-time period level. The model determines

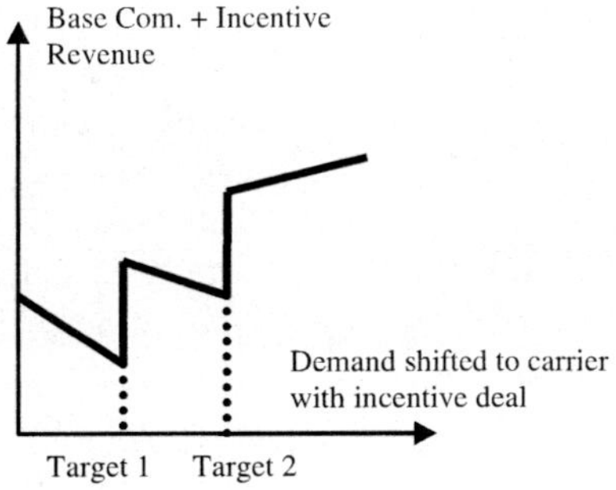

Fig. 4.14 Total compensation as a function of carrier share

how to influence ticket sales via service fees and merchandising efforts such that Travelocity's air revenue is maximized.

The impact of merchandising efforts on ticket sales is captured via market level demand shift and growth variables. In particular, the shift variable $shift_{a,m}$ represents the number ticket sales shifted to or away from carrier a in market m. The growth variable $growth_m$ represents the change in total ticket sales in market m. The shift and growth variables are constrained by user-defined estimates of the effectiveness of merchandising efforts.

The model also considers the impact of service fees. Service fee variables ($svcFee_{a,m}$) represent the service fee that is charged for a ticket sale on airline a in market m. The impact of service fees on ticket sales is modeled using piecewise linear functions. The parameters of these functions are computed via scenario analysis using simulation and customer choice models.

1. $\max \sum_{a\in A} \sum_{m\in M} \left(\text{tix}_{a,m} + \Delta\text{tix}_{a,m}\right)\text{comBase}_{a,m}$
   $+ \sum_{d\in D} \text{comIncent}_d + \sum_{a\in A} \sum_{m\in M} \left(\text{tix}_{a,m} + \Delta\text{tix}_{a,m}\right)\text{svcFee}_{a,m}$
2. $\text{shiftMin}_{a,m} \leq \text{shift}_{a,m} \leq \text{shiftMax}_{a,m}, \quad \forall a \in, \ m \in M$
3. $\text{growthMin}_m \leq \text{growth}_m \leq \text{growthMax}_m, \quad \forall m \in M$
4. $\Delta_{a,m} = \text{shift}_{a,m} + \text{growth}_m \text{share}_{a,m} + \sum_{oa\in A} \Phi_{a,oa,m}$
   $\left(\text{svcFee}_{oa,m}\right), \forall_a \in A, \quad \forall m \in M$
5. $\Delta\text{tix}_{a,m} = \Delta_{a,m} - \text{share}_{a,m} \sum_{oa=A,\, oa\neq a} \frac{\text{shift}_{\text{oa,m}}}{1-\text{share}_{oa,m}}, \quad \forall m \in M, \ a \in A$
6. $\text{comIncent}_d \leq \text{bigM} \sum_{k\in L(d),\, k>l} \delta_{d,k} + r_{d,l} \sum_{m\in M} \left(\text{tix}_{a(d),m} + \Delta\text{tix}_{a(d),m}\right.$
   $\left.+\text{tixToDate}_{a(d),m}\right), \quad \forall d \in D, \ l \in L(d)$

Where:

$\text{tix}_{a,m}$	Forecasted ticket sales on airline *a* in market *m*
$\Delta\text{tix}_{a,m}$	Targeted change in ticket sales on airline a in market m
$\text{comBase}_{a,m}$	Base commission received for a ticket sale on airline a in market m

conIncent_d	Incentive commissions collected on deal d
$r_{d,l}$	Per ticket incentive commission paid under deal d when level l is reached
$\text{shiftMin}_{a,m}$	Upper bound on the number of ticket sales Travelocity can shift away from airline a to another airline in market m
$\text{shiftMax}_{a,m}$	Upper bound on the number of ticket sales Travelocity can shift toward airline a from another airline in market m
growthMin_m	Upper bound on the number of ticket sales Travelocity can decrease total sales in market m
growthMax_m	Upper bound on the number of ticket sales Travelocity can increase total sales in market m
$\text{tixToDate}_{a,m}$	Number of tickets already sold on airline a in market m
$\text{share}_{a,m}$	Expected market share of airline a in market m
$\Phi_{a,oa,m}(svcFee_{oa,m})$	Function that denotes the change in ticket sales for airline *a* in market *m* given the service fee for airline *oa*. The function is used here only for simplicity of notation

The relationship is modeled explicitly via a piecewise linear in the model used by Travelocity:

A	Set of all airlines
D	Set of all incentive deals
$L(d)$	Set of all performance levels of incentive deal d
$a(d)$	Airline associated with incentive deal d
$bigM$	A large number

Please note that constraint set (4) captures the first order impact of merchandising and service fees on ticket sales. Constraint set (5) models second order effects due to share shift interdependencies. The underlying assumption is that second order effects are a function of current market shares.

The model can quickly become large and complex given that large agencies serve several 10,000 markets and have performance deals with multiple carriers. However, generating the inputs of the model is even more complex. It involves not only traditional demand forecasting but also correct modeling of demand relationships and dependencies.

Smith et al. (2007) report that the deployment of the supply management model allowed Travelocity to more aggressively pursue and perform on override deals: in the first 2 years after deploying the model Travelocity's override payments increased by \$6.3 M and \$5.3 M, respectively. During the same time Travelocity moved from network-wide to over 100 network-wide and market-specific override agreements.

4.4 OR Modeling Opportunities

In Sect. 4.3 we discussed OR techniques used today in air travel distribution. In our view, a great deal of effort is still needed in this field. As we have said earlier, the multi-constituency aspect of air travel distribution creates opportunities to increase performance and efficiency. An airline's distribution strategy determines which content to sell in which channel at which price, time and cost. For a GDS, the strategy narrows down to which content to obtain (at which price), how it is compensated when the product is sold, and which agencies should participate. For an agency, the distribution strategy defines how it obtains the relevant content (i.e., direct-connect vs. GDS, GDS A vs. GDS B) and how it gets compensated. Clearly, the more compensation GDS's (and agencies) require, the higher the distribution cost for airlines. Also, the fewer the content in a GDS, the less relevant (and unattractive to agencies) it becomes.

In this section, we provide a modeling framework to the conflict caused by non-aligned strategies of the various constituents. Section 4.4.1 suggests an approach to the content management problem. Section 4.4.2 proposes an approach airlines can use to determine the right distribution channels for its content. Section 4.4.3 provides a method for GDS's to evaluate the profitability of travel agencies.

4.4.1 Content Management

A key problem faced by airlines, GDS's, and travel agencies alike is that of content management. For airlines, it is deciding which content to make available in which channel, given the cost of distribution. For GDS's (and agencies), it is deciding whether to reduce its booking fee (commissions for agencies) to get access to some content, typically low fares. We have developed a framework to address this problem.

When a specific product (content) becomes unavailable in a given channel, a diversion of demand ensues: customers who intended to buy this product have to either buy a different product in the channel, buy in a different channel or make no purchase at all, i.e. abandon. Customers who do not abandon make up the diverted demand. We distinguish three types of diverted demand:

- *Channel switching*: Customers choose other outlets to buy the product
- *Sell-ups*: Customers buy a more expensive product offered in the same channel by the carrier in the same channel
- *Displacement*: Customers purchase comparable products from a different carrier

A customer's choice between channel switching, sell-up, and displacement is related to the customer's brand preference and price sensitivity[2]:

---

[2] Other attributes (e.g., schedule quality) also affect the level and type of diversion.

- Buyers with no or little brand preference will purchase a competitor's product.
- Buyers with strong brand preference (e.g., due to frequent flyer points) will either switch channels or sell up, depending on their sensitivity to price.

Suppliers need to estimate channel switching, sell-up and displacement rates when evaluating the net revenue impact of withholding certain content from a given channel. Similarly, distributors, i.e. GDS's and TA's need to estimate these rates when evaluating the net revenue impact of having versus not having certain products in their channels.

In the following we present a logit model that can be used to estimate demand diversion rates. Without loss of generality, let's assume that we have a universe consisting of two airlines[3] and one distributor. Each airline (A and B) sells two fare products, Low and High (represented respectively by L and H) both through the distributor and their own direct-channel at the same price. Let's also assume that the distributor has a different booking fee for each product. We will analyze the effects of airline A withholding their low fare product from the distributor while still making it available in its own direct channel.

If airline A withholds its low fare product from being distributed in the GDS, the following events may occur (possibly simultaneously):

- The demand D_L of the low fare product on the direct channel of airline A increases—some customers currently booking via the GDS will switch to the direct channel[4]—and the demand G_L of its low fare product in the GDS channel will be zero (i.e., the product is no longer available); this is referred to as *channel switch*
- The demand G_H of the high fare product of airline A in the GDS channel will increase—some customers will switch to the high product instead of switching channels; this is referred to as *sell-up*
- The demand of the low fare product of airline B in the GDS increases—some bookers will settle for this alternative instead of switching channels; this is referred to as *displacement*
- Marketing expenditures of airline A increase—additional costs of attracting new customers to its direct-channel
- Finally, the GDS's passive bookings increase—agencies that switch channels sometimes manually replicate the bookings in the GDS to facilitate their workflow and integration with back-office systems; GDS's typically charge a fee[5] to the airline

The diverted demand results from (1) channel switch, (2) sell-up and (3) displacement; assuming no leakage of demand, we can express the diverted demand in terms of fractional quantities that add up to one $\theta_1 + \theta_2 + \theta_3 = 1$ and $\theta_i \leq 1$,

---

[3] In the general case, we could have one airline versus all other airlines.

[4] In the general case, channel switch also means diversion toward other booking sources (e.g., GDS's or travel agencies with direct connections).

[5] Passive booking fees are typically less than 10% of regular booking fees.

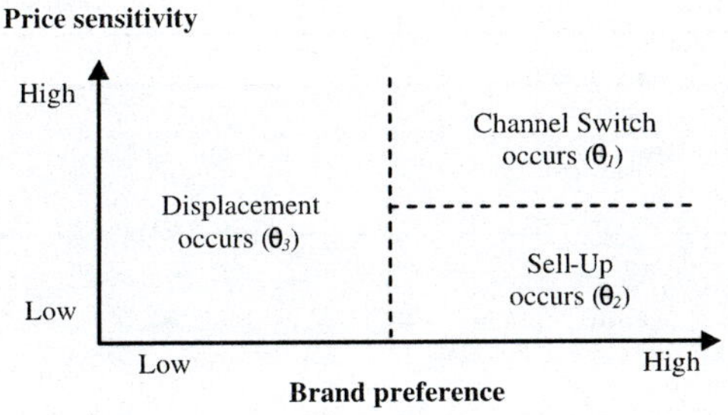

Fig. 4.15 Diversion due to differentiated content considering brand and price

where θ_1 is the fraction of demand shifted away from the channel, θ_2 the fraction of demand shifted to higher fares and θ_3 the fraction of demand shifted to other low fare products (e.g., from the competition). These relationships are portrayed in Fig. 4.15.

There are several methods that could be used to evaluate θ_1, θ_2 and θ_3, including customer surveys and case studies of similar actions by in the past. Another approach is logit based modeling of actual sale sessions to evaluate brand preference and price sensitivity, and to predict changes in customer behavior in case of product withdrawal. We will focus on this method in the next section.

4.4.1.1 Logit Estimation

Let's define a utility function of customer choice where brand and price are the determinant factors. This utility function[6] could be represented as $U_{ij} = \alpha_1 \text{ price}_i + \alpha_2 \text{ brand}_j + \varepsilon_{ij}$ where α_1 represents the price sensitivity, α_2 the brand preference and ε_{ij} represents unobserved characteristics of the utility value. Please note that this model assumes that customers are indifferent between the two available sales sessions. Using actual sales sessions where customers chose an airline (i.e., A or B) given a price level (i.e. P_L or P_H), this model can be calibrated using logistic regression. The probability of a choice is determined by $P(i,j) = \frac{e^{U_{ij}}}{\sum_{ij} e^{U_{ij}}}$. For example, when the choices available are (1) airline A at price P_L and (2) airline B at price P_H, the probability of choosing the first option is

$$P(A,L) = \frac{\alpha_1 p_L + \alpha_2(\text{brand} = \text{A})}{[\alpha_1 p_L + \alpha_2(\text{brand} = A)] + [\alpha_1 p_H + \alpha_2(\text{brand} = B)]}$$

[6] This function could be extended by capturing other variables such as schedule parameter (e.g., departure/arrival day and time, number of stops, flight time), and multiple carrier and product type; in that case a multinomial logit model could be used.

Table 4.6 The table below shows how θ_k ($k = 1, 2$ or 3) is derived from $P(i,j)$

	Case 1	Case 2	Case 3
Choice 1	Price = P_L Brand = A	Price = P_H Brand = A	Price = P_H Brand = A
Choice 2	Price = P_L Brand = B	Price = P_L Brand = B	Price = P_L Brand = B
	$\theta_1 = p(A, L)$	$\theta_2 = p(A, H)$	$\theta_3 = p(B, L)$

4.4.1.2 Airline Content Management

Without loss of generality, let's assume that we make the same modeling assumptions as above. In particular, we consider a universe consisting of two airlines[7] and one GDS. Each airline (A and B) sells two fare products, Low and High (represented respectively by L and H) both in the GDS and their own direct-channel at the same price. Let's also assume that the GDS has a different booking fee for each product.[8] The three distinct streams of diverted demand, denoted θ_1, θ_2 and θ_3 are as follow:

θ_1 probability that customer chooses airline A at price P_L in different-channel
θ_2 probability that customer chooses airline A at higher price P_H
θ_3 probability that customer chooses airline B at lower price P_L

Airline A before and after gross profit[9] can be expressed as follow:

$$\prod_{\text{before}} = (P_L - c_L)A_L + (P_H - c_H)A_H + P_L D_L + P_H D_H - m_A(D_L D_H)$$

$$\prod_{\text{after}} = -c'_L PB'_L + (P_H - c_H)A'_H + P_L D'_L + P_H D_H - m_A(D'_L D_H)$$

where

A_i Airline A's GDS bookings
D_i Airline A's direct-channel bookings
PB_i GDS Passive bookings
P_i Product price
c_i GDS fees
m_A Airline A's maketing cost
i product type L or H

---

[7] In the general case, we could have one airline versus all other airlines.

[8] GDS's tend to have a flat fee structure within product lines (air, hotel, car, …).

[9] Gross profit represents the revenue minus the cost of distributing the product.

Let D'_L, the new bookings on the direct-channel, be $D_L + \theta_1 \cdot A_L$ (the "before" bookings in the direct-channel plus a portion of bookings previously made in the GDS); similarly, we can express A'_H, the new bookings of the high fare product on the GDS, as $A_H + \theta_2 \cdot A_L$. θ_1 and θ_2 represent a fraction of the diverted demand; $0 \leq \theta_1 + \theta_2 \leq 1$ assuming no new demand is created, the extra demand in the direct-channel and the demand created for other products in the GDS is equal to the demand loss in the GDS. The passive bookings PB'_L can be expressed as a fraction of the direct-channel bookings, $\beta \cdot D'_L$ where $\beta \leq 1$. Let c'_L (passive booking fees) be a fraction of the regular fees, say c_L/K; it's worth noting that $P_i \gg m_A$ for any i and $P_i \gg c_j$ for any i-j pair[10]. We can also express both D_L and A_L as fractions of an overall market demand M_L ($M_L = D_L + A_L$), with D_L as $(1-\delta_L) \cdot M_L$ and A_L as $\delta_L \cdot M_L$ where δ_L is simply the GDS share of the total demand for airline A's low fare product ($0 \leq \delta_L \leq 1$).

The change in gross profit of airline 1 is:

$$\Delta \prod = \left\{ \left[\theta_2 P_H + \theta_1 P_L + c_L \left(1 + \frac{\beta}{K} \right) \right] - \left[\theta_1 \frac{c_L \beta}{K} + P_L + \theta_2 c_H + \theta_1 m_A \right] \right\} \delta_L M_L - \frac{c_L \beta}{K} M_L \tag{4.1}$$

We conclude from Eq. 2.1 that the change in profit is (a) positively correlated to θ_1 (measure of channel switch) since $P_L > \frac{c_L \beta}{K} + m_A$ and (b) positively correlated to θ_2 (measure of sell-up) since $P_H > C_H$. Clearly, as θ_1 and θ_2 increase, bookings in the direct-channel increase and sell-up bookings also increase.

$$\Delta \prod = \left\{ \left[\theta_2 P_H + \theta_1 P_L + c_L \left(1 + \frac{\beta}{K} \right) \right] - \left[\theta_1 \frac{c_L \beta}{K} + P_L + \theta_2 c_H + \theta_1 m_A \right] \right\} \delta_L M_L - \frac{c_L \beta}{K} M_L \tag{4.2}$$

By re-writing equation (2.1) as (2.2); it becomes clear that the market share δ_L does not determine future profitability (i.e., whether it makes economic sense to withhold content) but rather the amplitude of the change in profit; when δ_L is small (i.e., most bookings go through the direct-channel) there is little economic incentive to withhold the content; conversely, if δ_L is large (i.e., most bookings go the GDS) then the airline needs to estimate θ_1 and θ_2 before making a decision.

---

[10] Marketing cost per booking and booking fee are typically a small fraction of the fare amount. For a typical network carrier, these costs amount to 5–10% of fares, and are in decline.

4.4.1.3 Inventory Control and Distribution

For airlines, an alternative to estimating demand diversion (and eventually decide whether to withhold content) is to integrate the decision-making process in the revenue management process, and control the availability of its product not only by product type, but also by distribution channel. So, instead of considering gross revenue in the revenue-mix (or discount-allocation) model,[11] the airline could consider gross-profit as the objective function. The revenue-mix model could be re-written as

$$\begin{aligned} \text{Max} \quad & \sum_{i,k} (P_i - c_{ik}) \cdot E[\min(D_{ik}, X_{ik})] \\ \text{s.t.} \quad & \sum_{i,k} X_{ik} \leq \text{capacity}_{j,\forall j} \\ & X_{ik} \geq 0, \ \forall k \end{aligned} \tag{4.3}$$

where P_i is the price of product i, c_{ik} the cost of distributing product i in channel k, D_{ik} and x_{ik} are respectively the demand and allocation of seats to product i distributed in channel k, $capacity_j$ the capacity for each flight segment j where each (i, k) combination using flight segment j appears in the constraint, $E()$ the expected value,. Clearly, as (P_i-c_{ik}) takes on varying values equation 4.3 becomes more appropriate as a revenue-mix model. We should also note that forecasting becomes more difficult because of increasingly smaller demand values. However, that can be mitigated using approximations; for example, one may use the aggregated values D_i (the total demand of product i) and x_i (the total number of seats allocated to product i) in the revenue-mix model and use δ_k (the historical market share of channel k) to estimate the channel demand D_{ik} and the allocation x_{ik} using $D_{ik} = \delta_k D_i$ and $X_{ik} = \delta_k X_i$ where $0 \leq \delta_k \leq 1$ and $\Sigma \delta_k = 1$

4.4.1.4 GDS Content Management

For GDS's, the fare removal process discussed above has obviously different implications. The before/after gross profit of the GDS can be expressed as follow:

$$\prod_{\text{before}} = (c_L - m_G)A_L + (c_H - m_G)A_H + (c_L - m_G)B_L + (c_H - m_G)B_H$$

$$\prod_{\text{after}} = c_L' PB_L' + (c_H - m_G)A_H' + (c_L - m_G)B_L' + (c_H - m_G)B_H$$

where

A_i Airline A's GDS bookings
B_i Airline B's GDS bookings
PV_i Passive bookings

[11] See Chap. 3

c_i GDS fees
m_G GDS marketing cost
i product type L or H

As in the previous section, we can rewrite certain variables; A'_H can be expressed as $A_H + \theta_2 \cdot A_L$ and B'_L can be expressed as $B_L + \theta_3 \cdot A_L$. The passive bookings PB'_L can be again written as a fraction of bookings lost to the direct-channel, $\beta \cdot \theta_1 \cdot D'_L$ and c'_L (passive booking fees) as a fraction of the regular fees (c'_L/K).

The change in profit can be written as[12]:

$$\Delta \prod = A_L \left\{ \theta_2 (c_H - c_L) - \theta_1 \left[\left(1 - \frac{\beta}{K} \right) c_L - m_G \right] \right\} \tag{4.4}$$

Clearly, it is in the best interest of the GDS for θ_1 to be as small as possible; in fact if $c_H = c_L$[13] then its profitability depends solely on the value of θ_1 (i.e., zero or non-zero). This is an indication that a multi-level booking fee structure may be valuable to GDS's. Also, the profitability is directly proportional to A_L (airline A's low fare bookings on the GDS). The term $\left(1 - \frac{\beta}{K}\right) c_L - m_G$ can be best interpreted as "booking revenues minus incentives"; for carriers and/or markets where m_G is quite high (as a proportion of c_L) the GDS actually loses less in terms of gross profits.

4.4.1.5 The MFN Clause

The *Most Favored Nation* clause in GDS participation contracts requires an airline to provide the GDS its lowest available fare (and inventory) at all times. In exchange, the GDS reduces its booking fee. What is an MFN worth to a GDS? The alternative to an MFN is the airline withholding its low fare content. We could use the analysis above to answer this question.

The GDS reduces its low fare booking fee c_L in order to avoid channel switching (to the direct-channel or other GDS's) in case the airline decides to withhold its low fare product. The GDS's revenue at risk is $c_L \theta_1 A_L$ where c_L is the booking fee of the low fare, θ_1 the fraction of demand that could get diverted and A_L the volume of low fare booking. If the GDS agrees to a fee change of Δc_L, the change in its revenue will be $\Delta c_L A_L$. The neutral change of revenue is $\Delta c_L A_L = c_L \theta_1 A_L$ which implies that[14] $\Delta c_L = c_L \theta_1$.

---

[12] See Appendix for derivation.

[13] Usually, GDS's have the same fee for all products within the same category (air, car, hotel, etc....).

[14] In case the GDS charges the same fee (i.e., $c_L = c_H$), the revenue at risk is $\Delta c_L (A_L + A_H)$ implying that $\Delta c_L = c_L \theta_1 \frac{A_L}{A_L + A_H}$.

The booking fee change is directly proportional to c_L and θ_l. If $\Delta c_L = c_L\theta_1$, the GDS clearly benefits from the MFN clause by reducing its fee by Δc_L. Otherwise, the GDS is better off without; in which case, the loss of revenue for the GDS is $c_L\theta_1 A_L$.

4.4.2 Airline Channel Management

A carrier's long term distribution objectives can be expressed as volume targets by channel and market. In this context the carrier decides in which channels to participate and at what level. In this section we present a modeling approach that could be used to address this problem.

On the most abstract level each channel is characterized by variable and fixed distribution costs, expected demand, and expected ticket revenue. Variable distribution costs can differ by market. Expected demand and ticket revenue are usually given on the market level. If we assume perfect and fixed demand and yield forecasts and if we assume that there are no channel interactions then the channel problem can be formulated as a simple LP.

A very similar LP formulation has been studied in the context of airline Revenue Management (see Chap. 3). The RM model even offers extensions that deal with demand uncertainty. These models are useful in this scenario as well since channel and market-specific demand is available only as a forecast. The reader is referred to (see Chap. 3) for a review of related models.

Two important extensions of the basic model capture a carrier's ability to stimulate demand for a channel and to impact channel economics.

A carrier can increase demand in the direct-channel (own web site) by heavily promoting its web site. Demand stimulation can be modeled in our LP formulation by allowing the violation of the expected demand constraint in exchange for paying a penalty. In reality though, the relationship between demand and cost of stimulating the demand is likely to be nonlinear though. One can approximate the nonlinear relationship with a piecewise linear function or by modeling the relationship explicitly. The piecewise approximation has the advantage that the model can still be used using mixed-integer programing techniques or even a sub-gradient approach—assuming that the underlying function is concave.

Carriers often impact channel economics by withholding low fare content from certain channels. Withholding low fare content results in upsells (i.e. customers buying a more expensive fare) and demand loss (i.e. customers not buying at all or buying tickets on competing carriers). To model content withholding we increase demand for higher fare classes by market-and-channel specific upsell rates and zero out demand for the content that is withheld.

So far our modeling ignores channel interactions that occur in reality. If a carrier, for example, stimulates demand for a given channel then it will likely win bookings from competitors in the same channel and, potentially, stimulate new demand for the market. However, some of the incremental bookings will also come from other channels in which the carrier competes. The effect of bookings moving from one channel to another is referred to as channel shift. Probably the best example is the direct-channel. If the carrier starts promoting its web site then it will likely see a pick-up in direct bookings. These bookings are made by customers who would have otherwise booked on another carrier's web site, booked via a travel agency (online or offline) or would not have booked air travel at all. The customers who would have otherwise booked via a travel agency have been shifted to the direct-channel. Similarly, promoting a certain destination may generate not only incremental demand but also shift demand from another market. This phenomenon is observed primarily in leisure markets where travelers choose from a set of similar destinations. Carriers have to account for these effects when modeling the impact of channel stimulation on the bottom line. In particular, a carrier has to estimate a set of coefficients $\beta(src, m_{src}, tgt, m_{tgt})$ that represent how much demand is diverted from source src, market m_{src} to generate a single unit of demand for target tgt, market m_{tgt}. The carrier can estimate the coefficients using, for example, historic data, market experiments and/or customer choice modeling. Given the set of β-coefficients the channel interactions can be incorporated into our model.

A similar (though reverse) effect can be observed when a carrier chooses to withhold content from a channel. Let's consider the case where the carrier withholds the lowest fare class from a channel to increase its yield in the channel. Some of the customers who would have booked the lowest fare class will book competing carriers in the same channel, others will move to other channels. Yet others will upsell, i.e. book one of the carrier's more expensive products.

As with demand stimulation, the carrier needs to estimate the rate at which customers choose the different options given a unit increase in revenue in a given channel. In particular, the carrier needs to estimate a set of coefficients $\gamma(src, m_{src}, tgt, m_{tgt})$ that represent the change in demand for channel src, market m_{src} given a unit increase in yield in channel tgt, market m_{tgt}. Also upsell results, in general, in generating a disproportional large number of "cheap demand", i.e. demand that is associated with low fare classes at the target. The carrier will have to account for this by adjusting the expected yield in the destination channel. Combining all extensions we arrive at the model shown below:

$$\text{Max} \sum_{c \in C} \left(\Delta_c \delta_c + \sum_{m \in M} \left(\left(rev_{c,m} - \cos t_{c,m} \right) s_{c,m} + rev^{\text{inc}}_{c,m} s^{\text{yield}}_{c,m} \right) \right)$$

$$- \sum_{\text{s.t. } c \in C, m \in M} \left(d^{\text{stim}}_{c,m} \cos t^{\text{stim}}_{c,m} \right)$$

$$AS1 \leq Cap$$

$$s_{c,m} + (1 - u_{c,m})s_{c,m}^{\text{yield}} \leq E[Dmd_{c,m}] + d_{c,m}^{\text{stim}} - \sum_{tgt \in C, n \in M} \beta(c, m, tgt, n) d_{tgt,n}^{\text{stim}} + \sum_{tgt \in C, n \in M} \gamma(c, m, tgt, n) s_{tgt,n}^{\text{stim}}$$

$$\forall_c \in C, \quad m \in M$$

$$s_{c,m} \leq \delta_c \quad \forall c \in C, \quad m \in M$$

$$d_{c,m}^{\text{stim}} \leq UB_{c,m}^{\text{stim}}$$

$$s_{c,m}^{\text{yield}} \leq UB_{c,m}^{\text{yield}}$$

where

δ_c	is the binary variable that is 1 if the carrier participates in channel c
$s_{c,m}$	is equal to the number of ticket sales through channel c in market m
$E[Dmd_{c,m}]$	is the expected demand for channel c, market m
K	is a large number
A	is a matrix that maps channel and market-specific demand to itineraries. Each entry $a_{l,m}$ represents the rate at which market m uses capacity of leg l. Please note that this assumes that all channels sell different itineraries at the same rate. One can model different itinerary utilization by replacing the expression *AS1* with $\sum_c A_c S_{c,*}$, where A_c represents consumption rates specific to channel c
Cap	is the vector of leg capacities
C, M	represent the sets of all channels and markets, respectively
Δ_c	represents the net of fixed revenues and costs of participating in channel c,
$d_{c,m}^{\text{stim}}$	is the amount of demand stimulated for market m in channel c
$\cos t_{c,m}^{\text{stim}}$	is the cost of stimulating one additional unit of demand for market m in channel c
$UB_{c,m}^{\text{stim}}$	is the most additional demand that can be stimulated for market m in channel c
$rev_{c,m}^{\text{inc}}$	is the average incremental revenue that is collected per upsell in market m and channel c
$s_{c,m}^{\text{yield}}$	are the number of upsells in market m and channel c
$u_{c,m}$	is the upsell rate, i.e. the ratio of demand that buys a higher yielding ticket when the lower fare class is closed stimulated in market m and channel c
$UB_{c,m}^{\text{yield}}$	is the upper bound on successful upsells in market m and channel c

The biggest challenge in solving the problem at hand lies in the accurate estimation of the coefficients that capture channel interdependencies. Shopping and booking data and customer choice modeling are best suited to develop accurate estimates. Carriers also have to account for the fact that coefficients change as channels become more or less important over time and as other distribution models arise.

Another type of uncertainty that we have not touched on yet concerns the distribution costs. This uncertainty is due to the uncertainty of future contracts with distributors and other unknowns. To account for this uncertainty carriers can employ, for example, a scenario based stochastic modeling approach.

As mentioned in the previous chapter, a strategic alliance can be viewed as a distribution channel of its own. The revenue collected from the channel is equal to the portion the carrier receives as either marketing or operating carrier, depending on its role. We can estimate the impact of a code share on demand by first computing for the base case, i.e. without code shares:

$$d^0_{c,m} = \sum_{i \in I} d^0_{c,m}(i)$$

Then we estimate demand for the case with code shares:

$$d'_{c,m} = \sum_{i \in I} d'_{c,m}(i)$$

Next we compute how much of the new demand uses itineraries that are operated and marketed by the carrier:

$$d''_{c,m} = \sum_{i \in I} d''_{c,m}(i)$$

The differences in demand and revenue can be reflected in our channel model. First the demand for each channel c in which the code shares are sold is reduced from $d^0_{c,m}$ to $d''_{c,m}$. Second, a new channel c' is added for each channel through which the code share is offered. The market demands in the channel are given by $d_{c',m} = d'_{c,m} - d''_{c,m}$. The yield associated with a channel c' is defined as $r_{c',m} = (\omega\alpha_{\text{oper}} + (1-\omega)\alpha_{mkt})r_{c,m}$, where α_{oper} and α_{oper} are the portion of revenue collected on flights operated and marketed by the carrier, respectively, and where ω is the portion of code share demand that uses flights operated by the carrier.

Partnerships with non-air travel suppliers can be modeled in a similar way. Each of these partnerships is represented by two distribution channels, one representing the content sold through the partner's outlets and a second representing package sales through the carrier's direct channels. Both new channels have an impact on the carrier's existing direct and indirect channels. We leave it up to the reader to develop methods to estimate these impacts.

The reader may recall from the previous section that frequent Flyer programs also play a very important role in an airline's strategic distribution planning. It is difficult to model frequent flyer programs since their impact on sales and yields is complex. Frequent flyer programs generate incremental traffic and yields for the airlines by giving away low opportunity cost inventory. The correlation between incremental traffic and yield on the one hand and compensation (i.e. free tickets or upgrades) on the other hand depends on many factors, such as customer loyalty, a carrier's air and non-air supplier alliances and also competitors' strength in the market place. We encourage the reader to think about how frequent flyer programs could be reflected in an airline's channel selection model.

4.4.3 Travel Agency Valuation

In the competition for agency bookings and shrinking share of the GDS channel, the question about incentives and rebates a GDS should give a travel agency for a booking becomes crucial. GDS's have both fixed and variable costs, and cannot increase their capacity to service agencies in short notice or indefinitely. Hence, a GDS has to estimate the expected gross profit associated with signing a travel agency at a certain incentive or rebate schedule. If the GDS's capacity is tight then this profit has to be compared to profits associated with signing other potential agency subscribers.

The gross profit of a GDS can be written as $(b - r) \cdot B - m \cdot M - p \cdot P$, where b is the booking fee collected from airlines, r is the remittance to the agency (booking incentive), B is the number of bookings, m is the cost of system messages (availability, sell and cancelation messages sent internally or externally), M is the number of messages, p is the unit cost of processing and P is the amount of processing required. Without loss of generality M and P can be expressed as multiples of B.

4.4.3.1 A Graphical Approach to Agency Valuation

The following graph (Fig. 4.16) shows how agencies can be positioned on either side of a break-even line to assess their profitability. If we ignore fixed costs, the determination of the profitability of an agency depends on which side of the line it lays. By selecting the right booking incentive payment an unprofitable agency could be turned into a profitable one. This is shown in the graph by the distance between the point representing the agency and the break-even line; by reducing the incentive, the agency moves closer to the profitability line.

Instead of using the break-even line, one could also generate different lines to manage margins. To achieve a (pre-determined) margin, every agency in the mix needs to lie below the specified line, without cost or revenue consideration.

This graphical solution assumes there are no capacity constraints on the GDS side, i.e., any agency with any booking volume and shopping requirement can be satisfied. In reality, this is hardly the case. Contractual agreements between agencies and GDS's are typically 3–5 years long and capacity/technology decisions are made for a much longer time frame; changing (some of) the technology

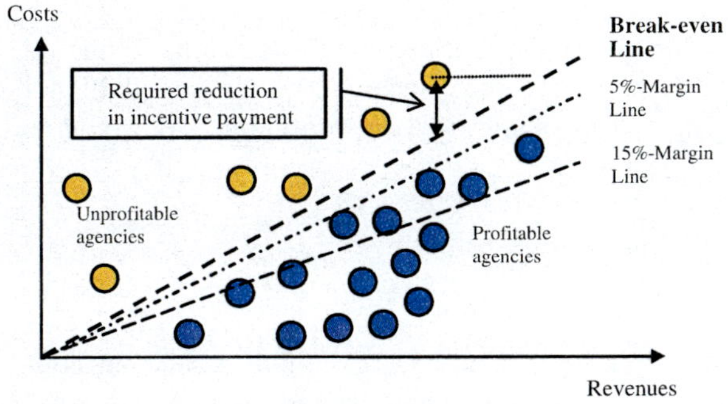

Fig. 4.16 Agency valuation breakeven curve

can be prohibitively expensive. We will describe a model that considers capacity constraints.

4.4.3.2 A Reduced-Cost Approach to Agency Valuation

Agencies make volume commitments and GDS's provide booking incentives as well a minimum service guarantee. For the GDS, this comes down to a demand vs. supply (or capacity) problem, which can be posed as follow:

$$\begin{aligned} \text{Maximize} \quad & \sum_i (b_i - r_i - m_i\mu_i - p_i\pi_i) \cdot X_i \\ \text{subject to} \quad & \sum_i \mu_i \cdot X_i \leq \text{message\_capacity} \\ & \sum_i \pi_i \cdot X_i \leq \text{processing\_capacity} \\ & 0 \leq X_i i, \ \forall_i \end{aligned}$$

where μ_i and π_i are respectively the multipliers to convert messages to bookings and processes to bookings for agency i, and x_i is the capacity allocation for agency i – b_i, r_i, B_i, m_i and p_i are defined as before but at the agency level.

By solving this allocation problem, one assigns the available capacities to each agency based on its profitability. An agency with low costs (i.e., low incentives or low capacity multipliers) will receive a higher allocation of capacity (up to its total demand); whereas an agency with high costs (i.e., high incentives or high capacity multipliers) will receive a lower allocation of capacity (possibly zero).

Suppose an agency gets no allocation (i.e., $x_i = 0$); a sensitivity analysis of its coefficient in the objective function $b_i - r_i - m_i\mu_i - p_i\pi_i$ will reveal the *minimum* change that will make it *profitable*. Since b, m_i, p_i, μ_i and π_i are all constant values, the change will have to come from r_i (agency incentives). Reducing the incentive by that amount increases its allocation to a non-zero value; further reductions will increase its share of the allotted capacity (until its total demand is fully satisfied).

This method also helps when making capacity purchase decisions; a by-product of the optimization is the shadow-price of the constraints. Using the shadow-price (or bid price) a GDS can decide when to purchase additional capacity and determine at forecast at the same time the profitability of doing so.

4.5 Future

In this section, we present five broad areas of research that could dramatically change the way airline tickets are bought and sold. These new areas include: dynamic pricing, risk inventory, auction-based revenue models, price unbundling and product bundling, and customer-direct distribution.

4.5.1 Dynamic Pricing

Traditional revenue management deals with demand uncertainty by using multiple, discrete price levels and inventory allocations; it provides a mechanism to alter prices based on remaining capacity, market conditions and anticipated demand. Prices of competing services are usually not explicitly considered; it is assumed that their effects are reflected in the overall market conditions.

Since the mid 1990s, price transparency has improved dramatically; travelers can instantly get price quotes from all airlines in matter of seconds. A low (or high) price level from any airline can significantly alter the demand for all airlines competing in a market. Consequently, it is becoming imperative for airlines to be able to: (1) gauge market conditions in (near) real time and (2) change price instantaneously.

GDS's (and travel agencies) could provide the facility where this new type of competition occurs by either: (1) communicating to all airlines the specific requirements of the buyer (e.g., preferred departure time and airline) prior to displaying the alternatives (schedule and price) or (2) changing prices based on general business policies. For example, airline A with a connecting service could choose to always price 10% lower than airline B in a given market when certain conditions are met (e.g., weekend travel) or offer a business class seat as an alternative to paying full fare in coach. Both approaches would necessarily be highly integrated with airline RM systems. In practice, only a few such tools have been implemented by the airlines, but dynamic pricing is gaining much attention in the industry and is a rapidly emerging field.

Two different types of dynamic pricing are described in Isler and D'Souza (2009). The first method discussed involves Internet channels whereby customers fill out a simple form that is used by an OR model to estimate that customer's price elasticity, and 'customized' prices are returned to the display. A second approach appropriate to GDS channels is described by the authors as a blend of both pricing and revenue management concepts. It uses traditional bid-price flight inventory controls coupled with a real-time evaluation of the fare less its dilution risk (to higher valued fare products) based on the estimated price elasticity; the authors indicate that a wide range of different factors can be used in assessing the current price elasticity (including competitive conditions). This second approach has been implemented by Swiss International Air Lines.

A third type of dynamic pricing involving real-time computation of the optimal price using current competitive conditions and customer choice models is described in a patent application by Ratliff (2003); an improved approach and a more comprehensive theoretical foundation are provided in a working paper by Gallego and Hu (2007). In this approach to airline dynamic pricing, the maximum expected revenue methodology described in Sect. 4.3.1 is applied given current competitive low fare search results; essentially a line search optimization is performed to find the marketplace maximum revenue for the airline whose fares are being optimized (while holding competitor fare levels constant). In addition to this marketplace revenue maximizing price, a separate, second price value must be calculated to determine the point at which the resulting demand matches the available flight inventory capacity (Gallego and van Ryzin 1994); loosely speaking, this price equates to the flight marginal revenues (i.e. sum of bid prices) which are, in turn, used to determine fare availability. To find the optimal price point, the marketplace maximum revenue price is compared against the sum of the flight bid-price control values on the flights considered, and the optimal price = maximum (marketplace revenue maximizing price, sum of flight bid prices). A proof of this optimality condition is offered in Gallego and Hu (2007). The basic idea behind this formula is that the airline should determine its optimal available price(s) explicitly considering current competitive offerings in terms of both price and schedule attractiveness, but the airline should never price below its flight marginal revenue (i.e. sum of bid prices). A working version of this type of dynamic pricing was built by two of the authors and deployed on a test basis.

For airlines, there are numerous benefits to these types of dynamic pricing system compared to traditional airline pricing and distribution. It provides more flexibility in market segmentation through less rigid segment fencing and the ability to charge different prices. More importantly, it allows for below-the-radar pricing changes without attracting competitive responses and the ability to stimulate specific market segments without increasing the risk of down-selling other segments.

However, many challenges need to be addressed; for example, maintaining price parity between channels where dynamic pricing takes place and their direct-channel where no competitive product is sold. Also, antitrust and price discrimination issues need to be considered; although existing laws allow current practices whereby two customers pay different fares depending on when they book, it is not entirely clear that a pure form of dynamic pricing will be permissible. Furthermore, although many practitioners favor the use of straight price matching of competitors to maintain marketplace competitiveness, simulation studies by d'Huart and Belobaba (2009) showed that application of such simple rules ignoring RM considerations often result in worsened revenue outcomes. Also, although (Gallego and Hu 2007) indicate otherwise, there are still widespread concerns among airline pricing analysts that the benefits of dynamic pricing to airlines will be marginal at best—i.e. if all airlines participate, it could turn into a zero-sum game. Further research in all of these areas is critical for the future success of dynamic pricing in the industry.

All things considered, the authors firmly believe there are many potential revenue benefits from the use of dynamic pricing technologies by airlines. However, the implementation of such tools should be carefully considered to minimize the potential risks involved, and special attention should be paid to post-implementation monitoring of such systems to ensure they are performing as expected.

4.5.2 *Risk Inventory Models*

Airlines usually shoulder the risk of unsold inventory; in this section we explore two models that allow for risk dispersion between the airline and other market players.

4.5.2.1 Options Market Model

Airfares generally consider airlines' operating costs. When operating costs increase significantly (fuel, labor, airplanes, etc..), airlines pass on the cost to consumers. For travel wholesalers, tour operators and corporate travel departments, sharp price increases could be devastating. To mitigate the risk of increasing airfare or reduction in capacity, these businesses could purchase options to hedge against rising airfares. Similarly, GDS, which usually rely on booking volume, could also hedge against higher fares that drive volumes down. On the opposite end, airlines can also hedge against decreasing airfares and/or lower travel volume.

Pre-paying air tickets at a fixed price is not a novel concept. Many airlines (or alliances) offer discounted miles or coupon-based travel when buying in advance. For example, American Airlines' AAirpass program offers fixed-rate pre-paid miles for travel since 1981 (source at end). Similarly, many airlines allow foreign travelers to purchase flight coupons at a reduced rate to visit any destination within a country or region. However, these pre-paid services are designed by the airlines, and the prices are set by them.

To be viable, an option market for airfares would have to be outside of the control of airlines, and allow all constituents of the industry to freely purchase and trade options in an open market. The uncertain nature of both the supply (airline capacity) and demand (passenger travel), the lack of total price transparency and relative small number of airlines in any given market makes this model difficult to implement.

On a more limited scale, Continental Airlines has implemented a consumer-direct version of option pricing when booking through its website (TravelPulse 2010). This process allows customers to hold an airfare at a specified price for some period of time prior to actually making the booking. Although there is a small service fee involved, option pricing is attractive to some customers because it mitigates the risk of unfavorable price changes.

4.5.2.2 Seat Allotment Model

An agency with significant buying power (agency/buying consortium, travel wholesalers, tour operators or corporate travel departments) can contract and hold a block of seats—referred to as allotments—at a preferential price. Allotments are contracted for a specific travel season and unsold seats generally released back to the airline at a pre-determined time (usually days or weeks) before departure, with or without penalty. Airlines sells seats this way for specific destinations (vacation resorts, tour destinations, etc.); extending it to every market is more challenging because it could dilute airlines' control of prices.

The volume, price, season, release date and penalty of the allotments are negotiated between the airline and the agency. For airlines, an allotment contract reduces the risk of unsold seats; for agencies, it provides a price advantage over their competition. To avoid competition with its own product, airlines typically require the seat to be bundled with other travel products (e.g. a cruise or hotel room).

To maximize revenue (and minimize risk) the airline must set its price, volume and penalty level given the capacity, forecast and historical performance of an agency. For agencies, market prices, demand forecast and airline brand are key in determining its strategy.

4.5.2.3 Auction-Based Distribution

When third-parties sell a ticket, airlines pay a booking fee. For GDS-based agencies, the airline pays the GDS, which may remit a portion to the agency. For GDS-bypass agencies, the airline pays the agency directly. Booking fees are a pay-per-sell paradigm not dissimilar to the pay-per-click model popularized by Internet search engines. Booking fees are flat and are either imposed by the GDS or negotiated with the airlines. During the shopping process, products from all airlines are displayed using a schedule attribute (departure time, number of stops, etc..) or price, giving equal chance to all products.

The essence of the pay-per-sell mechanism is neutrality; the GDS does not necessarily care about which airline is booked; there are no incentives for the GDS to do otherwise. And the value of the booking is not reflected; an airline with lots empty seats may value the booking more than another. Allowing airlines to raise/

lower the booking fee based on their demand and supply would result in a more efficient marketplace.

An auction-based, pay-per-sell business model may be the solution. It eliminates the lengthy (and sometimes difficult) negotiations that airlines and GDS's conduct every few years; the likelihood that either party gets stuck in non-advantageous contracts is eliminated. Airlines would bid for placement based on the importance of a particular city-pair and/or departure period. When flights are (expected to be) full airlines would bid smaller amounts or not all; when entering new markets, they would bid higher amounts. GDS's can have set minimum bids and tailor their response screen to show more itineraries from airlines with high bid and fewer (or no) itineraries from low bid airlines.

The cost of distribution becomes an integral part of the value the airline places of the booking. Airlines would automate the bidding process by linking their revenue management objectives to the new auction-based pay-per-sell fee; empty flights (low load factors or inventory bid price) nearing their departure date will result in higher pay-per-sell bids, whereas full flights would bid the minimum acceptable GDS fee (or none). The change will be significant for GDS's and airlines; it will require GDS's to maintain an up-to-date (standardized) view of the inventory of all airlines; and for airlines to share inventory information and create the internal mechanism that trigger low vs. high bids.

4.5.3 Price Unbundling and Product Bundling

Historically, airline tickets provided access not only to a seat, but also free transport of luggage (within quantity and/or weight limits), free meals, etc... Customers were segmented using aircraft compartment (e.g., first, business, economy) and fares rules and restrictions. In the past few years, we have witnessed both price unbundling and product bundling, creating a challenge in distribution and new segmentation mechanisms.

Price unbundling (also known as à-la-carte pricing) consists of reducing the product to its parts, which can be separately purchased by customers. Some parts are purchased before/after the flight (e.g., ground transport to/from airport, expedited/premium in-airport transfer, priority check-in, departure/arrival lounge access, priority boarding, luggage, no-fee changes) and others for in-flight consumption (e.g., advance seat selection, premium seating, meals and drinks, entertainment, telecommunication and Internet access, duty-free merchandizes). Low-cost carriers have pioneered price unbundling, but the concept has been adopted by full-service airlines as well. Table 4.7 below shows some examples of actual fees charged to passengers.

On the other hand, product bundling is when several items are grouped together to create a new product. The intention is to provide a clear differentiation, greater value and enhancing the airline's offering. The resulting simplification is supposed

Table 4.7 Airline fee schedule

		American Airlines (domestic service) ($)	Delta Airlines (domestic service) ($)	British Airways (international service) ($)	Singapore Airlines (international service) ($)
Baggage fees	First checked	25	25	0	0
	Second checked	35	35	60	0
	Third checked	100	125	140	N/A
Amenity fees	Internet access	4.95	4.95	N/A	N/A
	Meal	3.29	2	N/A	N/A
Seating fees	Seat selection	0	0	30	0
	Premium seating	N/A	N/A	N/A	50

Source Sabre Flight Fee Explorer (www.exploreflightfees.com)

Table 4.8 Air Canada product bundles

		Tango	Tango plus	Latitude	Latitude plus	Business
Frequent Flyer Program	Miles multiplier (%)	50	100	100	100	150
	$ spent online (mi/$)	0.33	0.5	1	1	1
Seat pre-selection		$15	Free	Free	Free	Free
Change fee		$40	$40	$0	$0	$0
	Last minute	$150	$50	$0	$0	$0
Refundable				Yes	Yes	Yes
Upgradable				Yes	Yes	Yes
Priority check-in/luggage					Yes	Yes
Lounge access	Air Canada Lounges				Yes	Yes
	Partner Airlines Lounges					Yes

Source AirCanada.com

to provide a clear segmentation of the market. Table 4.8 above shows the five product bundles used by Air Canada in the Canadian domestic market.

Both price unbundling and product bundling present several distribution challenges; the most notable is that shopping becomes more complex—product comparison becomes difficult. In the case of product bundling, packages from competing airlines are not always comparable; and in the case of price unbundling, the unbundled parts may be dissimilar or may be priced differently—customers often need to know the quantity and weight of their luggage months before traveling in order to get an apples-to-apples comparison.

The main challenge, however, remains with the airlines. Which ancillary products should be presented during shopping without overwhelming the customer? How should ancillary items be priced? How does customer valuation change? What's the profit-maximizing bundle in a given market? How are forecast revenue affected? What's the revenue maximizing inventory strategy? Should ancillary revenue be part of the inventory decision?

4.5.4 Customer-direct Distribution and Social Media

Airlines have historically relied on travel agents to deliver customers; as we have discussed earlier, Internet and rapid technology changes have provided airlines with more distribution options and outlets. The ability to deal directly with the customer during the research and shopping phase of travel—as opposed to after the booking has occurred—creates opportunities for airlines to generate more revenue and potentially lower costs.

Internet search tools (e.g., Google, Bing!) and metasearch sites (e.g., Kayak, Farechase, Mobissimo, etc…) deliver customers directly to airlines, allowing them to create tailored shopping experiences; airlines actively use these service providers. Another on-going effort is to use social media sites (e.g., Facebook, myspace, twitter, etc.) to facilitate shopping. Given the knowledge and information these sites dispose of, it is easy to deduce that airlines could create custom-made products to suit the needs of customers, even when privacy concerns are taken into consideration. These sites often possess profiles information (likes/dislikes, previous vacation/trip experience, family situation, mood, social habits, etc.) which could be used to generate custom-made trip offers for individuals or groups; for example, individuals with large families could receive vouchers for free bags or meals; people living in certain geographical areas could be incentivized to visit other areas (e.g., Florida residents to visit New York in Winter); these offers could be more targeted than fare sales available to the general market, be under the radar of the competition, and could be geared to generate a pre-defined number of sales lead.

For airlines, one of the difficulties is how to value customers delivered by Internet search tools or social media sites. Would distribution costs increase without a commensurate increase in revenue? Would the newfound customer information be exploitable? Would targeted offers work? Could customer choice models be used to better understand these shoppers?

Airlines use search engines to deliver customers, but to our knowledge, no airline has publicly embraced social media as an effective means of distribution; it is also clear that many are considering it. Delta Airlines and EasyJet are already bookable from Facebook; similar trends are likely to continue throughout the industry.

References

Adler K (2003) Controversy Surrounds 'Screen Scrapers'. New York Law Journal. http://www.nylj.com, July 24, 2003

AdWords (2009) Wikipedia, The free encyclopedia. http://en.wikipedia.org/w/index.php?title=AdWords&oldid=310724927. Accessed 29 Aug 2009

Air Canada (2011) http://www.aircanada.com/en/news/oneway/index.html

Airline (2009) (December 4). Wikipedia, The free encyclopedia. http://en.wikipedia.org/w/index.php?title=Airline&oldid=329744688. Accessed 8 Dec 2009

Airlines Reporting Corporation Web Site (2009) History section http://www.arccorp.com/aboutus/about_history_milestones.html

Associated Press (2005) Users confuse search results and ads. Survey: adults online naïve about how search engines work; MSNBC website; January 24, 2005; http://www.msnbc.msn.com/id/6861158/

ATPCO Website (2009) http://www.ATPCO.net/ATPCO/abouthis.shtml

Ben-Akiva M, Lerman S (1985) Discrete choice analysis: theory and application to travel demand. The MIT Press, Cambridge

Boyd A (1998) Airline Alliances; OR/MS, October 1998. http://www.lionhrtpub.com/orms/orms-10-98/boyd.html

Brunger W (2008a) The impact of the internet on airline fares: Part II: Understanding the 'Internet price effect'. Weatherhead School of Management, Case Western Reserve University. April 2008

Brunger W (2008b) The impact of the internet on airline fares: customer perspectives on the transition to internet distribution. J Revenue Pricing Manag 8(2–3):189

Cannizzaro M, Quinby D, Carroll B, Sileo L (2004) PhoCusWright's U.S. online travel overview—sixth edition. In: Burka K, Schetzina C (eds) PhoCusWright inc., Sherman, USA

Cisco Company Website (2009) X.25 overview. http://www.cisco.com/en/US/docs/internetworking/technology/handbook/X25.html

Computer Reservations System (2009) (November 24). Wikipedia, The free encyclopedia. http://en.wikipedia.org/w/index.php?title=Computer_reservations_system&oldid=327625895. Accessed Nov 30 2009

Consumer Reports Website (2004) http://www.consumerwebwatch.org/pdfs/global-concerns.pdf

Copeland D, Mason R, McKenney J (1995) Sabre: the development of information-based competence and execution of information-based competition. IEEE Ann Hist Comput 17(3):30–55

D'Huart O, Belobaba P (2009) Impacts of matching competitor RM seat availability; AGIFORS cargo and revenue management study group meeting. Amsterdam, The Netherlands

DeMarcken C (2003) Computational complexity of air travel planning. MIT Lecture Notes, Fall 2003. http://www.demarcken.org/carl/papers/ITA-software-travel-complexity/ITA-software-travel-complexity.pdf

Dunham MH (2002) Data mining: introductory and advanced topics. Prentice Hall, Englewood Cliffs

Expedia Inc (2009) (November 27). Wikipedia, The free encyclopedia. http://en.wikipedia.org/w/index.php?title=Expedia,_Inc.&oldid=328258433. Accessed Nov 30 2009

FrequentFlyer.com Web Site. History of Loyalty Programs; http://www.frequentflier.com/ffp-005.htm

Frequent-flyer Program (2009) (November 25) Wikipedia, The free encyclopedia. http://en.wikipedia.org/w/index.php?title=Frequent-flyer_program&oldid=327825373. Accessed Nov 30 2009

Galileo Website (2009) http://www.galileo.com/galileo/en-ch/about/History/

Gallego G, Hu M (2007) Dynamic pricing of perishable assets under competition. Technical report. Department of industrial engineering and operations research, Columbia University

Gallego G, van Ryzin G (1994) Optimal dynamic pricing of inventories with stochastic demand under finite time horizons. Manage Sci 40(8):999–1020

Gilliland Sam (2000) Strategy and the travel value chain. http://www.sabreairlinesolutions.com/news/pdfs/001002_Gilliland.pdf

Granados N, Gupta A, Kaufmann R (2004) Towards a theory of the move to unbiased electronic markets: lessons from the air travel industry. Information and Decision Sciences Department, University of Minnesota, September 30

Grossman D (2006) (April 17) Get plugged into meta-searching. USA Today online, Travel section. http://www.usatoday.com/travel/columnist/grossman/2006-04-14-grossman_x.htm

Grossman D (2010) (June 8) Airline fees a nightmare for corporate travel planners. USA Today online, Travel section. http://travel.usatoday.com/experts/grossman/2010-06-08-airline-fees-travel-planners_N.htm

Harteveldt H (2004) Why leisure travelers book at their favorite sites. Business view research document; Forrester Research; June 24, 2004

IATA Annual Report (2008) International air transport association (http://www.iata.org/), 64th Annual General Meeting, Istanbul, Turkey, June 2008, p 10

IATA Website (2009) http://www.iata.org/about/

IBM Website (2009) http://www-03.ibm.com/ibm/history/exhibits/vintage/vintage_4506VV3072.html

Isler K, D'Souza E (2009) GDS capabilities, OD control and dynamic pricing. J Revenue Pricing Manag 8(2–3):255–266

Karcher K (1996) The four global distributions systems in the travel and tourism industry. Electron Markets 6(2):20–24

Kleit A (2009) Competition without apology: market power and entry in the deregulated airline industry. CATO Institute website: http://www.cato.org/pubs/regulation/regv14n3/reg14n3-kleit.html

Lettovsky L (2006) The full cost of airline distribution. *Ascend magazine*, Issue 2006/2, December 2006, pp 70–73. http://www.sabreairlinesolutions.com/ascend/issues/ascend_2006_issue2.pdf

Lewis PH (1992) Booking with a computer. *The New York Times*; Nytimes.com; January 12, 1992. http://www.nytimes.com/1992/01/12/travel/booking-with-a-computer.html?sec=travel

Linden G, Smith B, York J (2003) Amazon.com recommendation: Item-to-item collaborative filtering. IEEE Internet Comput 7(1):76–80

Porter M (1985) Competitive advantage: Creating and sustaining superior performance. The Free Press, New York

Moe W (2006) An empirical two-stage choice model with decision rules applied to Internet Clickstream data. J Marketing Res 43(4):680–692

Moe WW, Fader PS (2004a) Capturing evolving visit behavior in Clickstream data. J Interactive Marketing 18(1):5–19

Moe WW, Fader PS (2004b) Dynamic conversion behavior at e-commerce sites. Manage Sci 50(3):326–335

Mullaney TJ (2004) Design your own discount getaway. Business week online. Personal business section; March 8

OAG Website (2009) http://www.oag.com/oagcorporate/history_of_OAG.html

Online Payment Services (2010) http://www.business.gov/manage/finances/online-payment-services.html

Orbitz (2009) (November 24) Wikipedia, The Free Encyclopedia. http://en.wikipedia.org/w/index.php?title=Orbitz&oldid=327759349. Accessed Nov 30 2009

Parker G (2007) (September 6) Revenue management forum website; http://rmforum.blogspot.com/2007/09/branded-fare-distribution.html

Parziale L, Britt D, Davis C, Forrester J, Liu W, Matthews C, Rosselot N (2006) TCP/IP tutorial and technical overview (8th edition). IBM Redbooks, IBM international technical support organization, December 2006

Peterson K (2009) Finding fees a tricky game even for seasoned fliers. Reuters, http://www.reuters.com/article/companyNews/idUKTRE51A7WW20090211. Accessed Feb 11

Phillips RL (2000) E-commerce: Revenue management at the speed of light. White paper, TALUS Solutions, Waterstone, Atlanta

Post D (2010) Variable opaque products in the airline industry: a tool to fill the gaps and increase revenues. J Revenue Pricing Manag 9(4):292–299

Priceline.com (2009) (November 20) Wikipedia, The free encyclopedia. http://en.wikipedia.org/w/index.php?title=Priceline.com&oldid=327009279. Accessed Nov 30 2009

Publication number: US 2003/0191725 A1; Filing date: Mar 26, 2003

Rao B, Smith B (2006) Decision support in online travel retailing. J Revenue Pricing Manag 5(1):72–80
Rao B, Zouaoui F (2009) Dynamic pricing of opaque airline tickets. J Revenue Pricing Manag 8(2–3):148–154
Rappa M (2006) Business models on the web. Managing the digital enterprise website. http://www.digitalenterprise.org/models/models.html
Ratliff R (2003) Approaches to availability processing: the increasing problem of shopping. AGIFORS revenue management study group meeting. Honolulu, Hawaii, June 2003
Ratliff R (2004) Competitive fare shopping for YM. AGIFORS revenue management study group meeting, Auckland, New Zealand, March 2004
Ratliff R, Walker A, Smith B, Brice T (2003) Availability based value creation method and system. U.S. patent application, U.S. Patent Office, Application number: 10/397708
Regan K Yahoo goes for overture in $1.6B deal. E-Commerce Times; July 14, 2003; http://www.ecommercetimes.com/story/21813.html
ReserVec (2009) (August 25) Wikipedia, The free encyclopedia. http://en.wikipedia.org/w/index.php?title=ReserVec&oldid=309954428. Accessed Aug 25 2009
RM Forum (2007) Branded fare distribution, Revenue Management Forum, September 6, 2007 http://rmforum.blogspot.com/2007/09/branded-fare-distribution.html
Robinson S (2002) Computer scientists find unexpected depths in airfare search problem. Published July 3, 2002. http://siam.org/pdf/news/456.pdf. Accessed June 12 2006
Sabre (2009) (computer system). (August 12) Wikipedia, The free encyclopedia. http://en.wikipedia.org/w/index.php?title=Sabre_(computer_system)&oldid=307605852. Accessed Aug 12 2009
Sabre Holdings Website (2009) http://www.sabre-holdings.com/aboutUs/history.html
Sabre Research (2006) Market driven retail engine—Air One pilot, Sabre Holdings Internal Report, March 2006
Sabre Travel Network website (2009) http://www.sabretravelnetwork.com/home/about/history.htm
Shelton D (1995) Alaska Airlines Sells Tickets Online. CNET News, December 28, 1995. http://news.cnet.com/Alaska-Airlines-Sells-Tickets-Online/2100-1023_3-200847.html
SITA Website (2009) http://www.sita.aero/content/facts-and-figures
Smith B, Guenther D, Rao B, Ratliff R (2001) E-commerce and operations research in airline planning, Marketing and distribution. Interfaces 31(2):37–55
Smith B, Darrow R, Elieson J, Guenther D, Rao B (2007) Travelocity becomes a travel retailer. Interfaces 37(1):68–81
Southwest Airlines website (2009) http://www.southwest.com/about_swa/airborne.html
Taylor B (2003) The great availability caching fallacy. ACP TPF Today J, September 2003
The Computerworld Honors Program (2009) http://www.cwhonors.org/CaseStudy/viewCaseStudy2009.asp?NominationID=273&Username=sAb273
The Economist (2005) Flying from the computer, September 29th 2005, Economist.com website (taken from the Economist print edition)
The New York Times (1981) 22 September 1981; Section D (Financial desk), Page 4, Column 5. http://www.nytimes.com/1981/09/22/business/company-news-american-to-offer-fixed-fare-passes.html
Travelocity (2009) (November 20) Wikipedia, The free encyclopedia. http://en.wikipedia.org/w/index.php?title=Travelocity&oldid=326988638. Accessed Nov 30 2009
Travelport Website (2008) http://www.travelport.com/sg/news_events/news_releases.aspx, April 14, 2008
TravelPulse Website (2010) http://www.travelpulse.com/continental-unveils-farelock-service-to-let-buyers-lock-ticket-prices.html, December 13, 2010
United States government role in civil aviation (2009) (October 19). Wikipedia, The free encyclopedia. http://en.wikipedia.org/w/index.php?title=United_States_government_role_in_civil_aviation&oldid=320859381. Accessed Nov 30 2009

U.S. Department of Transportation (2004) Part III, Computer reservations systems (CRS) regulation: final rule. Office of the secretary, 14 CFR Part 255, January 7, 2004

Vinod B (2007) Seat availability: alignment with the revenue management value proposition. J Revenue Pricing Manag December 2007, p 315–320

Vinod B (2009) Distribution and revenue management: origins and value proposition. J Revenue Pricing Manag 8(2–3):117–133

Wang T, Gal-Or E, Chatterjee R (2009) The name-your-own-price channel in the travel industry: an analytical exploration. Manage Sci 55(6):968–979

Westermann D (2005) Realtime dynamic pricing in an integrated revenue management and pricing environment: an approach to handling undifferentiated fare structures in low-fare markets. J Revenue Pricing Manag 4(4):389–405

Wright C, Groenevelt H, Shumsky R (2008) Dynamic revenue management in airline alliances. Working paper, Tuck School of Business at Dartmouth; Social Science Research Network Electronic Paper collection. http://ssrn.com/abstract=1105135. Accessed 11 Feb 2008

Chapter 5
Crew Management Information Systems

Diego Klabjan, Yu-Ching Lee, and Goran Stojković

5.1 Introduction

The airline industry is notorious for large swings between profitability and losses, following economic cycles closely. While economic cycles are getting less pronounced, on the contrary, profitability and losses of the airline industry are increasing in magnitude. These up-and-down swings were particularly observed in the last decade following the dot-com bubble. On top of the recession in 2000, the airlines were hit hard by the terrorist events of September 11. This was a defying moment in the industry.

Many large US carriers have drastically cut their costs through changes in business processes and renegotiating labor contracts (some through bankruptcy protection while others barely stayed out of it). Standard flight banks at hub airports were de-peaked into a more spread-out schedule. Pricing was substantially simplified to become more appealing to customers. Important cost reductions were obtained through labor renegotiations. Pilots are the highest paid unionized employees[1] and were the first target of such negotiations. Crew management

[1] http://www.bls.gov/oco/cg/cgs016.htm

D. Klabjan (✉)
Department of Industrial Engineering and Management Sciences,
Northwestern University, Evanston, IL, USA
e-mail: d-klabjan@northwestern.edu

Y.-C. Lee
Department of Industrial and Enterprise Systems Engineering,
University of Illinois at Urbana-Champaign, Urbana, IL, USA
e-mail: ylee77@uiuc.edu

G. Stojković
Jeppesen, Englewood, CO, USA
e-mail: Goran.Stojkovic@jeppesen.com

C. Barnhart and B. C. Smith, LLC (eds.), *Quantitative Problem Solving Methods in the Airline Industry*, International Series in Operations Research & Management Science 169,
DOI: 10.1007/978-1-4614-1608-1_5,

information systems, including the underlying decision support, played an important role since they provide both parties with an accurate and prompt assessment of negotiated rules. Such systems became indispensable. In the past they were mostly used for traditional planning, but today they bear a much higher importance. Unions are becoming a more respected partner in an airline's decision making process. At selected airlines the pilot unions are even involved in the selection process of components of crew management information systems.[2]

Crew cost remains the second largest expenditure behind fuel cost and thus the airlines spend significant resources for maintaining low crew cost. Manual decision making in the crew scheduling process such as constructing pairings or crew itineraries and later in the timeline rosters or bidlines has been mostly replaced by sophisticated and automated computer-based management information systems. Due to the complex nature of the airline business, it is a very challenging and daunting task to produce satisfactory decision support systems for constructing pairings and other entities. Solution methodologies for these types of problems started several decades ago and they have been evolving since then. Academicians, software vendors, and research and development groups within airlines frequently produce new methodologies to satisfy the ever-growing needs. With network capacity expansions and fleet consolidations, the problem sizes keep increasing. On the other hand, to further decrease the overall cost, the airlines started having a more holistic view of their entire operations and processes. The interest for systems that consider the crew cost beyond its historical boundary of crew is growing, e.g., taking into consideration the impact of crew scheduling to aircraft routings and fleeting. Such calls also require new models and the underlying solution methodologies with a substantial crew component.

This chapter has two main sections, each drilling into the aforementioned aspects. We start by describing basic models and the underlying state-of-the-art algorithms and solution methodologies. In the second part, we discuss the underlying information systems and the software arena. We end the introduction with an overview of the various business processes employed by many airlines.

5.1.1 Business Processes

While every airline has its own processes and organizational names, most of them follow the depicted processes and terminology outlined in Fig. 5.1. Long-term fleet and manpower planning consists of making strategic decisions with respect to the number of aircraft and the fleet decomposition, and cockpit crew manpower planning.

---

[2] Panel at AGIFORS Crew Management Study Group, New York City (2003): Preferential Bidding Task Force; Brett Wilkie, Manager Crew Planning, IT, US Airways and Charles Mayer, ALPA.

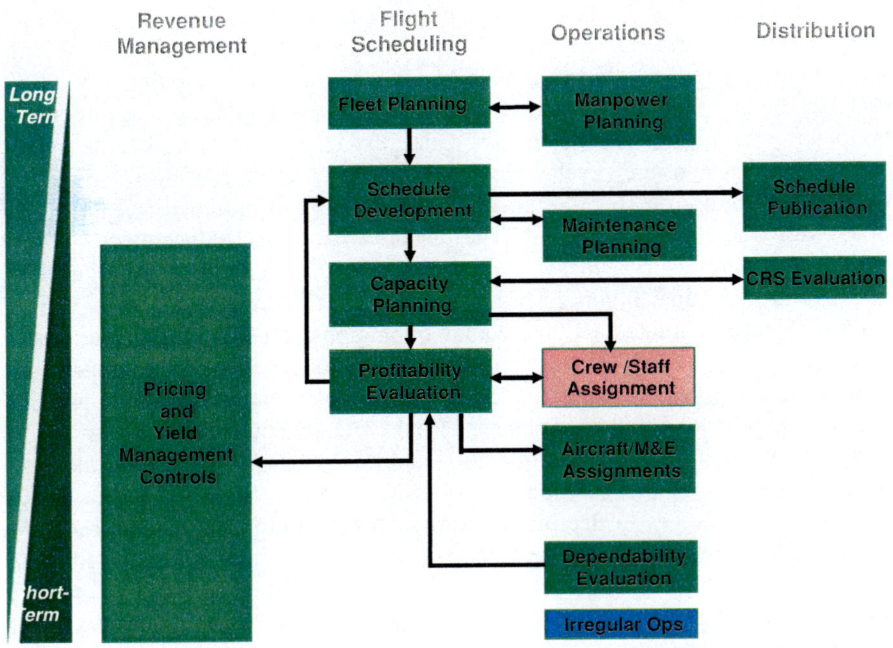

Fig. 5.1 Typical business processes (M&E = maintenance and engineering, CRS = customer reservation system)

In fleet planning, considerations such as the airline's mission (e.g., Southwest has a single fleet that allows relatively simple and efficient operations), aircraft utilization, route structure, cargo/passenger mix, etc., are taken into account. Long-term crew decisions revolve around the adequate staffing level in each fleet and requirements of crew training resources such as instructors, training facilities, and simulators.

The schedule development phase typically starts 12 months before the day of operations and lasts up to 9 months. In the first phase the airline establishes the service plan, which is the set of flights to operate in a given market. The service plan is either daily for domestic operations or weekly for international, long-haul services. The marketing group considers several factors such as traffic forecasts, status of competing carriers, internal resources, and marketing initiatives. Marketing initiatives are approved by the upper management and involve decisions such as entering a new market. The designed service plan typically does not divert substantially from the current schedule. Following the service plan, the scheduling group generates a detailed flight schedule, i.e., a flight departure and arrival time. The flight schedule has to obey a set of operating constraints, e.g., maintenance planning, and given generic resources such as the number of aircraft. The schedule is then published. Next is capacity planning or fleeting. In fleeting, an equipment type is assigned to each flight subject to available resources such as the number of aircraft. The goal of the fleet assignment model is to maximize profit. The schedule together with the seat capacity is then input to the computer reservation system. The produced fleeting solution is also evaluated with a profitability evaluation model and potential improvements are fed back to the schedule development group for possible minor adjustments. Once the equipment types are assigned, aircraft routing and crew scheduling follow. In aircraft routing, also called maintenance routing, a specific tail number or aircraft is assigned to each flight subject to maintenance constraints. The objectives are usually incentives such as through revenue and robustness. The goal of crew scheduling is to assign crew members to individual flights in order to minimize the crew cost and maximize various objectives related to contractual obligations, quality of life, and crew satisfaction. Crew schedules have to satisfy complex regulatory and contractual rules. Potentially, crew planners detect unfavorable connections and provide feedback to schedule and fleet planners. The crew scheduling process typically starts 3 months before the day of operation and it is constantly updated until a few weeks before the day of operations.

Only minor changes to fleeting, aircraft routes, and crew schedules are made during the last few weeks before the day of operations. If preferential bidding is used, approximately 1 month before the day of operations, crews bid for their monthly crew assignments and only minor changes such as two-way trip swaps are performed in the last few weeks. An alternative form of creating monthly work schedules of an individual crew member is rostering, where rosters are constructed simultaneously for all crew members by considering individual preferences, regulatory rules, and the direct cost.

Throughout the strategic planning processes, pricing and yield or revenue management are actively involved. In revenue management, the airline controls the seat inventory by adjusting fare prices, setting overbooking limits, and making decisions at any given time about selling particular fare classes on a given passenger itinerary.

The actual day of operations scheduling, also called execution management, consists of making final minor adjustments to the flight schedule (e.g., adjust arrival times based on the daily wind forecast), executing the preplanned schedule (e.g., file the flight plan) and rescheduling for irregular operations or disruption management. The latter is carried out by operations controllers, which are typically located in the airline operations control center. Most frequent sources of irregular operations are weather, unscheduled maintenance, congestion, crew unavailability, security problems, etc. Disruption management is composed of three processes. When an irregular operation occurs, first the aircraft are rerouted, which is called aircraft recovery. In this stage in addition to rerouting the aircraft, decisions on delaying and canceling flights are made. Next is the crew recovery process, where crews are assigned new crew itineraries. The controllers can use original, standby (crews available at the airport but not on active duty), and reserve crews (crews on-call but performing off-work activities such as being at home). At the end is the passenger reaccommodation process, where passengers are rerouted to alternative itineraries. Clearly the new schedule must conform to all regulatory

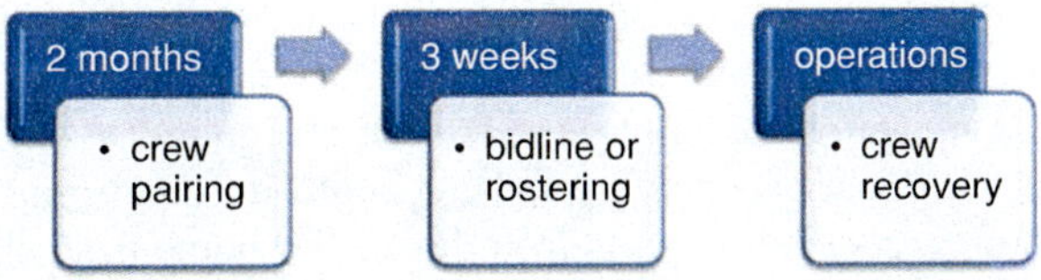

Fig. 5.2 Main crew processes

and contractual rules. While the airlines often impose more stringent rules in planning, in operations they typically use precise rules. Contractual rules for operations are usually different from those in planning.

The basic crew processes on the timeline are summarized in Fig. 5.2.

5.2 Optimization in Crew Management

Limited and costly crew resources make optimization within an airline's system crucial. From the crew standpoint, a fleet is a group of aircraft that can be operated by the same qualified cockpit crew. On short-haul flights, modern aircraft must typically carry two cockpit crew members: a captain and a first officer. Some older aircraft must also carry a flight engineer.

As pointed out earlier, in a typical planning process, the fleet or capacity assignment problem is performed before any crew assignments. Once a fleet is assigned to each flight—mainly with the objective of satisfying market demand—crews are then scheduled. The cockpit crew problems are fleet specific, i.e., a pilot can only be assigned to a certain fleet that he or she is qualified to fly. Compared to the cockpit crew, the cabin crews are more flexible because they are cross-fleet trained. Due to their large numbers and wider range of qualified flights, the cabin crew management problems are much more elaborate; however, their total cost represents a lower portion of the overall expenses.

The crew assignment problems have always defied the capabilities of our algorithmic knowledge and software science. They had and still do define the state of the art. The first difficulty results from the sheer size and nature of airline operations. A large carrier employs hundreds of costly crew members and it operates more than a thousand flights per day. The second challenge comes from complicated workload rules governing crew operations. Statutory rules are defined, for example, by the Federal Aviation Administration (FAA) in the US, to ensure flying safety. On top of these rules, an airline usually imposes its own rules for crew management and operations. Some of them result from crew union requirements, while others are imposed to provide additional operational safety and flexibility. These additional rules are generally more restricted, yet they can be violated in emergency or exceptional cases in operations (but not in tactical planning), while the statutory rules of the government agencies do not have such flexibilities. The workload rules are multidimensional. To explain the complexities of these rules, we start by introducing basic concepts of duty periods and pairings or crew itineraries.

A duty period is formed by a succession of flights and the required *sit times* between two consecutive flights. Duty periods form the basis of pairings, which are sequences of duty periods with a required overnight rest or *layover* between two consecutive duty periods. A multi-duty pairing typically spans several days. A crew member departs from his or her home base airport or domicile to start the pairing, and later comes back to the same home base after several successive duty periods to complete the pairing. Each pairing is followed in a monthly assignment by a rest period of several days.

Several workload rules pertain to each one of the following entities: flight, duty period, or pairing. For example, there are rules for the maximum numbers of duties in a pairing, the minimum time of rest between two consecutive duties, the maximum number of flying in a duty, and the maximum elapsed time of a duty and pairing. Some rules are expressed by the flying and rest time. For example, the complicated 8-in-24 rule imposed in the US states that whenever there are 8 continuous hours of flying time within a 24 h period, extra rest time called the compensatory rest is required.

The cost of a duty and pairing is traced to the contractual salary obligations. In the U.S., crew members are not paid by a fixed monthly salary, but instead, are paid with respect to the total flying time, total elapsed time, and the minimum guaranteed pay. The crew cost is usually measured in minutes. The following expression is often used to calculate the crew cost associated with duty period d:

$$b_d = \max\{f_d \cdot elapsed, fly, guarantee\},$$

where

b_d	the cost of duty period in d minutes,
f_d	a constant between 0 and 1 to give a discount to the total elapsed time of d,
elapsed	total duty elapsed time,
fly	total flying time, and
guarantee	a minimum guarantee per duty in minutes

Similarly, the crew cost associated with pairing p is usually computed as follows:

$$c_p = \max\left\{f_p \cdot TAFB, n \cdot guarantee, \sum_{d \in p} b_d\right\} + \sum_{\substack{\hat{d} \in p, \bar{d} \in p \\ \hat{d} \to \bar{d}}} e\left(\hat{d}, \bar{d}\right),$$

where

c_p	the cost of pairing p in minutes,
f_p	a constant between 0 and 1 to give a discount to the total elapsed time of pairing p,
TAFB	time-away-from-base (TAFB) time or the total elapsed time of the pairing,

n numbers of the duty periods in pairing p,

$d, \hat{d}, \bar{d}$ duty periods comprised in pairing p; notation $\hat{d} \rightarrow \bar{d}$ represents that duty period $\hat{d}$ is followed by duty period $\bar{d}$, and

$e(\hat{d}, \bar{d})$ extra cost for the layover between $\hat{d}$ and $\bar{d}$, e.g., per diem and lodging expenses.

An efficient crew management system should not only aim to fulfill the workload rules, but also assign crew members in an economical way to avoid unnecessary expenses. Because of the relatively high payment of crews, especially the cockpit crews, an efficient crew management system can significantly increase the profits of an airline. On the other hand, well-designed scheduling requirements and suitable workload can also increase the welfare of the employees.

5.2.1 Models

In this section, the prevailing mathematical models applied in crew management are surveyed. Based on their functional requirements within the business processes, the models are categorized into strategic, tactical, and operational or execution models. Strategic and tactical planning models, as the names suggest, are for long-to medium-term planning to optimize the use of all resources under a predicted most likely situation. They are usually employed from a few months to a few weeks before the day of the operation. On the other hand, operational planning is short-term planning mostly resulting from irregular events (the cancellation or delay of flights). Operational planning is to smooth out the schedule on the day of operation, to handle contingencies, and to recover the schedule. The main concept behind operational planning is not to additionally disrupt the original schedule too much, but to quickly return back to the original tactical plan. Optimization is particularly challenging in this case since the solutions need to be provided instantaneously, i.e., in real-time.

5.2.1.1 Strategic and Tactical Planning

The strategic and tactical planning regarding crew management comprises the crew pairing and crew assignment problems. In a typical crew management process, the pairing problem is solved first and then the assignment follows. The segregation of these two stages makes the problems easier to solve even though it might produce suboptimal solutions. Both crew pairing and crew assignment model formulations are of the set partitioning type. In these types of models, each element in the set must appear in one, and only one, subset. The model formulations are similar, yet the goals and the scope of problems do not overlap. We next introduce the crew pairing model and two types of crew assignment models: rostering and preferential bidding.

Crew Pairing

In the crew pairing problem, each flight has to be covered by a specific pairing. The process of selecting duty periods to form a pairing is referred to as pairing generation. The complicated workload rules and the pay structure yield many constraints and restrictions on pairings. Due to the excessive number of all pairings, it is impractical to enumerate all of them. One way to overcome this difficulty is to only enumerate selected pairings rather than all of them. The methodologies for generation are discussed in detail later.

The crew pairing problem differs slightly in the US from a typical problem in the rest of the world. Flights in the US are usually flown everyday, except on weekends or selected special days. Therefore, the pairing problem is solved first for the *daily* problem under the assumption that all flights are operated everyday. In the solution, some pairings contain flights, which are not operated on some days. These are referred to as *broken* pairings. To mend the broken pairings, the *weekly* exceptions problem is solved. In this step, the flights which are operated less than or equal to 4 days per week are covered together with all flights from broken pairings. At the end of this process, flights in the entire weekly cycle are assigned exactly to one pairing.

Outside of the US there is a larger variability of flights from day to day. As a result, the above two-stage process is not amenable. In such a situation, the entire weekly problem including all of the flights in a week needs to be solved at once.

In addition, airline companies tend to modify the flight schedule every season. Whenever a new flight schedule is constructed, a transition pairing problem between the two cycles of flight schedules is required. This problem is called the *dated* problem because each flight in such a problem is associated with a particular date.

Sometimes the weekly exceptions problem is infeasible. In this case it is allowed to send the crews as passengers on additional flights, usually to allow them to go back to their crew base for later dispatch. This action is referred to as *deadheading*.

The objective of the crew pairing problem is to minimize the pay-and-credit, which captures the cost above the cost accounting for the flying time. The model formulation is as follows:

$$\begin{aligned} \min \sum_{p \in P} c_p y_p & \\ \sum_{p: i \in p} y_p = 1 \quad & i \in F \\ y_p \in \{0,1\} \quad & p \in P \end{aligned} \tag{5.1}$$

where

P the set of all possible pairings,
F the set of all flights,
C_p the cost of pairing p, and
$i \in p$ represents that flight i is included in pairing p.

The decision variables are

$$y_P = \begin{cases} 1 & \text{if pairing } p \text{ is selected} \\ 0 & \text{otherwise} \end{cases}$$

The only constraints present impose that each flight must be covered by exactly one pairing. In practice, several additional constraints are included such as balancing pairings across crew bases. The model is also different for long-haul flights due to different crew requirements for certain flights, e.g., some flights require a single captain and two first officers, while others require a single captain and three first officers.

Rostering

The rostering problem is to assign individual crew members to a set of pairings. These pairings are obtained in the crew pairing phase and span a given time horizon, typically corresponding to a month. The output of rostering is a month of work or *roster* for each individual crew member. Each pairing in the time horizon has to be covered by a group of crew members as specified by the underlying fleet. Typical requirements and constraints with respect to a roster include the rest time interval between two consecutive pairings, the training and vacation periods, and the crew members' personal desires. Each roster is explicitly constructed for each individual crew member and it takes into account as many individual preferences as possible. Similar to pairings, regulatory bodies and union rules impose additional restrictions on a roster such as the total flying time within the time horizon.

In rostering, all rosters are constructed simultaneously. The following is a formulation of rostering, (Gamache and Soumis 1998):

$$\begin{aligned} \min \sum_{k\in K}\sum_{s\in S^k} c_s^k x_s^k & \\ \sum_{k\in K}\sum_{s\in S^k} \gamma_p^s x_s^k \geq n_p & \quad p \in P \\ \sum_{s\in S^k} x_s^k = 1 & \quad k \in K \\ x_s^k \in \{0,1\} & \quad s \in S^k, \quad k \in K \end{aligned} \tag{5.2}$$

where

K the set of all crew members,

S^k the set of rosters compatible with crew member k,

P the set of pairings (output from crew pairing optimization), training assignments, and open times to be covered,

c_s^k the cost of assigning roster to crew member k,

$$\gamma_p^s = \begin{cases} 1 & \text{if pairing } p \text{ is covered by roster } s \\ 0 & \text{otherwise,} \end{cases}$$

n_p the required numbers of crew members for pairing p,

and the decision variables read

$$x_s^k = \begin{cases} 1 & \text{if roster } s \text{ is assigned to crew member } k \\ 0 & \text{otherwise.} \end{cases}$$

The first constraint ensures adequate coverage for each specific pairing, while the second constraint ensures that each crew member is assigned to exactly one roster. The objective cost coefficients c_s^k capture the individual preferences, seniority, and possible direct or indirect costs incurred by the airline.

Preferential Bidding

An alternative business process to rostering is preferential bidding. In rostering all rosters are constructed at once and then handed over to the crew members. On the other hand, in preferential bidding, individual lines are constructed one by one again by taking personal preferences into account and guaranteeing that subsequent bidlines can be constructed (solely based on feasibility requirements). The process is usually driven by seniority, i.e., the most senior crew members are the first to specify their preferences and bids. Nowadays, the bidding process is usually carried out online. An individual crew member states his or her own preferences, such as the desire for specific pairings, specific time intervals in a day, particular regions, or expected workloads, through an interface of the preferential bidding system.

Let us assume there are m crew members labeled based on their seniority. The first crew member has the highest priority, while the last member has the lowest seniority. Given crew member k, the lines for all higher seniority crew members 1, ..., $k-1$ have already been constructed. The remaining available tasks (pairings, other open time, training) for crew member k are not the entire set of tasks but they correspond to the residual tasks. After all crew members receive specific lines, there might be some tasks that remain uncovered. We refer to the leftovers tasks, which include uncovered pairings, as the open time line, or simply open time. We also introduce a fictitious crew member labeled $m + 1$, which corresponds to covering open tasks, if such tasks exist. The open line is always assigned to crew member $m + 1$. The open time might actually require more than one crew member in which case several lines consisting of open time are being assigned to crew member $m + 1$. This is interpreted and implemented as each line of open time requiring a different crew member (not in the original pool of crew members). The main idea behind the model is to consider only the personal preferences of the current crew member k, while guaranteeing that all remaining tasks can be covered by yet to be assigned crew members $k+1, \ldots, m, m+1$. The following formulation for preferential bidding for crew member k is given by Gamache et al. (1998):

$$
\begin{aligned}
\max \sum_{s \in \Omega_k^k} & c_s^k y_s \\
\sum_{s \in \Omega^k} & a_{ps} y_s = 1 \quad p \in P^k \\
\sum_{s \in \Omega_e^k} & y_s = 1 \quad e = k, \quad k+1, \ldots, m \\
\sum_{s \in \Omega_{m+1}^k} & y_s \le N \\
\sum_{s \in \Omega_{m+1}^k} & d_s y_s \le H \\
& y_s \in \{0,1\} \quad s \in \Omega^k
\end{aligned}
\tag{5.3}
$$

where

Ω_e^k the set of residual tasks available for crew member e, $k \le e \le m+1$, $\Omega^k = \cup_{e=k}^{m+1} \Omega_e^k$,

P^k the set of residual tasks,

c_s^k the preference of line s when assigned to crew member k,

$$a_{ps} = \begin{cases} 1 & \text{if line } s \text{ includes pairing } p \\ 0 & \text{otherwise,} \end{cases}$$

d_s the duration of open line s,

N the maximum number of open time lines, and

H the maximum duration of open time lines,

and with decision variables

$$y_s = \begin{cases} 1 & \text{if line } s \text{ is selected} \\ 0 & \text{otherwise.} \end{cases}$$

This model is applied m times for every crew member following the seniority list and the corresponding residual set is updated each time. The objective of the residual problem is to maximize the score, which corresponds to maximizing the employee's satisfaction. The first constraint ensures that exactly one line covers each residual task. The second constraint ensures that every crew member receives exactly one line. The third and fourth constraints concern the open time lines by restricting the total number and durations of such lines, respectively. The purpose of these constraints is to limit the number of open tasks and open lines. At the very end, open lines are assigned to reserve crews.

In the US, yet another process, called bidding, is often employed. Under bidding, first all of the bidlines are constructed simultaneously, similar to rostering.

All of them are then presented to the crew members, who bid on them. The bidding process is based on seniority, i.e., the most senior crew member bids first.

Additional Considerations

Other than the high-level models given above, additional important considerations need to be taken into account in a decision support system.

Long-haul Versus Short-haul Networks

There are important differences in operations and the underlying network between long- and short-haul legs. Typically, international flights are long-haul, while domestic are short-haul legs. First, the network structure is different. In the U.S., domestic flights are usually operated under a *hub-and-spoke* network. Big market airports with high passenger flow act as *hubs*, while smaller airports serve as *spokes*. The hubs are connected with each other, while a typical spoke is connected with an adjacent hub. Most passenger itineraries connect at hubs and thus the network exhibits many connections. International, long-haul flights, on the other hand, usually operate back-and-forth. In such a network there are fewer possible crew (and passenger) connections, which has pros and cons. On the one hand, it implies fewer pairings and rosters or bidlines. On the other hand, due to fewer opportunities to create pairings, deadheads are required, leading at the same time to many more overall flights to consider.

Cockpit crews have different regulatory restrictions on long-haul flights. Crews sometimes are required to be upgraded on a long-haul flight. For example, an original two-member cockpit crew on a domestic flight is upgraded to a four-member team for safety reasons due to a much longer airborne time on long-haul flights. Crew members alternate on duty with one another during long flights.

Finally, the cost structure of a long-haul pairing is often dominated by the time-away-from-base time, which is considered to be the overall cost of a pairing.

Crew Splitting

The rostering model already exhibited typically applies separately for cockpit and cabin crews. For each of them, a separate optimization model is solved. Cockpit and cabin crews are usually treated separately due to the large discrepancy of the underlying requirements. Even pairing schedules are usually distinct. Cockpit crew members are rarely treated in these systems as individuals, but more as a team or crew. This convention is not only beneficial for cooperation in the cockpit, i.e., from the business processes standpoint, but it also reduces the size of the problem. Moreover, cockpit crews are grouped by the fleet family due to the training requirements, while cabin crew members are grouped by aircraft configurations, e.g., number of aisles, and geographical regions.

Balanced Crew Schedules

When constructing crew schedules, it is desirable that the workloads are balanced among the crew members. A situation where one crew member is assigned a long overtime duty, while the remaining crew members are assigned merely minimum guaranteed duties, is not desirable. To construct balanced schedules, we can either penalize in the objective function unbalanced assignment or impose additional balancing constraints.

Moreover, it is desirable for the crew schedules to also be equally balanced with respect to crew bases. The number of pairings originating at one crew base must be almost equal to the number of pairings originating at a different crew base. Sometimes the balance is not with respect to the number of pairings, but, e.g., the flying time. To reach this goal, crew base balancing constraints are often added to the crew pairing model. They impose that the total workload comprising of the pairings originating from each crew base must fall within an upper and lower bound.

Robustness

Besides the actual direct crew cost, robustness is gaining importance in modern crew software. The main scope of robustness is to consider the actual day of operations and the fact that disruptions occur, necessitating schedule changes. The aforementioned models assume that the schedules will be operated as planned, but this is seldom the case. The goal of robust crew schedules is to potentially produce suboptimal schedules with respect to the planned flight schedule, but the crew schedule would fare better in case of disruptions (Shebalov and Klabjan 2006; Yen and Birge 2006; Ehrgott and Ryan 2002). A non-robust schedule might yield highly increased crew cost at the end of a month mostly due to irregular operations. For example, a crew originally assigned to a cancelled flight in operations must be rescheduled. On the other hand, it is extremely undesirable to cancel a flight due to the crew's unavailability resulting from its late inbound flight.

It is also risky to push workload and safety rules to their absolute limits in planning because such a strategy would not allow any flexibility in case these bounds are reached in operations. The crew schedule would not have any buffers and thus it would be very fragile. A common strategy is to penalize in the models the patterns and bounds with a high probability of breaking during irregular operations. Other issues such as the allocation of reserve crews and the establishment of workload rules used in planning are also relevant to robustness and should be carefully considered.

5.2.1.2 Operational or Execution Planning

The main part of operational planning related to crew is the so-called recovery problem. Recovery is invoked in the case of unscheduled events such as a cancellation or delay of a flight in the case of inclement weather. In many situations, the underlying flight schedule is changed from the original one. The effects of a cancellation or delay are not only applicable to the underlying flight but can easily cascade throughout the flight network. For larger disruptions it is not uncommon that the snowball effects roll over to the operations during the next day. In general, it is not desirable to modify unaffected flights, but often it has to be done in order to obtain an acceptable and implementable low-cost solution. The general strategy of the recovery problem is to reschedule the part being directly affected by disruptions while maintaining the remaining portion of the original schedule as much as possible.

An important distinguishing feature of the recovery problem is the fact that it is an execution problem and thus solutions have to be produced in real time. As a result, different solution methodologies are usually employed. From the prospect of modeling, there are still rules confining the workload and connections. Labor contracts also dictate a different cost, which is usually the maximum of the planned pairing cost and the cost of the actual realized pairing. The constraints regarding safety or workload rules may be looser than those in strategic and tactical planning. Different airline companies may have different policies confronting an emergency situation. From the prospect of human resource, the reserve crews are available and should be included in the model.

The crew pay in the recovery problem will be recalculated for the new schedule based on the strategy mentioned before, which accounts for the flying time, elapsed time, etc. On the other hand, it requires that the crew payment for that month is at least the same as that of the original schedule. In addition, extra costs associated with cancelled flights including passengers' compensation or extra operational costs to return crew back to the crew bases, must be taken into account. One of the most important components in the recovery crew cost is to return all crews back on time as soon as possible. Several airlines drive their crew recovery decisions exclusively based on the back-on-time performance indicator while imposing the maximum number of reserved crews and deadheads.

The following is the pairing recovery model for fleet type e, (Lettovsky 1997; Lettovsky et al. 2000):

$$\begin{aligned}
\min \sum_{k\in K_e}\sum_{p\in P_k} c_p x_p + \sum_{l\in L_e} f_l y_l + \sum_{k\in K_e} q_k v_k & \\
\sum_{k\in K_e}\sum_{p\in P_k} \beta_{pl} x_p + y_l = 1 \qquad & l \in L_e \\
\sum_{p\in P_k} x_p + v_k = 1 \qquad & k \in K_e \\
\text{x } binary, y\, binary & \\
v \geq 0 &
\end{aligned} \tag{5.4}$$

where

K_e the set of crews available for fleet type e including reserve crews,

P_k the set of pairings that can be served by crew $k \in K_e$,

L_e the set of flights to be covered by crews available for fleet type e,

c_p the cost of pairing p,

q_k the cost of returning crew k to the crew base (or, in other words, not assigning a pairing to crew k),

f_l the cost of canceling flight l,

$$\beta_{pl} = \begin{cases} 1 & \text{if flight } l \text{ is covered by pairing } p \\ 0 & \text{otherwise.} \end{cases}$$

The decision variables are

$$x_p = \begin{cases} 1 & \text{if pairing } p \text{ is assigned to a crew} \\ 0 & \text{otherwise} \end{cases}$$

$$v_k = \begin{cases} 1 & \text{if no pairing is assigned to crew } k \\ 0 & \text{otherwise} \end{cases}$$

$$y_l = \begin{cases} 1 & \text{if flight } l \text{ is canceled} \\ 0 & \text{otherwise} \end{cases}$$

The objective of the recovery model is composed of three parts: the cost of rescheduling and reassigning crew, the cost of canceling a flight, and the cost of returning a crew back to the crew base. The first constraint ensures that each flight is either covered by a crew or cancelled. The second constraint captures the fact that crew cannot 'multi-task.' A crew is either reassigned to a single pairing or returned to the crew base for a possible later dispatch.

Other than the basic concepts described before, there are some other issues worth noting regarding the recovery problem.

The Original Schedule and 'Back-on-Time' Criteria

In the presented recovery model, a feasible solution comprises possible new pairings, i.e., pairings not originally in the monthly schedule. During recovery, it is desirable to retain the original schedule as much as possible. To take care of this point, in practice, a penalty measure determining the difference between the original and recovered schedules is often added to the objective function. A crew schedule that is less perturbed has a higher priority of being selected.

A deviation of this employed by some airlines is to recover the schedule as soon as possible. To this end, a new variable T is introduced capturing the last point in time where the crew schedule is changed. The objective function then simplifies to min T.

Evaluating a Solution

In modeling, the goal is to capture as many details as possible while remaining tractable. Nevertheless, it is virtually impossible to capture all possible events and circumstances such as the ever-changing operating environment. For this reason a human decision maker must interact with a recovery system. Due to this relation between the system and the decision makers, it is hard to evaluate the final solution. Unpredictability, such as the exact arrival of a snowstorm or its duration poses a major challenge in the evaluation effort. On a different front, estimation of cost coefficients such as the cost of cancelling a flight, typically involves capturing passenger goodwill, and the reputation of the carrier. In addition, the network effects are hard to isolate and quantify. The resulting solution is sensitive to such estimations and thus accurate cost estimations are difficult to obtain.

Crew Splitting

In practice, to gain additional flexibility, most airlines allow crews to be split in recovery. A crew arriving on a flight can be split so that a first officer continues on one flight while the captain resumes his or her duties on a different flight. The aforementioned model does not allow crew splitting since it deals with pairings, which are building blocks of crews. A similar model can be designed by capturing individual rosters of each crew member instead of pairings.

5.2.2 Solution Methodologies

Behind the hood of an optimization-based crew management system are mathematical optimization models with sophisticated solution methodologies. The crew management models yield instances with large numbers of variables and substantial complexity. Developing robust solution methodologies for these models is a continuing research topic. An acceptable solution methodology must excel in solution quality, accuracy, ease of use, robustness, and efficiency.

In the framework of crew management, rows and columns have associated meanings. The rows correspond to subjects, which in crew management are the flights, duty periods, or pairings. The subjects must be contained in containers, e.g., the flights, duty periods, or pairings must be covered. The columns are associated with feasible containers. The goal is to choose from all possible containers so that the selected containers include all of the subjects. We have to choose selected combinations of flights, which are duty periods, to cover all the flights; or we have to choose from combinations of duty periods, which are pairings, to cover all the duty periods. The process of finding possible containers or columns is called *column generation*.

The most popular algorithm to solve this problem consists of the following steps. First, we solve the *restricted master problem*, which considers only of a portion of the columns rather than all of them. Next, we solve the *subproblem*, which aims to identify or generate the columns that could improve the current solution. The basic idea applied in this process is to find the columns with the lowest reduced cost based on the dual variables (known also as shadow prices in economics). The columns obtained from the subproblem are added to the restricted master problem. The previous steps are repeated until we obtain a satisfactory solution or we have a provably optimal solution. We refer to this algorithm as *delayed column generation*.

It is worth noting that there is a difference between the so-called *master problem* and the restricted master problem. The master problem contains all possible variables, while the restricted master problem only contains a subset of them. Moreover, there are challenges in the step of subproblem solving because it involves sifting through a huge number of columns. Typically the subproblem is modeled as a network problem.

To illustrate the network concept, let us focus on pairing generation. The network is constructed in such a way that every possible pairing is a path in the network (but every path does not necessarily correspond to a pairing due to the various rules governing the feasibility of a pairing). Then the pairing generation problem becomes the shortest path problem, or a more complicated variant of the multi-constrained or multi-label shortest path problem. The nodes in the network could be the origins or destinations of flights or duty periods, called the flight or duty period network, respectively. In a flight network, the first family of arcs in the network links the departure node with the destination node of each flight. The origin node of a flight is also connected from an artificial source node of a crew base if the flight departs from the crew base. Similarly, the destination node of a flight is connected to an artificial sink node of a crew base if the flight arrives at the crew base. The remaining connecting arcs allow crew connections between two flights. They must capture either sit connections within a duty or rest connections between two consecutive duties in a pairing.

In a duty period network, flights are substituted by duty periods. Each duty period yields two nodes and the associated arc linking them. Connection arcs connect two duties that can be subsequent in a pairing.

There is a tradeoff between the two networks. The duty period network has many more nodes and arcs than the flight-based network. It could provide a severe strain on computing resources. The flight network can be easily represented even on low-end computers, yet it does not capture as many rules as the duty period network. The duty period network by construction embeds all of the rules directly pertaining to duties and thus it has a smaller number of paths that do not lead to a pairing.

In roster or bidline generation, nodes correspond to origin and destinations of pairings and arcs to legal rest periods between two consecutive pairings.

To capture legality requirements and for proper cost accounting, we introduce a label to track various resources at each node. The label, which is a multidimensional vector, contains attributes, such as the total elapsed time, the number of accumulated flights, or the cost that has been accumulated up to the node under consideration. The goal is to find the shortest (cheapest) path with all of the constraints satisfied. At an intermediate node, we usually keep track of several partial paths that connect to the node. Each path is associated with exactly one label. Technical details of the multi-label shortest path method are described in Sect. 5.2.2.2.

Similar to many other large-scale models, the methodologies used to solve the crew management optimization problems must be aligned with the development of computer architectures. For example, consider two different solution methodologies *A* and *B*. Methodology *A* takes 2 days to get an acceptable solution while methodology *B* requires 30 days to get an optimal solution using today's computing hardware. At this stage, methodology A is preferable for its efficiency. The developments in computer hardware such as multicore processing units may lead to methodology *B* becoming preferable over methodology *A*. This conceptual example illustrates that the resources at hand (here, the computing architecture technology) play an important role in determining the solution methodology.

On the other hand, different methodologies are more appropriate for solving different problems. Despite the superficial similarities of the presented models (they are all set covering problems with possible side constraints), each problem has its own characteristics stemming from column feasibility rules and the underlying scope. Before several models are to be solved, structures and characteristics of the models should be inspected and an adequate methodology must be carefully selected. In the following sections we present the methodologies and techniques used in solving large-scale crew related models. We also link them to the crew management problems outlined in the previous sections.

5.2.2.1 Basic Techniques in Large-scale Optimization

Solving mixed integer programs can be a daunting task, especially when the problems are large-scale. *Branch-and-price*, *Lagrangian decomposition*, and *Benders decomposition* are three prevailing techniques used in large-scale optimization. Each technique treats the problem from a different viewpoint. We introduce the basic ideas of these three techniques in this section, but explain the details pertaining to crew management in the subsequent sections.

Branch-and-Price

Integer programs are intriguing to solve since an explicit representation of the underlying feasibility set is impossible in a reasonable computational time. To circumvent this, we often relax the integrality restriction of selected variables and solve the resulting linear program instead, which is referred to as the linear programming *relaxation*. After the relaxation is solved, its feasible region is partitioned into two non-overlapping regions, or *branches*. The two branches are created based on a fractional variable in the optimal solution of the linear programming relaxation. This procedure, when applied repeatedly, is often denoted by a tree structure, where each node represents a linear programming relaxation and each child node represents the resulting branch. If an integer solution is obtained at a node, we record this solution, if it provides a better lower bound of the objective value, if we are maximizing, or upper bound, if we are minimizing.

The above method is called branch-and-bound. The disadvantage of the branch-and-bound algorithm, especially with a large-scale problem, is in the fact that often we have to visit a huge number of nodes and solve millions of linear programming relaxations. It is a method that, in the worst case, somehow enumerates all of the possible integer solutions and finds the best one among them. A variant of the branch-and-bound algorithm solves the linear programming relaxations using column generation at each node, which is preferable in a large-scale optimization. This method is called *branch-and-price* because we often refer to the step of solving the subproblem as *pricing* during the column generation process. The branch-and-price method focuses on efficient ways of solving the linear programming relaxations, while the next two methods take advantage of the special structure of the problem in a completely different way.

Lagrangian Decomposition

The Lagrangian decomposition method (Fisher 1985) decomposes the problem horizontally. All of the constraints are categorized into *hard* and *easy constraints*. Instead of solving the original problem with hard constraints, these constraints are relaxed and a penalty in the objective function is assigned, if they are violated. The problem we are now facing is called the Lagrangian relaxation problem. It has no hard constraints; but an additional penalty is introduced to the objective function, where each penalty coefficient or Lagrangian multiplier corresponds to a hard constraint. Lagrangian decomposition is an iterative procedure that aims to find a tighter bound at each iteration by adjusting the values of the multipliers. For example, if the original problem is a maximization problem, the objective value obtained from the relaxed problem is actually an upper bound of the true optimum. The Lagrangian dual problem is to find the lowest upper bound among all possible multipliers. In rare instances, the value of the Lagrangian dual equals the optimal value, but more often the Lagrangian dual bound is larger than the optimal value.

A subgradient method is often used to find the smallest upper bound of the relaxed problem. It is an iterative minimization problem over a piecewise linear convex function. The multipliers and step size are updated until a satisfactory upper bound is reached or a very small in norm descending direction is obtained. The encountered solutions may not be feasible and thus a separate heuristic needs to be designed that forces the solutions to satisfy the hard constraints (Lagrangian multipliers are used as guidance in this process). Lagrangian decomposition is especially powerful for the problems with several hard constraints linking together different parts of the model, while other constraints are relatively easy to capture.

Benders Decomposition

The Benders decomposition method (Benders 1962) decomposes the problem vertically. We categorized all of the variables into *linking* and *complicating variables*. This method takes advantage of the block diagonal structure of the problem. In the framework of Benders decomposition, the focus is on the so-called Benders master problem optimization. The Benders master problem is formed by projecting the complicated variables through the introduction of additional constraints in the original problem. At each iteration, first the Benders master problem is solved, which includes only the linking variables. Next we solve a problem, called the Benders subproblem, over the complicated variables while the linking variables are fixed. The optimal solution of the Benders subproblem is an upper bound, if the sense of the objective function is minimization, to the original problem. On the other hand, the solution of the Benders master problem is actually a lower bound. By adding a constraint resulting from the optimal solution of the Benders subproblem to the Benders master problem, the gap between the upper and lower bounds does not increase. The constraints added to the Benders master problem are called *Benders cuts*. The algorithm stops as soon as the gap becomes small enough.

A key requirement for using Benders decomposition is that the Benders subproblem has to be a linear program. The Benders decomposition methodology is especially powerful for mixed integer problems with integer linking variables.

5.2.2.2 Algorithms for Crew Management

In this section, we dive into the algorithmic techniques used in crew management. The already presented algorithms out-of-the-box yield unsatisfactory solutions and thus several custom-based strategies must be employed, e.g., branching decisions, pricing, etc. Elegant algorithms tailored to crew management focus mainly on two aspects. The first aspect is not to enumerate all the possible pairings, rosters, or bidlines, but merely enumerate those that have the potential to be in an optimal solution. The other aspect is to find an optimal (or near optimal) solution quickly. The techniques that follow focus on at least one of these two aspects.

Local Improvement Heuristics

We outline the local search heuristics only from a historical perspective. These were one of the first solution approaches employed by computer-based systems.

Nowadays they are outdated in comparison with more sophisticated algorithms with a more global view.

The algorithms require a starting feasible solution, which is then improved in subsequent iterations. In a typical local search algorithm for crew pairing optimization, a 'spot' corresponding to a subset of pairings of the whole problem at a given iteration is identified and the solution is improved spot by spot. For the selected spot of the incumbent solution, such as 10 pairings among the 100 pairings in the entire solution, we enumerate all possible pairings covering the flights contained in these 10 pairings. Next, a much smaller pairing problem is solved over newly generated pairings. The selected original pairings are replaced by the pairings in the solution of the smaller problem. If, for example, two spots are selected with non overlapping pairings, then two smaller problems can be solved concurrently in a given iteration.

A spot might be selected based on a time interval in the planning horizon, or based on stations, or even picked randomly. It is possible that the spot at one iteration overlaps with a subsequent spot in the next iteration. The local improvement heuristics have the advantage of solving much smaller integer programs even though they have to be solved many times.

Usually, it is not difficult to find a feasible initial solution (at least for large-scale instances). The pairing schedule operated in the previous season is an excellent reference point since the airlines do not change too many flights from season to season and year to year.

Branch-and-Price

Branch-and-price algorithms have a holistic view of the problem. They do not single out a smaller problem, but instead new columns are generated based on the entire network used in pricing.

To implement the branch-and-price algorithm for crew pairing, we first need to specify a suitable branching rule. A valid branching rule requires that the two branching decisions are mutually disjoint and compatible with the pricing algorithm. For example, standard variable dichotomy requires in one branch a pairing to be selected in the solution, while the other branch must disallow this pairing. This simple rule is intuitive but impractical because it is much more difficult to get solutions with the rules of "must not have" than the rules of "must have." In preventing a pairing to be selected, the pricing problem now becomes the problem of finding the second shortest path, which is a much more intriguing problem than the vanilla shortest path problem. A different branching rule (Ryan and Foster 1981) considers two flights simultaneously. On one branch, the two selected flights must be covered together by the same pairing and on the other branch, the two specific flights must be in two distinct pairings.

In the crew pairing application, we not only consider two specific flights, but also two consecutive flights in a pairing to make the strategy even more efficient. On one branch, the specific flights r and s must be covered consecutively by a pairing (the follow-on branch), and on the other branch, r and s must appear non-consecutively in pairings (they can either be in two distinct pairings or in the same pairing but not following each other). This strategy nicely balances the branch-and-price tree, yet it preserves the structure of the pricing algorithm. This branching rule is often called branch-on-follow-on.

In the follow-on branch, the two flights can be merged together and considered as a single entity. Within the flight network, this can be achieved by eliminating all connecting arcs from flight r except to flight s. In the duty period network, the same effect can be achieved by removing select duty periods and then fixing rest connections similarly as in the flight network. In terms of the constraint matrix, the two rows become identical and thus one of them can be eliminated. In the other branch we can take the same steps except that we remove the arc connecting the two flights. In this branch the constraint matrix remains intact, i.e., no reduction is possible.

If all flights depart at different times, the departure time becomes a distinguishing characteristic of flights. The branching rule based on departure times (Klabjan et al. 2001) is called the timeline branching. Based on this rule, we branch upon a specific flight r and a specific connection time t. On one branch, the flights immediately following r with the connection time of less than or equal to t are covered by the same pairing, and on the other branch, the flights immediately following r with the connection time longer than t are covered by the same pairing. For a flight schedule with equal departure times, we can either perturb the times slightly without harming the structure of the schedule or employ branch-on-follow-on when a 'fractional' r and t cannot be found. The timeline branching even more balanced the tree and it is also easily embeddable within the pricing network.

Besides creating balanced trees and compatibility with pricing, the branching decisions must guide the search of the tree toward finding a feasible integer solution resulting in an improved lower bound. A recent technique of selecting a good branching variable is called strong branching, and is today used in several commercial mathematical programming solvers. The basic idea is as follows. For a possible branching variable i and the two resulting branches, we perform k dual simplex iterations on each of the two branches, where k is a parameter. This yields two objective values f_i^0 and f_i^1 of the dual problems for the two branches. If f_i^0 and f_i^1 are large (for a maximization problem), the variable has a high priority for being selected for branching. We can use a combination of f_i^0 and f_i^1 as the score to evaluate a suitable variable. We perform the above steps on a set of candidate variables, and then choose the highest score variable within the set to branch on. The strong branching strategy should be combined with existing branching rules for crew pairing optimization (branch-on-follow-on or timeline branching).

Multi-label Shortest Path

We have already discussed that the subproblem or pricing is usually modeled as a network problem. Due to nonlinearities in the cost and restrictions on paths, the standard shortest path algorithms cannot be employed in pricing. Instead, a more complicated multi-label or resource constrained shortest path algorithm needs to be employed.

The multi-label shortest path algorithm relies on the network structure described at the beginning of Sect. 5.2. We demonstrate the basic principles by the example shown in Fig. 5.3. In the flight network case, nodes represent the origination/destination of flights and two types of arcs represent connections. The solid arcs represent the actual flights, while the dashed arcs represent ground connections, which is either the sit or layover connection between flights. The goal is to generate a pairing that begins and ends at the base station and has low reduced cost. To capture the dual prices, which correspond to flights, each flight arc in the network is assigned the matching dual value. The main idea is to proceed from one node to another, updating the labels along the way. Each node has several labels, each one associated with an underlying partial pairing. In our example, labels have four attributes, which are the reduced cost, total flying time, TAFB, and the number of flights. We have label (25, 3, 3, 1) at the destination node of flight *a* to represent the accumulated reduced cost of 25 units, total flying time of 3 units, TAFB of 3 units, and 1 operated flight. This label is obtained by traversing flight *a* from the label associated with the origin node of flight *a* (this label is not shown in the figure). After proceeding to the origin node of flight *b*, the accumulated reduced cost and the TAFB are updated and the label attached at this node becomes (28, 3, 9, 1). Such an update is based on the connection time between these two flights of 6 time units and the additional reduced cost contribution of 3 time units. Similarly, after proceeding to the destination node of flight *b*, the label is updated to (49, 6, 12, 2), which implies that the flying time of this flight is 3 time units. In addition, the crew cost-related contribution of this flight minus the associated dual price (the reduced cost contribution) of flight *b* equals 21 time units.

This procedure is then continued. The origin node of flight *d* yields label (69, 8, 19, 3), which is obtained after traversing 3 flights. Without additional enhancements, as presented, the method enumerates all possible pairings, which would result into computational intractability. If an attribute of a label exceeds one of the resource limits related to feasibility rules, then the label is discarded. In our example, if the maximum TAFB is 16 time units, then label (69, 8, 19, 3) is discarded. Some rules such as the non-replicated-flights-in-a-pairing rule, which is present in daily problems, are computationally challenging and should be checked only at the end after the pairing 'survived' all of the other feasibility rules.

An additional important enhancement is based on the notion of dominance. Upon visiting an intermediate node with several labels, one of them showing 'no competitiveness', we can discard this label since the path associated with it will never be in the optimal solution. The noncompetitive label is any label with all of its values dominated in every attribute by another label. In our example in Fig. 5.3, there are two possible legal paths at the origin of flight *d* and thus two labels. The first path consists of fights $a \to b \to c \to d$ and the second one of flights $e \to f \to g \to h \to d$. None of the two paths can be eliminated because one path has a longer flying time, a larger numbers of flights, and higher reduced cost, but the

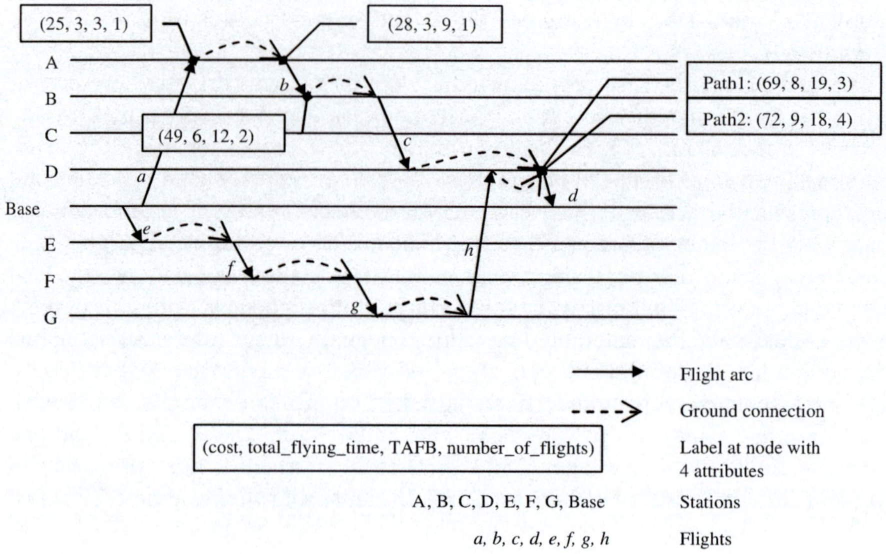

Fig. 5.3 Flight network and multi-label shortest path

other path has a longer TAFB. If label (72, 9, 18, 4) is changed to (72, 9, 20, 4), then it is dominated by the other one and thus it can be discarded. This is based on the observation that in this case every sequence of flights extending the bottom path can be appended to the top path to yield a feasible path with a lower reduced cost. It is important to note that the dominance strategy is not applicable to any rule. If there is a rule requiring that the total flying time in a pairing must exceed a certain value, then dominance with respect to the flying time is no longer applicable.

Our example relies on the flight network, but the application of multi-label shortest path in pricing is also applicable on the duty period network (see Barnhart et al. 1994; Lavoie et al. 1988; Vance et al. 1997).

Lagrangian-Based Methodology

While the crew problems are not really fitted for Lagrangian decomposition, they found success in some commercial implementations. For set partitioning problems usually all of the rows are relaxed and thus the selection of a Lagrangian multiplier plays an important role. The goal is to find multipliers u that lead to the unique solution of the Lagrangian relaxation problem, and this solution is also feasible to the original integer programming problem. We may take advantage of the special structure of the crew pairing problems with only 0 or 1 entries in the constraint matrix and the fact that the right-hand side is integral. With these characteristics, it is easy to obtain a solution to the Lagrangian relaxation problem. If the reduced cost of a variable is positive, the variable's value is 0; if the reduced cost is negative, its value is one. A difficulty arises in the case when the reduced cost is zero, and thus x can either be 0 or 1, and thus it creates ambiguity in finding a solution to the original integer programming problem. As a conclusion, the goal is to find multipliers u that lead to nonzero reduced costs.

We can apply a coordinate search algorithm and check feasibility of the original integer programming problem for every u iteratively. Unfortunately, a multiplier with the desired properties described above may not exist. One way to resolve this difficulty is by using the iterative approximation algorithm (Wedelin 1995). The main idea is to slightly perturb the coefficients of the objective function so that the optimality set does not substantially change and the uniqueness property holds. Instead of maintaining a single multiplier with each row, at each iteration, a constraint has two possible values u^+ and u^- corresponding to the multipliers. If the reduced cost of a variable in the iteration is positive, we increase u^+; if the reduced cost is negative, we decrease u^-. Values u^+ and u^- are controlled by two parameters κ and δ, where κ is the parameter controlling the level of distortion of cost coefficients, and δ ensures that the reduced cost is nonzero. If κ equals zero, there is no approximation or distortion.

To demonstrate the algorithm, let us assume that we are handling the minimizing problem of the form

$$\begin{aligned} \min \ & cx \\ Ax &= b \\ x_j &\in \{0,1\} \end{aligned} \tag{5.7}$$

where b is integral. The Lagrangian relaxation of this problem is

$$\min_{0 \leq x \leq 1} \ cx + u(b - x). \tag{5.8}$$

We define the reduced cost vector $\bar{c} = c - uA$, the index of rows by i, the index of columns by j, and an array s^i In array s^i each entry corresponds to a variable. The steps of the approximation algorithm are as follows, where we start with a random row.

Step 1: At current row (constraint) i, let

$$\begin{aligned} r^i &= \bar{c} - s^i, \\ r^+ &= \text{the } b_i'\text{th smallest element of } r^i, \\ r^- &= \text{the } (\mathrm{b}_i + 1)'\text{th smallest element of } r^i. \end{aligned}$$

Step 2: Choose $0 \leq \kappa \leq 1$ and δ based on empirical observations. Let

$$u_i^+ = u_i + \frac{\kappa}{1-\kappa}(r^+ - r^-) + \delta$$

$$u_i^- = u_i - \frac{\kappa}{1-\kappa}(r^+ - r^-) - \delta$$

Step 3: Update s^i by setting

$$s_j^i = \begin{cases} -u_i^- \text{ } if \text{ } r_j^i \le r^- \\ -u_i^+ \text{ } if \text{ } r_j^i \ge r^+. \end{cases}$$

Step 4: Update $\bar{c}$ by $\bar{c} = r^i + s^i$

Step 5: If all of the rows are visited, stop. Otherwise, go to the next row and repeat Steps 1 to 5.

These steps are repeated with κ gradually increasing from 0. Each time we loop through all of the rows, the corresponding x is computed based on the incumbent reduced cost $\bar{c}$. This corresponds to the reduced cost with respect to the Lagrangian multipliers $u_i = (r^+ + r^-)/2$. It is desirable that in each iteration the order of the rows is randomly selected.

In Table 5.1, computational results on real-world crew pairing problems are summarized, and a performance comparison between two algorithms is listed. From the experimental test, compared with the performance of the commercial solver CPLEX, the approximation Lagrangian algorithm performs very well in terms of the computational time on larger problem instances. Since crew management problems are largescale, this makes the algorithm suitable for such problems. On the other hand, the approximation algorithm provides an objective value that is often optimal or very close to being optimal. The approximation algorithm presented is the core of the optimization engine in the software provided by Jeppesen[3] (see Andersson et al. 1998).

5.2.2.3 Recent Advances

Three recent solution methodologies appearing in the last decade with substantial success in crew management are introduced in this section. They are the volume algorithm, primal–dual subproblem simplex, and constraint programming.

Volume Algorithm

The volume algorithm (Barahona and Anbil 2000, 2002), is an extension of the subgradient method. It can be used to solve a linear programming Lagrangian relaxation (5.8). In this section we focus on the case $b = 1$ since this is the usual set partitioning problem that is often confronted in the crew scheduling problems.

In the subgradient method, the current incumbent value is improved through an iterate search process. We update multiplier $\bar{u}$ in the direction of a subgradient, or based on the volume algorithm in the direction of a combination of a subgradient and the previous directions. The standard subgradient algorithm is simple,

[3] Carmen Systems was acquired in 2006 by Jeppesen.

Table 5.1 Computational results for crew pairing problems using CPLEX and the approximate Lagrangian algorithm (Wedelin 1995)

Problem	Size		CPLEX		Approximation algorithm	
	Rows	Columns	Obj.	Time	Obj.	Time
B727scratch	29	157	92,800	0.2 s	92,800	1.4 s
ALITALIA	118	1,165	5,017,500	1.0 s	5,017,500	11.0 s
A320	199	6,931	529,250	23.0 s	529,250	1 m 4 s
A320coc	235	18,753	565,000	2 m 12 s	565,000	5 m 2 s
SASjump	742	10,370	4,737,768	23 m	4,737,892	3 m 58 s
SASD9imp2	1,366	25,032	4,333,450	5 h 53 m	4,335,780	7 m 46 s
SASD9imp1	1,585	105,804	–	–	4,329,750	36 m 10 s

however, only the objective value and the value of the multipliers are obtained without providing a corresponding primal solution x. To obtain x, extra efforts are required, which are part of the volume algorithm. The main idea of the volume algorithm, which is given next, is to obtain an approximate primal value $\bar{x}$ by an updating process within the subgradient method. We denote the current objective value by z.

Step 1: Let $0<w<1$ be a parameter. Find an initial multiplier $\bar{u}$.

Solve (5.8) to obtain the current objective value $\bar{z}$ and solution $\bar{x}$.

Step 2: Make a step in the direction of a subgradient at $\bar{u}$.

Let the subgradient be $v^t = 1 - A\bar{x}$, the step size $s = \lambda \frac{UB - \bar{z}}{\|v^t\|^2}$, where $0 \leq \lambda \leq 2$, and UB is an upper bound of the optimal objective value.

Compute $u^t = \bar{u} + sv^t$.

Step 3: Solve (5.8) with u^t to obtain the new objective value z^t and solution x^t.

Step 4: Update the primal solution $\bar{x}$.

Compute $\bar{x} = \alpha x^t + (1 - \alpha)\bar{x}$, where α solves the one-dimensional problem:

$$\min_{\frac{w}{10} \leq \alpha \leq w} \|1 - A(\alpha x^t + (1 - \alpha)\bar{x})\|.$$

It is suggested to set $w = 0.1$, and if z^t is not increased by at least 0.1%, let $w = w/2$.

In addition, if $z^t > \bar{z}$ does not increase enough, let $\bar{u} = u^t$.

Step 5: If $\bar{z} = z^t$, let $t = t + 1$ and $\bar{z} = z^t$.

Step 6: Set $t = t + 1$ and go to step 2.

The volume algorithm provides an easier stopping criterion than the out-of-the-box subgradient methods whose stopping criterion is not robust and cumbersome to maintain. Since an approximate primal objective value is easily obtainable from the approximate primal solution $\bar{x}$, the gap between this approximate primal objective value and the lower bound $\bar{z}$ is used as a stopping criteria. If this gap is less than a given threshold, the procedure is stopped.

Table 5.2 Computational results of the volume algorithm, a combination of the volume algorithm and dual simplex, the pure dual simplex, and the primal–dual barrier algorithm on set partitioning problems (Barahona and Anbil 2000)

Problem	Size		Volume algorithm	Volume algorithm + Dual simplex	Dual simplex	Primal–dual barrier algorithm
	Rows	Columns	Time (s)	Time (s)	Time (s)	Time (s)
sp6	2,504	50,722	440	135	3,283	1,299
sp7	2,991	46,450	895	428	5,753	2,048

It is worth mentioning that if we unwind the approximate primal solution $\bar{x}$ in Step 4, it is a convex combination of $\{x^0, \ldots, x^t\}$if α is stationary:

$$\bar{x} = \alpha x^t + (1-\alpha)\alpha x^{t-1} + \cdots + (1-\alpha)^t x^0. \tag{5.9}$$

The most recent x^t has the largest contribution toward $\bar{x}$ while the previous vectors have lower contributions. To interpret the meaning of the primal solution $\bar{x}$ geometrically, we consider the space of $\bar{z}$ and $\bar{u}$. Given vector $(\bar{z}, \bar{u})$, the polytope in the space (z, u) is defined by all active constraints (at equality) with respect to $(\bar{z}, \bar{u})$ of $z + a_i u \leq b_i$ and $z \geq \bar{z}$. Vector $\bar{x}$ turns out to be the ratio of the volume between the active faces $z + a_i u = b_i$ and their projection at $z = \bar{z}$, to the total volume of the entire polytope in a neighborhood of $(\bar{z}, \bar{u})$– this statement is considered coordinate-wise with respect to coordinate i of $\bar{x}$. This result implies the name of the volume algorithm.

The volume algorithm has been used in a crew pairing system developed by IBM for US Airways and Southwest Airlines. In a crew scheduling set partitioning experiment with two instances, as Table 5.2 shows, the volume algorithm outperforms both the dual simplex and the primal–dual barrier algorithm. Moreover, a combination of the volume algorithm and dual simplex, which embeds the volume algorithm in column generation, reduces half of the computational time of the pure volume algorithm. In this combined algorithm, the dual values obtained from the volume algorithm are used to compute reduced costs and to select columns added to the restricted master problem, while dual simplex is applied to solve the restricted master problems.

Primal–dual Subproblem Simplex Method

The primal–dual algorithm dates back to 1956 (Dantzig et al. 1956). In the early days it was primarily used to exhibit polynomial algorithms for certain combinatorial problems. Later, the basic idea of the primal–dual algorithm was modified and embedded into many solution methodologies. The variant we present here is a primal–dual simplex algorithm combined with column generation. It is well fitted for solving large-scale linear programs.

To demonstrate the primal–dual simplex we assume the following primal–dual pair:

$$\text{(Primal problem)} \qquad \begin{array}{l} \min\ cx \\ Ax = b \\ x \geq 0 \end{array} \qquad\qquad \text{(Dual problem)} \qquad \begin{array}{l} \max\ \pi b \\ \pi A \leq c. \end{array}$$

In column generation for large-scale problems, only a subset of columns is considered at any point in time. Solution $\tilde{x}$ of the restricted master problem is always primal feasible (to obtain a complete primal solution x we simply append 0's to $\tilde{x}$). On the other hand, the corresponding dual solution ρ may not be feasible to the original dual problem. The key idea of the primal–dual simplex method (Hu and Johnson, 1999) is to improve the infeasible dual solution ρ by considering a convex combination of ρ and $\bar{\pi}$, where $\bar{\pi}$ is a given feasible dual solution to the original problem. We denote the new improved feasible dual solution as π', which is also used to select the desired columns that are included into the restricted master problem in the next iteration. The steps of the primal–dual simplex method are as follows:

Step 1: Obtain an initial feasible dual solution $\bar{\pi}$ of the original problem and initialize the restricted master problem with a subset of columns, potentially including artificial variables to make the restricted master problem feasible.

Step 2: Solve the restricted master problem with a subset of selected columns by any simplex-type linear programming algorithm.

Let $\tilde{x}, \rho$ be the corresponding optimal primal, dual solutions, respectively.

Step 3: If ρ is feasible to the dual problem, we stop. Pair $(\tilde{x}, \rho)$ is optimal.

Step 4: Let $\pi' = \theta\bar{\pi} + (1-\theta)\rho$, where $0 \leq \theta \leq 1$ is such that $\pi' A \leq c$ and $\pi' b$ is as large as possible. Value θ can be expressed as is expressed as

$$\theta = \max_{j, c_j^{\rho} < 0} \left\{ 0, \frac{-\bar{c}_j^{\rho}}{\bar{c}_j^{\pi} - \bar{c}_j^{\rho}} \right\}$$

where $\bar{c}_j^{\pi} = c_j - \pi A^j$, $\bar{c}_j^{\rho} = c_j - \rho A^j$ for column j and A^j is columnj of the constraint matrix. These are the underlying reduced costs.

Step 5: Construct the new restricted master problem. Preserve the basis columns in the restricted master problem but remove all other columns. Add the columns associated with the lowest reduced costs $c_j - \pi' A^j$ to the new restricted master problem.

Step 6: Set $\bar{\pi} = \pi'$ and go to step 2.

The primal–dual simplex method is efficient when the number of columns is much larger than the number of rows. In the computational experiment results summarized in Table 5.3, the primal–dual simplex reduces by half the time it takes using standard linear programming algorithms to solve linear programming relaxations of the crew pairing problems. If an incumbent primal solution is known in advance, the computation time for the primal–dual simplex can be further reduced albeit with some technical tricks. The primal–dual simplex method can

Table 5.3 Computational results for the primal–dual (P–D) simplex and primal simplex on linear programming relaxations of crew pairing (Hu and Johnson 1999)

Problem no.	No. of flights	No. of pairings	P–D simplex without an incumbent solution Time (s)	P–D simplex with an incumbent solution Time (s)	P–D simplex with an incumbent solution and improved initial dual Time (s)	Primal simplex with 1,000 columns added at a time Time (s)
1	61	3,754	0.8	0.9	0.7	–
2	144	48,574	7.6	7.4	5.85	12.6
3	174	209,935	30.0	25.7	20.3	36.0
4	202	2,335,782	155.6	–	–	–
5	382	2,000,000	58.7	39.5	–	65.0
6	930	2,000,000	277.0	194.0	–	–

also be parallelized by distributing the columns across various computational resources to solve several restricted master problems concurrently and dramatically reducing the overall execution time (see Klabjan et al. 2000). Another variant of primal–dual simplex (Barnes et al. 2002) uses a non-negative least squares problem to find a direction of the steepest ascent. This method shows its merit in extremely degenerate problems.

Constraint Programming-Based Column Generation

In a very dynamic world of crew management, the decision support components and the underlying algorithms must be very robust. Regulatory rules and, in particular, labor agreements often change. Each such change requires modifications to the business rules embedded in solution methodologies. The algorithms presented so far pose major challenges in this direction. It is very hard to design a very robust branch-and-price algorithm. Pricing is the 'sticking' point since it relies heavily on rules specifying legality of a pairing. Constraint programming is an interesting direction mostly focusing on ease of incorporating legality rules.

In a general mathematical program, the goal is usually to find an optimal solution within a feasible set. In constraint programming, the focus is not on optimality, but rather feasibility. In pricing, this amounts to switching from the goal of finding a column with the lowest reduced cost to the problem of finding a column with a reduced cost below a certain parameter. This can be cast as a feasibility problem: Does there exist a column with the reduced cost below the parameter? The main concept behind constraint programming is to find solutions by confining the relations between variables as long as the output solutions satisfy some desired properties. This type of a problem is also called a *constraint satisfaction problem* (CSP). A CSP is defined by a triple (X, D, C), where $x = \{x_1, x_2, \ldots, x_n\}$ is a set of variables, C is a set of constraints over variables in $x \in X$ each with a relation R, and $D = \{d(x_1), \ldots, d(x_n)\}$ is a set of values referred to as the *domain* that variables must belong to. A solution, denoted by v, which satisfies constraints in C, is a mapping, i.e., it is a value assignment to the variables. Furthermore, a solution v satisfies a constraint in C if and only if $v(x_i)$ is an element in the set of relation R of the constraint. Formally, the relation between v and (X, D, C) can be expressed as follows:

$$v: \text{X} \rightarrow \text{D such that} \quad v(x_i) \in R \subseteq C.$$

In order to solve large-scale crew management problems, the methodology recently developed by Fahle et al. (2002) and Junker et al. (1999) embeds constraint programming within column generation. Constraint programming has no effect on the overall branch-and-price scheme (or Lagrangian), it is merely incorporated as a pricing algorithm. The general framework of the *constraint programming-based column generation algorithm* (Junker et al. 1999) is as follows.

- *The subproblem*: A subproblem *SP* is a CSP where the set of the variables is denoted as (*y*, s, b). There is a one-to-one correspondence between the elements in *y* and those in the master problem. Let y_i denote a row and $y \in X$ a column. A solution $s \in \{\leq, =, \geq\}$ to the subproblem corresponds to a column (variable) b of the master problem. The constraint programming variable y_i in a subproblem is introduced to correspond to coefficient a_{ij} of variable x_j in constraint i in the master constraints, and $v_j(y_i)$ gives the value of a_{ij}. Variable z is introduced to correspond to the cost coefficient c_j in the objective function of the master problem, and $v_j(z)$ gives the value of c_j.
- *The master problem*: Given a set of solutions S to *SP*, for each solution $v \in S$ to the subproblem *SP*, which is a CSP, there is a corresponding variable x_v in the master problem. A constraint in the master problem for *SP* is specified by triple (y,s,b) A master problem, which is a mixed integer program, is specified by triple$(X, \text{D,C})$, where $M = \{mc_1, \ldots, \text{mc}_n\}$ is a set of master constraints of *SP*, $(Y, \text{ D, C})$ is a CSP, and $z \in X$ is a variable in *SP*. The objective function of the master problem is given as :

$$\min \sum_{v \in S} v(z) \cdot x_v.$$

- *Negative reduced cost constraint in the subproblem*: The goal of selecting columns with negative reduced cost can be achieved by introducing a negative reduced cost constraint in CSP. Let λ_i be the dual value for each master constraint $mc_i = y_i, s_i, b_i$. The goal is to search for a value of $\bar{y} = [\bar{y}_1, \ldots, \bar{y}_n]^T$ such that the negative reduced cost constraint specified by
  $\bar{z} - \sum_{i=1}^{n} \lambda_i \cdot \bar{y}_i \leq 0$, is satisfied, where $\bar{z}$ is the value of $\bar{y} = [\bar{y}_1, \ldots, \bar{y}_n]$
- *The master constraint*: A master constraint $mc_i = y_i, s_i, b_i$. can be in one of the following forms:

$\sum_{v \in S} v(y_i) \cdot x_v \leq b_i$ if $s_i = {}'' \leq ''$

$\sum_{v \in S} v(y_i) \cdot x_v = b_i$ if $s_i = {}'' = ''$, where i represents a row index and variable

$\sum_{v \in S} v(y_i) \cdot x_v \geq b_i$ if $s_i = {}'' \geq ''$,

x_v represents a selected column.

CSP can be embedded in a network-based shortest path problem to solve a subproblem. The application of CP on the rostering subproblem was proposed by Fahle et al. (2002) and is outlined next. In the shortest path CSP, we search for the shortest path in a directed acyclic network. It requires the network to be topologically ordered, which means that the nodes are ordered and labeled based on the embedded time. If there exists an arc from node A to node B, it therefore implies that B is higher in the order than A, i.e., it 'occurs' later in time. For crew rostering, the nodes in the network represent the tasks (pairings, open time, etc.) ordered based on the starting time. If the completion time of task t is smaller than the start time of task t', we add arc (t, t') to the network. The network also has a sink s' and a source s_c for each crew member c. Each path that starts at one of the source nodes and ends at s', and it satisfies all legality rules, represents a roster. Each arc has an associated cost denoted by $w_{t,t'}$ and representing the cost of assigning task t' immediately after task t. The assumption here is that the cost contribution of a certain task t is related to task t itself and the task t' that directly follows t. We also let $w_{s_c,t} = w_{t,s'} = 0$ be the cost associated with the arcs associated with the sources and sink.

The goal is to obtain the rosters (paths) with negative reduced costs in this acyclic network. Recall that there are two families of constraints in the master problem of rostering (see (5.2)). The first family of constraints ensures that tasks are covered by an adequate number of crew members, while the second type of constraints ensure that exactly one roster is assigned to each crew member. Let λ_t denote the dual value of the first family for task t, and let μ_c denote the dual value of the second family for crew member c. To capture the reduced cost, the cost associated with arcs is then substituted by $c'_{s_c,t} = -\mu_c$ for arcs $(s_c,$ t) and $c'_{t,t'} = w_{t,t'} - \lambda_t$ for arcs (t, t'). The negative reduced cost constraint becomes

$$-\lambda_c - \sum_{t \in X, t\prime \in X} \left(w_{t,t\prime} - \lambda_t\right) \quad < \quad 0,$$

where X is the set of all tasks that can possibly be included in a roster for crew member c.

Efficient computational techniques for embedding this negative reduced cost constraint into the constraint programming-based column generation can be found in by Fahle et al. (2002). An experiment based on real-world data shows that with constraint programming the search space is reduced, and so is the computation time. In an instance consisting of 65 crew members, 165 pre-assigned tasks, and

250 tasks to be assigned, it takes 3,162 s for constraint programming-based column generation to solve the problem to optimality.

5.3 State of the art in Information Systems

In the previous section we discussed the models and solution methodologies for various crew management problems. In this part we focus on the underlying data management systems and we review the offerings of the leading software providers.

5.3.1 Data Management Information Systems

The airline industry is constantly undergoing changes and the ability to react and adjust swiftly is imperative. The volume of collected and stored data is rapidly increasing and data management capabilities become not only a key for success but also of survival. Superior optimization applications and durable airline's data management information systems are key success factors in such an environment. Many legacy and startup airlines are realizing the importance of data management and resource planning, which is reflected in the recent investment in these fields despite, or maybe because of the current financial difficulties. Controlling the large number of vital data and its updates is the key to reducing costly data redundancies and misleading analytics, especially within enterprises like airlines that have multiple systems depending on the same information. Accordingly, data management is evolving from scattered databases toward consolidated data warehouses and lately toward integrated enterprise data management systems.

5.3.1.1 Data Warehouse

A modern data warehouse is more than just a big database. It also includes information management software to extend the value of the data by collecting and distributing necessary information and moving real-time insights closer to all interested users. By consolidating information as-sets and applying real-time analytics to turn information into intelligence, it is possible to deliver information in a selective and transparent way. Data warehouses are becoming much more real-time-oriented. Access to the data warehouse is becoming more and more a part of front-end transactions and is not anymore a traditional backend transaction. Traditionally, an airline data storage system consists of several databases or data marts organized around specific departments that are poorly connected with non-standardized and non-efficient data interfaces. Substantial customization, code

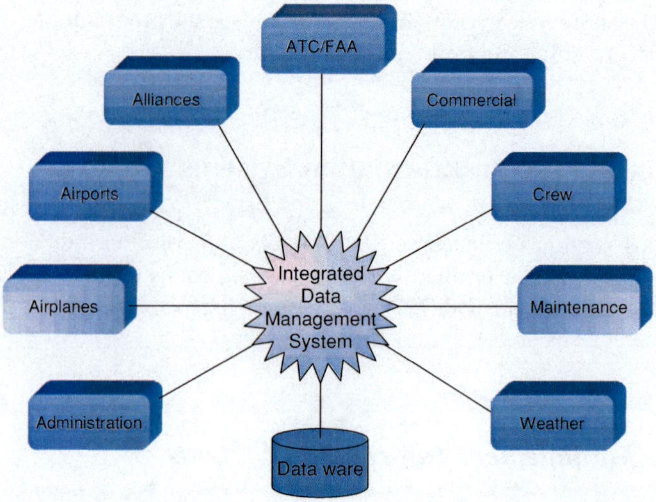

Fig. 5.4 An integrated enterprise data management system

changes, and testing across the entire system are required whenever even a small modification is needed in a department. The basic requirement of a modern airline data management system is that legacy data storage system needs to move toward a more efficient and better structured data warehouse. Such a system requires a good architecture that allows:

- Modularity—enhancement of one department's domain without impacting other domains,
- Migration—secure conversion from legacy,
- Integration—combining selected in-house and external systems and applications,
- Harmonization—data synchronization across different operational systems and data sources.

5.3.1.2 Integrated Enterprise Data Management System

A data management information system is quite often thought of as part of a data warehouse, but it actually sits on top of it. An integrated enterprise data management system also includes enterprise application integration capabilities as well as other services like user interfaces, data mining, re-porting capabilities, etc. Solid raw data and strong analyses are essential for making timely and appropriate business decisions and maintaining a competitive edge. An integrated solution, as the one shown in Fig. 5.4, should collect data from multiple systems into a consistent, accurate data warehouse.

Selected data is then presented in multiple business views, such as views for dispatchers, ATC managers, crew managers and maintenance coordinators to enable corresponding planners and managers to rapidly retrieve information during planning and operations and make more timely and informed decisions. Accounting and finance, route profitability, human resource, materials management, enterprise management, and customer relationship management should not be forgotten as a part of a comprehensive integrated data management system landscape.

An integrated data management system provides organizations also with the flexibility, reliability, and agility to respond to different vendors' software consolidation needs, to facilitate development of one integrated solution, and to reduce the overall cost of existing and potentially new data integration projects. Such a system needs to provide the following functionalities:

- *Connectivity*: Leverage all data, regardless of its source. With the "connected world" piping in data from all areas, connectivity is the foundation of enterprise data integration. For example, connectivity helps a crew tracking application to get in-flight aircraft positions by capturing and correlating data received from both airline systems and FAA sources.
- *Monitoring*: Profile, cleanse, and monitor data to assure standardized, consistent, and reliable information.
- *Capturing*: Capture and propagate data changes in real-time to assure data integrity, consistency, and credibility.
- *Processing*: Extract, transform, and load data from across the enterprise to create consistent and accurate information.
- *Data caching*: Enable different decision support tools a quick access to critical data. For example, data caching is used to enable crew managers in airline operations centers to quickly assemble views of the overall system status, delivering responsiveness that simply was not available from its relational data marts.
- *Data mining*: Help users to discover patterns, dependencies, and consistencies of huge data sets. It also allows queries and uses data across multiple systems without the physical movement of source data.
- *Master data management*: Quickly and reliably create a unified view of enterprise data from multiple sources.

5.3.1.3 Crew Data Management Information Systems

A fast access to the most current operational data drives the success of an airline's crew disruption management system. A crew data management system needs to be able to deal with such a dynamic environment in real-time. At the same time, this system should support the entire planning process, which is divided into long-term planning, short-term planning (scheduling and rostering), tracking and pre-operations crew schedule repairing. While speed is not crucial for planning data management except short term planning, it requires the storage and distribution of huge amounts of crew related data.

It is believed that a global recovery plan for the three resources (aircraft, crew, and passengers) should be obtained simultaneously due to their related nature. For example, if a flight is cancelled and the next flight is not covered by a crew or an airplane, a recovery plan for a crew would be of no use, if an aircraft is not recovered and positioned at the right place. The combination of sophisticated operations research techniques and an integrated data management system substantially increases the quality and accuracy of a crew decision support system. A comprehensive data management information system, in order to support crew planning and operations, needs to provide, among others, the following functionalities:

- manage all flight schedules, crew assignments, crew training, and crew time tracking,
- manage all key data including aircraft, airports, bases, limitations, and payroll policies,
- maintain all relevant crew data including contact information, qualifications, and payroll data,
- define and use multiple regulation templates,
- generate pairings and rosters for flight and cabin crew,
- use actual and future scheduled flights to check for crew feasibility,
- modify or swap pairings and rosters to clear feasibility conflicts, and
- drag-and-drop environment.

5.3.1.4 Case Studies

To name just a few, the following are examples of successful implementations and future plans for data warehouse upgrades and development of an integrated data management system.

Continental Airlines

Continental Airlines was proclaimed a winner in the 2003 Data Warehousing Award competition organized by the Data Warehousing Institute (TDWI) for their effort to streamline their data management and reporting practices across their entire enterprise. In the past, each department at Continental had its own approach to data management and reporting. The airline lacked a corporate data infrastructure that a broad range of employees could use for quick access to key insights about its business. Information was inconsistent across the different areas, and root causes were hard to identify.

In 1998, a Teradata-based data warehouse was implemented to bring consistency to data management and reporting at Continental. Five years later the warehouse consisted of 27 source systems including schedules, inventory, reservations, airline tickets, airline revenue flown information SOCC (airline operational system), One Pass (frequent flyer program), customer profiles and demographics, aircraft maintenance, alliance data, employee/crew payroll, and customer care. In addition, business rules are applied to the data at an enterprise level to derive additional information such as the true origin and destination of a trip. Even further, more than 1,000 users have fast and easy access to enterprise information for strategic and tactical decision support and full-spectrum business intelligence. The enterprise data warehouse demonstrates best practices in a number of ways.

- *Architecture*: Continental developed a model that simplifies joining different subject areas in real time and allows a single view of information.
- *Training*: Users receive extensive training so they have the skills to find the data they need and come up with enterprise-wide questions that could not have been answered before.
- *Transparency*: Profitability and loss are associated with any decision no matter how small.
- *Standardization*: A data warehouse steering committee works to standardize definitions and data metrics so that all users can use data from across subject areas.
- *Automation*: All transformation and loading processes are fully automated and automatically monitored. Only exceptional situations require human attention and intervention.

Continental reports multiple millions in cost savings and estimates that revenue has increased by multiple millions of U.S. dollars in 2002 through just four applications running on the enterprise data warehouse: revenue management, fraud detection, crew payroll, and customer relationship management.

Air France

At the AGIFORS 2006 Crew Management conference in Honolulu, Air France presented a plan to phase out their crew data management legacy system. The ambitious plan contains stable steps to assure a secure and smooth migration process through a period that is estimated to be approximately 10 years. The proposed migration suggests a modular transition process in order to allow stability during ups and downs of the airline industry. The goal was to define separate migration projects that are not dependent on each other. Each migration project is designed to bring value in order to pay for the phase-out. Projects may be jointly accomplished together with their sister company KLM, as well as with other airlines and providers. This is an ongoing initiative at Air France that has a high business priority, which becomes even more important with new airline operations center plans that are starting to shape up. The company's gradual transformation approach has already taken several steps toward this direction:

- continuous slimming down of the legacy data management systems,
- stop adding new data into the legacy data management systems,
- removing concepts that are better served by new data management systems,
- developing real-time pairings tracking,
- centralizing rules,
- defining business intelligence and production key performance indicators,
- developing manpower planning.

Air France believes that the integrated enterprise data management system should be owned and maintained by the company because it allows the evolution of the company's IT and represents a foundation of all business processes and decisions. Additionally, it facilitates the inclusion of decision support tools such as internally and externally developed operations research applications.

Delta Airlines

The airline crew data management system contains critical business and safety information and usually airlines are not ready to share such an asset, as seen in the above examples. However, Delta recently asked the U.S. bankruptcy court for permission to strike an outsourcing deal. According to the plan, Delta Air Lines will turn over the bulk of its back-office computer operations to IBM under a 7 year outsourcing agreement that Delta hopes will contribute to the savings it needs to successfully emerge from Chap. 11 bankruptcy. The airline wants IBM to assume operation of the computers and software that support Delta's customer reservations, business record-keeping, flight management, and maintenance tracking systems. Delta's agreement with IBM is the second major technology outsourcing deal involving a major U.S. airline. At the be-ginning of 2006, United Airlines inked a 10 year desktop and infrastructure management deal with Electronic Data Systems.

5.3.2 *The Software Market*

IT infrastructure and the corresponding complex software applications are an indispensable part of the airline industry today. Computerization is one of the key business areas for each airline. In the early days computerization was mainly oriented toward the automation of manual work and maintenance of basic databases, primarily the reservation system. Today, the main goals are optimization and integrated enterprise data management systems. Trends in the airline industry have created a market for firms capable of using technology to increase productivity and fully integrate operations management, while gradually phasing out existing legacy systems.

For more than 40 years, software providers have been offering solutions in areas such as revenue management, pricing, flight scheduling, cargo, flight operations, and crew scheduling. In the early sixties, Sabre Holdings, at that time part of American Airlines, transformed the airline industry through technological advancement by creating the first passenger reservations system. In these early days, airline software was mostly developed by airlines' IT departments themselves and by external consultants mostly coming from academia. Probably the first independent software provider with an established list of clients was SBS International, founded in the early seventies. The SBS' launch product was a rostering application with bidding capabilities, which is where the name SBS originated—Schedule Bidding System.

Depending on an airline or a software provider, crew management applications may have different structures. Two types of crew management applications are considered in this chapter in order to support two different processes, planning and operations. The main planning tools are manpower planning in the form of pairing and rostering applications, while the main operations tools are crew tracking and crew recovery applications. The following analysis of major crew software providers is focused around these main offerings and their optimization capabilities.

5.3.2.1 In-house Crew Management Software

In the past, many airlines had created large operation research and IT teams who developed, and are still developing their own crew management software applications. These were mostly big legacy carriers such as Air France, American Airlines, British Airways, Delta Airlines, Emirates, and United Airlines, to name just a few. Operation research resources are scarce and expensive and it is hard to keep a large operation research group in the environment where profit margins are very thin. Recently, some airlines decided to decrease their internal operations research and IT departments to steadily grow outsourcing and to look toward the best-of-the-breed operation research applications available on the market. There are many reasons behind this tendency, but one of the most important is to become lean and to cut costs through outsourcing. Delta Airlines recently even dissolved its own operation research group, Delta Technologies, in order to cut costs and to get quickly out of bankruptcy protection. However, to be competitive, a critical mass of more highly skilled operation research people able to cope with technology advancements is still required to be an integral part of a carrier.

Smaller airlines usually prefer buying a product that is close to satisfying their needs rather than of investing time and money in developing an application that will exactly fit their needs. Some airlines, like Northwest (acquired by Delta Airlines in 2008) and Southwest, are deciding to buy the source codes or pay contractors to develop for them strategically important applications that they have insufficient resources to develop and then employ their small operation research teams to support them. However, there are still some airlines such as American, Emirates, and United that are growing their own operation research teams.

While operation research team landscape is still shifting and adjusting, the most probable scenario for the future is that most airlines will continue buying well established and proven crew planning products like pairing and rostering, as well as the cutting edge technology required for disruption management applications. There are many atypical airline specific problems that will still need to be addressed using internal operation research expertise. The role of internal operation research resources is also crucial to performing data analysis, to analyzing planning and operations business flow, to understanding the external applications, and to helping with their selection, as well as to adjusting these applications to deal with changes, which are inevitable in the airline industry.

5.3.2.2 Software Providers and their Products

Today, there are several crew scheduling and rostering software providers offering a large number of crew management software packages, ranging from the long-term manpower planning through pairing and rostering, to the day of operations decision support systems. Airline crew management applications provide many benefits to their users. However, the quality and complexity of these offerings are not always the same and may include one or more of the following: graphical user interfaces, database services, reporting capabilities, web services, and simulation and optimization tools. Optimization tools may consist of simple heuristics or complex advanced mathematical methods. Nowadays, more powerful computers and sophisticated operation research methods enable significant improvements of the quality and speed of the software applications. Consequently, the airline crew management process becomes much easier and even modified in a way that obtained results are of a better quality and more quickly achieved. The user may play with different scenarios and solve them much closer to the actual operations. Making planning decisions closer to operations has indirect impact on the quality and robustness of these decisions by allowing for more accurate information to be taken into account.

In the following section we list, in lexicographical order, the major software vendors on the market and their main offerings. Some of them are major airline information technology software companies while others are specialized crew management software divisions within a bigger aviation or non-aviation company. Finally, there are a couple of small, independent companies that offer some components of an integrated crew management solution. While planning software packages have been available for quite some time and offered by most providers, more recent research and development efforts are focused on the operations.

AIMS Inc.

AIMS Inc. offering covers a wide range of applications including crew management tools such as manpower planning, pairing, and tracking. The crew pairing product is probably one of the most installed AIMS crew management products. Optimization capabilities of their crew management tools are debatable, but lower benchmark results are often compensated by the lower price of their products. They developed a solid customer base but it seems that it has been shrinking lately. Their web site is www.aims.aero with the head office registered in the United Kingdom.

Advanced Optimization Systems, Inc. (AOS)

AOS has been developing and implementing crew optimization software and solutions for the airlines since 1992. This small company has established a decent reputation and developed sophisticated optimization tools for pairing and line optimization. Despite very limited offering, they managed to develop a respectable customer base that includes small, medium, and large airlines. The company is based in New Jersey, US and their web site is www.aos.us.

Avient, an IBS Group Company

Avient provides solutions, services, and products to the global travel, transportation, and logistics industry. The company has over 15 years of experience in the design, development, and implementation of software solutions worldwide. Avient has built separate offerings for an established airline (Avient Velocity) and for a startup airline (Avient Take-Off). The Velocity product contains the pairing, rostering, tracking, and recovery modules. While tracking is their key product, it seems that their offerings do not include sophisticated optimization tools. This is probably the main reason for the modest airline customer base of Avient. The company is based in Berks, UK and their web site is www.avientsolutions.com.

Jeppesen, A Boeing Company

Elrey Jeppesen founded the navigation chart company in the early thirties. Today, the company is part of Boeing and represents one of the most comprehensive aviation software companies. Thanks to the recent acquisition of Carmen Systems AB and SBS International, now called Jeppesen AB, Jeppesen offers a comprehensive line of crew management products consisting of long- and short-term planning tools, operations tracking and decision support tools, administrative tools and web access services supported by an advanced rule engine and database services. With the acquisition of SBS the company acquired strong data management while Carmen's optimization capabilities are one of the best on the market. The combined Jeppesen airline customer base is one of the largest, including very small but also almost all of the largest airlines in the world. The range of operation research techniques implemented in their products spans from single business rule-based heuristics, to more advanced heuristics and meta heuristics, and to highly sophisticated mathematical programming methods. Also, as a result of the collaboration with academia—mainly the Massachusetts Institute of Technology and Georgia Institute of Technology—an advanced simulation engine is used in concert with their optimization tools in order to evaluate different scenarios and calibrate optimization parameters. Jeppesen is a global company with their corporate headquarter in Colorado, US. The airline crew management offices are based in New York, US and Gothenburg, Sweden. The company's official web site is www.jeppesen.com.

Lufthansa Systems

Lufthansa Systems, part of Lufthansa Group, is one of the world's leading IT service providers for the airline and aviation industry. It has around 4,550 employees in several locations in Ger-many and offices in 17 countries. Its portfolio addresses all airline business processes and includes services for passenger and cargo handling, flight operations, and aircraft maintenance and repair. As a system integrator with one of the state-of-the-art data centers in Europe, Lufthansa Systems covers the entire spectrum of IT services including consulting, application development and implementation, and reliable 24 h operation. The offering also includes a complete crew management product suite with crew tracking as the key module. Their optimization capabilities seem modest and they are mostly based on heuristics rather than on advanced mathematical programming methods. The company's headquarter is in Kelsterbach, Germany and their web site is www.lhsystems.com.

Kronos Incorporated

Kronos provides general workforce management applications. These tools help organizations staff, develop, deploy, track, and reward their workforce. As part of an acquisition of AD OPT Technologies in 2004, Kronos obtained Altitude division specialized in airline crew management. Kronos/Altitude's most known applications are the crew pairing and preferential bidding system and lately the bidline and manpower planning. The acquisition of Mercury in 2003 by AD OPT brought to the company a crew tracking application, but it seems that it did not generate too much interest in the market. The strongest point of Altitude's crew planning applications is the quality of the optimization engines. They are the first provider in this field that introduced and commercialized the branch-and-price optimization approach. Once the leading resource optimization provider, Altitude is losing lately its competitive advantage due to the lack of investment in research and development. Kronos headquarters is in Minneapolis, US, while the airline crew management software office is in Montreal, Canada. Their web site is www.kronos.com.

Navitaire

Navitaire Inc. provides reservations, direct distribution, revenue protection, decision-support, and passenger revenue accounting services to the airline industry through its hosted delivery model. With the latest acquisitions of Caleb Technologies Corporation and Forte, the product portfolio is significantly expanded by the crew management products. Among the key crew management software applications, acquired from Caleb, is the recovery suite system. Other applications comprise manpower, pairing, rostering, and tracking tools. Additional crew planning and tracking tools are supplied by the acquisition of Forte. The company's solid customer base includes several large airlines and some leading low-fare and midsize airlines. A wholly owned subsidiary of Accenture, Navitaire is headquartered in Minneapolis, US, with the crew management office in Austin, US. Its homepage is www.navitaire.com.

Navtech, Inc.

Navtech, Inc. was originally incorporated in the State of New York in 1981 and then reincorporated in the State of Delaware in 1987. Since then, together with their subsidiaries, Navtech developed a wide range of airline software offerings. While their focus is on charting and flight planning, over the last couple of years they have been investing in the development of crew planning products. The crew planning offering includes a pairing optimizer as well as preferential bidding system. Today, Navtech is an important airline software provider with a large customer base. However, their current optimization capabilities seem to be limited and their crew planning software customer base is still modest. Navtech's head office is in Monterey, US, while principle operations are based in Waterloo, Canada. Their web site is www.navtechinc.com.

Sabre Airline Solutions

Sabre, an ex-subsidiary of American Airlines, has been in business for more than 40 years. Today, it is a leading global company that provides airline products and services, reservations and departure control tools, as well as consulting services. Their airline crew management suite is an integrated end-to-end resource management solution designed to handle the airlines' needs. The suite addresses all phases of crew management operations from long-term crew resource planning to day-of-operations, more precisely manpower planning, pairing, rostering, web access, crew tracking, and recovery. Sabre's applications are very data processing oriented, but they have also developed a wide range of sophisticated operation research tools. Their optimization tools are state of the art and they are planning to stay a big player in this game. Their commitment to this objective is proven by their willingness to invest in its improvement, especially following the recent Boeing acquisitions. This Texas, US-based company has the largest share of the airline resource management market and is definitely the major airline information technology provider. Their web site is www.sabresolutions.com.

5.3.2.3 Providers Consolidation

Many industries are facing an increasingly competitive environment due to globalization, deregulation, increased complexity and consolidation. Consolidation through mergers and acquisitions is a major trend in many industries. A central goal of most mergers has been to improve investment returns through cost cutting, productivity gains, and economies of scale. The airlines industry is not an exception; the latest examples are the mergers of America West and US Airways, Delta and Northwester Airlines, and the looming merger of United and Continental Airlines. Their objectives are expanded routes, increased fares through reduced capacity, improved service and a more stable operating environment. In order to have more stable operating environments, airlines are looking more toward an integrated solution from one provider, which offers multiple modules instead of the best-of-the-breed isolated solutions that are hard to integrate. It is likely that software providers will follow this tendency. The latest acquisition trend confirms such a direction and it will not be surprising to see more acquisitions in the future.

A merger and acquisition activity should target a market presence increase, and fixed and operations costs improvements. Increased presence may be achieved by removing a competitor from the market, increasing the customer base, and offering a more complete portfolio of products. Costs can be reduced by removing redundant overheads, product overlaps, and reorganizing research and development to better use resources and reduce development costs. In recent years, several important crew software provider acquisitions have taken place. Boeing acquired a number of companies as part of its vision to offer airlines a complete suite of products and services in support of the aircraft they buy from them. In 1999, 2000, and 2001 Boeing acquired The Preston Group, Jeppesen Sanderson, and SBS International, respectively. In 2005, Carmen Systems AB acquired OpCom and just a year later Carmen was acquired by Boeing as well. Today, they are all part of one single family—Jeppesen, A Boeing Company. In 2001, Sabre Airline Solutions acquired David R. Bornemann Associates as part of its strategy to penetrate the small and medium-sized airline software solutions market. In 2004, they acquired the Stockholm-based company RM Rocade as well. In 2003, Adopt Technologies Inc. acquired Mercury Scheduling Inc. and in 2004 Adopt itself was acquired by Kronos Incorporated. In 2004, Navitaire acquired Caleb and in 2006 they added Forte to their list.

5.4 Conclusions

What all these acquisitions mean and how successful they are is hard to say and predict. It is hard to find any public data about the success of the above acquisitions. The available information shows mostly the market aspect of an acquisition and may create the wrong impression about the real financial outcome. The process of consolidation is beneficial if it facilitates increased efficiency on both sides, acquired and acquirer organizations, or at least for one that survives. While many of these transactions have a potential for efficiency gains, it seems that it is not systematically exploited. However, it seems that the current trend unveils a typical big-fish-eats-small-fish pattern where small independent providers get absorbed by larger companies. The feeling is that the ultimate objective of larger corporations such as Boeing, Lufthansa, and Sabre is not necessarily increased efficiency, but rather an expansion of the product and service portfolios in order to offer an enterprise crew management solution.

References

Andersson E, Housos E, Kohl N, Wedelin D (1998) Crew pairing optimization. In: Yu G (ed) Operations research in the airline industry. Kluwer Academic Publishers, Hingham, pp 228–258

Barahona F, Anbil R (2000) The volume algorithm: producing primal solutions with a subgradient method. Math Program 87(3):385

Barahona F, Anbil R (2002) On some difficult linear programs coming from set partitioning. Discret Appl Math 118(1/2):3

Barnes E, Chen V, Gopalakrishnan B, Johnson EL (2002) A least-squares primal-dual algorithm for solving linear programming problems. Operations Res Lett 30(5):289–294

Barnhart C, Johnson EL, Anbil R, Hatay L (1994) A column generation technique for the Long-Haul crew assignment problem. In: Cirani TA, Leachman RC (eds) Optimization in industry II. Wiley, New York

Barnhart C, Johnson EL, Nemhauser GL, Savelsbergh MWP, Vance PH (1998) Branch-and-price: column generation for solving huge integer programs. Operations Res 46(3):316–329

Benders J (1962) Partitioning procedures for solving mixed-integer programming models. Numer Math 4:238–252

Dantzig G, Ford L, Fulkerson D (1956) A primal-dual algorithm for linear programs. In: Kuhn H, Tucker A (eds) Linear inequalities and related systems. Princeton University Press, Princeton, pp 171–181

Ehrgott M, Ryan D (2002) Constructing robust crew schedules with bicriteria optimization. J Multi-Criteria Decis Anal 11:139–150

Fahle T, Junker U, Karisch SE, Kohl N, Sellmann M, Vaaben B (2002) Constraint programming based column generation for crew assignment. J Heuristics 8(1):59–81

Fisher M (1985) An applications oriented guide to Lagrangian relaxation. Interfaces 15:10–21

Gamache M, Soumis F (1998) A method for optimally solving the rostering problem. In: Yu G (ed) Operations research in the airline industry. Kluwer Academic Publishers, USA, pp 124–157

Gamache M, Soumis F, Villeneuve D, Desrosiers J, Gelinas E (1998) The preferential bidding system at Air Canada. Transp Sci 32(3):246–255

Hu J, Johnson EL (1999) Computational results with a primal-dual subproblem simplex method. Operations Res Lett 25(4):149–157

Junker U, Karisch SE, Kohl N, Vaaben B, Fahle T, & Sellmann M (1999) A framework for constraint programming based column generation. In: Principles and practice of constraint programming-CP'99, vol 1713, pp 261–274

Klabjan D, Johnson EL, Nemhauser GL (2000) A parallel primal-dual simplex algorithm. Operations Res Lett 27(2):47–55

Klabjan D, Johnson EL, Nemhauser GL, Gelman E, Ramaswamy S (2001) Solving large airline crew scheduling problems: random pairing generation and strong branching. Comput Optim Appl 20(1):73–91

Lavoie S, Minoux M, Odier E (1988) A new approach for crew pairing problems by column generation with an application to air transportation. Eur J Oper Res 35:45–58

Lettovsky, L. (1997). Airline operations recovery: an optimization approach. Ph.D. Thesis, Georgia Institute of Technology, Atlanta

Lettovsky L, Johnson E, Nemhauser G (2000) Airline crew recovery. Transp Sci 34:337–348

Ryan D, Foster B (1981) An integer programming approach to scheduling. In: Wren A (ed) Computer scheduling of public transport urban passenger vehicle and crew scheduling. Elsevier Sci B.V, Amsterdam, pp 269–280

Shebalov S, Klabjan D (2006) Robust airline crew pairing: move-up crews. Transp Sci 40:300–312

Vance P, Barnhart C, Johnson E, Nemhauser G (1997) Airline crew scheduling: a new formulation and decomposition algorithm. Oper Res 45:188–200

Wedelin D (1995) An algorithm for large-scale 0–1-integer programming with application to airline crew scheduling. Ann Oper Res 57:283–301

Yen J, Birge J (2006) A stochastic programming approach to the airline crew scheduling problem. Transp Sci 40:3–14

Chapter 6
Operations

Stefan E. Karisch, Stephen S. Altus, Goran Stojković and Mirela Stojković

6.1 Introduction

Operating an airline is an extremely complex undertaking. In order to successfully dispatch hundreds of aircraft, schedule thousands of air crew, operate up to one hundred thousand flights per month, and transport tens of millions of passengers per year, detailed planning and near-flawless execution are critical. On the day of operations, plans and schedules for many different resource areas need to come together for the airline to provide reliable service to its customers. No two such days are alike. Airlines not only vary their daily schedules to adapt to changes in demand, they also face uncontrollable external events, such as bad weather, that require them to modify the planned execution to restore the schedule. There are many examples of how well—or how poorly—airlines have dealt with such disruptions.

Ice and snowstorms in recent winters left passengers stranded on airplanes for up to 11 h, and caused havoc in the affected airlines' systems for several days. JetBlue Airways' reputation and financial strength were hit during the *Valentine's Day Storm* in February 2007 (Wikipedia 2009; JetBlue 2007), when the airline was forced to cancel more than a thousand flights due to operational issues during a

S. E. Karisch (✉)
Operations Research and Optimization, Jeppesen, a Boeing Company, Englewood, CO, USA
e-mail: Stefan.Karisch@jeppesen.com

S. S. Altus
Strategic Initiatives, Jeppesen, a Boeing Company, San Jose, CA, USA

G. Stojković
Enterprise Solutions, Jeppesen, a Boeing Company, Englewood, CO, USA

M. Stojković
Operations Research, Delta Air Lines, Eagan, MN, USA

C. Barnhart and B. C. Smith, LLC (eds.), *Quantitative Problem Solving Methods in the Airline Industry*, International Series in Operations Research & Management Science 169, DOI: 10.1007/978-1-4614-1608-1_6, © Springer Science+Business Media, LLC 2012

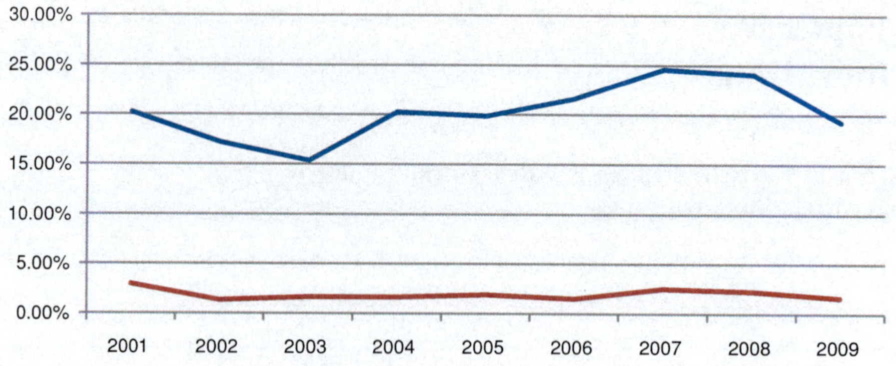

Fig. 6.1 Percentage of delayed (*blue*) and cancelled (*red*) flights of US carriers

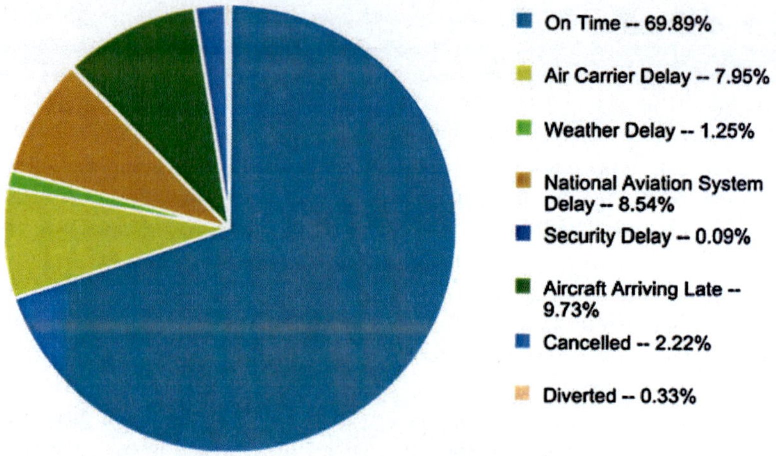

Fig. 6.2 Delays and cancellations June–August 2007

winter storm. Besides direct costs to airlines, such disruptions have a visible, and harmful, impact on passenger goodwill. Several such occurrences sparked an initiative to create an "Airline Passenger Bill of Rights", which in turn resulted in strict tarmac delay rules that went into effect in April 2010.

The summer of 2007 was the worst in aviation history when it comes to delays and cancellations. On-time performance was the worst on record (see Fig. 6.1), and almost one-third of all flights were significantly delayed or canceled. According to the Bureau of Transportation Statistics (2009), the 20 largest U.S. airlines had an on-time arrival rate of 71.7% in August 2007, down from 75.8% a year earlier. While subsequent years saw a significant reduction in capacity caused by record fuel prices first and then the global economic downturn, analysts expect the airline industry to exceed 2007 passenger levels sooner rather than later.

The diagram in Fig. 6.2 categorizes the delays during the summer of 2007 (Bureau of Transportation Statistics 2009). During these months, air carrier delays

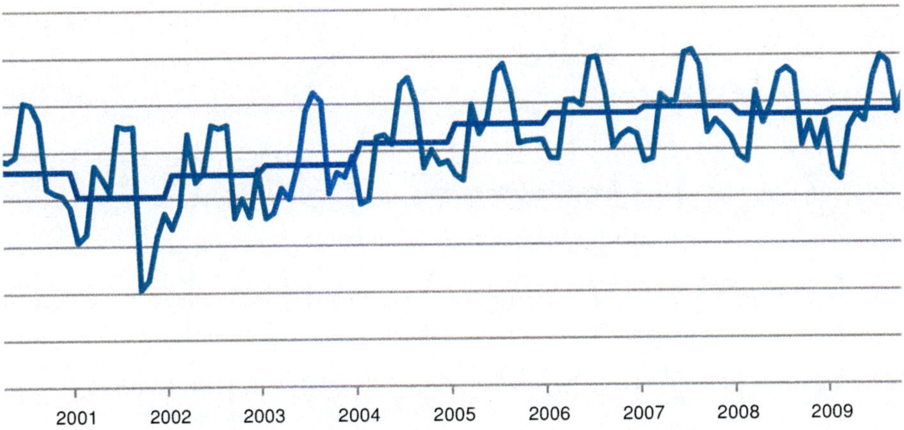

Fig. 6.3 Increase of load factor of US airlines

and the National Aviation System have been significant factors in the worst delays in US aviation history. As the figure depicts, a large portion of delays can be attributed to the National Aviation System and its limited capacity.

The Air Transport Association (ATA) estimated that, in 2009, 100 million system delay minutes generated nearly $6.1 billion in direct aircraft operating costs for scheduled U.S. passenger airlines; compared to 129 million system delay minutes in 2010 that generated nearly $10 billion in direct aircraft operating costs for scheduled U.S. passenger airlines (ATA 2010). Assuming $35.70 per hour the average value of a passenger's time, ATA estimated that 2009 delays have cost air travelers alone $3.3 billion.

With higher load factors seen in recent years (see Fig. 6.3), the execution of a dense flight schedule becomes even more delicate, since airlines lose flexibility in recovering from disruptions. Canceling flights becomes a less-attractive option when there is no space to re-accommodate passengers on other flights.

Many of the proposed solutions to delays and cancellations involve massive infrastructure improvements, but these have a long implementation timeline. More efficient operations by airlines—through improved decision-making leading to better allocation of aircraft, crews, and ground resources on the day of operations—makes an immediate impact in the interim, and will continue to provide value in an updated operating environment. Operations Research (OR) plays a central role in improved operational efficiency, through development of new policies and introduction of new tools towards this end.

Performance measurement is a key for ensuring that operational improvement indeed happens. The airline industry relies on Key Performance Indicators (KPIs), quantitative measurements that reflect the critical success factors of a system, organization, product, or solution. KPIs are usually associated with revenue or costs. A familiar KPI related to revenue is *Revenue per Available Seat Mile (RASM)*, while a typical cost-based measure is *Cost per Available Seat Mile*

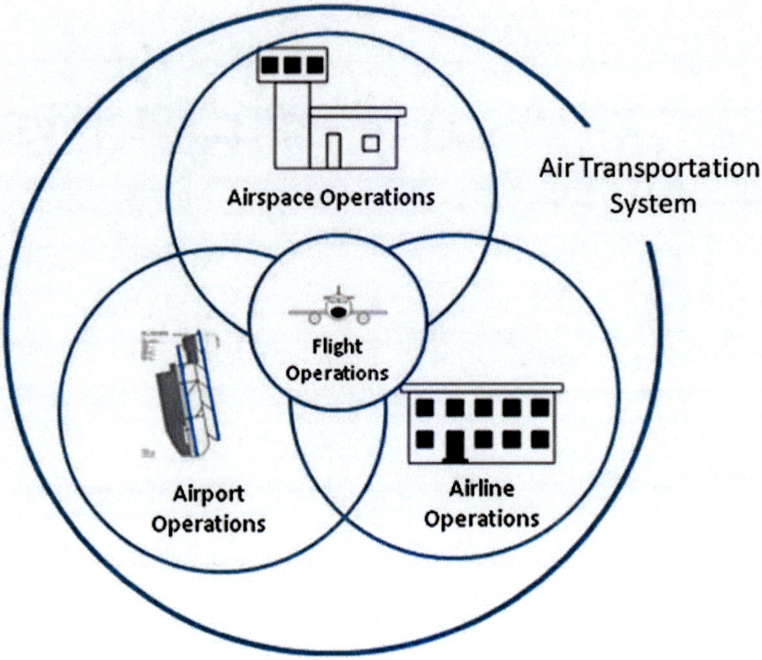

Fig. 6.4 The Air Transportation System as system of systems (© 2011 Jeppesen)

(CASM). Since the early days of commercial aviation, unit costs have decreased while unit revenue also decreased. Until the 1980s, improvements in aircraft technology contributed substantial costs savings, which masked the fact that other operational costs, especially crew costs, actually increased. In the more competitive post-deregulation years, airlines have turned their attention to these other costs. Jiang and Hansman (2006) compiled CASM and RASM data since deregulation in 1978, and report an average decrease of CASM by 2% and RASM by 1.7%. Their data reflects past and recent industry losses in times when RASM dropped below CASM. To combat these losses and further reduce costs, airlines now must address the more difficult aspects of operational efficiency.

6.1.1 Overview

This chapter focuses on the role of OR in improving the efficiency of an airline from an operational point of view. We view the airline as a system of systems, which itself is part of the larger Air Transportation System, as illustrated in Fig. 6.4. Theoretically, the Air Transportation System can be viewed as one large operational problem. In practice, it is decomposed into many interdependent, more manageable but still challenging sub-problems. These involve coupled complex

Fig. 6.5 Overview of airline business processes (© 2011 Jeppesen)

subsystems such as airlines, airports and airspace with their equipment and workforces, all with a single purpose, to facilitate the movement of people and goods.

Many resources in the Air Transportation System are shared and they connect different stakeholders directly or indirectly and impact their operations. Operational irregularities can therefore propagate easily from one stakeholder to another. For example, one airline's disruptions can impact other airlines as they share the same airport and its scarce resources, e.g., taxiways or runways. This emphasizes once more the importance of having a holistic perspective of the entire Air Transportation System. The ultimate objective of the Air Transportation System is to offer transportation services in a safe and efficient manner. The role of OR is to provide analytical and optimization tools that allow for efficient planning and operations through the overall system. This task is nontrivial since each subsystem has its own objectives and constraints, which are often in conflict with each other. While these subsystems are intricately linked, they have traditionally been looked at in isolation. Focusing on local optimality can neglect the global objectives of the entire Air Transportation System.

Going a level deeper into the Airline Operations subsystem, we have a similar breakdown into smaller but still very complex subsystems. In Fig. 6.5, we highlight the most important subsystems and related processes concerning revenue generation, air crew scheduling, aircraft scheduling, flight operations, navigation information management, and maintenance scheduling. While these subsystems are linked, they also have traditionally been looked at in isolation. The figure lays out a typical timeline and depicts when each business process is executed. All these processes are dependent on each other. For example, depending on booking levels, an airline might adapt aircraft and crew schedules, reassigning aircraft of different sizes to match demand, and changing crew assignments to match the new aircraft schedule.

All aspects that are not related to marketing or revenue management are considered part of Operations. The revenue generation processes are shown only for completeness reasons in Fig. 6.5 and are not in our scope. In this chapter, we will mainly address three areas of these processes:

- Flight planning
- Airline disruption management
- Ground resources and turnaround management

When we refer to *Operations* from now on, we will only refer to these three areas. As for the bigger system depicted in Fig. 6.4, we hence cover parts of Flight Operations, Airline Operations, and Airport Operations.

We will not explicitly address the topic of *Maintenance, Repair and Overhaul (MRO)* planning and scheduling but rather refer to Butler and Keller (2000), for an overview of this important topic. Crew and fleet scheduling is covered in other chapters of this book.

Safety is the top priority in operating an airline. Only after an airline guarantees that safety is provided can it consider other priorities such as operational efficiency. A commitment to safety is an integral part of an airline's culture, with a deep impact on its operations practices. An absolute priority for safety does not restrict the types of tools and practices airlines can use in the pursuit of operational efficiency, but it requires that specific hard constraints be included in models, to ensure that solutions are feasible within this culture.

To tackle the challenging task of improving operational efficiency to reduce costs, airlines have been early adopters of Information Technology (IT) and decision support. Limitations in early technology, however, have impacted business processes and are reflected in how airlines are structured and their operations fragmented. This fragmentation leads to a local view and, with decision support and optimization systems whose scope lies within a single department, to suboptimality with respect to global objectives. Further compounding the problem of organizational fragmentation, early research and Decision Support Systems (DSS) were based on simplifying assumptions to obtain tractable problems. Many of these assumptions are still prevalent in academic research. In the future, OR should lead the industry to overcome these limitations, and by combining the latest IT advancements and algorithmic development create holistic DSS for decision makers.

Our goal with this chapter is to present the Operations problem as a real world problem with its many facets and challenges, and motivate future research. The ultimate goal is to overcome the existing fragmentation and to solve an Integrated

Operations Problem, addressing all resource areas at once, and producing a safe and holistic, globally-optimal solution.

6.1.2 The Role of OR in Airline Operations

OR has been applied in the airline industry since as early as the 1950s, and the airline industry has been pioneering in adopting OR ever since. Barnhart et al. (2003) provide an extensive historical overview of the application of OR in the air transport industry. They state that there have been more than 1,000 OR papers published on air transport over the last 50 years; the different chapters of this book indicate the breadth of topics treated. The impact of OR has been weaker for operations than for planning, due to the relative immaturity of the technology and a recently recognized need for efficient, real-time decision making (which places extreme requirements on solution speed).

In the 1960s, operations and planning started out on the same level with respect to the application of OR. Planning seemed to be more tractable and attracted much more attention for a long time. This resulted in a large body of work in the area of airline planning and scheduling. Even in these domains, without the pressure to provide real-time solutions, researchers were forced to make simplifying assumptions, and some of these simplifications carried over to operations.

As the use of OR in planning and scheduling became widespread, analysts turned their attention to the day of operations and the gap between the plan and the actual operation. A natural approach is to modify the planning models to be more robust to disruption by outside events. This topic, "robust scheduling", constitutes much current research in planning and scheduling.

Cohn (2007) reviews this research and classifies it into groups. One approach is to incorporate knowledge from data analysis into robust scheduling. For example, Rupp and Holmes (2006) gather empirical evidence on cancellation patterns. The authors analyze data from the U.S. Bureau of Transportation for 35 million flights from 1995 to 2001. They found that in addition to weather, there are several reasons for flight cancellations, the most prominent among them airline hub size, concern over revenue loss, the day of the week, and the time of the flight. Hansen and Hsiao (2006) consider weather and other factors in predicting delay patterns. They formulate an econometric model of average daily delay. The model considers the effects of arrival queuing, volume, terminal weather, en route weather, and seasonal effects. They confirm that queuing has a greater delay impact in the morning than in the afternoon or evening and 31% of the total delay increase between early 2004 and early 2005 can be attributed to traffic growth.

In a more theoretical approach, AhmadBeygi et al. (2008) show the impact of individual delays on the whole system by deploying propagation trees. The paper investigates the relationship between planned aircraft and crew schedules and delay propagation whose understanding is a prerequisite for developing tools to build more robust airline schedules. Borndörfer et al. (2010) propose an approach to minimize the probability of delay propagations along aircraft rotations which allows for the construction of delay resistant schedules. Their method can be generalized to crew and integrated scheduling problems.

Another body of work looks at multi-criteria optimization to crew scheduling. Ehrgott and Ryan (2002) use multi-criteria optimization for crew scheduling to maximize robustness and minimize cost by keeping crew and aircraft together and by adding extra slack when aircraft changes are scheduled. Shebalov and Klabjan (2006) look at so-called "move-up crews" to introduce more recovery opportunities and provide more recovery options. Weide et al. (2009) focus on improving crew robustness by keeping crew and aircraft together whenever possible. This idea was first presented by Klabjan et al. (2002) and implemented successfully in practice at several airlines, see Larsson (2006). Gao et al. (2009) describe integrated airline fleet and crew robust planning.

Flight schedule re-timing is considered in the work of Lan et al. (2006) to improve passenger metrics, and of Stojković et al. (2002) and AhmadBeygi et al. (2010) to decrease the propagation of delays. Rosenberger et al. (2004) focus on robust fleet assignments by introducing a lot of short cycles to allow easy cancellations, while Smith and Johnson (2006) improve operational robustness through network simplifications, such as minimizing the number of fleet types assigned into and out of specific stations. Burke et al. (2009) investigate multi-objective approaches for schedule robustness and show through simulation on KLM data that schedule flexibility can indeed increase operational performance. Heinold (2008) defines a "maintenance reachability" robustness metric and creates plans that have greater opportunities for maintenance recovery when planned aircraft routings are disrupted. In proof of concept work, Dunbar et al. (2010) incorporate the ideas from AhmadBeygi et al. (2008), Lan et al. (2006), and Weide et al. (2009) and introduce a framework that integrates aircraft routing and crew planning to accurately calculate and minimize the cost of propagated delay.

No matter how robust the plan, however, the disruption input can always exceed its resilience. Furthermore, robustness and optimality are a trade-off, with a pareto family of solutions where the more robust a solution, the higher its cost. Assuming a more effective disruption recovery capability allows a lower cost planning solution to be chosen. Therefore, advances in robust planning have not reduced the demand for optimal day-of-operations solutions, which we describe in this chapter.

The three specific operational topics discussed in this chapter, namely flight planning, disruption recovery, and ground operations, have seen increases in research lately.

The previously-described survey by Barnhart et al. (2003) does not include a single reference to published research on flight planning. Several events in the last few years increased awareness of the importance of efficient flight planning. Increased fuel prices, augmented by pressure to reduce emissions, have brought more organizational focus and investment in airlines towards any method of reducing fuel. Other economic changes have directly impacted the flight planning problem itself. Air traffic control agencies have used a variety of means to deal with congestion, adding more restrictions in some areas while liberalizing airspace in others. These differences in the airspace structure increase the complexity of the path optimization search space, including adding dynamicity as airspace constraints are applied only during peak times. Further, as IT advances make larger, more complex problems more tractable, flight planning becomes a viable addition to large, multi-disciplinary optimization problems such as collaborative air traffic management and integrated recovery optimization. For these reasons, we expect continued growth in interest in flight planning as a worthwhile academic topic. We give a detailed overview of the topic in Sect. 6.2 of this chapter.

In operational recovery or disruption management, DSS are now available, but not yet widely used. The state of the art can be compared with the state of revenue management in the 1980s or crew planning in the 1990s. Disruption management systems still need to evolve from luxury item to a business necessity. One challenge in this context is the difficulty to support a real-time environment and to capture highly complex business requirements and processes in OR models.

The best known published application of OR in the field of operational recovery has been the work by Navitaire (then CALEB Technologies) and Continental Airlines which won the 2002 Franz Edelman Award for Achievement in Operations Research and the Management Sciences (Yu et al. 2003). The authors state that the deployment of this crew recovery tool for major disruptions provided Continental Airlines an estimated savings of $40 million after September 11, 2001. Continental was the first airline that was able to get back to a full schedule after the complete shutdown of the North American airspace due to the terrorist attacks.

Although very rudimentary, the earliest work on operational recovery has been focusing on aircraft recovery and was published by Teodorović and Guberinić (1984), Teodorović and Stojković (1990) and Jarrah et al. (1993). Crew recovery models started to appear in the literature in late nineties led by Wei et al. (1997) and Stojković et al. (1998). First models proposing an integrated way to address operational disruptions appeared approximately about the same time led by Teodorović and Stojković (1995) and Lettovský (1997). Soon after, Stojković and Soumis (2001), propose integrated airline operations recovery models. Another broad overview on irregular operations is provided by Ball et al. (2006). For a detailed overview of work in disruption management, we refer to Sect. 6.3 of this chapter.

The ground subsystem is an important and integral part of the Air Transportation System. There are many interdependencies between these subsystems in both planning and operations phases. An insufficient match of aircraft activities (i.e., demand) and ground capacities (i.e., supply) may create lengthy delays, potential cancellations and, generally speaking, an unstable transportation system. Especially during the day of operations, collaboration between subsystems is extremely important not only to ensure efficiency but above all to guarantee security. With increased traffic at airports worldwide, ground operations are becoming a critical part of the Air Transportation System. New airport terminals are now opened with equally new IT infrastructure enabling better data gathering, faster and more complex analytical capabilities as well as the use of more sophisticated OR models and decision support tools, all resulting in lower cost and improved operational efficiency. Once again, the role of OR is to overcome localized and suboptimal decision making and increase the overall efficiency of ground resources, including workforce and equipment. Sect. 6.4 of this chapter focuses on ground resources and turnaround management.

OR has been used for several decades to help airlines build better plans, a practice that is widely implemented and culturally accepted. Airlines still have to expand their focus from planned to realized benefits and OR has a big role to play, providing intelligent decision support based on advanced analytics and IT advances.

6.1.3 The Current Landscape and Challenges

To be able to implement efficient OR solutions, an appropriate IT infrastructure is required. Airlines have been early adopters of IT to better manage the complexity of their operations. But this early introduction has led to legacy IT infrastructure. While this infrastructure keeps airlines operating, it makes the introduction of new technology more challenging. There is substantial variation, however, between the practices and tools at different airlines, and the challenges they face in upgrading their capabilities. Some of the barriers to improvements are organizational or cultural, while others are driven by the legacy IT infrastructure itself.

Rogers et al. (2000) provide a still relevant overview of this dilemma for a supplier of IT and Decision Support Technology (DST). In their survey of airline operations personnel from a variety of airlines, they identified the following areas that would need to be addressed:

- Data: lack of accuracy, timeliness, availability, etc.
- Economic: lack of economic data and metrics to measure impact
- Integration: lack of ability to integrate databases, software applications, etc.
- Situational awareness: lack of big picture for local decision makers
- Decision support: lack of DST to satisfactorily address operational complexity

Many challenges stem from limitations of existing IT infrastructures. Legacy mainframe systems are being replaced slowly due to the immense efforts and costs involved. One can compare replacing a mainframe IT infrastructure at an airline with changing the engine of a car while driving down the road at high speed. There are new and promising technologies available, but it is difficult to measure its impact to an airline's bottom line, and the cost and risk of implementation are substantial. Because of the expense and risk, it is unpalatable to replace everything at once. Replacing individual capabilities requires integration and information sharing between modern and legacy technologies, which presents an additional challenge. It is widely recognized that startup airlines have an IT advantage over legacy carriers.

As biggest challenges and shortcomings, airline executives point out the lack of situational awareness across all areas during irregular operations. There are still thousands of relatively small tactical decisions that are made by employees across the system, with only a local view. Some of these decisions have large ramifications system-wide and could be improved with appropriate technology. For example, for a gate agent the question arises whether (s)he should wait, and if so, how long, for delayed connecting passengers. A ramp agent asks the same question about waiting for delayed connecting bags.

Access to relevant information is essential to achieve situational awareness and improve decision making. This requires communication technology that supports the retrieval of information anytime and anywhere. One of the challenges airlines face during an irregular operation is to communicate with crew. Even for a smaller airline with fewer than 100 aircraft, the airline would still have at least 1,500 crewmembers flying each day and only a limited number of crew trackers available to locate and communicate with crew during disruptions. Without the ability to communicate, a recovery solution to a major disruption is only a plan in computer memory that cannot be implemented. While the necessary communication technology is already available, it is not yet widely used, in parts also due to the legacy IT burden many airlines face.

Traditional communication between ground and aircraft such as the Aircraft Communications Addressing and Reporting System (ACARS) has been expensive, but today's availability of wireless technology onboard of aircraft open up new possibilities for communication and better operational decision making. The implementation of state-of-the art communication technology would not only provide airlines operational benefits but also drastically improve customer satisfaction, as passengers could also receive more accurate and timely information.

When applied on top of an existing IT infrastructure and linked to real-time information, OR can power comprehensive and user-friendly DST solutions in the hands of operational decision makers, being it front-line employees at the gate or back-office personnel in the airline's Operations Control Center (OCC). Being able to make the decisions based on accurate, comprehensive, and timely information undoubtedly improves the efficiency of an airline's operation, and thus the bottom-line results of the company.

Within the broader Air Transportation System, the same air-to-ground data-exchange capabilities are enabling enhanced aircraft path adjustments with collaboration between the aircraft, the airline's OCC, and Air Traffic Service Providers (ATSP). This real-time three-part collaboration is the key enabling technology for future air traffic concepts such as the Federal Aviation Administration's Next Generation Air Transportation System–NextGen (Federal Aviation Administration 2009) and Eurocontrol's Single European Sky ATM Research–SESAR (Eurocontrol 2009), which promise to increase capacity for airlines and reduce operating costs due to more holistic solutions to operational irregularities. These initiatives are expected to have significant impact on operational efficiency, increasing capacity while reducing operating costs. Incremental work to support this collaborative traffic management model is being introduced in the current environment in the form of Dynamic Airborne Re-Planning (DARP) in several oceanic regions (ASPIRE 2011), but the IT infrastructure, especially the lack of automated links between the OCC and ATSP, reduce the use of this technique by increasing workload.

There is still a lack of globally applied cost models, which make benchmarking of operational performance across the Air Transportation System and its subsystems difficult. Industry executives question whether current efforts to improve the efficiency of operations are aimed at the right places, without a more common view and clear measures of the return on investment. Along similar lines, researchers are still looking for ways to measure the value of a plan before it is executed. Commonly used economic models as basis for OR models will be a partial answer to this challenge.

No matter what systems or technology are put in place, their full benefits will not be realized unless they provide situational awareness through communication and connectivity, i.e., put timely information in the hands of the right people. A related question is how to display this information and guide the decision makers through an enormous amount of total data to the relevant information. The importance of visualization in DST is recognized by the OR community, but there has been limited published research in this context for airline operations. As Baumgarten (2009) states, the humans can process and hold only a limited amount of information at one time. The effectiveness of data visualization hinges on two things: the ability to clearly and accurately represent relevant information and how the human interacts with and processes this information. To address these aspects and ensure that decision makers have the right tools, user centric design and user experience have to be considered when developing OR based DST.

Related to usability, flexibility and adaptability are still lacking in many tools offered for airline operations, especially for the day of operations. Some suppliers offer software systems based on rules engines that allow the separation of the software core and a business rules layer. These systems are fully customizable, but there are few such installations for day-of-operations disruption management.

As a final topic of this introduction, we would like to address the dual topics of fuel price and environmental impact, which together have driven radically increased attention to fuel use. This has led to investments for saving fuel that range from operational analyses by third-party consultants to investment in new technologies and infrastructure, both in the air and on the ground. Fuel conservation is further driven by global policy-making on climate change. Even though aviation is responsible for only 2% of global CO_2 emissions today, it is often seen as a luxury item, and lacks the lobbying strength of some other carbon-creating industries; furthermore, its emissions may be more damaging to the atmosphere because of where they occur. In Europe, nearly all aircraft operators fall under the Emissions Trading Scheme to be put into place by 2012 (Ellerman and Buchner 2007). Although other emission species are a factor (Schwartz and Kroo 2009), most environmental policy is based on CO_2, which is directly proportional to fuel use, so emissions and fuel prices both drive fuel reduction by operators.

New aircraft are 70% more fuel efficient than 40 years ago, and further efficiency gains are forthcoming in aircraft being developed now. It can take years, however, for new aircraft designs to reach operational use, and decades for them to completely replace existing types. Therefore, there is significant advantage associated with optimizing the flight of current aircraft, including fuel and emissions fees within the total cost. This need has contributed to the resurgence of interest in flight planning optimization throughout the industry.

Summarizing the current landscape, OR has started to play an integral role in making the Air Transportation System more efficient. However, OR advancements need to be combined with an enhanced IT infrastructure. Both OR and IT need to embrace new technologies to create powerful and useful DST. Operators are slowly and painfully replacing legacy IT infrastructure, as they incorporate new technologies, such as mobile computing, which create many opportunities for providing better situational awareness to all stakeholders and decision makers. OR's role is both to enable and to leverage these opportunities, improving operational efficiency in the Air Transportation System and its subsystems through advanced analysis.

6.2 Flight Planning

6.2.1 Introduction

Before any flight can depart from its gate, there is a regulatory need to evaluate the planned operation, to ensure that all operational safety requirements are met. In addition, the operator must provide details of the intended operation, such as the departure time, route, altitude, speed, and capabilities of the specific aircraft operating the flight, to the relevant Air Traffic Control (ATC) service providers. The details of this requirement vary between different regulators, but always involve ensuring that enough fuel is being carried for foreseeable irregularities, such as the impact of forecast weather.

To meet this requirement, airlines use software to analyze the flight, calculating the fuel and time and comparing the operation against several specific regulations. This flight planning software produces legal documents that confirm the legality of the planned operation, give the crew operational details such as the route, altitudes, time, and speed to fly and amount of fuel required, and transmit the aforementioned information to ATC. In most of the world, dedicated staff, often known as dispatchers, prepare these documents. In much of Europe, the legal requirement for flight planning does not specify the use of dispatchers, and some flight plans are calculated autonomously.

For each flight segment, the airline typically knows the departure and arrival airports, scheduled departure time, and desired payload weight (where payload is passengers and/or cargo). They also have operational data describing the performance of the aircraft expected to operate the flight. Depending on the distance and other characteristics of the flight, other operational parameters may be specified, or calculated by the flight planning software.

While the requirement to prepare a flight plan is regulatory, the calculations used to meet this requirement provide an opportunity to optimize the way the flight will be operated, to minimize cost. The same detailed analyses needed to ensure regulatory compliance can be coupled to an optimizer, where the regulations, the underlying physics, and other requirements are constraints. Optimizing many of the operational variables not only reduces operating costs, but also reduces workload on the dispatcher, since any parameter chosen by an optimizer does not need to be determined manually.

In this section we describe the casting of flight planning as an optimization problem. A constant theme in the discussion that follows is the critical importance of the regulatory and aeronautical domains on the identification of relevant factors and the development of appropriate algorithms. As a general reference, we recommend Grover (1990), which, although out of date in terms of automation tools, and focused on strategic planning rather than day-of-operations, effectively describes the many concerns in the preparation of a flight plan. For more information about aircraft performance, we recommend Padilla (1996) and Airbus (2002).

6.2.2 Processes and Solution Approaches

6.2.2.1 Problem Formulation

There has been an overall reluctance in the literature to present a concise formulation of flight plan optimization. This may be because, as de Jong (1974) writes, "…the interdisciplinary nature of flight planning makes that the design of an overall system is not so simple as could be guessed from the optimal control formulation." Illustrating de Jong's point, McIntyre (1996) paints a grim picture of what happens when the optimization does not sufficiently consider the many practical issues that are critical constraints for the 4D path optimization through a 3-dimensional network where speed constitutes the fourth dimension. Nonetheless, we will develop a formulation that we believe captures the fundamental characteristics of the problem without overwhelming the reader with specific details.

6.2.2.2 Optimal Control Theory Approaches

One natural approach to the flight planning problem is to solve the 4D trajectory optimization problem with optimal control theory, using methods generally associated with Bryson (1969, 1998). The physical equations of motion are written as a system of differential equations, which are then numerically integrated, and control variables (defining the local speed and direction of flight) are optimized to minimize an objective function. Traditionally this objective function is often flight duration (time), but a version of (6.1) can be used with the same approach.

The most general formulation of the optimal control theory approach is to define a state vector, X, of the aircraft's 3D position and mass, and a general function F that defines the system dynamics:

$$\dot{X} = F(X, T, U, \varphi) \tag{6.1}$$

where T is the temperature, U is the wind vector, and φ is a vector of controls, which define how the aircraft is operated.

This approach was applied to airline flight plan optimization by de Jong (1974), who formulates the "Air Navigation Problem" using classical optimal control theory in 2 dimensions, assuming constant speed and altitude.

A recent implementation (Jardin and Bryson 2001) of this general approach used Neighboring Optimal Control (NOC), which minimizes an objective function to second order. Compared to classical optimal control theory, NOC is reported to have a substantially faster solution time, but to be less robust. For large variations in the wind field, it may not find the optimal path.

The optimal control problem can be converted to a static nonlinear optimization problem by the replacement of the equations of motion (which cause the dynamic nature of trajectory optimization) with constraints enforcing physical continuity. Instead of propagating the finite-difference solution (by directly calculating X for each timestep sequentially for each solution iteration), the X can be replaced by a set of new decision variables Z. When the trajectory is divided into k timesteps, this introduces k additional constraints,

$$Z_i = \dot{X}_i \tag{6.2}$$

The resulting problem is large, but the constraint matrix is sparse, so specialized sparse nonlinear programming algorithms (e.g., sparse sequential quadratic programming) can solve it efficiently. These "collocation methods" were explored by Hargraves and Paris (1987) for military and space applications, and adopted by Betts and Cramer (1995) for airline trajectory optimization. The authors are not aware, however, of any operational use of this method by an airline.

The primary disadvantage of these "trajectory optimization" approaches is that the paths they produce are not domain feasible, and require what de Jong calls "a complicated correction cycle" to fit the optimal route to the allowable route and altitude structure. In fact, due to specific airspace availability constraints, the optimal unconstrained path may lead to a non-optimal constrained solution, depending on the algorithm used to fit the optimal unconstrained trajectory to an allowable flight path.

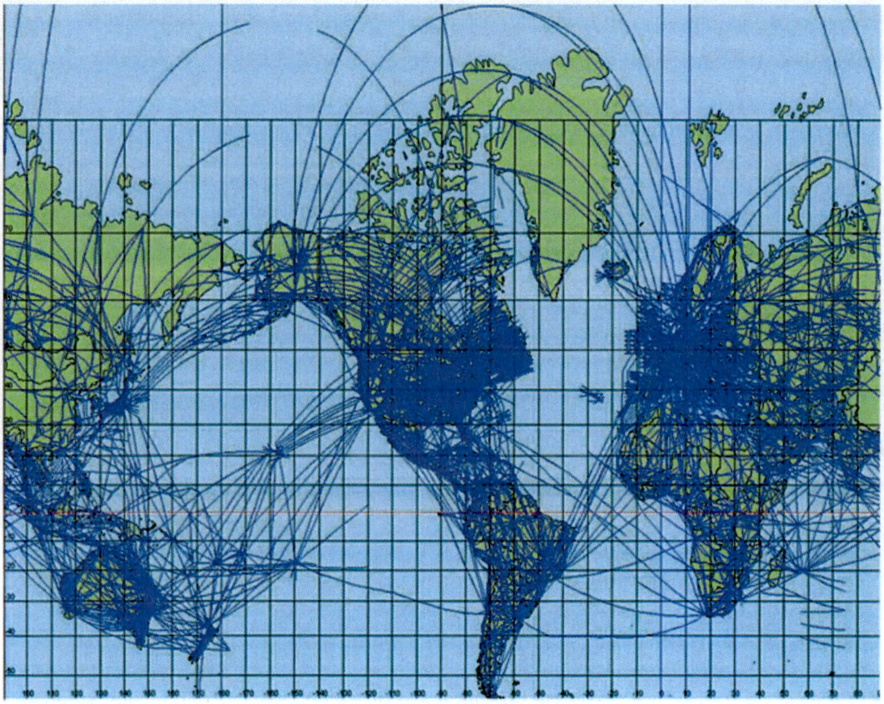

Fig. 6.6 The world's statically-defined airways (© 2011 Jeppesen)

6.2.2.3 Network Optimization Approach

Rather than optimal control, we choose network optimization as our basic formulation. Two distinguishing features emerge: the effort required to identify the set of nodes and edges, and the dynamic nature of the cost of an individual path (considering the other decision variables and constraints).

Over a large portion of the world, aircraft are limited to an airway structure similar to a road structure on the ground. Permanent points are defined, sometimes in conjunction with a navigational radio on the ground, sometimes just a specified latitude and longitude. Airways connect the points, although in some parts of the world it is permitted to fly directly between points not connected by airways. Figure 6.6 is a snapshot of the airways defined around the world in the mid-2000's.

A close examination of Fig. 6.6 reveals airways that have "dead ends", generally over water. In addition to the statically-defined airway structure, there are trans-oceanic airways published on a daily basis in high traffic density areas such as the North Atlantic. In other oceanic areas, aircraft fly unstructured routes between points defined by their latitude and longitude.

Some areas over land have moved to less-restrictive routing systems. In much of the United States, a regular grid of named points, the "National Reference

System" (NRS), has been introduced, and aircraft can fly routes defined as a series of these points, according to restrictions published by the Federal Aviation Administration (2003).

Further complications arise, however, from restrictions imposed by various authorities. Specific airways may be closed at certain times of day or day of week, on a regular basis. More complex restrictions on which airways or points a certain flight can use are imposed in the US by the FAA and in Europe by individual state ATC service providers (ATSP). In Europe, Eurocontrol's Central Flow Management Unit (CFMU) collects restrictions from the individual ATSP's and republishes them as the Route Availability Document (RAD) (Eurocontrol 2006), updated every 28 days, and the Conditional Route Availability Message (CRAM), currently distributed daily but soon to be more frequent. In the US, the dissemination of restrictions is less formal, consisting of some lists of routes (e.g., "Conditional Departure Routes" or "Playbook Routes"), ad hoc Letters of Agreement, and further restrictions published in Notices to Airmen (NOTAMS). These restrictions can be exclusionary or inclusionary, issued on a real-time basis, and even be dependent on other elements of a flight's route (e.g., a certain airway cannot be flown by aircraft using a certain other airway, regardless of origin or destination). Increasingly, ATC providers also issue restrictions based on real-time traffic congestion.

Certain airway segments may also be unavailable to a specific aircraft based on its capabilities or equipment. For example, some airways require aircraft to be certified to specific levels of Required Navigational Performance (RNP) or Reduced Vertical Separation Minimums (RVSM). Some, but not all, aircraft carry life rafts and/or life jackets, establishing three different maximum distances from land that an individual aircraft can fly. An airline's fleet of similar aircraft may have different levels of capabilities or equipment, either permanently (e.g., life rafts not installed) or temporarily (e.g., a certain sensor is out of service, allowing flight but suspending a navigational performance certification).

Together, the static airway structure, dynamic airway structure, and unstructured routing form a network of vertices and edges from the departure airport, known as the Point of Departure or POD, to the arrival airport, the Point of Arrival or POA. Development of the network from the various domains is a non-trivial information management problem, but mathematically straight-forward.

De Jong (1974) was the first to propose a network optimization approach based on dynamic programming. He describes the application of Bellman's shortest path algorithm (Bellman 1958) in the context of flight planning and points out its suitability in relationship with the solution of partial differential equations in a continuous system. De Jong highlights the advantages of dynamic programming to find the shortest path through the network over control theory approaches, which center mainly around the capability to better model complex airspace constraints. He however also points out the computational challenges inherent with the size of these networks, especially in his time. Today, industrial strength optimization models overcome these challenges by dynamically updating and sizing the network. Sophisticated network modeling can replace or approximate some of the resource constraints used in a classical dynamic programming model and therefore find a fine balance between the speed of calculation and the quality of a solution. At the same time, the deployment of vastly improved shortest path methods has made these network optimization approaches much more tractable so that iterative correction cycles are much easier and faster to accomplish.

In the following, we provide a high-level mathematical formulation of flight planning as a network optimization problem. First, we need to capture all the regulatory and operational constraints in the form of rules, which determine the network itself, i.e., whether two nodes can be connected by an edge, and restrict any feasible path through the network. The set Z of constraints involves the edges themselves, the state of the aircraft, and the departure time, t_D. The set of edges $E = \{e_j\}$, $j = 1, \ldots B$ that are available for a given flight cannot be predetermined, but must instead be evaluated against Z during the search for the optimal path, since some of the availability rules depend on what other edges are being traversed, or what specific time an edge is reached, which depends not only on t_D but also the path taken so far.

Each edge e_j also has set of available altitudes $H_{e_j} = \{h_\gamma\}$, $\gamma = 1, \ldots, G$, which may also vary by time of day, thus by the time the edge is reached.

Unlike many network-path optimization problems, the cost associated with each edge is not a simple value that can be pre-computed, and, in fact, the total cost of a given path is not a sum of the costs of the individual edges.

The overall objective function representing total operating cost, consists of four major sources of cost: fuel, time, overflight fees, and lost revenue due to spilled payload. The time costs can be further divided into costs based on flight duration, and costs based on absolute arrival time.

The fuel and duration costs are sums of the consumed fuel and elapsed time of each edge. Because these costs depend on the state of the aircraft at the initial node of the edge, they cannot be pre-calculated for each edge. The fuel, especially, cannot be accurately predicted without knowing the aircraft weight, since the fuel flow is a nonlinear function of weight. The speed generally varies with aircraft weight as well, and the total fuel and time depend on the forecast temperature and winds, which vary with the absolute time when the aircraft arrives at the initial node of the edge.

The objective is to find the optimal 4-dimensional path from origin (POD) to destination (POA) in a 3-dimensional network where speeds V_i, $i = 1, \ldots n$ on each edge constitute the fourth dimension. The cost function of the flight plan optimization problem can be formulated as follows:

$$\text{Minimize } C_F(W_D - W_A) + C_D(t_A - t_D) + C_A(t_A - t'_A) + \sum_{r \in FIR} C_N(r) + \sum_{k \in K} C_P(P_k) \tag{6.3}$$

where:

C_A: costs based on the difference between actual and scheduled arrival times (e.g., due to airspace/airport congestions, noise reduction policies or airport closures)
C_D: costs based strictly on duration of flight (e.g., maintenance or lease rates)

C_F: cost of consumed fuel including price of fuel per weight unit, typically varying by POD
C_N: navigation charges based on FIR r, where flight is taking place
C_P: cost of not being able to carry revenue-generating payload element k (spill costs)
W_D: weight of aircraft at departure
W_A: weight of aircraft at arrival
P_k: value of specific payload element k spilled
t_D: time of departure
t_A: time of arrival
t'_A : scheduled time of arrival
r: index for specific Flight Information Region (FIR)
k: index for specific payload element spilled

The constraints include weather related restrictions, speed and altitude limitations, fuel constraints, and many more. Before we highlight the components of the cost function in more detail and some of the main constraints, we introduce an important concept called *cost index*.

6.2.2.4 Cost Index

For modern aircraft, it is typical to allow an onboard computer to vary the local velocity subject to a scalar parameter CI, the cost index, which is an assumed ratio of the cost of time to the cost of fuel. A higher CI corresponds to higher speeds and constitutes one of five independent variables that determine the actual speed. The other four are altitude, weight, temperature, and wind velocity. The flight plan optimization will then correctly predict the speed at any point in the flight; the speed is necessary to accurately calculate the fuel and time.

Figure 6.7 shows the variation of CI from 0 (minimize fuel, time costs are nil) to 999, creating a Pareto front of minimum-fuel solutions for each possible arrival time.

The definition of CI assumes a purely linear cost of time. As we will discover in further discussion of the objective function (6.3), this is not the case. Thus, the CI, rather than being a pre-determined parameter, is a decision variable in flight plan optimization.

Both Boeing and Airbus recommend that a single CI value be used for the entire flight. The speeds for individual segments will vary as the five independent variables determining CI change, but a single scalar decision variable can be used for speed optimization for the entire flight.

6.2.2.5 Flight Operating Costs

The objective function (6.3) is a linear combination of several individual costs. Since they share a common unit (currency), no weighting is needed. While the

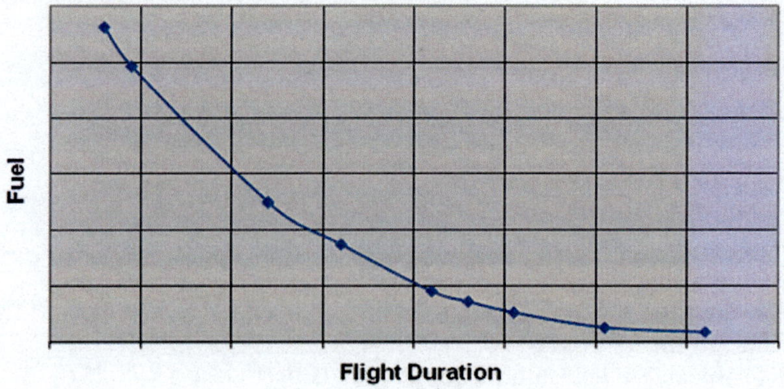

Fig. 6.7 Varying the cost index creates a Pareto front, a family of minimum-cost solutions trading time against fuel (© 2011 Jeppesen)

objective function is a linear combination of the four terms, the terms themselves are not linear with respect to the decision variables.

6.2.2.6 Fuel Cost

The fuel cost term $C_F(W_D - W_A)$ is straight-forward at first glance—the amount of fuel burned during the flight, multiplied by the price per unit. Several operational details, however, impact the implementation.

The fuel required for a flight is determined in units of mass (kg or lb). Fuel is sold, however, in units of volume (usually US gallons or liters). Unlike water, jet fuel varies in density, both due to temperature and specific formulation of the fuel, so the fuel density must be known to determine the cost of the fuel required for a specific flight.

Fuel prices vary from airport to airport. Each airline has its own contracted price at specific airports. These prices, which are generally considered competitively sensitive information, must be made available to the flight planning system to calculate the fuel costs.

There is often a waiver of taxes on fuel for international flights. Consequently two flights operated by the same airline and being fueled at the same airport may have different gross fuel prices. Furthermore, the definition of "international flight" may apply to a domestic segment that continues internationally, provided that certain criteria linking the segments are met. Eligibility for the tax waiver may vary from day to day for a specific domestic flight segment.

A common practice regarding fuel costs involves the carrying of additional fuel on flights departing from airports with cheaper fuel, to use on subsequent flights from airports with more expensive fuel. The resulting higher aircraft weight leads to higher fuel burn. The flight plan calculation must evaluate whether the advantage is greater than the cost of the additional fuel burn. In fact this practice,

Table 6.1 Time-based costs for a flight

Source	Costs based on duration	Costs based on arrival time
Aircraft	• Time-dependent maintenance costs • Hourly lease rates	• Subsequent flights • Operational flexibility
Crew	• Direct pay (some operators)	• Subsequent flights • Operational flexibility
Payload		• Re-accommodation • Compensation

known as "fuel tankering" or "fuel ferry," also leads to other cost increases, such as the increased wear on engines and brakes of the higher takeoff and landing weights, and possibly higher spill costs, which must be included to obtain the optimal decision.

Finally, in the present formulation we could include emissions-based costs in the fuel costs. The first implementation of emissions costs is the European Union Emissions Trading Scheme (EU-ETS), described in European Parliament and Council of the European Union (2009). This presents a new problem in the fuel cost calculation, because each operator has a certain emission allowance, and is required to purchase additional allowances or offsets for fuel burned beyond their allowance. The calculation of the cost contribution towards emission allowance or offset purchase for a single flight is difficult to estimate. One solution is for airlines to maintain high-fidelity estimates of their total fuel use, and an estimate of the margin by which they will exceed their allowance, and increase the fuel price for each flight by a corresponding amount.

6.2.2.7 Time-Based Costs

Of the four terms in the objective function (6.3), the time-based costs,

$$C_D(t_A - t_D) + C_A(t_A - t'_A) \tag{6.4}$$

are the most complex, the most difficult to model correctly, and the most challenging in their mathematical nature.

The first term of (6.4) is related to the elapsed time of the flight (the duration, independent of the actual departure or arrival times), while the second term is related only to the arrival time. The costs based on arrival time are further decomposed into those that are linear in elapsed time between actual and scheduled arrival and discrete costs that occur when the arrival is within a specific time interval. While there are direct costs of longer flight or later arrival, there are also indirect costs that should be taken into account but are more difficult to express explicitly.

Table 6.1 summarizes different costs associated with the duration or arrival time of a flight, each of which we then discuss in more detail.

6.2.2.8 Time-dependent maintenance costs

Aircraft must have regular maintenance at specific flight-time intervals. This results in a linear cost of time expressed in units of currency per flight hour. Since this depends on flight duration, independent of arrival time, it is part of the first term of (6.4).

6.2.2.9 Hourly Lease Rates

Some aircraft and engine lease rates are per flight hour, rather than calendar days. This results in a linear cost of time expressed in units of currency per flight hour. Once again, this cost is included in the first term of (6.4).

Some lease terms have flight time thresholds, beyond which additional costs are implemented. These are more difficult to model in the context of an individual flight, but if the total flight time of the aircraft or engines is estimated and compared with the lease terms, these can be expressed as an additional amount of currency per flight hour.

6.2.2.10 Subsequent Flights

If the current flight arrives late enough to impact the departure time of the next flight the aircraft is required to operate, there are costs associated—either for delaying the next flight, for swapping flights with another aircraft (reducing the optimality of the aircraft routing), or for using a spare aircraft. This applies identically to crew; their next flight may be delayed, crews may be swapped, or a reserve crew may be used. Even if the aircraft or crew are not scheduled to operate another flight immediately, a late arrival can impact their next flight after planned maintenance of the aircraft or required rest time for the crew.

Depending on the detailed consequences of aircraft or crew arriving too late for their next scheduled flight to depart on time, these costs may be in the second term of (6.4).

6.2.2.11 Direct Pay

Some, but not all, crews are paid based on their total flight time. This could be linear total time, or it could be a minimum scheduled time for the flight with additional time above that paid in currency per hour, which complicates the formulation of this term. In fact, some crews have increasing rates beyond certain time thresholds, further complicating the model.

6.2.2.12 Re-Accommodation

If passengers or cargo arrive too late for their connecting flights, there may be a cost associated with providing alternate means for them to reach their destination. The information management problem for these costs is complicated by different re-accommodation policies, with corresponding different costs, for different value-levels of payload (e.g., different elite levels of frequent flyer, different fare class of passenger or cargo). These costs, which are in the second term of (6.4), are discussed further in Sect. 6.3.

6.2.2.13 Compensation

Some passengers or cargo receive direct compensation for delayed arrival at their final destination. As with re-accommodation, the actual costs vary with the value-level of the payload component. These costs are in the second term of (6.4).

6.2.2.14 Operational Flexibility

There is additional cost associated with accrued flight hours and/or late arrival of aircraft and crews due to the resulting lack of operational flexibility. These costs could be immediate, in the inability to swap aircraft or crew to cover for another late incoming flight, or could be delayed, in adding constraints to a subsequent aircraft or crew assignment optimization because the aircraft or crew in question have fewer available hours.

Two key aspects of the time-dependent costs impact practical solution strategies for flight plan optimization. One is the amount of detailed data, which change on a daily basis, such as the levels of compensation or re-accommodation for individual payload elements. In fact, connectivity data for aircraft, crew, and passengers are highly dynamic, since they are coupled to other flights whose status can change. The solution quality, therefore, depends on the level of integration with other systems throughout an airline that track these data. The second, related, aspect is the fact that some of the costs are based on the duration of the flight, while others are based on the actual arrival time. This sometimes provides a trade-off between discrete flight plan options, where a departure delay reduces flight duration and fuel, but results in a later arrival time. Correctly evaluating this trade-off demands real-time integrated data.

6.2.2.15 Overflight Charge Costs

Most countries fund their ATC agencies (whether public or private) by charging fees to fly through their airspace. The formula for the fees can vary depending on whether the flight is departing or arriving in the country in question, or overflying it without landing. In (6.3), we separate the overflight charges,

$$\sum_{r \in FIR} C_N(r)$$

into two components: one based on distance flown within a specific ATC area or *Flight Information Region (FIR)*, and one, often called "unit charges", consisting of a flat fee for entering the FIR.

A guide to worldwide overflight charges is published by IATA, formerly in book form, but now as an interactive online tool (IATA 2010).

The significance to flight plan optimization is that the charges vary greatly from country to country, and different routing options for a single flight can have different overflight charges based on which countries they overfly (or how much distance they spend overflying each). A longer route, with higher fuel and time costs, may be optimal if it avoids expensive airspace.

6.2.2.16 Spill Cost

Spill cost refers specifically to lost revenue when a certain plan for operating a flight reduces the maximum payload that can be carried below the revenue-generating payload available to the airline. If a flight is limited by the total takeoff weight of the aircraft, increasing the fuel required reduces the payload that can be carried. If the route, profile, and speeds that minimize the other three components of the cost increase the fuel load and thereby reduce the payload that can be carried, the cost of lost revenue must be included and may change the optimal solution.

Our model of spill cost in (6.3),

$$\sum_{k \in K} C_P(P_k)$$

accounts further for the fact that payload, whether passengers or cargo, is quantized into individual weight amounts, each of which may have a unique cost rate. $C_P(P_k)$ determines the cost if payload element k is left behind to allow a 4D path requiring more fuel.

6.2.2.17 Fuel and Time of Flight Constraints

The fundamental calculations driving flight plan optimization are the determination of fuel and time of a flight.

One physical characteristic of flight that drives the implementation of flight planning systems is the variation of speed and fuel flow (which determine operating cost) with altitude, temperature, and the weight of the aircraft. Figure 6.8

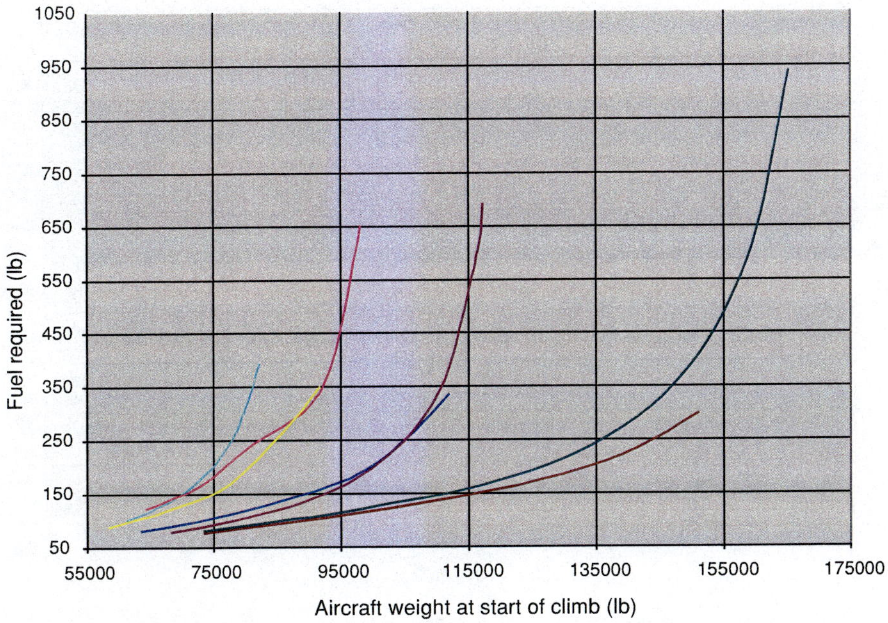

Fig. 6.8 Variation of fuel required for a 1,000 foot enroute altitude change with aircraft weight (© 2011 Jeppesen)

shows the variation with respect to aircraft weight of the fuel required for a 1,000 foot enroute climb. The curves shown are from a variety of different aircraft types, from multiple manufacturers and technology generations, but have a similar nonlinear nature.

Figure 6.9 shows the variation in total aircraft weight over long- and medium-range flights. The significance of Figs. 6.8 and 6.9 together is that the calculation of aircraft weight is dynamic—the fuel flow at an arbitrary point in the flight cannot be determined without a detailed calculation of the flight between that point and a boundary point (departure or arrival depending on which is known), and it is impossible to fix both the arrival and departure weight independently. This limits the algorithms that can be used to optimize the flight plan.

Another relevant aspect of aircraft performance is the weight limitations on an aircraft, which may involve trading payload for fuel. This is taken care of by limits to the takeoff and landing weights for a flight to maximum values. The takeoff weight of the aircraft may be limited by its structure, a certification for the specific airframe by the airline. More likely, however, the takeoff weight is limited by some combination of the departure airport (3D runway geography, obstacles near the airport), meteorological conditions at departure time (wind, temperature, and runway surface contamination such as rain or snow), and condition of the aircraft (tire and break wear, inoperable systems such as steering or automatic braking). Evaluating this limit requires a complex calculation involving real-time data.

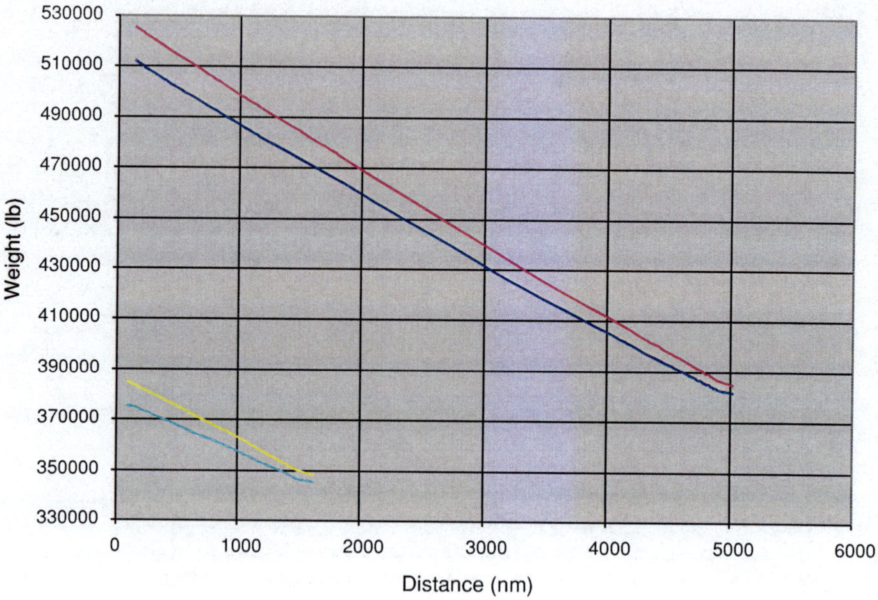

Fig. 6.9 Variation of aircraft weight over long- and medium-range flights (© 2011 Jeppesen)

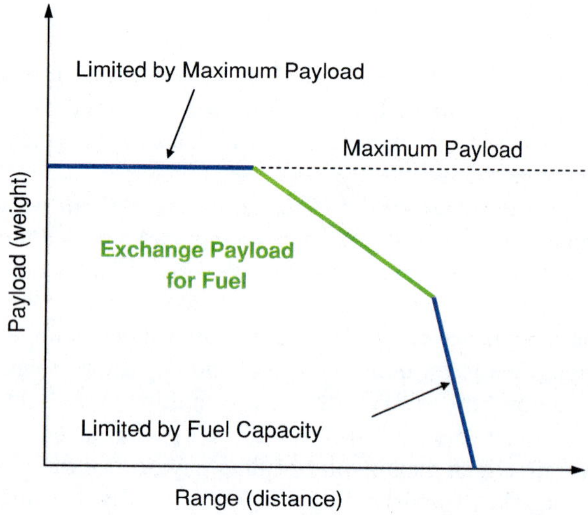

Fig. 6.10 The payload-range diagram illustrates the relationship between the distance of flight and the amount of payload that can be carried (© 2011 Jeppesen)

Figure 6.10 shows the payload that can be carried by an aircraft at a specific maximum takeoff weight as a function of the flight distance (labeled as "range"). Up until a certain distance, the aircraft's maximum payload can be carried. At some point, however, the aircraft reaches its maximum takeoff weight. To fly further, payload must be removed, to be replaced by more fuel. This tradeoff continues until the aircraft's maximum fuel capacity is met; beyond this point, modest gains in range are possible by reducing payload further, because the weight of the aircraft is reduced, so fuel consumption decreases.

It is the reduction in payload with increased light distance that gives rise to the "spill cost" term in the objective function (6.3). A longer distance route that might be optimal with respect to time-based and overflight fee costs may, because of the higher fuel required, reduce the payload that can be carried.

6.2.2.18 Weather

The characteristics of flight, especially climbing and cruising flight, depend heavily on both winds and temperatures.

The fuel flow for a given weight, speed, and altitude depends additionally on the ambient air temperature, which can very over a range on the order of 50 degrees Celsius.

The optimal route is heavily dependent on the wind, since the ground speed (which determines how quickly the aircraft actually reaches its destination, and how much fuel it burns in doing so) is a simple vector addition of the airspeed (the speed of the aircraft relative to the air, created by the aircraft itself) and the wind. On any given day, the wind-optimal route (shortest path in the air) can be up to 10% longer than the great-circle "shortest path" over the ground.

The aviation industry is a driver of weather forecasting technology, and development of new wind and temperature models will likely continue indefinitely. Currently the primary sources of worldwide wind and temperature forecasts are the United States National Weather Service and the United Kingdom Meteorological Office. These forecasts form a data set of so called vector wind velocity constraints.

Currently, most (if not all) operational flight planning systems accept the forecasts as deterministic and fully accurate. The contingency fuel constraint is partly intended to account for uncertainty in the forecast winds and temperatures. As forecasts have become more accurate, contingency fuel requirements have been reduced. In Europe, operators can base their contingency fuel on a statistical analysis of planned vs. actual fuel burns, which implicitly accounts for the accuracy of forecast weather. Tino et al. (2009) explored the impact on aircraft trajectories of uncertainty in forecast winds and temperature. While their motivation was trajectory prediction for air traffic management, extension of their work to airline flight planning could identify ways to more explicitly account for uncertainty in forecasts.

Besides the actual forecasts used for operational flight planning, historical data are required for marketing studies and route planning. When considering a city-pair that may be flown, or considering which aircraft type to use for a city-pair, it is useful to know how long the flight will take, how much payload can be carried, and how much fuel will be burned for a given aircraft type. The calculation of percentile wind-components (i.e., winds that are less favorable than $n\%$ of historical days during a specific month or season) is not completely straightforward because of the presence of the jetstream, a narrow band of very strong eastbound winds in the altitude range where airliners fly. Wind-optimal routes almost always avoid the jetstream when flying westbound, and will divert substantially to “surf” it flying eastbound. Without special handling, a high percentile wind field would assume the jetstream does not exist for eastbound flights and is “everywhere” for westbound flights, which is unrealistic. A widely-used method for handling this delicate issue was first proposed by Sawyer (1950).

6.2.2.19 Maximum Speed and Altitude

The model requires constraints that limit the combination of speed and altitude for each edge. In general, the speed for each available altitude is chosen to minimize cost. The altitude, speed, and temperature then provide a maximum weight based on aircraft data; if the aircraft weight W_i exceeds this maximum weight, the altitude in question is infeasible, and another altitude must be chosen.

Further limits on altitude or speed are based on inoperable or degraded systems on an individual aircraft. Real-time data regarding the condition of the aircraft, and a database of rules defining the limits associated with each configuration deviation, are necessary to accurately implement these constraints. A further limit on aircraft speed is based on forecast turbulence, defined here as a change in vector wind velocity with respect to altitude.

6.2.2.20 Fuel Constraints

The model also requires a constraint that enforces additional fuel, beyond that planned to be consumed during the flight, required by regulatory agencies and individual airline operational policy.

In most cases, the amount of additional fuel required depends on the conditions of the aircraft at the end of flight—at a minimum its weight, but for the percentage fuels also its final cruise altitude and the temperature there (thus, the detailed location). This means the total required fuel for the flight can only be iteratively solved whether calculated backwards from the POA or forwards from the POD, and depends on the profile and route selected, further complicating the overall optimization of a flight plan.

The constraint for the required additional fuel is usually broken down into several terms. The first term is a percentage of the planned trip fuel while the second term is a fixed amount. The third term is a required duration flown at a calculated fuel-flow rate, defined at a specific aircraft state, usually at the end of cruise, just before descent. The final term is additional fuel for operational requirements, such as forecast ATC delays or assumed tactical path deviations due to forecast convective weather.

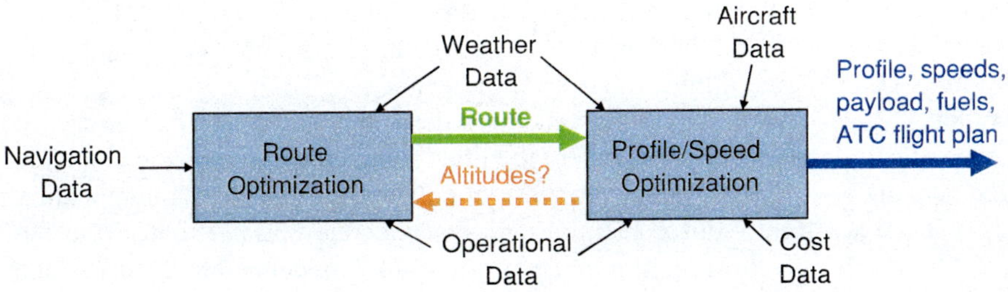

Fig. 6.11 Decomposition of flight planning optimization into two subproblems (© 2011 Jeppesen)

Other fuel regulations deal with the loss of an engine and/or pressurization either far from diversion airfields (e.g., over the ocean) or in regions of challenging terrain. A further constraint on fuel is to limit the total departure fuel to less than the volumetric capacity of the aircraft's fuel tanks. Here one explicitly uses the fuel density, which varies by POD and ambient temperature.

6.2.2.21 Sequential 2D Network Solution Approach

The approach described above is challenging from a computational point of view and hence quite often for practical reasons, a second approach is used. This approach decomposes the 4D path optimization problem into two 2D subproblems: Route Optimization and Profile/Speed Optimization. The first subproblem is a network optimization problem to solve the 2D route or ground track. The second subproblem determines the altitudes, speed, payload, and departure fuel. This approach is illustrated in Fig. 6.11.

In the Route Optimization sub-problem, the network from POD to POA is determined, with constraints from ATM, traffic, severe weather, and aircraft capabilities applied. The wind forecast (2.10) is used to find a wind component for each edge in the network, and an aircraft speed is assumed to determine an equivalent air distance for each edge. This equivalent air distance is the cost for each edge. The optimal 2D route is then the minimum-distance path through the network. This approach is described by de Jong (1974), who lists several potential network optimization methods that can be used.

As previously described, all current routing options, including oceanic unstructured routing (which in practice is constrained to specific intervals of latitude and longitude), occurs between specific locations that form a grid. For areas of unconstrained airspace (consistent with proposed future air traffic management concepts), an artificial grid can be created, as is done by de Jong (1974) and Wilson et al. (1997). Hence, using network optimization to find the horizontal route is universally applicable.

Once an optimal route is found, the Profile/Speed sub-problem determines the altitudes, speeds, and departure payload and fuel for the route.

Since only discrete altitudes can be flown, the altitude profile consists of a series of level sections with altitude changes between them. The altitude changes are commonly called "step climbs" and "step descents".

While the location of step climbs and step descents can be made arbitrary, in practice most operators choose for them to either commence or complete at a node of the network (i.e., a named point) for ease of crew execution and communication with ATC. If we follow this constraint, and consider the optimal route from the Route Optimization to be a set of $n + 1$ points $i \mid i=0,\ldots,n$, defined by their latitude and longitude, and the $n$great-circle segments between them, then at each point i we must decide whether to continue at the same altitude, climb to a higher altitude, or descend to a lower altitude, for the transit to point $i+1$. Let us place a node x_i^a at each point i at each allowed altitude $a|a \in A$, where A is the set of allowed altitudes. Then, for each segment between i and $i+1$ we can define edges connecting all the nodes x_i^aat point i at each allowed altitude a to all the nodes x_{i+1}^aat point $i+1$ at each allowed altitude a. Defined this way, the profile optimization problem could be solved as a network optimization problem. Several details of the domain, however, constrain the set of methods that can be used.

An essential feature of profile optimization is that the cost to climb or descend to, and then fly at, a certain altitude, depends on the weight of the aircraft at the decision point (recall Fig. 6.8 and its associated discussion). Thus, the optimal altitude at which to fly from i to $i+1$ cannot be determined without knowing the weight at x_i^a (i.e., it is dependent on the path taken from *POD* to current position i). This means the cost of each edge in the network profile optimization profile cannot be determined in advance, but must be calculated during the optimization process.

A limit to further decomposition of the optimization is that speed optimization, which is non-linear, is highly coupled to profile optimization, so must take place simultaneously, or iteratively. In practice, it is effective for the Profile/Speed Optimization sub-problem to consist of an outer iterative optimization of speed, with the profile, payload, and fuel optimized for each speed in an inner algorithm.

The speed optimization itself should use CI as the decision variable. The CI is varied using a single-variable nonlinear optimization technique (line search) to find the total cost of the flight, where for each candidate CI the profile is optimized. The discrete nature of the profile optimization introduces discontinuities in the derivative of CI with respect to cost, and in the range of arrival time as a function of CI, so a robust line-search technique must be used. Lidden (1992) discusses tactics for dealing with such discontinuities. Further discontinuities are introduced by the time-based elements of the objective function (6.3), especially those based on arrival time. Altus (2010) discusses a heuristic algorithm where discontinuities in total cost as a function of CI are calculated, intervals that might contain the global minimum are identified, and independent line searches are carried out on each of these intervals.

The primary difficulty with the sequential 2-D approach is that the winds vary by altitude, so an altitude for each network edge must be chosen to use as the basis of wind application in the Route Optimization sub-problem. There are several approaches to determining which altitudes, and therefore which winds, to use in Route Optimization. In addition to an altitude for winds, an aircraft speed must be chosen, since the effect of wind on the air distance varies depending on how fast the aircraft is flying. In any event, the altitude used for each edge should take into account the available altitudes for that edge, including any known restrictions.

The simplest approach is to use a heuristic rule to select altitudes. Depending on the quality of the heuristic, this can be very effective. For a scheduled operator with extensive historical data on a specific airport pair, the typical altitude for a specific aircraft type and payload may be fairly certainly known.

An extension of the heuristic method is to actually consider multiple heuristics, and generate a route for each one. For example, a heuristic rule could be to assume flight starting at one altitude and step-climbing to the next higher available one, but at three different lateral points (expressed as a distance from the POD), generating three different lateral routes. Then each of the routes is used as input to the Profile/Speed Optimization sub-problem, and the best solution is identified.

A third approach is to use a reduced basis method—a lower-fidelity model that is faster, but less accurate—of the profile/speed optimization to approximate the optimal profile within the Route Optimization sub-problem. The Profile/Speed Optimization sub-problem is then run to refine the profile and speeds and accurately calculate the fuel and time. The reduced basis model might use an estimated departure weight and not ensure feasibility with performance constraints such as required contingency fuel.

Finally, the two sub-problems can be run iteratively—a heuristic provides the original altitudes to use in route optimization, an optimal profile and speeds are calculated, route optimization is re-run using the optimal profile and speeds from the first iteration, and Profile/Speed Optimization occurs again. Two iterations is nearly always sufficient for convergence. As with all iterative subspace optimizations, however, it is possible to converge to a non-optimal solution by getting "stuck" in the wrong region of the search space.

Any of these methods can help find the optimal route as part of the optimal 4D path. The choice of which to use depends on the actual intended use (e.g., average frequency, length of routes, prevalence of ATC restrictions and non-homogeneous overflight charges in the main area of operation), and the implementation details with respect to the specific subspace optimizations being used. In general, however, the reduced basis method is most accurate, and deals best with taking overflight charges and 4-dimensional no-fly areas (e.g., severe weather) into account in finding the optimal lateral route.

6.2.2.22 4D Network Solution Approach

A natural extension of the network optimization approach described in the previous section is the creation of an extended network with nodes at each lateral waypoint at each available altitude for the airspace in question. In this manner, the two 2D problems are combined to a single 4D problem that can be solved directly.

This approach was described by de Jong (1974), with numerical examples, and is also used by Sorensen and Goka (1985) and Wilson et al. (1997). A further extension of this general approach is the specialized algorithm described by Tang and Mukherjee (2000).

As with profile optimization in the sequential 2D approach, the cost to fly the next segment from *i* to *i+1* depends on the path taken from *POD* to the current position *i*, so the cost for each edge cannot be pre-determined, but must be calculated during the optimization process. Obviously, the dimension and complexity of the problem solved with this approach is significantly higher than the previous one, which significantly increases the computation time. However, by using 'smart' heuristics to reduce the feasibility space without significantly deteriorating the quality of the solution, this approach is showing superior results.

6.2.3 Flight Planning Tools

While a few of the largest airlines use internally-developed tools for flight planning, most use a tool developed and maintained by a third-party system provider. Due to the highly specialized nature of the domain, most of the vendors who provide such systems have diverse aviation operations portfolios.

The primary goal of these systems is to ensure legality of the planned flight with respect to the relevant regulations, but they all optimize flight plans to provide additional value to the airlines that use them. Some of the implementation details that differ between the different systems are as follows:

- The extent to which they can be integrated with real-time data from other operational systems—whether from the same vendor, internally developed by the airline, or provided by a third-party.
- The fidelity of the cost model used as the objective function.
- The extent to which they can be integrated into user interfaces that airline operational staff use for other tasks.
- The ability to quickly and flexibly re-optimize the flight plan subject to real-time data updates, both before and after departure.
- The extent to which, and effectiveness with which, they model complex, dynamic, airspace restrictions such as RAD, CRAM, and FAA Letters of Agreement.

Whether the solutions any real system provides to the optimal flight plan problem are truly globally optimal depends as much on the quality of the data—on cost, operational constraints, aircraft characteristics, etc.—as on the algorithms used to solve the problem. Furthermore, users may elect to fix certain decision variables, either for the preference of the flight planner or flight crew, lack of confidence in constraint modeling, or faster computational performance.

Whether flights are re-optimized after departure depends largely on organizational issues for the individual airline and the airspace in which they are being operated. The number of flights an airline dispatcher is responsible for at a time varies greatly, allowing varying amount of attention to be devoted to individual flights. In many areas, airline-initiated reroutes are not possible, or possible only by pilots directly requesting them from air traffic controllers by voice, which is inefficient. Future air traffic management concepts, such as NextGen (FAA 2009) and SESAR (Eurocontrol 2009), should increase the practicality of post-departure re-optimization. While many proponents of such concepts assume that new trajectories will be determined by pilots, the detailed cost model (6.3) available to ground-based flight planning systems will likely keep airline operations centers involved in the re-optimization process.

6.2.4 Conclusion

For some time airlines have considered flight planning a commodity, a necessity for which different solutions had no special value. During this time period, published work pertinent to flight planning was usually in the context of small-scale trajectory optimization for onboard computer flight management systems, or studying the interaction between flight planning and other topics, such as advanced ATM concepts.

The rise in fuel prices and premium on operational efficiency, as well as the completion of other cost savings possibilities, have caused airlines to re-consider improved flight planning as a cost saving possibility.

In addition, general advances in OR and IT capabilities have introduced the possibility to include flight planning in bigger, multidisciplinary optimization efforts.

Furthermore, as the regulatory and operating environments continue to evolve, flight plan optimization must evolve with them, because problem formulations are highly coupled to the mathematical nature of the domain.

We therefore expect to see continued refinement of the methods described in this section, and development of new ones, over the next several years.

6.3 Airline Disruption Management

6.3.1 Introduction

Airlines manage their resources according to a published flight schedule that is optimized from a revenue standpoint. However, the execution of the airline schedule is rarely regular. Irregular airline operations require quick and efficient corrective actions. The systematic way of dealing with irregular operations is referred to as airline disruption management.

Airline disruption management is a tremendously complex and challenging process. Airlines typically spend a couple of months carefully planning for the operations to come. However, once in operations, the recovery time is compressed to minutes. Although a relatively short recovery window slightly simplifies the task, efficiently solving very complex multi-resource problems within minutes is still an almost unattainable task.

Any event during the day of operations that interrupts the smooth execution of the airline's scheduled activities is considered a schedule disruption. The disruptions may be classified, according to their sources, as internal and external.

Internal disruptions originate within the airline and appear in the form of aircraft and crew unavailability. Aircraft become unavailable due to flight schedule changes or mechanical problems, while crews become unavailable due to unforeseen events, such as sickness, strikes, and other unplanned absences, as well as a result of flight schedule modifications.

External disruptions originate in the outside world and are most often caused by inclement weather, airport problems or ATC-related problems.

Both internal and external disruptions are further categorized as minor or major. Minor disruptions occur regularly, on a daily basis. Crew members calling in sick, or a few flights getting delayed or cancelled, are minor disruptions. Crew strikes, or hub closures due to inclement weather, are major disruptions. The average number of major disruptions varies from airline to airline, but usually ranges from 10 to 20 annually.

Although a day with no irregular operation is very hard to imagine, airline disruption management had not attracted much attention until relatively recently, when deteriorated financial results forced airlines to shift their focus from minimizing the planned (scheduled) costs to achieving the minimal realized (operational) costs. That shift resulted from awareness that cost-efficient planning for an ideal, disruption-free world often leads to poor performance when the plan comes to its realization, as little slack is available for absorbing deviations from the planned schedules. Tight plans are very vulnerable to disruptions which tend to extend far beyond the events that originated them. In extreme cases, even minor operational disruptions may lead to high operational costs when the original disruptions propagate through the tightly planned schedules. The ripple effects of the initial disruptions, which result in an increased volume of aircraft, crew and passenger problems, may be reduced by adding more slack during the planning process. However, a robust schedule, which contains slack to absorb operational disruptions, leads to a cost inefficient planned solution and a needless increase in the airline's costs when the operations run smoothly.

Regardless of the degree of robustness of a planned schedule, disruptions during the day of operations are inevitable. Better understanding of airline irregular operations and their implications on overall financial results, in combination with efficient airline disruption management, are the keys to improving airline performance.

Although the objectives of any recovery process are similar, recovery results are airline specific. One example is the type of network an airline operates. The hub-and-spoke networks, mostly operated by larger carriers, are very sensitive to irregular operations at their hubs that easily spread throughout the network thus creating major disruptions. Point-to-point networks are less sensitive to major disruptions, but are more vulnerable to the minor ones. While quite a large number of possible recovery options in a hub-and-spoke network allows for finding an efficient solution to a given problem, considering all those options while solving a problem in real time may significantly slow down the solution process. Point-to-point networks offer a very limited number of recovery options and therefore allow for a quick response, but in turn exhibit very low solution efficiency. The business processes in place, which can be quite different from one airline to another, is another example. While some airlines insist on keeping together the entire scheduled crew during the rescheduling process, others allow splitting crews to achieve better recovery efficiency. Proposing flight delays to the operation controllers whenever necessary to cover the input flight schedule by the required crew complement (crew members of various ranks required to staff the flight) is another option viewed as indispensable by some airlines and as unthinkable by others.

6.3.2 Processes and Solution Approaches

6.3.2.1 Planning Versus Disruption Management

Planning and operations differ in many aspects, the most important being the following:

- Problem scope: global vs. local
  While during the planning phase all available resources and information are considered to build a solution from scratch, the recovery phase aims at repairing the plan at a local level, considering primarily the parts of the system affected by disruptions. A recovery solution extends the already executed part of the plan and is extended by the future planned activities.
- Input data quality: static vs. dynamic data
  While the data used for planning purposes are static and therefore highly reliable, the recovery actions are taken in an uncertain, dynamic environment using information subject to constant change.
- Solution time: solving ahead of time vs. reacting in real-time
  Planning problems are solved well in advance of their execution and hence are not very time-sensitive. On the contrary, the recovery processes require making quick decisions in real-time.
- The objectives: minimize total planned cost vs. minimize the deviations from the plan
  While in the planning phase the objectives are defined in terms of cost minimization and efficient utilization of resources, during the recovery phase an additional goal is to minimize the modifications to the plan.

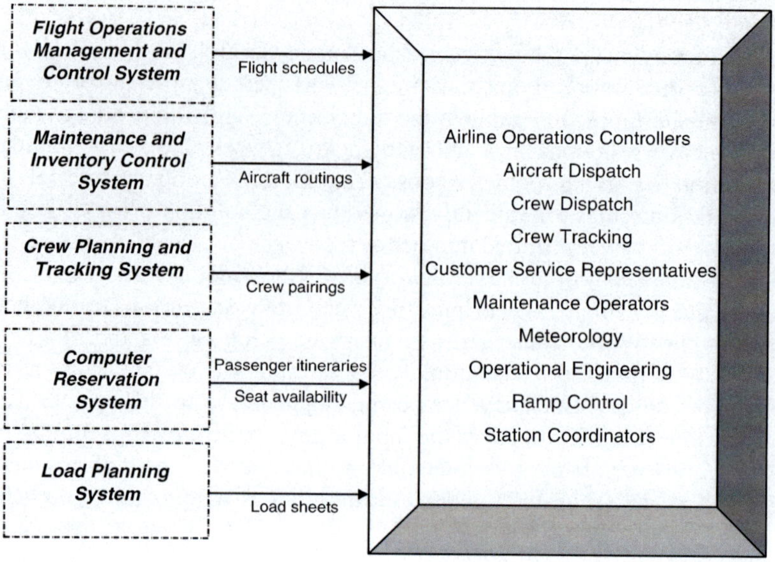

Fig. 6.12 Information flow within OCC

6.3.2.2 Disruption Management Organization and Airline Business Process

Schedule execution on a daily basis is managed through an Operations Control Center (OCC) (alternatively referred to as System Operations Control (SOC), Airline Operations Center (AOC) or Airline Operations Control Center (AOCC)). The OCC is the heart of the operations. Its purpose is to safely and reliably achieve the airline's operating goals on a daily basis. Some of the operating goals are to maximize system on-time performance and the number of scheduled flights actually flown, to preserve scheduled connections and, in the case of a hub-and-spoke network structure, to maintain hub bank integrity (i.e., to ensure that flights with highly connected passenger itineraries arrive and depart in a way that preserves these connections). In order to meet these goals, the OCC takes actions such as canceling and delaying flights, rerouting crews and aircraft, inserting extra flight segments and ferries (flights that do not carry passengers but reposition aircraft for subsequent flights), and uses other operational tactics with the ultimate goal to provide the best possible flight schedule to the passengers. Even though the airlines are traditionally organized in functional silos, each supporting silo is represented in the OCC at all times.

The information flow within an OCC, as portrayed by Clarke et al. (2000) from Sabre Research Group, is shown in Fig. 6.12. More details about the organization of an OCC are available in Kohl et al. (2007).

The traditional OCC decision process is hierarchical in nature, with aircraft, crew and passenger recovery problems being solved in that order.

Aircraft recovery decisions, such as flight cancellations and delays, equipment changes, and other applicable actions, often result in the disruption of crew schedules. The operation controllers, who monitor and coordinate airline activities with respect to schedule problems or any other potential problems, communicate the changes in the original flight schedule to the crew trackers. The latter are responsible for checking the feasibility of the modified flight schedule from the crew perspective and applying the corrective actions to fill the open crew positions in the altered input flight schedule. Internal crew disruptions, such as sicknesses or other unplanned absences, no-shows, or late reporting for flights, cause additional disruptions in the original crew schedule.

After taking appropriate actions to recover disrupted crew schedules, crew trackers communicate their solution to the operations controllers. When the flight schedule cannot be completely covered, they also reveal the set of uncovered flights and recommend actions that would help solve the problem from the crew perspective, such as delaying or canceling specific flights.

The operations controllers may either accept the proposed actions, or come up with a different set of delayed and cancelled flights, and ask the crew trackers to re-evaluate the new alternative from the crew perspective. It may take several iterations and a significant amount of time before a satisfactory, or at least a feasible, solution from both the aircraft and crew perspectives is found. The process becomes extremely time-consuming in case of major disruptions where even finding a feasible holistic solution for both aircraft and crew becomes a real challenge.

Once there is agreement upon the final flight schedule, modifications are communicated to the customer service representatives, who examine the feasibility of the original passenger itineraries and take corrective actions to repair those that are broken.

In spite of the hierarchical nature of the recovery process, airline operations controllers, crew trackers, and customer service representatives are in constant communication throughout the recovery cycle. Their collaborative efforts are crucial when assessing the feasibility and impact of possible decisions. Research on airline recovery performed to date and various developed and deployed recovery tools reflect this traditional, sequential business process.

6.3.2.3 Aircraft Recovery

Aircraft recovery, also known as operations or schedule recovery, is hierarchically the first recovery problem that an airline faces in the presence of irregular operations. Solving the aircraft recovery problem consists of modifying the original flight and aircraft schedules to compensate for the input schedule disruptions. The schedule modifications and aircraft rerouting decisions are usually made simultaneously.

The main goals of an aircraft recovery process are to minimize the deviations from the planned schedule and allow the airline to resume normal operations as fast as possible at the lowest possible cost. Minimizing the total deviation from the planned schedule is crucial not only because the flight and aircraft schedules are optimized in the planning phase and input as such to the recovery problem, but also because of the adverse effects that flight and aircraft schedule deviations may have on the other planned areas within an airline. A time span, within which the modifications to the airline's schedule are allowed, referred to as recovery period, is a mean of controlling the speed at which the disrupted airline returns to the normal operations. While the recovery period usually starts when a disruption occurs or is made known, the airline's target time to restore normal operations determines the end of the recovery period. The length of a recovery period depends on the size of the input disruption and the amount of time available to recover from it. While a full and cost-efficient recovery from massive disruptions requires a longer recovery period, the amount of time available to produce a recovery plan usually imposes a practical limit on the size of the recovery period, with shorter periods allowing for faster, but more costly, solutions.

Various actions, such as flight delays and cancellations, flight diversions (flying to an alternate airport), over-flying (flying to another scheduled destination) and ferry flights are employed on a daily basis to cope with operational disruptions. The resulted, altered flight schedule must comply with all applicable rules and restrictions, such as airport curfew (hours during which flights are not allowed in or out of a specific airport).

Aircraft rerouting is required whenever an original aircraft rotation (sequence of flights and maintenance activities assigned to an aircraft) becomes infeasible due to flight schedule modifications. The original rotations are modified by swapping flights between aircraft and by inserting new flights added to the flight schedule into the existing aircraft rotations, or by building new aircraft rotations for the unused available aircraft. Both the modified and new aircraft rotations must respect the aircraft maintenance requirements. In addition, aircraft balance requirements must be satisfied throughout the aircraft rerouting process, with the scheduled aircraft types properly positioned at the end of the recovery period. The aircraft balance requirements influence, among others, flight cancellation decisions.

When canceling flights becomes the only viable option, or the most promising of the available options, flights that form round trips that originate at hubs or major airports, rather than isolated flights, are removed from the original flight schedule. This approach creates temporary spare aircraft, referred to as virtual spares, which can help in solving other aircraft problems. It also allows the airline to cut off the disrupted part of the network and hence to prevent the disruption from propagating throughout the entire network. The end result is higher operational reliability, better aircraft utilization, and lower recovery costs. Canceling a single flight creates an aircraft surplus at the departure airport of the cancelled flight by leaving the originally scheduled aircraft unused. At the same time, it creates an aircraft shortage at the destination airport of the cancelled flight and, as a result, makes it impossible to fly the downstream flights that the original aircraft was scheduled to operate. Flying those unassigned flights would require bringing in a spare aircraft of the same type as the scheduled one. Being a very expensive resource, spare aircraft are available in very limited quantities and at a limited number of locations, usually only at major hubs. Creating an aircraft shortage at an airport with no spare aircraft available results in incremental recovery costs resulting from ferrying a spare aircraft from another airport.

No matter how good, an aircraft recovery plan, if taken in isolation, may have disastrous consequences on the crew and passengers. For example, although canceling flights may be a very attractive option from the aircraft recovery perspective, flight cancellations may create havoc with rerouting passengers. An efficient aircraft recovery strategy must therefore account for the downstream costs and impacts on crew and passengers, as well as on the gate utilization and ground workforce, just to name a few. Although the extent to which these effects are captured is model specific, a level of sophistication keeps improving over time in general.

In their pioneering work, Teodorović and Guberinić (1984) solve a problem caused by the unavailability of a single or multiple aircraft using flight retiming and swapping options. They use a branch and bound procedure to find a new daily airline schedule that minimizes the total passenger delay.

Several years later, Teodorović and Stojković (1990) proposed an extension of the previous work. They formulate the problem as a mixed integer linear program and solve it using a dynamic programming heuristic. The primary objective of their model is to minimize the total number of cancelled flights. The secondary objective is to minimize the total passenger delay on the preserved flights in case the number of cancelled flights is identical.

Subsequently, Teodorović and Stojković (1995) proposed an approach that sequentially creates crew and aircraft reroutes and allows for using aircraft swaps when an aircraft routing is maintenance infeasible (i.e., the assigned aircraft cannot complete the scheduled set of flights due to maintenance limitations).

Jarrah et al. (1993) propose two separate network flow models to cope with aircraft shortages: a flight delay model that only allows for retiming flights, and a flight cancellation model which only allows for canceling flights. Aircraft swapping and ferrying is allowed in both the flight delay and flight cancellation models. The objective function minimizes the cost of delaying or canceling a flight that depends on the total number of passengers on the flight, number of passengers connecting to other flights, possible down-line flight cancellations and the lost crew time. Revenue loss associated with passengers lost to another airline and passenger ill-will costs are also accounted for.

Using a basic space–time network representation, Yan and Yang (1996) incorporate, for the first time, flight cancellations, delays and ferry flights in a single model. Their work is further extended by Yan and Lin (1997) and Yan and Tu (1997). Yan and Young (1996) model the aircraft recovery problem as a minimum cost flow model with side constraints and use the Lagrangian relaxation with a subgradient method to solve it. Their objective function maximizes the airline profit obtained by subtracting the cost of delays and cancellations from the airline revenue. Maintenance schedule, crew restrictions and passenger connections are not taken into consideration.

Cao and Kanafani (1997a, 1997b) extend the delay model of Jarrah et al. (1993) to address both cancellations and delays simultaneously using a 0–1 Quadratic Programming model.

Argüello et al. (1997) (Gang Yu served as a Chairman and CEO of CALEB Technologies from 1995 to 2001) formulate the aircraft rerouting problem as a resource assignment model, where resources correspond to aircraft:

$$\text{Minimize} \sum_{k \in Q} \sum_{j \in P} d_j^k x_j^k + \sum_{i \in F} c_i y_i \tag{6.5}$$

subject to

$$\sum_{k \in Q} \sum_{j \in P} a_{ij} x_j^k + y_i = 1, \quad \forall i \in F \tag{6.6}$$

$$\sum_{k \in Q} \sum_{j \in P} b_{tj} x_j^k = h_t, \quad \forall t \in S \tag{6.7}$$

$$\sum_{j \in P} x_j^k = 1, \quad \forall k \in Q \tag{6.8}$$

$$x_j^k \in \{0,1\}, \forall j \in P, k \in Q \tag{6.9}$$

$$y_i \in \{0,1\}, \forall_i \in F \tag{6.10}$$

where:

F: set of flights indexed by i
P: set of feasible aircraft routes indexed by j
Q: set of available aircraft indexed by k
S: set of stations indexed by t
a_{ij}: binary coefficient equal to 1 if flight i is in route j, 0 otherwise
b_{tj}: binary coefficient equal to 1 if route j terminates at station t, 0 otherwise
c_i: cost of canceling flight i
d_j^k: cost of assigning aircraft k to route j
h_t: number of aircraft required to terminate at station t
x_j^k: binary variable equal to 1 if aircraft k is assigned to route j, 0 otherwise
y_i: binary variable equal to 1 if flight i is cancelled, 0 otherwise

The objective function (6.5) minimizes the total cost of assigned aircraft routes and flight cancellations. Constraints (6.6) are the flight covering constraints. Constraints (6.7) are the aircraft balance constraints. Constraints (6.8) are the aircraft utilization constraints. The binary assignment variables are defined by (6.9) and (6.10). A greedy randomized adaptive search procedure (GRASP) is used to rebuild aircraft routes disrupted due to aircraft groundings and flight delays during a day of operations. In the first, solution construction phase, a greedy evaluation function and a randomized selection method are used to iteratively construct a feasible solution. The feasible solution is further improved in the second, local search phase. The time-band approximation scheme, detailed in Argüello (1997), is employed to obtain a lower bound of the problem. The time-band scheme consists in partitioning the recovery horizon into uniform length time-bands and then aggregating all activities within a time-band into a single activity. The resulting mathematical model, which is an integral minimum cost network flow problem with a set of flight cover side constraints, is solved to produce a lower bound.

The model presented by Thengvall et al. (2000) is an extension of the work by Argüello et al. (1997). The objective function minimizes the deviation from the original schedule. Aircraft planners are allowed to specify preferences related to the recovery operations.

Thengvall et al. (2001) extend the work of Yan and his co-authors (Yan and Yang 1996; Yan and Young 1996; Yan and Lin 1997; Yan and Tu 1997) and solve the aircraft recovery problem for multiple fleets. Their model allows for cancellations, delays, ferry flights, and aircraft swaps (inter-fleet swaps) and substitutions (intra-fleet swaps). The resulting model is an integer multi-commodity network model with side constraints. The paper presents three multi-commodity network-type models: a pure network with side constraints, a generalized network, and a pure network with side constraints in which the time horizon is discretized. The objective function in the first two models includes an incentive to maintain as much of the original aircraft routings as possible. Minimizing the sum of cancellation and delay costs is the objective of the third model. The first model outperforms the other two in regards to both solution quality and computation times on the set of test scenarios. The model is later used to test the bundle algorithm proposed in Thengvall et al. (2003).

Stojković et al. (2002) propose a model that modifies flight schedules based on planned crew transfers and rest periods, aircraft maintenance and passenger connections.

Rosenberger et al. (2003) present an optimization model that reschedules legs and reroutes aircraft by minimizing rerouting and cancellation costs. Their aircraft recovery model is a set packing problem in which each flight leg either belongs to exactly one route, or is cancelled. The proposed model is validated using a simulation of airline operations.

Hang (2007) presented a fast metaheuristic aircraft recovery algorithm developed by Carmen Systems (now a part of Jeppesen). The algorithm protects crew connections and pre-assigned maintenance activities, and allows for generating multiple structurally different solutions within seconds.

None of the models, with the exception of one described in Rosenberger et al. (2003), allows for including aircraft maintenance constraints. Therefore, an aircraft recovery decision that implies the infeasibility of the aircraft planned maintenance inevitably invokes an aircraft maintenance recovery action, covered in the following section.

6.3.2.4 Special Instances of Schedule Recovery Problems

Airline Response to Ground Delay Programs (GDP)

When severe weather or other irregularities reduce the capacity at or around an airport, the traffic service provider employs special techniques to reduce traffic flow to match capacity. One such technique that is common in the United States is the Ground Delay Program (GDP). As described in Chap. 7, the GDP employs an allocation scheme called ration-by-schedule (RBS). In RBS, an individual airline has access to specific arrival times ("slots") at the airport in question. The FAA assigns flights to slots using a "first-scheduled, first-assigned" approach that preserves the scheduled arrival sequence. However, preserving the original flight sequence may not be the best alternative for an airline because the involved flights exhibit different levels of vulnerability to incurred delays—the costs associated with passenger, crew, and aircraft connectivity vary greatly between flights.

To give the airlines better control over their economic performances, a collaborative decision making (CDM) program was implemented in 1997. The basic concept of the program consists of allowing an iterative exchange between the airlines and the FAA. The FAA's first-scheduled, first-assigned slot reallocation is the first iteration. Upon the reception of the new slots the airlines are given the opportunity to reassign some of their flights to the allocated arrival slots using flight substitutions and cancellations. Canceling selected flights into the affected airport frees up the corresponding arrival slots that can be used by flights with later controlled arrival slots but earlier scheduled arrival times. Flight cancellations of low priority flights, so called smart cancellations, are tactically used to reduce delays of higher-priority flights.

While much of the literature on GDP, CDM, and ATFM (Air Traffic Flow Management) slot allocation in general is from an Air Traffic Management point of view, the results demonstrate the advantage to the airline of optimally assigning flights to slots. For example, Andersson et al. (2003) explore the benefits of increasing communication and collaboration between airlines and air traffic controllers during the arrival process at hub airports. The estimated operational improvements from providing airlines with more accurate landing time assessments (first study), and from allowing airlines to re-sequence their incoming traffic (second study), are measured by passenger delay minutes. The results of the second study indicate that the ability of an airline to influence the sequence of its arriving aircraft provides significantly higher potential benefits than the provision of accurate arrival time predictions addressed in the first study.

Given that each year there are several hundred GDPs enacted by the FAA, a smart flight canceling tactic can significantly improve an airline's on-time rating. According to the US DOT (Department of Transportation) on-time performance metrics, a flight is considered to be on time if its actual arrival time is within 15 min of the schedule arrival time. The on-time percentage is calculated as the proportion of all on time arrivals. Cancelled flights are excluded from this computation. Thus, cancelling some flights has added benefit in the overall optimal solution of delays and cancellations.

An airline can further reduce the impact of a GDP by using flight substitutions to allocate early arrival slots to those flights where delays would have the most negative impact. Traditionally, slot reassignment is done manually by traffic specialists. Being constantly under pressure to respond to numerous requests, a manual process concentrates on minimizing arrival delays at the GDP airport, with at most very limited attention to down-line effects of the implemented actions, such as delays of subsequent flights (aircraft and crew rotations), illegal crews, and passenger misconnections.

The problem of assigning inbound flights to the allocated slots is a significant combinatorial problem, further complicated by the presence of multiple objectives. An efficient and timely approach to the problem necessitates using an optimization engine. The goal is to help traffic specialists find a combination of flight cancellations and slot reassignments that minimizes the airline's losses and maximizes its on-time performances, while respecting various input constraints. Some of the possible objectives are to:

- Maximize on-time arrival performance for both inbound and outbound flights
- Minimize cancellations
- Preserve passenger and crew connections
- Minimize the impact of scheduled aircraft maintenance and crew legality limits

A GDP optimizer can be used as a standalone tool or coupled with an aircraft recovery optimizer that minimizes the down-line consequences of the GDP delays and cancellations. If fast enough, the tool allows for testing different scenarios and saving their solutions for a future fast implementation. Due to uncertainty in weather predictions, an anticipated GDP sometimes fails to materialize, or has a reduced or expanded scope. Delaying solving and implementing the GDP solution as long as possible allows for avoiding unnecessary cancellations and delays, putting an additional premium on solution speed.

Vasquez-Marquez (1991) addresses schedule disruptions due to the ground holding or ground delay programs. A quick tour-building heuristic is proposed to find the most appropriate sequence of reallocations. Given the negative impact that the slot reallocation has on the crew and gate schedules, the proposed heuristic is viewed as a decision support tool that makes it possible for operations managers to evaluate the effects of a number of different cancellations before implementation.

Luo and Yu (1997) consider the same problem with a different objective function. Their model is based on the observation that no departure will be delayed more than any arrival will be. While the minimization of maximum delay among all flights is achieved by retaining the original landing sequence, the reassignment of the arrivals may be needed to minimize the maximum delay among departures. The authors present a fast heuristic for a restricted version of the problem.

Niznik and Green (2001) described two models developed by American Airlines. Although the models differ in the objectives and the required solution techniques, their overall goal is to reduce the impact of a GDP on airline operations and passenger experience.

The basic model, formulated as a standard assignment problem, assigns landing slots to arriving flights such that the number and amount of flight and passenger delays, as well as the number of flights with arrival delays over 15 min, are minimized.

The advanced model incorporates dependencies of departure flights on arrival flights in terms of crews and maintenance when assigning landing slots to arriving flights. The objective of the advanced model is to minimize the number and amount of departure and arrival delays. The problem is formulated as a generalized assignment problem, which is solvable using standard integer programming solvers. The incorporated dependences between arrivals and departures complicate the underlying mathematical structure and cause solution times to dramatically increase. To cope with unacceptable solution times, a heuristic approach is used. It consists of generating multiple sets of (flight, arrival slot) assignments, evaluating each of them independently, and finally choosing the best subset of assignments so as to cover all flights.

The technical considerations of a production application at American Airlines were discussed by Salam and Niznik (2005). The results of a bipartite matching problem for a single airport GDP were presented. Fast solution times were reported for scenarios with neither intra-fleet swaps nor flight cancellations allowed.

Gumireddy and Ince (2002) presented a prototype developed by US Airways. The underlying model ensures crew availability and equipment balance, with neither crew nor equipment substitutions permitted. In order to solve a multi-objective problem, weighted objectives based on changing priorities were used. The methodology adopted to calibrate the weighted cost function was presented by Ince (2009). The problem constraints were modeled either as hard constraints, or soft constraints with variable penalties. Both priorities and penalties were cost driven. The problem was formulated as a standard assignment model and solved through multiple runs with hierarchical objectives. The constraints used are:

- Every flight must be assigned to exactly (or at most) one slot;
- A slot may have a flight assigned to it or not;
- The slot time cannot be earlier that the estimated time of arrival;
- The slot time cannot be earlier than the flight scheduled time of arrival;
- Most binding constraint from crew, passengers, maintenance and operational requirements;
- Constraints on objectives from previous runs.

The prototype was tested on several GDPs. Important savings in regards to all considered metrics and very fast solution times were reported.

Stojković and Soumis (2004) presented an exact optimization model that simultaneously adjusts the flight schedule (using delays and cancellations) and re-routes aircraft and crews, while minimizing the resulted aircraft, crew and passenger costs. The passenger delays are minimized and passenger connections are preserved to the maximum possible degree. The set of constraints is represented by:

- Airport: capacity;
- Flight: maximal delay, aircraft type;
- Aircraft: connection time, maintenance;
- Crew: qualification, duty duration, connection time.

That holistic approach used produces globally feasible solutions. Several scenarios, created by delaying flights arriving at an airport, reducing airport capacity (GDP), or both, were solved in under a minute. The solution quality is superior (in many regards) to the traditional, sequential approach.

Aircraft Maintenance Recovery

The schedule of which individual aircraft will operate each specific flight is often called the "maintenance schedule" because it is driven largely by the hard constraint of required maintenance. The plan each day is for aircraft requiring maintenance to end at a station where that maintenance can be performed. Planned maintenance work orders differ both in maintenance location (any maintenance station or at a specific location, hangar or a line maintenance location) and time requirements (on a specific day or on a range of days).

The aircraft recovery models have a partial view of maintenance issues, limited to keeping intact an aircraft scheduled for maintenance within a considered recovery period. That approach is restrictive because a potentially better solution is excluded from consideration. Instead of preserving the planned maintenance activities, some other valuable options may be exploited, such as:

- Swapping two aircraft due for maintenance the very same night if the maintenance work for each aircraft can be accomplished at the new overnight station;
- Swapping an aircraft whose planned maintenance can be legally postponed with a maintenance-critical aircraft (a maintenance critical aircraft is an aircraft in close proximity to the schedule maintenance);
- Conduct premature maintenance of an aircraft when there is unused maintenance capacity, freeing up the planned maintenance slot for another maintenance-critical aircraft, etc.

To solve the aircraft maintenance recovery problem the following common input data are needed:

- Set of aircraft to be routed and their maintenance requirements,
- Flight schedule, and
- Capacity of the maintenance stations in terms of manpower and hangar slots available for consumption each day.

Different levels of sophistication can be built into the model, starting with a least complicated model that simply assigns aircraft to fleet-compatible maintenance stations. Parts and labor availability, and evenly distributed workload among the maintenance stations, are considered in more advanced models.

The potential benefits of a maintenance recovery tool are not only improved aircraft decisions, but also improved aircraft routing productivity and potentially higher aircraft utilization between two scheduled maintenances.

Herrera (2005) discusses the maintenance routing problem and the approaches considered at Southwest Airlines. A tactical aircraft routing optimizer, which serves as a basis for a day-of-operations recovery, generates 7-day aircraft routing solutions for the entire published schedule. The solutions satisfy maintenance requirements for all aircraft while assigning work to maintenance stations such that maintenance capacities and capabilities are respected. The maintenance routing problem is formulated as a set-covering problem with side constraints which include flow balance, work order coverage and capacities of the maintenance stations. Two solution approaches were tested: a branch-and-price approach, where a column corresponds to a flight string (Barnhart et al. 1998), i.e., a maintenance feasible routing from a maintenance station to a maintenance stations, and a multi-commodity network flow model in its closed form. Promising initial results and fast solution times are reported for both solution approaches. While it takes hours to find a manual solution, the optimization approaches tested take less than 40 min to accomplish the same task.

The presentation by Heinold (2008) extends the previous work. The most frequent overnight routine maintenance checks, performed every 7 days, were considered. Over-the-day operational swaps happen after a 7-day plan is built. Roughly 25% of aircraft operated by Southwest get swapped every day. Day 7 aircraft (due for maintenance at the end of the day) are carefully rerouted, but rerouted Day 6 aircraft (due for maintenance on the following day) finish the day dispersed throughout the network. At the end of the day the optimizer applies aircraft swaps to bring Day 6 aircraft to appropriate maintenance stations the following night. Due to the previous over-the-day swaps the optimizer may be unable to successfully reroute the Day 6 aircraft using the overnight swaps. Only then are additional over-the day swaps, which may be undesirable, considered. To minimize their number, Southwest uses probability analysis to improve the planned routes, and thus makes the maintenance schedule more robust. The design goal is to minimize the number of over-the-day swaps needed to get Day 6 aircraft to a maintenance base by the end of the following day, while balancing work at the maintenance bases. This approach is a good example of the relationship between day-of-operations optimization and effective robust schedule optimization.

6.3.2.5 Crew Recovery

Disruptions to planned crew schedules are inevitable. In addition to internal crew problems (unplanned crew absences, down-line crew unavailability), the hierarchically preceding aircraft recovery decisions (flight cancellations, delays, diversions and aircraft swaps) may result in additional crew disruptions. Regardless the disruption's origin, the final outcome is the unavailability of crews at the airports where they are needed at the time when they are needed.

The crew recovery process consists in applying various actions to repair, at the minimum cost, the original crew schedule while minimizing the number of flights that cannot be operated due to lack of sufficient crew on board. Crew recovery actions include, but are not limited to, rescheduling already scheduled crews or deploying reserve crew members.

Crew rescheduling consists of swapping the original crew assignments between crew members. Reserve crew members are available throughout the airline's network at specific airports known as crew bases. Reserve crews are either on-call reserves sitting at home, or reserves standing by at airports for an immediate deployment if need be. Recovery from a disruption when a sufficient number of reserve crew members is available is relatively effortless. However, reserve crews are a limited resource. Aggressively using reserves at the beginning of the month typically results in their shortage as the month progresses. An efficient crew recovery when only a very limited number or reserves (or no reserve at all) is available necessitates highly creative and complex solutions unlikely to be generated by humans.

Crew trackers are advised of the irregular operations by several different means: the involved crew members, the operations controller, or via the crew management system messaging screen or printed messages. Whether a disruption results from the aircraft recovery process or is crew driven, the crew trackers are responsible for determining how to crew the uncovered flight segments in a most appropriate way. They are also responsible for proposing solutions to different what-if scenarios submitted by the operations controllers in order to help them assess the crew impact on different irregular operations possibilities. The solutions are generated either manually or with a help of a recovery tool.

What makes the crew recovery process significantly different from the aircraft recovery is a human component to consider when building a recovery plan in response to a crew disruption. Any proposed solution must comply with the numerous and very complex crew legality rules and regulations. Those comprise the rules used for the planning purposes and additional rules that regulate the recovery process. The rules are separated into three large groups: mandatory governmental rules (such as Federal Aviation Regulations—FAR in the U.S. and Joint Aviation Requirements—JAR in Europe), contractual rules that further restrain the governmental rules, and airline-specific business rules aimed at improving the quality of life of crew members. Although different sets of rules apply to cockpit and cabin crews, those rules are either safety-related or quality-of-life-oriented. The safety-related rules restrict the amount of working and flying hours between mandatory rest periods, whose minimum duration is also regulated. The maximum duration of a crew pairing, also known as the time away from base, is one of the rules not only safety-related but also aimed at protecting the quality of life of the crewmembers. The majority of the quality-of-life-oriented rules reflect the airline-specific crew recovery business process. As an example, they prescribe the order in which different crewmember groups, such as disrupted, undisrupted and reserve crews, are introduced into the solution process. Further sorting is based on the factors such as a crew member's status, position and seniority. The order may change depending on the specific characteristics of a problem to be solved.

Those rules also facilitate the task of the crew trackers when manually dealing with crew disruptions by reducing the number of considered alternatives to a reasonable size that a human can handle.

During the planning phase the anonymous crew pairings (a crew pairing is a sequence of flights separated by rest periods, worked by a crew leaving and returning to the same crew home base) that cover the set of given flights are built first, and then awarded to the available crew members in the form of monthly schedules. Once awarded, a crew pairing becomes personalized, i.e., specific to a particular crew member. Regardless of the approach used to build a monthly crew schedule (rostering, bidlines or preferential bidding), the final result of the crew scheduling process is the set of personalized monthly schedules for the considered group of crew members.

The crew disruption management process, also referred to as the roster maintenance problem, thus consists in modifying the personalized crew pairings by preserving the integrity of the entire personalized monthly blocks to the greatest degree possible (a personalized monthly block, or roster, is a sequence of pairings, rest periods, annual vacations, transition activities, reserve blocks, desiderata, and other periods of unavailability activities carried out by one employee in one month). The main crew recovery objectives are to cover in a most cost efficient way the set of given flights, while remaining as close as possible to the original personalized crew schedules. Besides the crew location, the information about the recent flying history (already flown portion of the crew members' monthly schedule) and the future scheduled activities must be continuously available for each crew member. The up to date crew information is necessary to generate a legal and implementable recovery plan. While past crew activities must be used when checking the legality of the recovery plan, the future activities are used both for the legality purposes and as the quality of life measures. The legality concerns result from the fact that some future scheduled activities, such as training or vacation periods, must remain assigned to the crewmembers in any recovery solution. The latest possible return time to the home base of a modified pairing, which depends on the scheduled return time, is another example of using the recent history for legality purposes. The future scheduled activities are also used as a quality marker when reassigning crew members. To increase the crew satisfaction, crew trackers make every attempt to reschedule crew members such that their modified pairings remain as close as possible to the original ones.

In addition to returning to normal operations as soon as possible, and with as little changes as possible to original crew schedules, minimizing the incremental crew cost is another important objective of the crew recovery process. Crew rules state that the crews must receive the higher of the planned pay and the pay associated with the realized schedule. While the crews removed from their original assignments due to the irregular operations are still paid for them, the reassigned crews receive either the scheduled pay when the reassignment makes them work less, or the new higher pay otherwise. In addition to the latter one, the incremental cost also results from higher than guaranteed reserve utilization, additional hotel rooms, and meals and airline tickets when crews deadhead (or position) on other airlines, just to name a few.

Bearing in mind that the crew cost represents one of the largest components of the direct operational cost of an airline, usually exceeded only by fuel cost, it is evident how important cost-effective crew recovery solutions are for the well being of an airline. The manual crew recovery process is highly constrained by the prescribed steps that a crew tracker follows when solving one disrupted pairing at a time. The computing and algorithmic powers of decision support tools allow stepping away from that highly constrained procedure, and produce more efficient recovery plans in return.

Much work has been published on the traditional crew recovery problem, where flight schedules and aircraft rotations are fixed.

Teodorović and Stojković (1995) present a dynamic programming based approach that minimizes the crew ground times using the first-in-first-out principle.

Wei et al. (1997) formulate the problem as a system-wide multi-commodity integer network flow model where a commodity corresponds to a crew, and solve it using a heuristic depth-first branch and bound search algorithm. Their objective is to minimize the total cost of the reassigned pairings. The proposed approach starts with a preprocessing phase where each of the disrupted patterns is repaired by finding a path that brings the disrupted crew back to their home base. The flights left uncovered at the end of that phase are gathered in a set of uncovered flights, which represents the starting node of the search tree. A new node is created by assigning an uncovered flight to a crew that looks suitable for covering it. A node is pruned if the total number of modified pairings generated from the root node until the considered node is reached is greater than the number of modified pairings associated with its parent node. The solution process continues by generating new alternatives and solving the problem again until the set of uncovered flights becomes empty, in which case a solution to the given problem is found and the process terminates. A negative cost shortest path algorithm is used to find a path between a given pair of the start and end points within a pairing. The authors provide a comprehensive framework for crew management during irregular operations, but do not report computational results for a real-size instances (the largest instance solved comprises 18 disrupted crew pairings).

The proposed solution process is feasibility rather that optimality driven. A very first solution with no uncovered flight left cannot be further improved (no matter how bad it may be). In addition, the planned crews are treated as inseparable units. Legality checking is performed at two levels. The more frequently violated rules are checked upon a pairing has been generated. The less frequently violated rules are checked only once a solution to the whole problem has been found. The individual crew recovery decisions are made in an independent way.

Yu et al. (2003) discuss the implementation of a crew recovery decision support system at Continental Airlines. The implemented model is a refined version of the model from Wei et al. (1997):

$$\text{Minimize} \sum_{k\in K}\sum_{p\in P_k} c_p x_p + \sum_{f\in F} u_f y_f + \sum_{k\in K} q_k z_k + \sum_{f\in F} d_f s_f \tag{6.11}$$

subject to

$$\sum_{k \in K} \sum_{p \in P_k} a_{fp} x_p + y_f - s_f = 1, \quad \forall f \in F \tag{6.12}$$

$$\sum_{p \in P_k} x_p + z_k = 1, \quad \forall k \in K \tag{6.13}$$

$$x_p = \{0, 1\}, \quad \forall k \in K, \quad \forall p \in P_k \tag{6.14}$$

$$y_f = \{0, 1\}, \quad \forall f \in F \tag{6.15}$$

$$z_k = \{0, 1\}, \quad \forall k \in K \tag{6.16}$$

$$s_f = \{0, 1, 2, \ldots\}, \quad \forall f \in F \tag{6.17}$$

where:

F: set of active flights f
K: set of available crew k (active crew and reserves)
P_k: set of pairings p that crew k can serve within the recovery window
c_p: cost of assigning pairing p to crew k
u_f: cost of leaving flight f uncovered
d_f: cost of deadheading each crew on flight f
q_k: cost of leaving crew k unassigned
a_{fp}: binary coefficient equal to 1 if flight f is covered by pairing p, 0 otherwise
x_p: binary variable equal to 1 if pairing p is used in the solution, 0 otherwise
y_f: binary variable equal to 1 if flight f is uncovered, 0 otherwise
z_k: binary variable equal to 1 if crew k is unassigned, 0 otherwise
s_f: number of deadheads on flight f

The objective function (6.11) minimizes the total cost. Constraints (6.12) are the flight covering constraints. Uncovered flights and crew deadheading are penalized in the objective function. Constraints (6.13) are the crew assignment constraints. Crews with no reassignment are penalized in the objective function. Constraints (6.14) to (6.17) describe the binary variables used in the model. Crew legalities are enforced by the crew pairing generator. Continental Airlines received the 2002 Franz Edelman Award for Achievements in Operations Research and the Management Sciences for developing and implementing, in collaboration with CALEB Technologies (later acquired by Navitaire), a DSS for crew recovery. The optimization engine described in the paper produces feasible solutions that comply with legality requirements and honor crew quality of life. Continental Airlines estimates savings of approximately US $40 million from using the implemented system to cope with four major disruptions in 2001.

Treating the planned crews as inseparable units is a simplification that greatly reduces the complexity of the problem, but also has an adverse effect on the quality of the solution: to generate a legal modified pairing the limits of the most constrained crewmember are imposed on the entire crew. Stojković et al. (1998) propose an adapted version of the traditional crew pairing model to handle the individual crew pairings. Modifications to a crew roster are limited to a single pairing from the recovery period (pairing in progress or the subsequent pairing if the last flight of the pairing in progress is delayed, making the rest period prior to the subsequent pairing insufficient). All flight segments from the modifiable pairings that are completely within the recovery period are subject to reassignments, while the other flight segments (completely or partially outside the operational period) are frozen and thus must remain unchanged. Their recovery model allows for crew splitting (each crew member is considered separately) and therefore improves recovery efficiency over the previous models. The problem is mathematically formulated as a set partitioning type problem, and solved using a branch-and-price algorithm (column generation method embedded in a branch-and-bound search tree).

The advantage of the proposed branch-and-price approach is that it can be applied to small, just as to large, operational problems where a priori enumeration of feasible pairings may be impossible given the tight solution time constraints.

Medard and Sawhney (2007) (Carmen Systems, now part of Jeppesen) expand the framework introduced by Stojković et al. (1998) by allowing changing more than one pairing in each of the disrupted crew rosters. Within a given (possibly sliding) recovery time window they apply a generate-and-optimize technique that incorporates both pairing construction and pairing assignment in one step. The approach used is flight based rostering subject both to pairing and rostering rules and additional multi roster requirements (requirements imposed on groups of crew members). While the parts of the roster before and after the disruption are fixed, flights within the recovery window are de-assigned from disrupted crew and a few other crew ("helper "crew). Newly scheduled flights may also be added to the problem. Pre-assignments within the recovery window, such as days off and training duties, must remained untouched by the recovery actions. The resulting integer programming model is the flight based rostering model that incorporates all the pairing and rostering dimensions of the recovery problem. Two solution approaches are tested: greedy enumeration of legal rosters using depth-first-search (DFS) and reduced cost column generation (COLGEN) where the k-shortest path algorithm is used on a duty-based network to find attractive legal rosters. In both cases roster legality is checked using a black box legality module. Both generation methods are tested for a single recovery time window on instances for single and multi base problems. To keep the computational time of the DFS approach reasonable, the total number of rosters generated using DFS is regulated by parameters specifying the search characteristics. While the number of enumerated rosters is restricted to the smallest possible number that enables DFS to find an operationally feasible solution, no restrictions need to be imposed on COLGEN which, unlike DFS, does not suffer from combinatorial explosion. The computational results confirm that the number of columns required by the reduced cost column generator is much less than the number required for the enumerative depth first search approach. The COLGEN approach outperforms the DFS approach in regards to the solution quality on small to medium size instances (up to 500 crew members). This is mostly because the search width parameters for the enumerative approach were chosen to be small enough to avoid combinatorial explosion. Still, increasing the search width dramatically increases the computational time, especially when deadheads need to be added (to solve one such instances DFS generated more than four hundred thousand columns and still could not find the optimal solution). However, too much computational time is required to set up the network used to generate reduced cost attractive rosters for large problems.

Lettovský et al. (2000) develop a set covering approach to select the best among the generated crew pairings for small to medium size disruptions. They assume that a crew is inseparable for the length of the pairing. To efficiently solve the problem, they use several different techniques that reduce the complexity and size of the problem, and apply several preprocessing techniques to extract a subset of the schedule for rescheduling. New pairings are generated by enumerating feasible continuations of partially flown crew pairings, using a fast crew pairing generator. Crew legalities are checked on the fly so that only legal crew pairings are generated. The pairing generation method is designed to minimize the modifications to the original schedule. Efficient storage and data access are achieved by using appropriate data structures. Computational results using a schedule from a major air carrier are presented. While the proposed methodology efficiently solves medium-sized disruptions to the crew schedule, it cannot handle large-scale disruptions within an acceptable running time. To cope with the problem the authors propose to use a sliding window to isolate smaller portions of the schedule for rescheduling and solve them sequentially, one at a time.

Nissen and Haase (2006) present a crew recovery formulation and solution method adapted to the needs of European airlines, where fixed crew salaries are predominant. Their duty-based formulation with a new type of resource constraints is solved using a branch-and-price methodology. The model's capability to produce quick solutions to different recovery scenarios is reported in the paper.

In addition to published research, airlines have been internally developing crew recovery decision support tools. Abdelghany et al. (2004) describe a recovery tool developed by the R&D department at United Airlines. The tool, designed for a hub-and-spoke network structure, proactively recovers crew problems ahead of time using a rolling horizon to consider flights in chronological order of their departure times. In each of the consecutive assignment problems the objective is to cover a maximum number of flights while minimizing flight delays and the deviation from the original crew schedule. Despite the missing information about aircraft and flight attendants, and a heuristic solution approach used, the authors report a high level of the users' satisfaction. Deregnaucourt (2004) describes a crew recovery application developed by the IT and OR departments at Air France. The application consists of three modules: Graph module which generates alternative crew pairings for each of the input disrupted crew pairings which respect a subset of legality rules; Legality Control module which ensures full legal compliance of preserved alternative crew pairings; and LP (Linear Programming) module which selects the set of best alternative pairings from the set of fully legal crew pairings provided by the Legality Control module. The application allows the crew trackers to solve both what-if and real disruption scenarios, and update crew rosters with the rebuilt crew pairings. Several examples of the tool's successful use at Air France are presented.

Numerous presentations at AGIFORS and INFORMS conferences were devoted to vendor products. Only a couple of them are cited here. Shaw (2005) presented a column generation based crew recovery approach developed by Sabre. The model allows for modifying multiple pairings within a given crew roster, while keeping the crew members planned to work together unsplittable during the recovery phase. The objective function minimizes the impact of disruptions on the planned operations and limits the down-line ripple effect. A tactical crew recovery model by Gao and Kalyta (2006), developed by Sabre in collaboration with Georgia Tech, allows for recovering crew duties, with an option to keep an entire crew together or split it. A greedy diving heuristic based on a scoring mechanism and rolling horizon, is employed to solve the problem. Each flight is assigned a flexibility index. The flights are partitioned into groups of predefined size such that the least flexible flights belong to the first group and the most flexible to the very last group. The problem is then solved sequentially, starting with the least flexible group.

Efficiently selecting offline deadheads, when needed, from a giant set of all available flights from the Official Airline Guide (OAG) is a topic that has not attracted much attention in the crew recovery literature. The deadhead selection for the long-haul crew assignment problem, as proposed by Barnhart et al. (1995), is not directly applicable to the dense short-haul schedules because of the huge number of potential pairings. Lettovský et al. (2000) categorize deadheading options with respect to crew recovery and provide computationally inexpensive deadhead selection heuristics. However, their pairing generator is able to produce pairings only when selected deadheads are given a priori. Although a careful algorithmic design and its efficient implementation would allow for on-the-fly offline deadhead selection, reliability of the published flight schedules during irregular operations and the inaccessibility of the real-time seat-availability data for flights operated by other airlines are still open issues. While deadheading crews on the flights operated by the disrupted airline (online deadheads) is one of the recovery options available in most recovery models, deadheading on other airlines (offline deadheads) is still usually restricted to the set of deadheads built into the planned crew pairings.

6.3.2.6 Passenger Recovery

The last step of the traditional airline disruption management process is to develop a passenger re-accommodation plan for disrupted passengers. A disrupted passenger is a passenger whose booked itinerary is broken and therefore impossible to realize. A passenger itinerary becomes broken when at least one of its flight legs is canceled, when the connection time between its consecutive flight legs becomes too short due to flight delays, or when a booked seat becomes unavailable due to equipment downsizing. Passenger recovery consists in rebooking disrupted passengers on selected alternate itineraries. Displacing undisrupted passengers to better accommodate disrupted ones is never allowed. Due to the reduced system slack—with lower capacity and higher load-factors—described in the chapter introduction, re-accommodation of passengers from operational disruptions has become substantially more challenging.

Passenger recovery is the most data intensive of all the airline recovery processes. A non-exhaustive list of input data includes the following: passenger information from the PNR[1] (itinerary, fare/cabin class, passenger attributes such as frequent flyer status, special requirements, etc.), operational flight schedules, aircraft and seat configurations, seat availability on online flights and flights operated by partners and other airlines, minimum connection times, airline specific rebooking business strategy (procedures, policies and rules), passenger re-accommodation costs (cost of flying on another airline, ground transportation, lodging and food cost), and passenger goodwill costs. Many of these data elements are dynamic, further complicating the passenger recovery process, which calls for efficient information handling, such as high performance database access, exception management, and effective information presentation.

Traditionally, disrupted passengers are rebooked manually or with limited help from DSS, whose complexity and level of sophistication vary from one system to another. The simplest, semi-automated DSS only generate options for the disrupted passenger itineraries. They require a human to select a specific rebooking for a given disrupted passenger from the set of generated alternatives. More advanced systems automatically select rebooking for disrupted passengers based on the airline's rebooking policies. The least sophisticated automated systems rebook a disrupted passenger on the very next flight with the same origin and destination as the original flight, regardless of the flight capacity and downstream connections. Further manual interventions are needed whenever rebooking results in flight overbooking or passenger misconnections. More sophisticated tools rebook disrupted passengers to their final destination while repeating the airport sequence from the booked itinerary or without necessarily visiting the same intermediate airports.

The rebooking process is either sequential or simultaneous. Greedy, sequential processes rebook passengers one by one, following a predefined rebooking order. The passenger rebooking order is typically based on the perceived passenger value to the airline, expressed through passenger attributes such as the frequent flyer status, cabin booked, ticket price paid, etc. The sequential rebooking processes are not only extremely time consuming and frustrating to both customer services and passengers, but do not achieve efficient results.

[1] Passenger Name Record (PNR) is a record in the database of a computer reservation system that contains the travel record for a passenger, or a group of passengers traveling together.

Both manual approaches and simple, heuristic DSS are restricted to a very narrow, local view of the problem. A broader, holistic view of the problem is achieved through an optimization based, simultaneous passenger recovery process. The optimization-based systems assist passenger services and airport agents in making quicker and better, value-based decisions using a variety of options unavailable to the simple approaches. The global view of the problem allows for increased rebook rates thanks to a better utilization of the limited seat inventories. The efficiency is achieved through simultaneously considering all selected passengers who compete for the same available seat inventories. Although particularly focusing on high-value passengers, these systems improve the solution quality across all disrupted passengers by intelligently selected among the comparable options for the high-value passengers. The end result is the reduced overall re-accommodation costs and the increased passenger satisfaction.

The major weakness of the automated rebooking systems is that a passenger's real preferences are unknown at the time the problem is solved. The metrics used to assess the quality of a newly generated itinerary are defined by the airline and hence do not necessarily reflect the passenger's real preferences, which may vary from one situation to another. The quality of a rebooking option is measured relative to the booked itinerary. The typical key solution quality metrics are the discrepancies in the arrival times, flight cabins and the number of connections between the booked and rebooked itineraries. However, the system optimum (best solution from the airline's perspective) does not necessarily coincide with the user optimum (best solution from the passenger's perspective). As a consequence, some passengers may refuse their rebooked itineraries. This situation is highly undesirable from the company's perspective, especially when the refusal comes from a high-value passenger. A simple option generator is of great help in such a situation. However, the number and quality of the rebooking alternatives decrease over time, because the inventory of available seats gets smaller as more passengers are re-accommodated. As a result, by the time a rebooking gets refused there may be no seats left to generate valuable alternatives taking into account passenger's real preferences.

The automation of the rebooking process, coupled with the use of efficient optimization models, is the key for an efficient passenger re-accommodation process and better customer service quality. An efficient passenger recovery system needs to be modular (option generator, business rules and solver modules), configurable and scalable (small and large disruptions).

For an airline to select, implement, and actually use an automated passenger re-accommodation system, it must demonstrate a fast response time and provide high-quality solutions. The solution speed greatly influences the rebooking success rate in a highly dynamic operational environment typical for massive passenger disruptions. Though feasible when selected, a rebooking itinerary may become infeasible due to modified operational data. The amount of operational changes and therefore the rebooking rejection rate is directly proportional to the solution time. While an efficient data management is critical for the fast passenger recovery, solution performance is critical for the system to be competitive.

Provided solutions must maximize passenger satisfaction and minimize the incremental cost to the airline—the extra cost on meals, accommodation, transportation cost (other modes or carriers), and all the unplanned cost related to the new passengers itineraries. The extent to which an airline approximates real preferences is reflected in increased passenger satisfaction.

Clarke (2005) presented an optimization based passenger recovery tool built by Sabre to assist passenger services and airport agents to re-accommodate passengers disrupted due to schedule changes. The alternative travel itineraries for the disrupted passengers are built using only online or alliance partner flights. To reduce problem dimensions, disrupted passengers are aggregated into groups. A passenger group is formed by aggregating disrupted passengers with common disrupted location (airport) and destination. The predefined passenger re-accommodation priorities and business rules, set by the airline, are incorporated in order to protect high-value passengers. The passenger re-accommodation priority is expressed through a passenger value index, based on the relevant passenger and itinerary characteristics stored in a PNR. Business rules set by the airline, passenger loss to other carriers or modes of transportation, and the auxiliary cost components are taken into consideration when generating and pricing out alternative passenger itineraries.

The passenger re-accommodation problem is mathematically formulated as a multi-commodity network flow problem, where a commodity corresponds to a passenger group:

$$\text{Maximize} \sum_{i \in I} \sum_{p \in P(i)} f_i X_{ip} \tag{6.18}$$

s.t.

$$\sum_{i \in I} \sum_{p \in P(i)} \alpha_{if} X_{ip} \leq CAP_f, \quad \forall f \in F \tag{6.19}$$

$$\sum_{p \in P(i)} X_{ip} \leq PAX_i, \quad \forall i \in I \tag{6.20}$$

$$X_{ip} \geq 0, \quad \forall i, p \tag{6.21}$$

where:

X_{ip}: number of passengers for itinerary i assigned to path p
F: set of all operational flights f
I: set of all potential origin–destination itineraries i
P: set of all feasible paths p in the residual network
$P(i)$: subset of paths that can be considered for itinerary i
f_i: passenger value (revenue, cost) for itinerary i
CAP_f: capacity of aircraft assigned to flight f
PAX_i: total number of passengers booked on itinerary i
α_{if}: binary coefficient equal to one if itinerary i contains flight f, zero otherwise

The objective function (6.18) maximizes the overall passenger re-accommodation value. Constraints (6.19) limit the number of passengers assigned to flight f to the availability capacity of aircraft operating the flight. Constraints (6.20) limit the number of reassigned passengers on itinerary i to the total number of passengers booked on the itinerary. Constraint (6.21) defines the non-negative model variable that counts the number of passengers for itinerary i assigned to path p.

The initial concept was developed by Clarke et al. (1999). Productization of the model prototype was done in collaboration with Air New Zealand.

The problem is solved using either a greedy heuristic or an optimization-based model. Only the best alternatives are generated for each passenger group. This simplification, although beneficial to the overall solution time, has an adverse effect on the solution performance. Best alternatives typically comprise flights which are very attractive to a number of passenger groups with common disrupted location (and possibly some other passenger groups). However, limited seat inventory on the most attractive flights usually does not allow for accommodating all of the disrupted passengers competing for them. It follows that using some less attractive alternatives is inevitable to increase the overall rebooking rate.

Green and Ande (2005) presented a passenger recovery tool developed by American Airlines that assigns, in a timely manner, alternative itineraries to disrupted passengers. The objective is to maximize the total number of rebooked passengers while assigning the best options to the most valuable customers. The tool consists of an optimization model and a score algorithm. The score algorithm calculates the score of each alternative passenger itinerary based on the itinerary's characteristics such as the itinerary's departure and arrival times, the number of stops and cabin upgrades and downgrades. The optimization model is an assignment model that maximizes the total score of all passenger re-accommodations. Only online flights are used when building alternative passenger itineraries for disrupted passengers. The presented optimization tool had shown improvements over the manual rebooking process in regards to the use of limited seat availability across the entire American Airline flight network.

Alves (2006) presented a passenger re-accommodation tool built by Jeppesen and applied at AeroMéxico. The goal of the tool is to efficiently rebook disrupted passengers while minimizing the passenger perceived delay and associated costs. Once again, getting and handling the large amount of information is quoted as the biggest challenge on the road to automating the passenger rebooking process. Alves and Vaaben (2009) compared different sequential re-accommodations strategies with a holistic, simultaneous re-accommodation model using real life passenger itineraries from a mid-size North American airline. The holistic model was not only able to find better solutions for the high value passengers, but also significantly increased the solution quality for the low value passengers (18% improvement over the sequential approach). In addition, the severe passenger delays were successfully avoided.

Gabteni (2010) presented a large scale passenger reaccommodation model developed by Amadeus. He uses a multi-commodity network flow model formulation coupled with Floyd-based network pruning to control the number of alternative routes generated between two airports. The proposed model is tested on 32 public instances. Good performances are reported for large instances: problems with up to 32,000 disrupted passengers were solved within 5 min.

To be more successful, passenger recovery systems still need to overcome a number of practical challenges. The problems of data integration, with the typical airline IT landscape of legacy (stand-alone systems that were never intended to work in a multi-disciplinary fashion) was described in the chapter introduction. We reiterate here, however, that the data required for efficient recovery reside in disparate systems, and this forms a further challenge to practical application of effective passenger recovery optimization.

To find passenger-valued rebooking alternatives (alternatives which match passengers' needs/desires), an automated passenger recovery process requires information about the passengers themselves, beyond what is typically stored in a PNR. One example is whether a group of passengers on the same PNR is separable and, if so, into how many subgroups. Such information is valuable, for example, if non-related subgroups of passengers are booked as a group by a travel agent because of a common vacation destination. Another example is information about an international passenger's visa status to enter specific countries. This information may be crucial when considering rerouting options. Other information, such as how time-flexible a passenger's travel plan is, or whether keeping the same cabin plays a more significant role than the arrival time at the final destination, illustrates the type of questions that would be very useful from the airline's perspective when evaluating different rerouting options for a disrupted passenger. The information about a passenger's tolerance of small planes (regional jets, turbo props, etc.) is yet another type of information that would be very useful to have. Given some passengers' strong aversion towards small planes, any reroute that requires a passenger to fly the whole or a portion of the new itinerary on a small plane, when the original itinerary did not require that, has a high probability to be refused by a passenger. At the same time, the information is very hard to get because of the sensitivity to gathering of contingency preferences. Asking passengers for extensive information to be used in recovery may be interpreted as a lack of reliability, so such requests must be carefully considered.

To make meaningful decisions, an automated passenger recovery process requires the precise operational schedules and seat inventories for the airline itself, its alliance partners and other airlines. Getting accurate and timely information during a major disruption, even for the airline's own flights, is a highly challenging task in itself because of the highly dynamic, constantly changing environment. Any disruption that requires rebooking on other airlines (OA) usually results from major weather issues, which impact all the airlines operating to and from the affected regions. Using publicly-available schedules and seat inventories for OA makes little sense in a disrupted environment, where the operational flight schedules are unreliable (many cancellations and delays), and the number of available seats constantly changes due to ongoing rebooking activities. Getting accurate OA information in a timely manner is still an unresolved issue.

Another data challenge arises in building itineraries with non-standard connecting times. Standard minimum connection times are commonly used when building alternative passenger routes. Using precise airport layout information, expressed through minimum passenger and baggage connection times between pairs of gates, increases the feasibility of generated itineraries. Even though the passenger connection time between two adjacent gates may be very small, a significant baggage connection time may be required when the gates are on the opposite sides of the terminal building with no under-terminal passage between them. A reroute option with a tight connection from the passenger perspective may thus become infeasible when baggage is taken into the picture.

Contacting the rebooked passengers is yet another challenge, especially when the disruption occurs before the start of the booked trip. Even though the wide spread use of mobile devices nowadays allows for exploring and using new options, such as text messaging and voice mails, reaching each individual passenger in time is still a highly challenging task, especially when the number of disrupted passengers is high.

6.3.2.7 Integrated Recovery

Airlines are traditionally organized in functional silos. Dividing the whole business process into smaller and thus manageable functional pieces was necessary from an historical perspective. Creating functional silos splits the large, intractable recovery problem into smaller, computationally tractable sub-problems traditionally solved in a sequential manner. In that sequential solution approach a solution for an upstream silo becomes a fixed input data for downstream silos, thus reducing the solution flexibility for each subsequent sub-problem. The price to pay is the lack of a holistic view that makes the traditional, sequential recovery process globally suboptimal even though locally optimal solutions may exist within each silo.

Technological (computer technology), methodological (mathematical modeling) and software advances have made it possible to more efficiently tackle complex, large size problems whose resolution was previously unthinkable, and hence to start revamping the traditional business process. The very first step towards blurring the line between the silos and commencing the integration of the previously disconnected pieces consists in understanding and structuring the activities and information across the silos. Stojković (2008) proposed an Airline Business Process Modeling concept as a way to map airline processes and build cross-functional views. Realizing cross-functional views is the first step towards gaining control over airline planning and recovery processes. Better insight into data and processes helps integration processes and therefore leads to better operational results.

The idea behind the integrated recovery is to replace the traditional sequential, and thus sub-optimal, recovery process by a process that simultaneously tackles the aircraft, crew and passenger recovery problems.

A sequential recovery process is based on using integrated data across traditional silos and a messaging system that allows for exchanging suggested schedule changes between specialized recovery modules and evaluating their impact on the corresponding operational areas. Even a partial integration, which consists of integrating the most relevant constraints from other recovery areas into a dedicated recovery module, helps improve overall solution quality.

A fully integrated, simultaneous recovery process, which concurrently considers the information of all major recovery components (aircraft, crew, and passengers), would allow recovery decisions that are better balanced across the resource areas than sequentially generated solutions. Simultaneous recovery implies a mathematical model able to incorporate all relevant objectives and constraints from the resource areas simultaneously considered. The major obstacles to developing such a system are the huge volume of integrated data to be considered and manipulated in real time, and the level of complexity of such an integrated optimization model. This approach is generally considered computationally intractable at current computing technology and mathematical programming levels.

In his doctoral dissertation Lettovský (1997) discusses the fully integrated airline recovery problem (aircraft, crew and passengers). He formulates the problem as a mixed-integer linear model and proposes a Benders' decomposition algorithm to solve it. The master problem, which is the Schedule Recovery Model, provides a cancellation and delay plan. Aircraft and Crew Recovery sub-problems are then solved for each aircraft type. A flight with no aircraft or crew found is cancelled. Finally, the Passenger Reaccommodation sub-problem reroutes disrupted passengers using the residual flights. The model's huge dimensions make it computationally intractable and therefore unusable for a real application.

Petersen et al. (2010) confirm that solving the airline integrated recovery using the Benders decomposition framework proposed by Lettovský (1997) is not a viable option due to the large problem scope and identify the Crew Recovery module as a bottleneck. To make the problem tractable, they propose delivering a solution which is globally optimal with respect to passengers, locally optimal with respect to crew, and feasible with respect to aircraft. The integrated approach is tested against the sequential approach on five scenarios. The results show that the integrated approach produces better solutions than the sequential approach for two scenarios and equally good solutions for the remaining three scenarios in regards to the passenger and flight delays. Even though the integrated approach takes longer to solve the problems, the reported solution times are still comparable to the solution times of the sequential approach. Some future improvements are listed in the presentation.

Papers on partially integrated recovery are more abundant. Stojković and Soumis (2001) presented the first pilot recovery model that allows for simultaneously modifying, as necessary, existing flight departure schedules and planned individual work days (duties), while keeping original aircraft itineraries unchanged. The objectives of the model are to minimize the number of cancelled flights, the total delay of all flights, and the number of pilots with down-line problems caused by the recovery actions. The problem is formulated as an integer nonlinear multi-commodity network flow model with time windows. Flight precedence constraints preserve the feasibility of the input aircraft itineraries. The model thus implicitly preserves scheduled maintenance activities. The problem is solved by a branch-and-bound algorithm, where the lower bounds are derived from an extension of the Dantzig-Wolfe decomposition principle (a.k.a. branch-and-price method). All test problems are successfully solved in very short computational times. The advantages of simultaneously considering crew recovery and flight re-timing are tested in comparison with the traditional sequential approach, where the flight schedule is fixed first and thus given as an input data to the crew recovery problem. The results show the superior solution quality of the integrated approach over the traditional one with respect to the number of uncovered flights, with no increase in the number of pilots with the next-day down-line violations.

Stojković and Soumis (2005) is an extension of the previous work. They introduce a general form of the multi-crew rescheduling problem with interchangeable positions (most cabin crew positions). The multi-crew problem without interchangeable positions may be decomposed into one problem per position and solved by the approach presented in the previous paper (cockpit crew). Synchronization constraints are introduced to ensure the same departure time for a flight that belongs to several personalized new duties, where the number of duties is equal to the number of crew members assigned to the flight. The superior solution quality of the integrated approach over the traditional one is demonstrated again, as well as quick computing times for small and medium-size problems. Potential model improvements and applications are proposed in the paper.

Stojković et al. (2002) present a real time flight retiming model that preserves planned aircraft itineraries and crew routes. The new schedule is obtained by reducing flying, ground service, maintenance, or passenger transfer times within the given limits. The costs of time reductions, elements of the crew costs, and passenger inconvenience are included in the objective function. The model is intended for minor disruptions, as a first attempt to restore normal operations by delaying some flights without modifying existing aircraft itineraries and crew rotations, before resorting to much more expensive changes. The dual of the proposed model, which turns to be a network model, is solvable in real time given its linear behavior with respect to problem size.

Cordeau et al. (2001) propose a planning model that can also be used at the operational level to update aircraft routes and crew pairings following schedule disruptions. The model incorporates aircraft maintenance constraints, as well as the most important crew scheduling constraints. A solution approach based on Benders' decomposition is proposed to handle the linking constraints that impose minimum connection times for crews that depend on aircraft connections. Integer solutions are computed using a heuristic branch-and-bound method. The integrated approach demonstrates significant cost savings in comparison with the sequential planning process. The authors suggest using the same model to address hub-and-spoke network structures with a higher saving potential. Flight re-timing is seen as another possibility to further reduce crew costs.

Rosenberger et al. (2003) expand their aircraft recovery model to minimize crew and passenger disruptions. To achieve a more accurate estimate of delay costs they perform separate estimations of delay costs for disrupted and non-disrupted passengers and crews, and include those in the objective function of the extended model. They evaluate their approach using a stochastic model by Rosenberger et al. (2002) and compare the results of their extended model with those of the simplified model. They report significant reductions in passenger inconvenience and disruptions achieved with the extended model. The flipside is the deterioration of on-time schedule performance and the increased number of flight cancellations.

Stojković and Soumis (2004) analyze the advantages of integrated optimization for recovery under airport capacity reduction. They compare the integrated approach to two traditional approaches: the sequential solution of individual disrupted flights in chronological order, which models the way humans solve the problem, and (partially) integrated approaches. The metrics used for the comparisons are the number of cancelled flights, the number of delayed flights, the average delay, the number of modified aircraft and crew routes, and extra crew work hours. The simultaneous optimization approach is verified as far superior to the manual sequential approach in respect to all considered metrics and the total response time. Clear benefits of the integrated recovery are shown using a real scenario (airport capacity reduction). Three recovery approaches were tested: the traditional one that takes into account only aircraft, a partially integrated version that simultaneously considers aircraft and passengers, and the fully integrated aircraft, crew, and passenger recovery. The fully integrated system exhibits a significant improvement in the overall solution quality over the partially integrated ones. This result clearly demonstrates that increasing the number of resources simultaneously considered produces a solution that better satisfies all involved parties and minimizes the overall recovery cost.

Bratu and Barnhart (2006) report similar results using an aircraft recovery model that allows for delaying and canceling flights to minimize the total recovery cost. The costs of re-accommodating disrupted passengers and crews, re-routing aircraft, and canceling flight legs are included in the cost function. Using leg-based aircraft routing decision variables, instead of route-based ones, allows them to generate recovery solutions for aircraft, crew, and passengers simultaneously. Maintenance requirements are handled by fixing the routes for maintenance-critical aircraft. The obtained solutions exhibit a reduced number of disrupted passengers and passengers required to overnight at a destination other than the planned one, and reduced total delay minutes of disrupted passengers. As an expected consequence, the total delay minutes of non-disrupted passengers and the airline's on-time performance suffer.

Kohl et al. (2007) describe an integrated sequential recovery (ISR) developed by Carmen Systems (now Jeppesen) in collaboration with British Airways (BA) and the Technical University of Denmark, as part of the Descartes (DEcision Support for integrated Crew and AiRcrafT recovery) project. ISR uses the dedicated solvers for single resources (aircraft, crew and passenger recovery optimizers) as black boxes and an intelligent messaging system that integrates them. While only aircraft and crew recovery modules are able to suggest schedule changes to other subsystems, each dedicated recovery module is capable of evaluating the impact of a schedule change on its corresponding area of operations. To generate schedule change suggestions that are good from the overall business perspective, the aircraft and crew recovery modules consider constraints and objectives from other areas. Demonstrated improvements in the quality of decision making for the system that integrates data spanning multiple resources (ISR) are claimed although not quantified in the paper. Another system that determines solutions in an integrated fashion, using concurrent access to available information for each of the resources, is discussed as well. The system is referred to as a tailored integrated recovery (TIRS). An initial TIRS version that builds a time-band network (based on Arguëllo (1997b)) both with respect to aircraft and crew has been developed. To simplify the problem a full crew was considered as one unit. A basic heuristic local search, based on simulated annealing, has been implemented, but no computational results were presented.

6.3.3 Airline Disruption Management Tools

6.3.3.1 Design Requirements

The major benefit airlines seek from a disruption management tool is to realize productivity gains by mitigating the impact of irregular operations through faster and more efficient return to normal operations. Recovery tools combine both the algorithmic and computing powers to achieve that gain. To be justifiable, a recovery tool needs to prove its capability to deal with all kinds and volumes of disruptions in a better and faster way than humans do. The solutions need to be not only cost efficient, but also logical and applicable from the airline's perspective.

To be considered as valuable, a recovery tool has to demonstrate fast response times, ability to consider a wide scope of possible recovery actions, capability to consider a wide range of scenarios, potential to generate a most effective and least disruptive recovery solution, and capacity to present multiple meaningful solutions on request.

The most important tangible benefits expected from a recovery tool are:

- Reduced recovery times during irregular operations,
- Improved on-time performances, and
- Revenue protection from avoided canceled flights.

The major intangible benefits are:

- Increased passenger goodwill, and
- Improved control over airline operations.

In addition to production mode, a disruption management tool should allow the user to employ it in what-if mode. What-if mode lets the users create and solve several plausible scenarios in anticipation of irregular operations. Selected scenarios and the corresponding solutions may then be saved for potential future use. That makes users prepared for an oncoming event ahead of time and smoothes the users' workload distribution, especially in case of major irregular operations. If one of the scenarios materializes, the related solution is immediately deployable. The ultimate benefit is a better and faster decision.

Although the key objective of any recovery tool is to promptly produce minimum-cost solutions which remain as close as possible to the original schedule (flight, crew, maintenance schedule), minor and major disruptions require different solution approaches dictated not only by different dimensions of the problems, but also by different user expectations, primarily in regard to the solution time and quality.

Minor Disruptions

Users have very high expectations of a recovery tool when it is employed to solve a minor problem. The reason is that small problems are solvable manually in a quality and timely manner. To be valued by the end-users, a recovery tool invoked to solve a minor disruption has to be at least as good as them in both quality and response time aspects. To produce a quality solution, the tool must incorporate the user's knowledge when selecting the best among the concurrent options, especially when it comes to choosing among several alternatives with the same or similar costs. Offering the user a choice between multiple high-quality alternative solutions further increases the attractiveness of a tool and increases its chances to be embraced and therefore employed on a daily bases. The multiple solutions may be generated with no user intervention (batch process) or through an interactive and iterative process between the user and the tool.

It follows that heuristics and local searches, when properly designed and implemented, are very well suited for solving minor disruptions due to their ability to always generate a feasible solution to a problem, capacity to imitate a human way of thinking when solving problems, quick response time to small-scale problems, flexibility in terms of using a variety of constraints and objectives, and ability to generate multiple solutions for a given problem.

However, although the automation of a manual process adds certain value to the disruption management process by significantly reducing the recovery time, a recovery tool that automates without improving the manual solution process is an investment only partially justified.

Major Disruptions

Dealing with major disruptions implies different objectives and poses different challenges to a recovery tool. Major problems are hardly solvable by a human in a timely and efficient manner. Due to the limited capacity of humans to deal with a complex problem in a holistic way, the users decompose a large-scale problem into small, solvable pieces. The most common approach is to chronologically order the problems and solve them one at a time, starting with the earliest one. Although the process makes it possible for humans to deal with otherwise unsolvable problems, it usually leads to inefficient solutions due to the isolated views of the whole picture. A recovery tool needs to demonstrate its ability to deal with the complex problems in a holistic and timely manner. The major characteristics of the heuristic approaches and local searches, which are beneficial for solving minor disruptions, become disadvantages when it comes to handling complex, large-scale problems. Heuristic approaches are, in general, ill-suited for solving massive problems because, like humans, they consider one action at a time. Because the solutions they produce tend to be very similar to what humans are able to produce manually, there are no obvious benefits of using them for solving large disruptions. In addition, their ability to generate multiple solutions has little practical value for large disruptions. The first reason is that the quality of the generated solutions cannot be guaranteed. The second one is that the volume of information makes it difficult for a human to compare solutions to a complex problem in a timely manner. In fact, the time needed to perform such a comparison could easily surpass the time needed to generate the solutions themselves. Although associating some quality metrics with each of the solutions could help, it is unlikely to be sufficient in steering the user attention toward the most acceptable solution.

Solution approaches based on mathematical programming are better suited to solving complex problems because of their capability to preserve the global view of the problem, and produce complicated and interesting solutions that humans are unlikely to generate. Their main disadvantages are that a feasible solution within a defined timeframe cannot be guaranteed, solution time may be sensitive to details of model formulation, numerical models may be difficult to modify, and solutions may be less intuitive to the user.

6.3.3.2 User Acceptance

Although the process of accepting a new tool is influenced by numerous factors, experienced users generally do not accept a new DSS for the following reasons:

- Difficulties associated with the calibration of user-controllable parameters
  Most of the parameters need to be carefully calibrated and fixed as such, especially optimization-related parameters, which users do not have the expertise to tune. Only a limited number of parameters should remain on an adjustable-setting base according to the user's needs.
- Challenges associated with data transfers
  The problems to be solved, as well as the generated solutions, need to be transferred, with little or no user intervention, from the main system (aircraft, crew, passengers) and back to it. Otherwise the recovery tool becomes unattractive because of the additional time required to manually input the problems to the recovery tool, and even less attractive when a solution needs to be transferred back

into the main system. That extra time may well be many times longer than the actual time needed to solve a problem, especially for major disruptions.

- Solution formats or GUI views
  Although it may seem trivial, issues such as the look and feel of the GUI or the formatting of the solution, if different from what the user is accustomed to, further reduce the likelihood of a tool to be accepted regardless its solution speed and quality.
  Baumgarten (2009) (Sabre) presents a comprehensive insight into data visualization whose goal is to make information understood, not just available. He defines the effectiveness of data visualization by its ability to clearly and accurately represent information and by the user's ability to interact with it to figure out what the information means.

No matter how good a recovery tool is, the users must be given some time to get used to it.

6.3.3.3 Common Usage

To fully benefit from a disruption management tool the users need to embrace the following ideas:

- Solve simultaneously multiple, ideally all, existing problems at a given time
  Using a tool to solve a single problem at a time is a typical example of poor utilization of a recovery tool that results in no obvious benefit over the manual mode because of the same limited view of the situation that prevents the tool to exercise its real strengths. The widely accepted "time-buy approach", which solves the urgent problems first leaving more distant problems for some later times, is greedy by nature and frequently results in cost inefficient solutions because of its incapacity to consider the whole picture. The recovery actions employed to solve the problems close in time may prevent the tool from finding an efficient solution for the time-distant problems, even though some other equally good alternatives for the close problems would result in a better solution for the distant problems. The inefficiency is expressed through caused, but avoidable, flight delays and cancellations down the line, and the increased crew and passenger cost. Stepping away from that tradition requires changing the users' mindset. The end-users need to get familiar with the idea that only when given an opportunity to use its global-view capability to the full extent can a tool produce anticipated superior quality solutions.
- Deploy the tool closer to the problem materialization
  Solving a problem in its early phase and immediately implementing generated solution usually results in needless modifications and implies further adjustments as the irregular operations progress. Postponing the decision time implies higher reliability of input information and better data quality in general, which in turn improves decision quality. To achieve it the users must embrace the previous idea of simultaneously solving all existing problems and also trust the

tool from the solution time, quality and reliability perspectives. The tool's proven ability to return quality solutions in a timely manner makes users comfortable with the idea of waiting longer before initiating the solution process. That wait is beneficial for the overall solution quality because of the increased reliability of the information used to make a decision. However, humans are very sensitive to waiting times. Even though it may take several hours for a human to solve a large problem, a few-minute solution time may be perceived as too long by the user sitting at the computer and waiting for the solution to come back. The reason is not only a negative perception of waiting times in general, but also the lack of confidence in the tool's capability to produce adequate solutions. If the returned solution is unsatisfactory, the user needs to either rerun the application or manually adjust the solution before deploying it. However, it may be impossible to apply either action if no time is left to do that.
Baumgarten (2010) (Sabre) provides very functional observations on the importance of communicating progress and estimating completion during optimization. He states that "as developers of systems that require time, we need to be aware of the human perception of time and create user interfaces that mitigate the negative effects of waiting". The following six facts about the psychology of waiting together with the appropriate actions, where applicable, are identified:

1. Fact: Occupied time feels shorter than unoccupied time
   Action: Visualize and quantify optimization progress
2. Fact: People want to get started
   Action: Consider workflow changes to start processing sooner
3. Fact: Uncertain waits seem longer than known, finite waits
   Action: Quantify optimization progress
4. Fact: Anxiety makes the wait seem longer
   Action: Give users more control, more information, and more visibility
5. Fact: Unexplained waits seem longer than explained waits
   Action: Display optimization summary statics while processing
6. Fact: The more valuable the service, the longer customers will wait

- Use the tool on a daily basis instead of uniquely using it for major irregular operation
  Small disruptions occur on a regular basis. The users are accustomed to manually deal with them regardless of a decision support tool they may have at their disposition. However, the overall cost of recovering from minor disruptions (low single-problem recovery cost, but very high problem frequency) may exceed, on an annual basis, the total recovery cost from major disruptions (high single-problem recovery cost, but only few times a year). The main reason for a less frequent utilization of recovery tools when responding to minor disruptions is that small problems can be solved in a timely manner by humans. However, the user's recovery decisions are not cost-oriented. A good solution from the user's perspective is a solution easy to implement, the easiness of

implementation being measured by the required number of phone calls or text messages. Changing the users' mentality by shifting their attention to the cost component of the recovery decision is the challenging task left to the airline management. The transformation is definitely worth the efforts.

6.3.4 Conclusion

Although several recovery suits and stand-alone recovery products have been available on the market for quite some time, there is still no industry-wide recognized and accepted recovery tool. Reasons for that can be found on both the airline and vendor sides.

Some of the issues on the airline side are lack of availability of the input data needed to achieve required solution quality, as well as fuzzy definitions of the business objectives and constraints, especially on the more granulated level, which harms solution quality.

On the vendor side, one of the major challenges is to successfully adapt existing tools to different airline business models. Unlike the fairly standardized planning process, airline-specific recovery rules and processes make it very difficult to build "one size fits all" tools. The conditions under which a tool is invoked may be quite different as well, with some airlines using the tool specifically for major crises, and others on a daily basis regardless the size of the problems. Some third-party tools were initially developed for a specific airline customer. If the implementation approach was not flexible, it could be difficult to use these tools off-the-shelf for other airlines. Reusing such a tool for a different airline may require substantial redesign. Adapting even highly customizable tools to a specific real-time environment can be a painful process for both vendor and airline, not only because of the tool's characteristics, but also because user expectations are often complex and difficult to capture in detailed specifications. Long-term dedication and willingness to adjust to meet a client's success criteria are the keys to a wider utilization of recovery tools. Another highly important issue is a very limited insight into the implications of the recovery decisions on the overall airline business results. Even though the economies from using an optimization-based recovery may be significant, as reported in Yu et al. (2003), they are very difficult to quantify.

There is still no consensus in the airline community on how to measure the quality of a recovery plan. Establishing the set of metrics, better known as Key Performance Indicators (KPI), is a very important step toward promoting the use of the recovery tools. KPIs can be used to predict the operational and financial performances of an airline, assist in comparing different recovery policies and tools, and increase the situational awareness of the airline management.

An intuitive way to evaluate airline plans under different operational conditions and different recovery procedures is to simulate them by a model of airline operations. SimAir and MEANS are two frameworks developed to this end in academia.

SimAir, a modular airline simulation of the daily operations of an airline described in Rosenberger et al. (2002), was developed at the Georgia Institute of Technology to evaluate a plan's robustness and recovery policies in a random environment. SimAir emulates an Operational Control Center and comprises three modules: the Event Generator Module, which generates random ground time delays, additional block time delays, and unscheduled maintenance delays; the Controller Module, which determines when a disruption prevents the flights from flying as scheduled; and the Recovery Module, which is activated by the Controller Module when a disruption occurs. SimAir also allows for external recovery tools to be seamlessly integrated.

The MIT Extensible Air Network Simulation, MEANS, is another modular, event-based simulation tool that tracks aircraft, crew and passengers through a network of queues that capture capacity constraints at airports (Clarke 2004). It includes, among others, an Airline Module, composed of the scheduling and recovery sub-modules. The recovery sub-module determines airline actions in response to input schedule disruptions. The built-in recovery module can be replaced with any recovery tool whose performance can then be evaluated. MEANS can also be used as a platform for research projects aimed at determining an airlines' KPIs, as described in Riedel and Clarke (2005), Karisch (2004) and Tiourine and Chiraphadhanakul (2009).

Dealing with operational problems in an uncertain, dynamic environment using information subject to constant change is still a challenge. Efficiently solving a problem when input information evolves following a sequence of unforeseen disruptions necessitates efficient data integration and conflict management. Without being able to capture the dynamics of a problem it is difficult to fully benefit from using a disruption management tool.

Yet another challenge in implementing recovery tools lies in end-user acceptance. Airline operations staff understands that their decisions are based on a wide variety of inputs, which they learn to balance through experience, and are justifiably suspicious of efforts to automate them. While recovery tools are meant to complement the users, not replace them, the overall economic climate in the airline industry impacts attitudes towards all new technologies. Delicate, yet aggressive, education of user groups is required to encourage users to embrace the use of recovery tools.

The use of DST allows the user's role to evolve from a tense, labor and nerve-intensive search for a feasible solution, to a higher level that consists in controlling the recovery process and making better decisions. The process control is achieved through applying the user's knowledge and expertise when using and calibrating the user's configurable parameters to drive the solution process in the right direction, depending on the problem-specific characteristics, as well as towards selecting the best of the offered alternatives in respect to hardly quantifiable factors. Using a recovery tool to cope with operational disruptions means combining the best of two worlds: the algorithmic and computing power, on one side, with a human brain's capability to successfully deal with fuzzy, ill-defined notions, and synthesize them into "human expertise" often hardly quantifiable and thus almost impossible to integrate into a recovery algorithm, on the other side. The ultimate goal is to achieve a unique synthesis of human intelligence, industry expertise and innovative technology that provides decision-makers with the knowledge to act, enabling them to make better decisions.

Although the best application of a recovery tool is seen in conjunction with human expertise, as part of an interactive and iterative process, lots of up front work is required from management to prepare the end-user group and point out the benefits of using the new system. If the job is not properly done, instead of looking for ways to make the system work, the users find ways to reject it.

In the current airline market regional carriers quite often operate more flights and transport more passengers than the main carrier. Traditionally, the recovery decisions for the main carrier and its regional carriers are produced independently with minimum information exchanged. Due to unsynchronized decisions, the overall recovery results are sub-optimal for the involved airlines and passengers they serve. An effective recovery strategy must simultaneously take into account both the main and regional carrier(s) in order to achieve the best possible solution. Efficiently solving integrated recovery problem for the main carrier and its regional affiliates is a future challenge for both the airlines and the operations research community.

Pachon (2007) (joint work of Navitaire and Continental Airlines) describes the integrated recovery problem and the related technological challenges, using as an example Continental Airlines as a main carrier and Continental Express as its regional carrier. Both carriers operate hub-and-spoke networks and fly approximately the same number of flights daily. Company-specific operational rules and strategies that must be respected, coupled with the need to access data for both carriers, were cited as the major OR challenges in approaching the integrated recovery. Gaining acceptance from both carriers, because the individual solutions, when taken in isolation, may look better than the integrated one (especially for the regional carrier), was pointed out as another significant challenge. Some preliminary results, obtained using a sample of scenarios of different types and sizes, showed that significant savings could be achieved by making an integrated decision. The following benefits of the integrated approach are reported based on a sample of 100 scenarios of different types and sizes: an average reduction in number of cancelled passengers of 12%; an average reduction in number of delayed passengers of 5%; and an average reduction in number of delayed flights of 1%. Compared to the model for the main carrier only, the integrated model shows an increase of approximately 20% in number of variables and an increase of approximately 74% in number of constraints. The corresponding solution time only grew by a modest 10%.

In conclusion, notwithstanding numerous attempts over the past years to introduce effective airline disruption management models and tools, they are still regarded as luxury items. To recover from schedule disruptions most airlines still rely, at best, on humans equipped with operations and crew tracking systems that provide status information, but do not suggest solutions. Proving business necessity of using disruption management tools and providing tools that achieve the expectations of airlines are some of the unresolved, but highly challenging, issues the airline OR community still faces.

6.4 Ground Resources and Turnaround Management

6.4.1 Introduction

The scope of this section is not only to show the OR landscape but also to provide a glimpse into the business logic of airport operations. An airport can be characterized as a system containing multiple subsystems such as runways, taxiways, gates, passenger and cargo terminals, and ground access complexes, as shown in Fig. 6.13. Each of them has its own objectives and constraints, which are often conflicting. However, they all need to be working in a synchronized manner so that safety is guaranteed and efficiency is provided.

The capacity of an airport is determined by the capacity of the most restricting element of the airport—the capacity bottleneck. In most cases, this is the runway system of the airport. However, the bottleneck element can also be the taxiways, passenger terminal including check-in counters, immigration and security lines, gate availability, or some other part of the airport subsystems. This section focuses on the role of OR to increase the system's throughput, reduce bottlenecks, and increase overall efficiency of the ground resources, including workforce and equipment, located mainly in or closely around the passenger terminal area.

As shown in Fig. 6.14, a passenger terminal has its airside and its groundside, also called landside. Usually the airside includes the parts of an airport directly involved in the arrival and departure of aircraft or simply beyond the security check line. The groundside is everything else in a passenger terminal. Ground resources optimization is applicable in both of these areas.

Keeping an airport's capacity lower than it is designed for would definitely increase operational robustness and ensure the absence of serious delays. However, this is not a practical solution since it would result in underutilization of the airport's resources. On the other hand, high utilization of resources would increase the throughput and the profitability of the airport but it could create an unstable environment where small disturbances may lead to big delays and cancellations. The effect of those disruptions may be experienced not only at the given airport but also propagate through the entire Air Transportation Systems, as mentioned in the introduction of this chapter. Consequently, it is very important to find a fine balance between capacity and resource utilization. Unfortunately, it may be difficult to calculate these costs especially in an environment where airport management involves different stakeholders with their own interests and preferences.

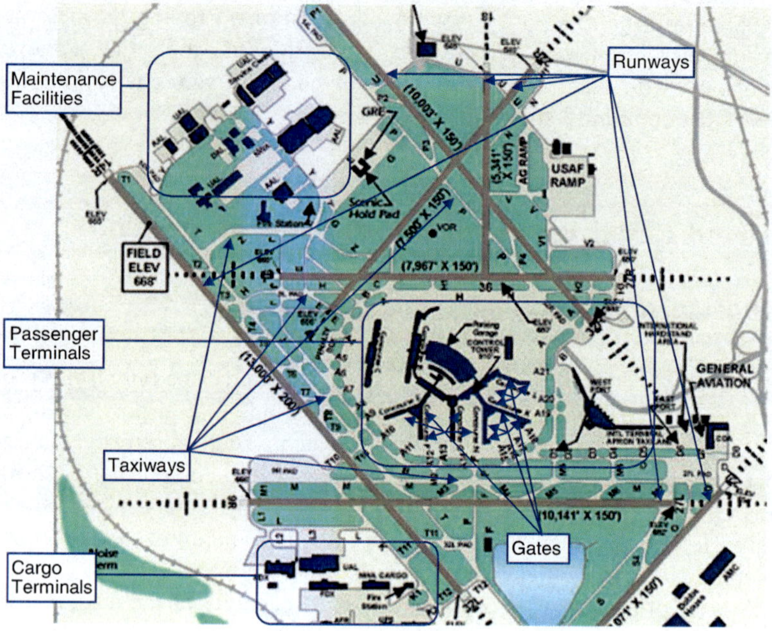

Fig. 6.13 Airport layout

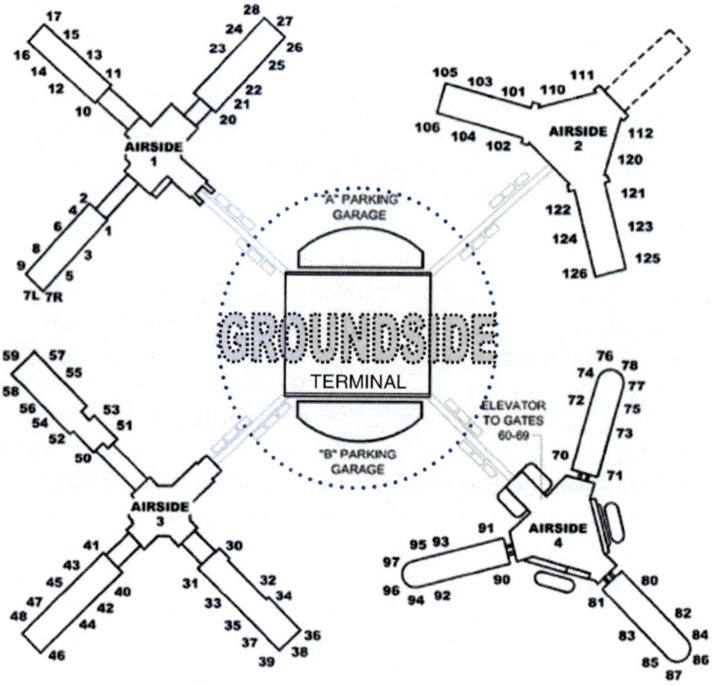

Fig. 6.14 Airport's airside and groundside

Ground resource management is one of the more complex and time-critical processes in the airline industry. Multiple parties work on the same object, where even only a small delay in a single activity can create havoc on the work-schedules of everybody else, leading to a substantial waste of resources and multiplied delay of the entire system. Such interdependencies, the time-criticality, and natural factors such as weather, highlight the importance of careful planning, coordination, and flexible management of all activities and resources needed to serve an aircraft, its crew, and of course passengers.

A widespread and officially accepted classification of ground resources and turnaround management does not exist in the literature. Some authors define ground resources as the airline ground staff servicing an airplane on the ground between two flights. Others include all staff at the airport that supports one or more airlines to operate at the airport. Some people also include ground equipment and facilities required at the airport. Here, under the term of ground resource management we consider all of the above. The following are some examples of ground staff resources and equipment: check-in agents and counters, gate agents, ramp agents, baggage conveyors, baggage handling vehicles and staff, aircraft cleaning staff, passenger assistance agents, fuelling, catering, push-back towing and cargo loading/unloading equipment and staff, deicing services, air traffic control staff, and security and border control counters and agents. Due to its specific nature, airplane maintenance is often treated as a separate problem and usually is not a part of the ground resource management applications.

As per the U.S. Transportation Department's (2004) Bureau of Transportation Statistics 2004 Report, in 2001 there were 5,286 public-used airports in the US serving almost 600 million passengers. About 445 million passengers were served by 72 airports. An additional 130 million passengers were served by 124 smaller airports while the rest was served by small airports scattered all around the country. In 2002 around 55% of all airline passengers were served by the 20 biggest US airports.

The number of passengers and the required number of air and ground resources, including workforce and equipment, is huge and represent a logistical challenge, especially for hub airport authorities. To avoid propagation of inefficient operations at one airport on the whole transportation system, it is critical to understand the interdependencies between airports. Hub airport have an even bigger impact, and to illustrate the challenge, Fig. 6.15 depicts the layout of hub airports in the US in 2009, as provided by the Bureau of Transportation Statistics (BTS) on the U.S. Department of Transportation (2010a) web site.

The same web site provides the BTS T-100 Market data showing that in the last decade a record number of passengers (835,436,440) on all airlines serving all US airports were transported in 2007.

Chin et al. (1997) found that in 1994, 74.5 percent of total delay hours were on the surface in the form of gateholds, taxi-out delays, or taxi-in delays. They state that the vast majority of these surface delays are incurred while waiting in the departure queue, and to a lesser extent, arrivals waiting for gates. This statement is still valid and at some airports such as La Guardia in New York, ramp and taxiway

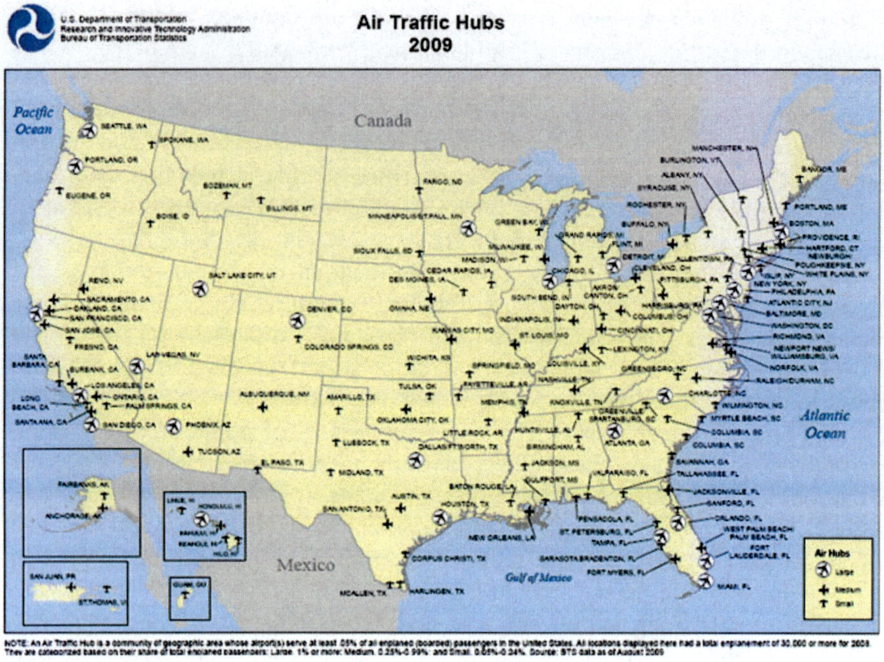

Fig. 6.15 US Hub Airports

congestion causes huge taxi-out and taxi-in delays. Sometimes, especially at big hub airports with many connection flights, passengers sit on the ground waiting for a gate to be free so that their airplane may pull-in and deplane. Sitting on an airplane watching other aircraft pulling out and knowing that one of them is their connecting flight is very frustrating for the passengers and directly influences the overall passenger satisfaction with an airline.

In well publicized incidents in the winter of 2006–2007, passengers were stuck on airplanes on the ground for up to 11 h (Wikipedia 2009). A more recent incident occurred in summer 2009, when 47 people were trapped aboard a regional jet with faulty lavatories for seven hours. As a consequence to such extreme delays, the US DOT imposed a maximum of three hours of tarmac delay, effective on April 29, 2010. While many airlines argued that such a rule would increase cancelled flights or extreme delays due to crews losing their ability to complete a flight once they returned to a gate, BTS data (2010b) show that in August 2010, compared to August 2009, cancellations were constant, and delays decreased. In August 2010 there was only one flight with a tarmac delay over three hours, and in October 2010, for the first time ever there were no flights with such a delay (DOT 2010c).

As reported by Wolf (2010), in April 2011 even foreign airlines may become subject to similar tarmac delay rules. This decision would come as a consequence of the snowstorm that hit the New York area during the days before New Year's

Eve 2010. Hundreds of passengers were stranded on planes, a majority of them on international flights. Among others, five Cathay Pacific Airways flights were sitting on the ground for at least 4 h on Dec. 29, 2010, and one was on the tarmac for almost 11 h. More than 1,100 people were on board these aircraft. While the airline is pointing to insufficient gate availability as a cause of this excessive tarmac delay, government officials are insisting that airlines should take a pro-active approach and delay/cancel rather than fly into an airport like JFK in weather conditions that result in tarmac gridlock. In reality, for such long flights, weather forecasts are not very reliable. As the December 2010 winter storm shows, the availability of ground support staff, including customs and immigration officers, is critical for successful operations at the airport as well. Furthermore, their availability also depends on the local ground transportation systems. Adding new fees or penalties would only burden airlines by increasing costs, which would eventually be passed on to passengers, and will not prevent similar events in the future.

Airport disruptions vary depending on traffic volume, airport layout, and runway configurations. These disruptions are usually caused by bad weather, low visibility, approach area congestion, taxiway congestion/closure, deicing, technical problems, infrastructure bottlenecks, staff absences or shortages, accidents, gate unavailability, unplanned aircraft maintenance, and the maintenance of airport and airspace navigation devices.

Highly efficient airline schedules and poorly planned ground resource schedules can become a source of disruption as well. The optimization of 'productive' resources, such as airplanes and air crew, does not always take into consideration the 'non-productive' ground service resources. Such an approach results in tight ground connection times which are very sensitive not only to schedule changes but also to variations in ground service resources during the day of operations. Combined with inefficient ground resource planning and high operational costs, an unstable operations environment may be created. The workload of unbalanced ground resources is difficult to optimize and vulnerable to demand variations during operations. In such an environment, even feasibility of schedules is hard to achieve when a disruption occurs. A holistic view on both the planning and the operation phase in ground resources allocation and its coordination with other resources is important. It is essential to deploy appropriate models and tools in each of these phases as well as to develop the linking constraints between them in order to create an integrated optimization approach.

Ground resources are often owned and managed by different stakeholders, such as airlines, airport authorities, and government agencies such as the FAA and TSA, and Customers and Border Protection (for immigration), see also Fig. 6.16. Authorities such as the FAA are usually on the airside of the airport, while security and immigration authorities belong to the groundside. Airline and airport authorities usually operate on both sides. They all have in common a flight schedule that needs to be carried out and passengers that need to be served.

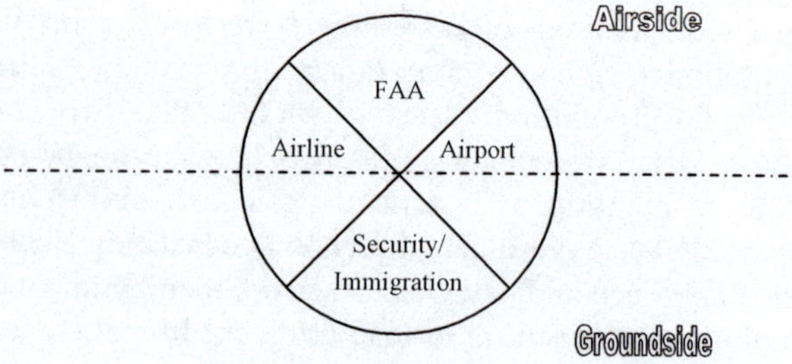

Fig. 6.16 Ground resource stakeholders

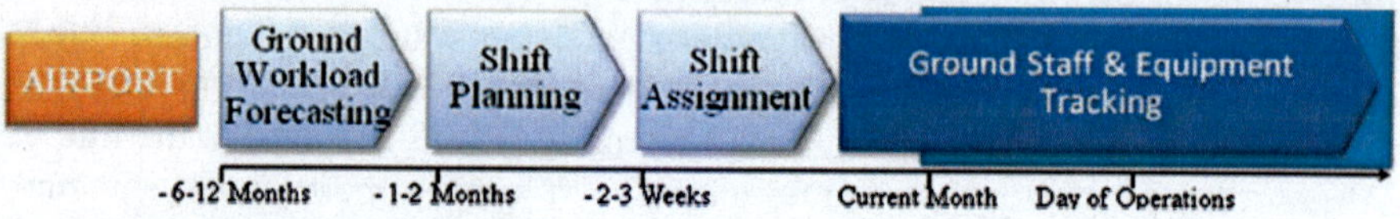

Fig. 6.17 Ground resource scheduling (© 2011 Jeppesen)

6.4.2 Processes and Solution Approaches

6.4.2.1 Ground Resource Management

Ground resources management is a multistage process that can be separated into phases: *long term planning, midterm planning, short term planning, maintenance of the plan,* and *execution of the plan,* i.e., operations as well as *post operations analysis*. Long term planning is a necessary first step in ground management due to long recruitment and training cycles for some positions.

Ground resources are required to perform tasks that are closely related to the flight schedule. Early equipment and manpower planning processes need to start well in advance of the final schedule being published. They include forecasts of labor and equipment requirements based upon estimated flight schedule and passenger, baggage and cargo volumes.

Once the flight schedule is published, usually a year to six months in advance, midterm planning takes place. As presented in Fig. 6.17 (extracted from Fig. 6.5), midterm planning is also called Ground Workload Forecasting. At this point, it is possible to determine the ground staff and equipment workload required to cover the given schedule, i.e., to determine tasks, their start/end time and qualification requirements for a given scheduling period. It is also possible to add a demand estimation component at this stage that allows the workload to be adjusted based on historical data.

Around a month or two in advance, short term planning or Shift Planning takes place. In order to determine a set of valid shifts that covers the resource requirements at minimal cost, a set of shift patterns is often generated first. With the schedule and resource information updated, shifts are populated with tasks generated for each ground resource type. The objective is to cover the workload profile for the given scheduling period, and to maximize the number of tasks assigned to a fixed set of shift patterns while honoring constraints such as minimum time between tasks, task precedence and shift/task qualification.

A couple of weeks before the start of the scheduling period, the Shift Assignment or Rostering process takes place. At that point roster lines are built, i.e., the shifts previously generated are assigned to resources taking into account government, company and union regulations such as minimum rest times, maximum overtime and other limitations, while aiming at balancing the work schedule. A roster line includes work shifts, trainings, meetings and absences such as days-off, vacation, and sick-days. While overtime is allowed, an objective is usually to minimize overtime while still covering all required tasks. Similar to air crew scheduling, rosters may contain lines assigned to specific ground staff, anonymous lines available for bidding, or personalized lines that include staff's preferences, known as preferential bidding. The shift generation and rostering phases may shift back and forth in order to adjust to changes occurring during the planning period and to cover all tasks that are still uncovered.

Usually, one distinguishes the transition period between the planning and operations processes. The transition or pre-operations process starts once the planning phase is completed. At that point, maintenance of the published rosters takes place, accounting for schedule updates, shift changes, absence management and rosters repair. The objective at this stage is to keep the integrity of the plan and allow for a safe and efficient start of operations.

On the day of operations, the transition phase is over and real-time roster management is required. Real time information is constantly received from external systems. Examples of such real-time updates are schedule modifications (departures/arrivals/cancellations), staff clock-in/clock-out, airport congestion, bad weather, and technical problems. Schedules associated to these changes may experience increasing operational costs and even infeasibilities in some cases. An operations management tool with staff and equipment tracking, data management, alert mechanism and disruption recovery capabilities is required at this stage to manage disruptions and allow recovery in a timely and efficient manner. Such a tool should also contain an administration module supporting post operations analysis, collecting and publishing operations details, payroll calculation and report writing.

In its complexity, ground resource management is comparable to air crew management. Solution methodologies used to address these resource scheduling problems are similar. However, there are important differences between the two problems regarding the nature of the tasks that need to be assigned, the underlying business rules, and the business objectives. Flight crew need to satisfy linking constraints that guarantee the geographic continuity between two successive flights. Ground staff does not have such constraints since everything is happening at the same location. Compared to air crew, the tasks for ground resources are shorter and have less restrictive constraints, resulting in a larger solution space and hence a larger number of combinations to create a single line of work. Due to the combinatorial nature of ground resource problems they are often hard to solve, especially for large hub airports. Additionally, ground tasks are very interdependent, concentrated around the same airplane, facility or ground equipment. This interdependence contributes additional difficulty in achieving robust solutions which explains why the airport ground problem is harder to solve compared to other industries' ground personnel, such as shop and restaurant employees. Like air crew, ground personnel are usually heavily unionized and a complex set of business and contractual rules are imposed. On the positive side, ground tasks are usually more periodic, with the highest load around the airport's peak periods when the most of inbound and outbound flights are taking place. Consequently, it is possible to decompose the problem into smaller time periods that are easier to solve.

Inappropriate ground resources planning may have a negative effect on the day of operations. Planning and operations are tightly related and all benefits of optimized schedule planning may be easily lost if they are not aligned with the operational objectives. As already mentioned, strategic and tactical planning decisions are based on assumed ideal operating conditions, which rarely occur in practice. For that reason more robust planning is required to protect plans at least from minor and some medium size disruptions.

A big challenge in ground resource management is the high fluctuation in workforce demand. Demand fluctuation may be planned fluctuations, such as sporting events, or unforeseen fluctuations. An extreme example of unforeseen fluctuation is security and border control at the airport. Routinely, this type of workforce is planned based on the published timetables and is very repetitive from day to day. However, after the bomb attacks in Madrid and London additional border control had to be carried out, which even included Intra-European (Schengen) flights, and it was critical to have such resources deployed quickly. After the July 2006 terrorist plot involving liquid explosives on North Atlantic flights, all hand luggage was forbidden on UK flights and liquids and gels were forbidden in hand luggage on North American flights. Many airlines and airport authorities, especially British Airports Authority (BAA), were unprepared for stricter security rules and a drastic increase of checked-in luggage. Another unforeseen fluctuation example is deicing service at the airports. When a snowstorm hits an airport, a huge amount of personnel is required for a short period of time. For example, snowstorms are very rare at Dallas/Fort Worth airport. However, as experienced by one of these authors, in January 1997 a snowstorm created great operational challenges, especially at the airport. The weather conditions required deicing of all airplanes departing that day, and the airport's modest deicing equipment and resources were not sufficient, which led to many cancellations as well as long queues of airplanes waiting for deicing. In some cases, departing passengers were stranded on airplanes for six or more hours. These incidents resulted in huge costs for airlines and airports as well as in the loss of a lot of goodwill of the traveling public.

We point out, however, that it is impossible to completely avoid potential disruptions. It would be more expensive to keep surplus staff for extensive periods in time while the probability to need it is very low, as the deicing example shows. Ground resources and turnaround management tools may help provide better visibility of the current situation, faster deployment of reserves or other contingency plans as well as better coordination with other elements of the system. However, the real value comes from OR capabilities of such tools by creating more cost/benefit balanced deterministic and stochastic models combined with forecasting capabilities.

Unfortunately, sometimes organizational issues such as business processes and real estate ownership create obstacles for an effective use of OR. For example, terminal and gate space is often restricted to or owned by particular airlines rather than being part of the common usage terminal system, especially in North America. By dedicating gates to one or two airlines, the entire set of gates is decomposed in smaller subsets. Like in any other optimization problem, having the optimal solution for each subsystem doesn't guarantee the optimality for the entire system, in this case the optimal gate assignment for the entire airport.

An important part of ground resources and turnaround management is airport gate assignment. The gate assignment problem may be executed in different planning phases. An initial gate allocation can be made once a flight schedule is published and the allocation is based on airline and airport preferences, contracts, and constraints. It may be used even during the flight schedule planning phase itself to evaluate different schedule scenarios by taking into account the costs of ground handling operations. However, gate assignment is usually solved on a daily basis at the beginning of the day of operations to determine the feasibility of gate-flight assignments such that total passenger connection times are minimized. Although flight and passenger data are quite accurate at that point, the provided solution is still based on a fixed scheduled arrival and departure time. For that reason, models that support dynamic airport gate assignment are crucial for efficient operations.

6.4.2.2 Solution Methods

The cost of ground resources is a significant component of an airline's fixed and operational cost. Yet, compared to flight crew, the costs associated with ground resources are rather low. This is probably the reason why OR models and practices have not been as widely applied in this area as they have been in flight crew scheduling. However, as operations margins become lower, passenger flow increases, and the integration of air transportation subsystems becomes widespread, more sophisticated solution methods will have to be implemented.

As discussed above, the ground resources optimization problem is a highly constrained and complex optimization problem. The majority of constraints come from government regulations and company and union agreements and these are also called hard rules. There are usually also soft rules, which correspond to 'best business practices' for how to build shifts and lines, but are not necessarily part of contractual agreements and do not cause any penalties if they are not respected.

These soft rules may be imposed by the user and, while they may be helpful to decrease the dimension of a problem that needs to be solved, they also may be too restrictive and exclude some cost effective solutions from the solution space. Adding to the complexity are different types of daily shift schedules, full-time and part-time workers, high fluctuation of required workforce demand from one day or period of the day to the next, and multi-skilled workforce.

There is an increasing market demand for staff scheduling and rostering solutions in general. Research in this area has been extensive over the last two decades and Ernst et al. (2004a, b) provide an extensive bibliography of achievements in this field. However, the amount of published airport ground resource optimization work in airline industry is not so large.

Brusco et al. (1995) describe the development and implementation of two modules designed to enhance ground personnel scheduling at United Airlines. The first module uses column generation to improve the selection of employee shifts. The second module, a local search heuristic based on Simulated Annealing, enables initial rosters to rapidly improve. Using data collected from 119 United Airlines stations across the U.S., they report that the incorporation of the modules resulted in potential annual cost savings of more than \$8 million.

Many mathematical models for personnel and equipment problems may be found in the literature such as quadratic assignment, multi-objective 0–1 integer, and multi-commodity network formulations. Equally, a large number of solution methods are used including linear, integer and dynamic programming, shortest path algorithm, column generation, Lagrangian and Benders' decompositions, branch-and-bound, or custom made heuristics. Some of them are using solution specific heuristic formulations that may be also found in the literature and further increase the number of possible formulations for this type of problems. Here, we use the set partitioning (SP) formulation to describe generic ground personnel and equipment problems.

Let L denote the set of all feasible lines, i.e., sequences of events; let E denote the set of events which need to be covered. SP contains binary decision variables x_l, which equal 1 if sequence $l \in L$ is selected and 0 otherwise, and continuous decision variables s_e, which equal the number of uncovered events $e \in E$. The cost of sequence $l \in L$ is denoted by C_l and the penalty for not covering event $e \in E$ is denoted by p_e. A cost is a generic term that may include a real (dollar) cost, and different synthetic costs like credits, preferences, satisfaction level, etc. The coefficient a_{le} is equal 1 if the sequence l covers the event e and 0 otherwise. Finally, n_e is the number of times that event $e \in E$ needs to be covered. The SP formulation for a ground resource scheduling and rostering model may be written as follows:

$$\text{(SP) Minimize} \sum_{l \in L} c_l x_l + \sum_{e \in E} p_e s_e \tag{6.22}$$

$$\sum_{l \in L} a_{le} x_l + s_e = n_e \quad \forall e \in E, \tag{6.23}$$

$$x_l \in \{0, 1\} \quad \forall l \in L, \tag{6.24}$$

$$s_e \geq 0 \quad \forall e \in E. \tag{6.25}$$

The objective function (6.22) seeks to minimize the total cost of sequences selected in the solution and the number of uncovered tasks. Constraints (6.23) are generalized set partitioning constraints, which ensure that each event e is covered exactly with n_e sequences. Slack variables s_e are added to allow events to remain uncovered. Similar surplus variables may be added to allow so-called over-coverage. Note that these surplus variables (6.24) can be continuous since the right-hand side values n_e (6.23) as well as the x_l variables are integers (6.25). The same model may be used to describe both shift generation and rostering problems. In a shift generation model an event represents a task and sequence represents a shift. In a rostering model, an event represents a shift and sequence represents a roster line.

While this model may be used to describe many aspects of ground resource and turnaround management problems, a disadvantage of the given mathematical model is a large number of feasible sequences, which correspond to a large number of sequence (path-like) variables. The purpose of the presented SP model is to illustrate the problem rather than to necessarily solve it this way in practice.

To be usable to solve real-world problems, the model needs many more constraints. Additional constraints may be added for modeling global constraints, which deal with combinations of sequences such as interdependent tasks, overlapping shifts, limits in the number of shifts starting simultaneously, and distribution targets of tasks among shifts. Also, there is plenty of so called "resource" or local constraints that influence the shape of a sequence and actually reduce the number of feasible sequences in set L. Local constrains comprise, e.g., of precedence constraints or max duty rules.

In practice, more or less sophisticated heuristics or complex mathematical programming methods are used. One practical example is provided by Dugan et al. (2002) from Sabre Holdings. The authors decompose the shift planning process as two stage optimization problem: a shift scheduling model builds empty shifts first and then a task assignment model assigns tasks to these shifts. Even further, they solve the shift scheduling problem through a sequence of six linear/integer programming models. Such a heavy decomposition brings a risk of getting far off from the global optimum, even though each single optimization step reaches its local optimum. The authors still report 25% savings compared to the previous heuristic approach used to solve this type of problems. The task assignment problem is formulated as an IP and solved with column generation. While this sequential approach is inherently heuristic, its practical value should not be neglected. It is capable of dealing with large scale problems, its solution time is very reasonable, and it takes into account a large number of hard and soft constraints that are challenging in practice.

In follow up work, Dugan et al. (2003), present the ground task assignment and model it as a maximum clique problem. This formulation helps in modeling many hard and soft constraints such as shift check-in/out times, black-out periods, partial coverage and floating lunches. The original max clique formulation, with fixed lunches, is resolved directly with an IP. However, by introducing the floating lunches into the model, the problem size is increased significantly. By using the column generation approach, the model with floating lunches almost double the solution quality without drastically increasing solution time.

Finally, the shifts are assigned to individuals in the ground shift assignment, or rostering, step. To solve this large scale, heavily constrained rostering problem, a simple heuristic can be used to build and assign cyclic roster where all employees of the same group perform exactly the same line of work, but with different starting times for the first shift or duty—also known as 'snake' lines due to their appearance in a Gantt view. This roster type is most applicable for cases with repeating demand patterns. While cyclic rosters are sometimes required by airline and/or union representatives, a much more efficient way of generating shifts and assigning them to staff over the rostering horizon is to use mixed integer programming techniques.

Often, union agreements require predetermined working patterns. Lusby et al. (2010) address the ground rostering problem with work patterns. They use a cutting stock based integer programming formulation combined with a decomposition approach based on column generation and variable fixing. The authors go even further and propose a model to build rosters straight after the Workload Forecast process. That way the Shift Planning process is avoided and a better global solution is found. They also introduce two robustness parameters into their model: one called *workload contingency* to be able to handle higher workload and the other called *slippage* to account for potential workload delays. Due to the large planning horizon, the problem is decomposed into smaller time intervals in order to be solved in a reasonable time. The computational results on three real-life instances and 10 artificial instances demonstrate the efficiency of the proposed methodology. Computational results, compared with existing solutions implemented by the airline, provide reportedly a higher level of robustness simultaneously with staffing levels reduced by 10–12%.

Presented model assumes that planned flight schedule is not going to change. To be able to build more 'operation friendly' shifts, Kraay and Safabakhsh (2007) from United Airlines forecast actual flight performance which is then used to build for more robust shifts and rosters.

Another aspect of check-in resource planning is presented by Campbell et al. (2007) from United. Improved customer processing time during check-in is achieved through the better forecast of passenger arrival patterns, reorganization of the check-in lobby positions, right staffing of check-in counters.

As mentioned above, airport gate assignment is important for efficient airline and airport operations. The mathematical models used in this area are quite different compared to the generic set partitioning model (6.22–6.25) presented above. For that reason, we present a multi-objective formulation, a version of the model originally proposed by Ding et al. (2004a, b). We follow the formulation proposed by Drexl and Nikulin (2008). In the following we describe the formulation to illustrate the complexity of the model, without going into too much detail. For more details, we refer to the references above.

Airport gate assignment is modeled here as quadratic assignment problem using the following parameters:

N = set of flights arriving at and/or departing from the airport;
M = set of available gates at the airport;
n = total number of flights, i.e., $n = |N|$;
m = total number of gates, i.e., $m = |M|$;
a_i = arrival time of flight i;
d_i = departure time of flight i ($ai < di$);
$w_{k,l}$ = walking distance for passengers from gate k to gate l;
$f_{i,j}$ = number of passengers transferring from flight i to flight j;
$u_{i,k}$ = normalized preference value of assigning flight i to gate k;
v_i = priority value for flight i.

Additionally, two dummy gates are introduced. Gate 0 represents the entrance/exit (check in/out) area of the airport. Gate $m + 1$ represents the apron where flights park when no terminal gates are available. If flights are assigned to the apron, then passengers are usually transferred by bus to the concourse. The binary variable $\pi_{i,k} = 1$ denotes that flight i is assigned to gate k, $0 < k \leq m + 1$, and $\pi_{i,k} = 0$ otherwise. Then the objectives can be expressed as follows:

$$\text{Minimize } z_1 = \sum_{i=1}^{n} \pi_{i,m+1} \tag{6.26}$$

$$\text{Minimize } z_2 = \sum_{i=1}^{n}\sum_{j=1}^{n}\sum_{k=1}^{m+1}\sum_{l=1}^{m+1} f_{ij} w_{kl} \pi_{ik} \pi_{jl} + \sum_{i=1}^{n}\sum_{k=1}^{m+1} f_{0t} w_{0k} \pi_{ik} + \sum_{i=1}^{n}\sum_{k=1}^{m+1} f_{i0} w_{0k} \pi_{ik} \tag{6.27}$$

$$\text{Minimize } z_3 = \sum_{i=1}^{n}\sum_{k=1}^{m+1} v_i u_{il} \pi_{ik} \tag{6.28}$$

Objective (6.26) addresses the number of flights that are not assigned to any terminal gate (i.e., to the apron). Objective (6.27) represents the total passenger walking distance. It consists of three terms: the walking distance of transfer passengers, originating departure passengers and disembarking arrival passengers. Objective (6.28) represents the total flight gate assignment preference value. Note that the last objective can be converted into a minimization objective by simply changing its sign. The set of restrictions is defined by the following system of constraints. The first constraint:

$$\sum_{i=1}^{m+1} \pi_{ik} = 1, \quad 1 \leq i \leq n$$

ensures that every flight must be assigned to exactly one gate including the apron. The second constraint:

$$\pi_{ik}\pi_{jk}(d_j - a_i)(d_i - a_j) \leq 0, \quad 1 \leq i,j \leq n, k \neq m+1$$

prohibits schedule overlapping of two flights if they are assigned to the same terminal gate. The last constraint:

$$\pi_{ik} \in \{0, 1\}, \quad 1 \leq i \leq n, \quad 1 \leq k \leq m+1$$

defines the variables to be Boolean.

Drexl and Nikulin (2008) are solving this formulation by Simulated Annealing.

Shebalov (2008) from Sabre Holdings proposes another multi-objective gate assignment formulation with a quadratic objective function. He also introduces robustness in his formulation by implementing safety buffers on arrival/departure times and maximizing the uniformity of the gate utilization.

More variants of these problems and other solution methods can be found in Babić et al (1984); Mangoubi and Mathaisel (1985); Vanderstraetan and Bergeron (1988); Xu and Bailey (2001); Dorndorf et al. (2007).

The above formulations are suitable for a planning version of the gate assignment problem. To be useful in operations, a dynamic gate assignment model needs to focus on minimizing the walking distance of passengers, i.e., the connection time. Maharjan and Matis (2011) are proposing a proactive gate reassignment model for hub and spoke operations to minimize passenger disruptions and (taxi and idle) fuel burn costs. To further increase convenience for passengers, the objective function may take into consideration the reduction of the passengers' luggage transfer time. Such a model also needs to take into account the stochastic aspect of operational changes as well as have the real time feeds related to the flight schedule, aircraft rotations, crew schedules, weather, ground staff, and equipment changes. Such a model needs to be re-optimized often to accommodate real time changes.

6.4.3 Ground Management Tools

There are various computer software packages for ground staff scheduling and rostering, ranging from spreadsheet implementations of manual processes to systems based on mathematical models using efficient optimal or heuristic algorithms. Ground resources management applications are able to handle constraints that a human-being cannot overview simultaneously. Labor laws, union regulations, consideration of security and border restrictions (e.g., Schengen/non-Schengen within Europe) in gate assignment, gating interdependencies (e.g., with "horse-shoe" terminal designs) are just a few examples of such constraints. At the same time, complex objective functions that include non-linear costs, penalties and preferences may be introduced in the optimization models. OR becomes especially important with the increasing scale of airline/airport operations.

While planning software packages have been available for quite some time, more recent research and development focuses on operations. We do not review software applications in this chapter. However, we list some of the main elements of such applications and present more dominant software providers on the market.

6.4.3.1 Business and IT Requirements

On April 6, 2004, The Greater Toronto Airports Authority (GTAA) opened Terminal 1 at Toronto Pearson airport as a part of the ambitious 10 years, $4.4 billion Airport Development Program, whose master plan is published on the GTAA's corporate web site: Greater Toronto Airports Authority (GTAA) Master Plan (2010). At the heart of its successful opening was new innovative IT system. As stated by McCoomb (2007), the president and CEO of GTAA, in 2007 International Airport Review, the IT group at the GTAA introduced the idea of a common-use IT infrastructure, unlike the traditional model where every airline and organization built and maintained its own infrastructure. Under this new model, the GTAA creates the same infrastructure across the entire airport campus, more than 30 buildings spread out over 4400 acres, allowing the centralized control and maintenance that should result with lower IT support cost. More than 50 airlines, a dozen agencies, and thousands of passengers would be able to access the system as allocated by the GTAA to facilitate the processing of passengers and baggage as well as other airport operations. In 2004, the airport handled 28.6 million passengers. The users of the system (airlines, airport and government employees) as well as passengers benefited from the IT infrastructure, through baggage handling, security systems and the check-in process. Airport operations have been streamlined, simplified and made more efficient. Money has been saved and training is easier. By introducing the common use philosophy, the GTAA was able to offer a cost reduction to all users and gain central control over the IT system. The GTAA's IT system was designed to be flexible, efficient and accommodating for future technology. The idea behind the common-use philosophy is that the airport, not the airlines, manages the network, telecommunications, video feeds, check-in counters, gates, etc. More details about the advantages of the common-use IT systems may be found in the Briody's (2004) article: Toronto's Common-Use Airport Infrastructure.

As described on the Guangdong Airport Management Corp. (GAMC) web site (2010), China's new Guangzhou Baiyun International Airport (GBIA) opened for business in August 2004. The Chinese Government gave Unisys the overall responsibility to provide the airport's IT backbone. The new airport is designed for a yearly passenger-handling capability of 25 million people, apron space for 66 aircraft and a freight-handling capacity of 1.1 million tons. As reported in Computerworld Honors (2005), the Guangzhou Airport Central Integration Information management system supports schedule management, daily operations management, resource allocation, centralized accounting and management reporting needs of a modern hub airport. Not only do travelers benefit but also airport operations are much more efficient. The airport personnel may receive the flight information in real time and agents can better service passengers with check-in information. Pro-active alerts control apron activity if critical tasks fall behind schedule and are likely to impact flight departure. This includes baggage and cargo loading and unloading, aircraft cleaning, loading meals, and refueling. Now GBIA provides on time flight operations, smooth ground handling, efficient resource management and an enhanced passenger experience at the airport. It promotes a decentralized airport operations model, in which the Airport serves as a Supervisor/Coordinator of Ground Handling companies, monitoring their operations and performance. Using the new operations resource management system, the airport is able to efficiently allocate resources such as aircraft parking stands, boarding gates and baggage belts. This has substantially reduced the time that incoming aircraft have to wait on the taxiway for parking slots to be assigned. As mentioned in Computerworld Honors (2005) report and the GBIA's own web site, passengers are able to quickly collect their bags at optimally assigned carousels, thus avoiding long queues and delays.

6.4.3.2 Solution Requirements

Ground resource management applications provide many benefits to their users. The required high transparency and maneuverability is supported by user friendly interfaces. Cost savings and accurate resource allocation are delivered by sophisticated optimization engines. Legality and flexibility need to be by reliable rules engines. Such applications need to maintain up-to-date flight information, plan resources for the long term and pre-operations, manage activities in real time, and provide relevant information to any party involved. There are some applications offering a simulation engine, with more or less sophisticated graphic display, that may be used as a ground situation display tool with look-ahead what-if capabilities. Such applications provide visualization of ground operations and detection of bottlenecks in the system. They are however not capable to offer a solution for a given situation without human intervention or some sort of the optimization tool.

To bring some structure in solution evaluation, metrics or KPIs, as mentioned earlier, may help. In ground resource management, KPIs for evaluating a product are usually cost oriented since ground services are traditionally seen as a cost center for an airline. KPIs differ depending on the organization, provider or user. Their importance may vary depending on the situation or location; for example, planning vs. operations, airside vs. groundside. It is crucial for evaluating a solution to better understand the nature of different indicators and select those that really align with the company needs. Examples of quantifiable KPIs that are usually used in benchmarking different solutions are:

- Solution cost
- Staff idle time
- Number of under covered tasks
- Amount of overtime
- Number of violated soft rules
- CPU time

Integrated ground resources management products would allow more transparency of ground planning and operations. By varying business rules, such as task start/end time, task and shift duration, pay rate, etc., it is possible to create different scenarios, and evaluate and compare them. Generally, more flexibility increases productivity and consequently lowers labor costs. Evaluation of necessary staffing size, estimation of the capacity of the ground handling system, and identifying terminal bottlenecks would be a basis for investment decisions and budget planning but also a part of short term planning and operational scenario analysis.

6.4.4 Conclusion

The products currently offered on the market are capable of managing airport ground resources and equipment. Usually, the planning process is fairly integrated stretching from long term to short term planning. Planning is based on a published flight schedule that represents a common denominator for all airline, airport and government staff and equipment schedules. The flight schedule implicitly connects the flight and ground planning processes within the Air Transportation System so that planning may be done in isolation for different disciplines. While this approach is good enough for long and short-term planning, it is often inefficient on the day of operations. As discussed earlier, delay at one airport may result in resource and equipment shortage at another. For that reason we see as potential future direction for ground resources and turnaround management products to develop an enterprise level integration of data and decision support tools.

While powerful and wider used optimization and better integration will be the main target of future applications development, there are still many small decisions that need to stay decentralized. The Airport Operations Center (APOC) is a crucial element of such an approach where all data are processed and collected in one hub which selectively distributes information to subscribed and authorized users. Guangzhou Baiyun International Airport promotes a decentralized airport operations model, in which the Airport serves as a Supervisor/Coordinator of Ground Handling companies, monitoring their operations and performance. The APOC allows transparency of ground planning, better control and better decision making, especially if supported by powerful integrated OR solvers.

More recently, like in other segments of the aviation industry, there has been a tendency to offer integrated end-to-end ground resources solutions for planning, operational and administrative processes at airlines, airports, ground handling and service companies. Such systems comprise several modules, which are available individually or as a package. An efficient staff and equipment planning tool can save an airline a substantial amount of money and improve at the same time the quality of life for employees. However, existing planning optimization products minimize the resource cost without taking into account consequences on the day of operations. Operations optimization, on the other side, tries to fix the problem as soon as possible without taking into consideration all cost criteria from the planning phase. We believe a future focus should be the development of models that combine planning and operations objectives: operations costs have to be included in the planning objective function and elements of planning a part of an operations objective function. Robust scheduling will significantly increase the reliability and stability of operations. With shift and resource plans more stable fewer delays would be produced by ground staff shortage on the day of operations.

Another research direction is the integration of the planning process. The way shifts are built can have significant impact on the quality of the rosters. Generating shifts and rosters simultaneously will produce better overall solutions at the cost of an increase in computing time. Better modeling, better performing solution methods and faster computers will make this tractable in the not so distant future.

Further improvement of technology may lead to important changes in the existing ground processes in both planning and operations. One example is the introduction of Radio Frequency Identification (RFID) units to enhance the traceability and disruption management of cargo, baggage and passengers. Another example comes from mobile computing technology, which enables automation of the check-in process by means of electronic boarding passes, increased productivity through smart phone and tablet applications, constant connectivity of ground personnel with the back office, GPS locator capabilities, etc.

Last but not least, the gate assignment problem and most importantly its dynamic version, play a critical role in integrated airport, airline and airspace operations. During the day of operations, there can be a large number of deviations to planned flight schedules and booked passenger itineraries. Even on a quiet day, gate pull-in and push-out time and passenger loads are fluctuating. Consequently, quick and efficient gate management is extremely important on the day of operations, and the capability to constantly and quickly re-optimize gate assignments based on actual data is crucial. An efficient and accurate data exchange between ground resource operations systems, disruption management systems and air traffic flow management systems represent a challenge by itself. However, the bigger challenge is the response time of optimization needed to provide gate reassignment in real time.

Another challenge for more efficient gate assignment is the fact that gates and facilities are often dedicated to or owned by specific airlines, which limits flexibility. In some cases it makes sense to have dedicated gates for security or business reasons, or in big hub airports. However, in many cases better global solutions form an Air Transportation System point of view may be obtained if scarce resources such as gates would not be limited.

6.5 Research Opportunities and Roadmap

As we described the history and current best practice in the application of OR to Airline Operations, we have also touched briefly on an ideal state, where a holistic, multidisciplinary optimization problem covers all problem domains and produces a globally optimal solution in which relevant aspects are distributed in real time to all decision makers. Progress towards such an ambitious goal must naturally be evolutionary and we now provide some ideas for researchers and practitioners as they chart a path towards this solution.

In Operations,[2] as well as in other parts of the Air Transportation System, decision makers have a growing need for situational awareness due to an increasing amount of data and constantly updated information. IT advances in information management and mobile computing connect an ever increasing number of data sources and make information available *when* and *where* users need it. OR provides insight and intelligence to transform this information into knowledge. Based on a modern and reliable IT infrastructure, OR enables the development of comprehensive and user-friendly decision support solutions. In the hands of both front-line employees and back-office staff, these solutions help users make decisions that improve operational efficiency and increase the bottom-line results of their companies. Communication capabilities are paramount, as all stakeholders need consistent and accurate information.

Figure 6.18 depicts a conceptual framework based on Systems Thinking principles. In this vision, IT and OR combine to transform data into knowledge. IT connects data sources and provides real-time availability of critical information. OR provides the necessary insight and enables decision support by providing decision-makers with exactly the information they need.

The combination of IT and OR will help overcome the current fragmentation of Operations and provide situational awareness to all stakeholders for optimal decision making. As we have emphasized throughout this chapter, an isolationist perspective contributes to suboptimal performance of the overall air transportation system. A significant value-add for all stakeholders will be the creation of intelligent information solutions that connect and integrate these subsystems. OR should strive to help deliver end-to-end planning and operations solutions for the whole system.

---

[2] Operations comprises Flight Operations, Airline Operations, and Airport Operations as defined in Sect. 6.1.

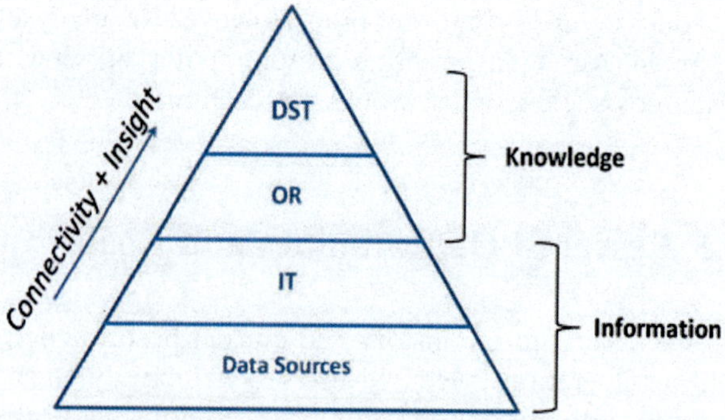

Fig. 6.18 The Information Technology, Operations Research, and Decision Support Technology triangle

Closing the loop between forecasting, planning, operations, and analysis constitutes an important area of OR research. As depicted in Fig. 6.19, integrated information management and decision support throughout the planning and operations cycle is critical to informed decision making. An important research area involves developing commonly accepted KPIs which truly indicate (or predict) a plan's performance or a decision's impact. As argued throughout this chapter, this understanding is important to eventually achieving minimal realized (operational) costs of a plan. Future research needs to clearly distinguish between a priori performance indicators and a posteriori performance measures. OR can help develop metrics used by airlines to measure operational performance and identify parameters that best correlate with operational performance. These KPIs will create an understanding of the functional relationship between the important performance characteristics of an airline.

The OR community can take a leadership role and work with practitioners to establish metrics. Once both financial and operational performance indicators are in place, these relationships can be better understood and managed, and OR methods can leverage them to predict or influence the profitability of an airline.

In this context, robust planning and scheduling is an area that requires more research attention. Being able to close the feedback loop from planning to operations will make robust scheduling practical. Simulation that allows "fast-forwarding" through possible implications will be an important component for valuing the robustness of specific decisions and, eventually, plans and schedules as well.

Each of the areas described in this chapter also offers unique research opportunities.

We consider *flight planning* a worthwhile academic topic. OR can help refine existing methods and develop new techniques to ensure that flight planning is not viewed as a commodity. More concretely, flight planning optimization by itself offers many opportunities as 4D models need to be further developed, and the reliability of cost models improved. Research into managing the complexity of airspace and other restrictions will allow flight planning optimization to further evolve. Related to airspace complexity, Performance Based Navigation and

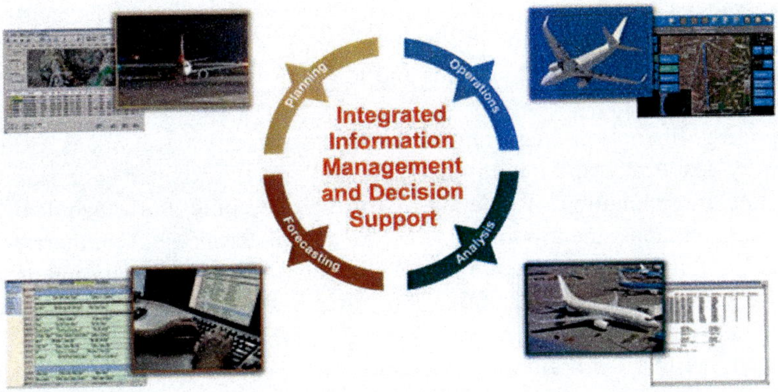

Fig. 6.19 Integrated information management and decision support for airline OPERATIONS (© 2011 Jeppesen)

Required Navigation Performance will add even more complexity to flight planning in the future while allowing for more efficient flight operations.

We also see opportunities in the integration of flight planning optimization into more complex, multi-disciplinary decision making. A concrete example is the integration of flight planning in disruption management as described by Vaaben et al. (2010). Huge opportunities will come from the current overhaul of the air transportation system by the FAA and Eurocontrol through NextGen and SESAR, respectively, as described below.

Disruption management still provides many research challenges. Tractable OR approaches for solving the holistic recovery problem still need to be developed, and they have to be practical in order for users to accept them. Especially when dealing with major disruptions that require complex recovery solutions integrating multiple resource areas, it is important to offer meaningful and intuitive information to users to avoid the rejection of a very viable approach. We also need to improve our understanding of how a series of minor disruptions is best managed. This not only requires a better handling of uncertainty in the decision making process, but advanced insight into how decisions might impact other disruptions as well.

More specifically, we see opportunities in several areas of disruption management. Improving optimization performance across the board will reduce response time to disruptions, allowing users to generate improved recovery solutions based on accurate and timely information. Providing multiple meaningful solutions that cover a wider scope will enable more powerful decision support. This, coupled with intuitive user interfaces, will empower stakeholders to be proactive, rather than reactive, in their decision making.

Of course, all of this requires improvements to data quality and availability. As pointed out in detail above, disruption management is data intensive, and it is essential for OR to deal with the challenges of real-time data. Related to this, OR models will also have to deal with imperfect problem statements and objectives that are often contradictory. Being able to calibrate OR models and decision support systems to specific airline needs, government regulations, and union agreements are important factors in the acceptance of optimization based DST by the end user. Additionally, OR systems and solutions need to cater for the dynamic nature of operations.

Integrating disruption management with other areas necessitates extensive research by academics and practitioners alike. We have to better understand how the existing fragmentation of Operations can be overcome, both technically and from a change management point of view. As described in Sect. 6.3, there is great potential for better, more efficient decision making through integration within one airline, within a group of airlines (e.g., mainline and regional carriers), or within the Air Transportation System in general (e.g., GDP).

We envision that disruption management can be used continuously at airlines, i.e., as part of "business as usual" for both minor and major disruptions. In fact, to truly become part of the regular business practice, the disruption management tool must be the primary operations control tool, used to track and modify the assignment of aircraft and crew to flights even in the absence of any disruption and assess the impact of potential modifications on the overall system. While there is evidence of the monetary benefit of applying OR in disruption management, there is still not a clear business case for large-scale investments into recovery technology due to the lack of common cost models and KPIs. Hence, quantifying the impact of recovery optimization is an important future research avenue.

Ground resource and turnaround management has many untapped opportunities as well. Although the fact that ground staff is not as expensive of a resource as air crew may have limited research in the past, the expensive consequences of delays resulting from inappropriate ground management are now recognized. We believe delay cost provides sufficient motivation for a general increase in research in this area.

A future research direction in ground resource management is the treatment of the topic across an airline's operation. Currently, ground resources are managed as a series of local problems. Integration of enterprise wide data and implementation of decision support tools using these data will allow improved decisions from a system-wide perspective. Another important research area is the development of OR approaches that cover both planning and operations to achieve better turnaround management. Other research opportunities comprise further work in ground equipment scheduling, ground staff scheduling, gate assignment, and turnaround time management. An important topic in ground resource management is balancing resources as it comes to workload fluctuations and the management of uncertainty.

To improve Operations in general, airlines have to keep being innovative and continue investing in new technologies that will help them gain a competitive advantage. We believe that innovation around wireless technology and connectivity in airline operations today can have a similar business impact on airlines, as did the introduction of revenue management in the 1980s. The connection of different information sources cannot be viewed as a pure IT problem; only once

OR gets deployed, can the true benefits of decision support for Airline Operations be realized.

Situational awareness can only be achieved through the application of advanced analytics. A research question is to determine what information a user needs, when and where it is needed, and how to take the underlying uncertainty into account to promote better decisions. OR will need to offer solutions that combine human intelligence, domain knowledge, and IT. OR analysts will also have to contribute to the important business process and organizational change management aspects when introducing these new solutions in a traditional operations environment.

We believe that there is an opportunity to apply *risk management* in airline operations more broadly. Risk management expertise will help regulators better manage oversight and lessen operational constraints, which will increase the value proposition of OR even further. In this context, Fatigue Risk Management (Romig and Klemets, 2009) in crew scheduling is a great example of how OR can help replace legacy flight time and duty time regulations with more accurate ones that are not only operationally advantageous, but actually improve safety levels. A related opportunity is to look at risk management to better manage uncertainty in operational decision making. As described in several places in this chapter, simulation is now being applied at some stages to better forecast the impact of decision alternatives and can also be incorporated into robust scheduling.

To be able to suggest a *high-level roadmap* for further research in Operations, it is important to understand the ongoing and wide-ranging changes in the air transportation system. NextGen aims to reduce congestion, both in the air and on the ground, by creating a new National Airspace System that will be implemented across the United States in stages between 2012 and 2025 (Federal Aviation Administration 2009). Its European counterpart is SESAR, which also aims to implement new technology to increase the capacity of the Air Transportation System (Eurocontrol 2009).

With NextGen and SESAR, pilots and dispatchers will be able to select their own, minimally-constrained, wind-optimal flight paths rather than follow existing airways. This freedom to find more efficient 4D paths will provide advantages to a wide variety of stakeholders, including airline customers, by lowering overall costs, increasing capability to reduce flight time as a recovery tactic, and making aircraft ground times more reliable.

All system decision makers in the air and on the ground will have access to the same information, transmitted via System Wide Information Management (SWIM). The increased volume of information will support collaborative decision making between air traffic management and airline operations, promoting better balance between demand and capacity. Improved weather and traffic forecasts, coupled with enhanced DSS, will minimize airspace limitations. The vision of NextGen and SESAR is to enable informed and better decision making through a holistic view of the Air Transportation System. As described in the chapter on Air Traffic Flow Management, the implementation of NextGen and SESAR is also creating many opportunities for the application of OR.

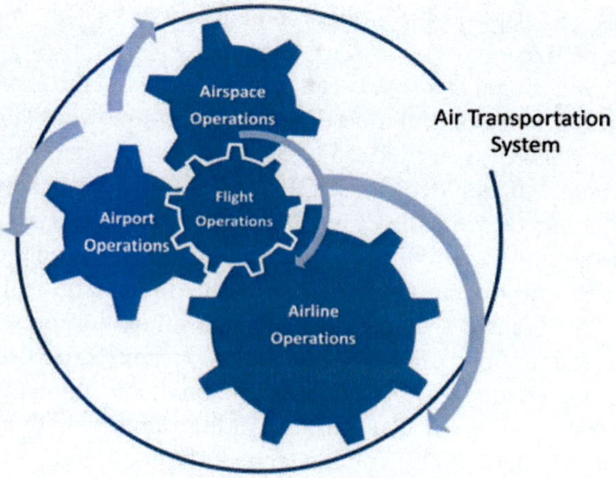

Fig. 6.20 Optimization as "grease" between air traffic management, flight operations, and airline operations

In a SWIM environment, where "everybody knows everything," the biggest challenge may be to find the pieces of information that really matter. OR and optimization can be viewed as the "grease" between Air Traffic Management, Flight Operations, Airline Operations, and other subsystems of the entire Air Transportation System, as depicted in Fig. 6.20. OR also keeps the smaller parts in each subsystem moving, as an ever increasing amount of data needs to be filtered and transformed into knowledge for decision makers.

Figure 6.21 depicts a *high-level research roadmap* with several research topics that build on each other from the bottom up (i.e., from the first level to the fourth level):

1. Specific research in flight planning, disruption management, and ground resource and turnaround management as described earlier
2. Advanced analytics to develop further insight into the Air Transportation System, its subsystems, and their interdependencies
   - Creation of common economic models and KPIs
   - Clearer problem statements and definitions of objectives
   - Robust scheduling by closing the feedback loop between planning and operations
   - Development of business cases that help accelerate investments into IT and OR
3. Integrated models and solution approaches
   - Integration of resources within an airline
   - Integration of operations within an airline group
   - Definition of the role for OR as part of NextGen and SESAR
   - Proposed changes of regulations of the Air Transportation System to improve safety and efficiency

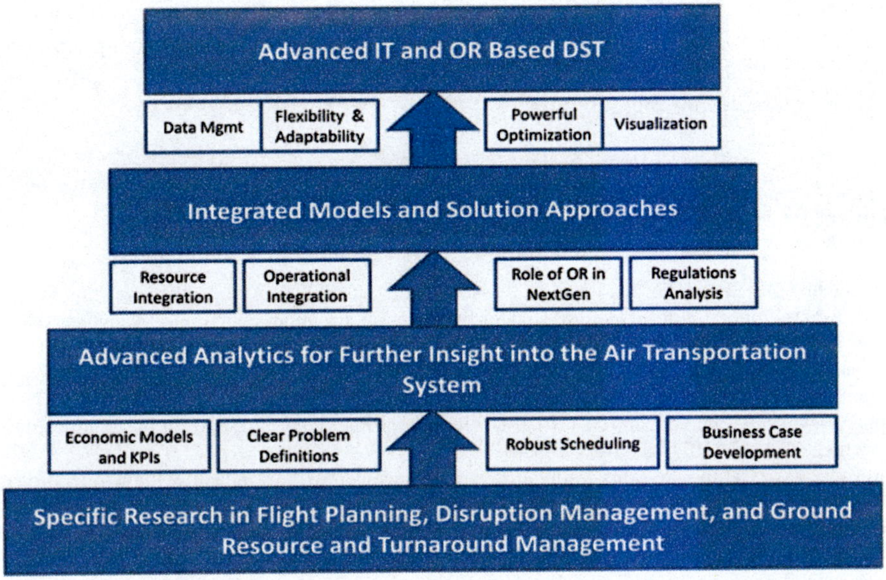

Fig. 6.21 High-level research roadmap

4. Development of smart DST that integrates the best of IT and OR
   - Comprehensive data and information management
   - Flexibility and adaptability
   - Powerful optimization models and approaches
   - Intuitive visualization of meaningful solutions based on user experience design principles

In this chapter, we presented Operations as a real world problem with its many facets and challenges, reviewed the application of OR for improving efficiency, and proposed future research directions for overcoming remaining challenges. Operators are slowly and painfully replacing aged legacy IT infrastructure and thereby starting to deploy mobile computing technology to achieve better situational awareness for all stakeholders and decision makers. OR is going to play an integral role in improving Airline Operations and the Air Transportation System, but it needs to be combined with an enhanced IT infrastructure. For the creation of intelligent and useful DST, OR needs to deal with the IT needs of the industry.

To further advance the state of airline operations, OR researchers need to "get their hands dirty" and overcome traditionally simplifying assumptions. Incremental improvements will help bring us closer to the global, holistic solution, as the industry cannot wait for the massive infrastructure investments that are underway to be completed. In other words, OR will need to play a central role in improving the Air Transportation System in its current form while, at the same time, help shape the future system. This is a unique opportunity to improve both the safety and efficiency of Operations without compromising one or the other. The ultimate goal is for OR to play a key role in eventually solving an Integrated Operations Problem, addressing all resource areas, and producing a safe and holistic, globally-optimal solution.

References

Abdelghany A, Ekollu G, Narasimhan R, Abdelghany K (2004) A proactive crew recovery decision support tool for commercial airlines during irregular operations. Ann Oper Res 127:309–331

AhmadBeygi S, Cohn A, Guan Y, Belobaba P (2008) Analysis of the potential for delay propagation in passenger airline networks. J Air Transp Manag 14(5):221–236

AhmadBeygi S, Cohn A, Lapp M (2010) Decreasing airline delay propagation by re-allocating scheduled slack. IIE Trans 42(7):478–489

Airbus (2002) Getting to grips with aircraft performance. Airbus industrie flight operations support & line assistance, January 2002

Air Transport Association (2010). http://www.airlines.org/economics/cost+of+delays/

Altus S (2010) Dynamic cost index management in flight planning and replanning. AGIFORS Airline Operations Study Group Meeting, Brisbane

Alves A (2006) Information management for passenger recovery. AGIFORS Airline Operations Study Group Meeting. Dubai

Alves A, Vaaben B (2009) Disruption management: passenger re-accommodation strategies. AGIFORS Airline Operations Study Group Meeting. Montreal

Andersson K, Hall W, Atkins S, Feron E (2003) Optimization-based analysis of collaborative airport arrival planning. Transp Sci 37(4):422–433

Argüello M, Bard J, Yu G (1997) A GRASP for aircraft routing in response to groundings and delays. J Comb Optim 5:211–228

Argüello M (1997) Framework for exact solutions and heuristics for approximate solutions to airlines' irregular operations control aircraft routing problem. Ph.D. Dissertation, Department of Mechanical Engineering, University of Texas, Austin

ASPIRE (2011) Best practice in Asia Pacific: Oceanic Flights. http://www.aspire-green.com/bestPractice/oceanic.asp

Babić O, Teodorović D, Tosić V (1984) Aircraft stand assignment to minimize walking. J Transp Eng 110:55–66

Ball M, Barnhart C, Nemhauser GL, Odoni A, Laporte G (2006) Air transportation: Irregular operations and control. In: Transportation: Handbooks of Operations Research and Management Science, North-Holland, Amsterdam, pp 1–64

Barnhart C, Hatay L, Johnson EL (1995) Deadhead selection for the long-haul crew pairing problem. Oper Res 43(3):491–499

Barnhart C, Boland NL, Clarke LW, Johnson EL, Nemhauser GL, Shenoi RG (1998) Flight string models for aircraft fleeting and routing. Transp Sci 32(3):208–220

Barnhart C, Belobaba P, Odoni AR (2003) Applications of operations research in the air transport industry. Transp Sci 37(4):368–391

Baumgarten A (2009) Back through the looking glass: data visualization and rich interaction. AGIFORS Crew Management Study Group Meeting. Las Vegas

Baumgarten A (2010) Communicating progress and estimating completion during optimization. AGIFORS Crew Management Study Group Meeting. Athens

Bellman RE (1958) On a Routing Problem. Quart J Appl Math 16:87–90

Betts JT, Cramer EJ (1995) Application of direct transcription to commercial aircraft trajectory optimization. AIAA J Guidance, Control, and Dyn 18(1):151–159

Borndörfer R, Dovica I, Nowak I, Schickinger T (2010) Robust tail assignment. ZIB-Report 10-08, Zuse-Institut, Berlin
Bratu S, Barnhart C (2006) Flight operations recovery: new approaches considering passenger recovery. J Sched 9(3):279–298
Briody D (2004) Toronto's common-use airport infrastructure. eWeek.com, 13 July
Brusco MJ, Jacobs LW, Bongiorno RJ, Lyons DV, Tang B (1995) Improving personnel scheduling at airline stations. Oper Res 43(5):741–751
Bryson AE, Ho Y-C (1969) Applied Optimal Control. Blaisdell Publishing Company
Bryson AE (1998) Dynamic Optimization. Pearson Education
Bureau of Transportation Statistics (2009) http://www.transtats.bts.gov/
Burke EK, De Causmaecker P, De Maere G, Mulder J, Paelinck M, Vanden Berghe G (2009) A multi-objective approach for robust airline scheduling. Computers and Operations Research, (In press)
Butler GF, Keller MR (2000) Handbook of Airline Operations. McGraw Hill, New York
Campbell C, Perry C, Rathi N, Safabakhsh N (2007) Enhancing customer check-in experience. AGIFORS Airline Operations Study Group Meeting, Denver
Cao J, Kanafani A (1997a) Real-time decision support for integration of airline flight cancellations and delays–part I: mathematical formulation. Transp Plan Technol 20:183–199
Cao J, Kanafani A (1997b) Real-time decision support for integration of airline flight cancellations and delays–part II: algorithm and computational experiments. Transp Plan Tech 20:201–217
Chin DK, Goldberg J, Tang T (1997) Airport surface delays and causes. NASA Contractor Report 201721
Clarke J-P (2004) MIT extensible air network simulation (MEANS). AGIFORS Airline Operations Study Group Meeting. Washington
Clarke M (2005) Passenger Reaccommodation. AGIFORS Airline Operations Study Group Meeting. Mainz
Clarke M, Lettovský L, Sylla A, Walker A (1999) Journey management passenger reaccommodation initiative. AGIFORS Airline Operations Study Group Meeting. Istanbul
Clarke M, Lettovský L, Smith B (2000) The Development of the Airline Operations Control Center. AGIFORS Airline Operations Study Group Meeting. Budapest
Cohn A (2007) Robust schedule planning and recovery: Current research and future challenges. INFORMS Annual Meeting. Seattle
Computerworld Honors (2005) Guangzhou Airport Central Integration Information management System: 2005 Computerworld Honors Case Study, http://www.cwhonors.org/laureates/transportation/guangzhou.pdf
Cordeau J-F, Stojković G, Soumis F, Desrosiers J (2001) Benders decomposition for simultaneous aircraft routing and crew scheduling. Transp Sci 35(4):375–388
de Jong HM (1974) Optimal track selection and 3-dimensional flight planning. Royal Netherlands Meteorological Institute, Report 93, De Bilt, Netherlands
Deregnaucourt B (2004) Crew pairings recovery at Air France. AGIFORS Annual Symposium. Singapore
Ding H, Lim A, Rodrigues B, Zhu Y (2004a) New heuristics for the over-constrained airport gate assignment problem. J Oper Res Soc 55:760–768
Ding H, Lim A, Rodrigues B, Zhu Y (2004b) Aircraft and gate scheduling optimization at oirports. Proceedings of the 37th Annual Hawaii International Conference on System Science, IEEE, Piscataway, NJ, 74–81
Dorndorf U, Drexl A, Nikulin Y, Pesch E (2007) Flight gate scheduling: State-of-the-art and recent developments. Omega 35(3):326–334
Drexl A, Nikulin Y (2008) Multicriteria airport gate assignment and Pareto Simulated Annealing. IIE Transactions 40:385–397
Dugan S, Kohli P, Ramak R, Smith B (2002) Shift scheduling optimization. AGIFORS Annual Smposium, Honolulu, HI

Dugan S, Kohli P, Ramak R, Smith B (2003) Task assignment model. AGIFORS Annual Symposium, Paris

Dunbar M, Froyland G, Wu C-L (2010) Robust airline schedule planning: Minimizing propagated delay in an integrated routing and crewing framework. Submission for 2010 AGIFORS Anna Valicek Award. http://www.agifors.org/award/submissions2010/MichelleDunbar_paper.pdf

Ehrgott M, Ryan D (2002) Constructing robust crew schedules with bicriteria optimization. J Multi-Criteria Decision Anal 11:139–150

Ellerman AD, Buchner BK (2007) The European union emissions trading scheme: origins, allocation, and early results. Rev Environ Econ Policy 1(1):66–87

Ernst AT, Jiang H, Krishnamoorthy M, Sier D (2004a) Staff scheduling and rostering: A review of applications, methods and models. Eur J Oper Res 153(1):3–27

Ernst AT, Jiang H, Krishnamoorthy M, Owens B, Sier D (2004b) An annotated bibliography of personnel scheduling and rostering. Ann Oper Res 127:21–144

Eurocontrol (2006) Route Availability Document RAD Users Manual. version 1.0

Eurocontrol (2009) SESAR. http://www.eurocontrol.int/sesar/public/subsite_homepage/homepage.html

European Parliament and the Council of the European Union (2009) Directive 2008/101/EC. Off J Eur Union, 13 Jan 2009

Federal Aviation Administration (2003) High altitude airspace redesign: Phase 1. Advisory Circular 90-00, 22 Sept. 2003

Federal Aviation Administration (2009) The FAA NextGen Site. http://www.faabuzz.com/2009/07/the-faa-nextgen-site/

Gabteni S (2010) Large scale reaccommodation. AGIFORS Airline Operations Study Group Meeting. Brisbane

Gao C, Kalyta A (2006) Tactical crew recovery during irregular airline operations. INFORMS Annual Meeting, Pittsburgh

Gao C, Johnson E, Smith B (2009) Integrated airline fleet and crew robust planning. Transp Sci 43(1):2–16

Greater Toronto Airports Authority (GTAA) Master Plan (2010). http://www.gtaa.com/en/gtaa_corporate/publications/master_plan

Green T, Ande M (2005) Automated passenger re-accommodation for real-time, day-of-ops passenger recovery. INFORMS Annual Meeting. San Francisco

Grover JHH (1990) Airline Route Planning. Science Press Blackwell

Guangdong Airport Management Corp. (GAMC) web site (2010). http://www.baiyunairport/english/about.jsp

Gumireddy L, Ince I (2002) Optimization for hierarchical objectives during ground delay programs. AGIFORS Airline Operations Study Group Meeting, Rome

Hang D (2007) Aircraft schedule recovery with crew connection protection. INFORMS Annual Meeting, Seattle

Hansen M, Hsiao CY (2006) Econometric analysis of US airline flight delays with time-of day effects. Transp Res Rec 1951:104–112

Hargraves CR, Paris SW (1987) Direct trajectory optimization using nonlinear programming and collocation. J Guid, Control Dyn 10(1):338–342

Heinold A (2008) Maintenance reachability at Southwest Airlines. AGIFORS Annual Symposium, Montreal

Herrera J (2005) STARR: Southwest Airlines Tactical Aircraft Routing and Recovery. AGIFORS Airline Operations Study Group Meeting, Mainz

IATA (2010) IATA Airport, ATC, and Fuel Charges Monitor. http://www.iata.org/ps/publications/Pages/aeronautics-charges-monitor.aspx

Ince I (2009) Lessons learned in implementing a slot optimizer during ground delay programs. AGIFORS Airline Operations Study Group Meeting, Montreal

Jardin MR, Bryson AE (2001) Neighboring optimal aircraft guidance in winds. J Guidance, Control, Dyn 24(4):710–715

Jarrah AIZ, Yu G, Krishnamurthy N, Rakshit A (1993) A decision support framework for airline flight cancellations and delays. Transp Sci 27(3):266–280

JetBlue Airways (2007) http://www.jetblue.com/about/ourcompany/flightlog/archive_march2007.html

Jiang H, Hansman RJ (2006) An analysis of profit cycles in the airline industry. 6th AIAA aviation technology, Integration and Operations Conference, Wichita, KA

Karisch S (2004) Monitoring and controlling key performance indicators. AGIFORS Annual Symposium, Singapore

Kraay D, Safabakhsh N (2007) Robust Airport Manpower Planning. AGIFORS Airline Operations Study Group Meeting, Denver

Klabjan D, Johnson EL, Nemhauser GL, Gelman E, Ramaswamy S (2002) Airline crew scheduling with time windows and plane-count constraints. Transp Sci 36(3):337–348

Kohl N, Larsen A, Larsen J, Ross A, Tiourine S (2007) Airline disruption management—perspectives, experiences and outlook. J Air Transp Manag 13:149–162

Lan S, Clarke J-P, Barnhart C (2006) Planning for robust airline operations: optimizing aircraft routings and flight departure times to minimize passenger disruptions. Transp Sci 40(1):15–28

Larsson T (2006) Operational Schedule Management. AGIFORS Airline Operations Study Group Meeting, Dubai

Lettovský L (1997) Airline operations recovery: An optimization approach. Ph.D. thesis. Georgia Institute of Technology, Atlanta

Lettovský L, Johnson EL, Nemhauser GL (2000) Airline crew recovery. Transp Sci 34(4):337–348

Lidden S (1992) Optimum 4-D guidance for long flights. Proceedings of the 11th AIAA/IEEE Digital Avionics Systems Conference 262–267

Luo S, Yu G (1997) On the airline schedule perturbation problem caused by the ground delay program. Transp Sci 31(4):298–311

Lusby R, Dohn A, Range TM, Larsen J (2010) An integrated approach to the ground crew rostering problem with work patterns. Demarks Tekniske Universitet—Management Engineering Department, Report 9

Maharjan B, Matis TI (2011) An optimization model for gate reassignment in response to flight delays. J Air Tran Manag 17(4):256–261

Mangoubi RS, Mathaisel DFX (1985) Optimizing gate assignments at airport terminals. Transp Sci 19(2):173–188

McCoomb L (2007) International Airport Review 1, http://www.gtaa.com/local/files/en/Corporate/Publications/External/InternationalAirportReview-Issue1-2007.pdf

McIntyre R (1996) The adverse impact of flight management systems on long range international airline operations. In: Proceedings of the 15th AIAA/IEEE Digital Avionics Systems Conference 359–363

Medard CP, Sawhney N (2007) Airline crew scheduling: from planning to operations. Eur J Oper Res 183(3):1013–1027

Nissen R, Haase K (2006) Duty-period-based network model for crew rescheduling in European airlines. J Sched 9(3):255–278

Niznik T, Green T (2001) Optimizing the airline response to ground delay programs. AGIFORS Airline Operations Study Group Meeting, Ocho Rios

Pachon J (2007) Managing integrated main and regional carriers' recovery decisions. INFORMS Annual Meeting, Seattle

Padilla CE (1996) Optimizing Jet Transport Efficiency: Performance, Operations, and Economics. McGraw Hill, NJ

Petersen J, Solveling G, Clarke J-P, Johnson S, Shebalov (2010) An optimization based approach to airline integrated recovery. AGIFORS Airline Operations Study Group Meeting. Brisbane

Rekik M, Cordeau J-F, Soumis F (2004) Using benders decomposition to implicitly model tour scheduling. Ann Oper Res 128:111–133

Riedel R, Clarke J–P (2005) Airline operations feedback control with KPIs. AGIFORS Airline Operations Study Group Meeting, Mainz

Rogers WH, Allen J, Hoyme K (2000) The airline operations center dilemma: Solving'day-of-operation'disruptions with greater economic efficiency. In: Butler G, Keller M (eds) Handbook of airline operations. aviation week group/McGraw-Hill, Washington
Romig E, Klemets T (2009) Fatigue risk management in flight crew scheduling. Aviat Space Environ Med 80(2):1073–1074
Rosenberger JM, Schaefer AJ, Goldsman D, Johnson EL, Kleywegt AJ, Nemhauser GL (2002) A stochastic model of airline operations. Transp Sci 36(4):357–377
Rosenberger JM, Johnson EL, Nemhauser GL (2003) Rerouting aircraft for airline recovery. Transp Sci 37(4):408–421
Rosenberger JM, Johnson EL, Nemhauser GL (2004) A robust fleet-assignment model with hub isolation and short cycles. Transp Sci 38(3):357–368
Rupp NG, Holmes GM (2006) An investigation into the determinants of flight cancellations. Economica 73(292):749–784
Salam R, Niznik T (2005) Ground delay program management at American Airlines. INFORMS Annual Meeting, San Francisco
Sawyer JA (1950) Equivalent headwinds: application of upper-wind statistics to air-route planning. Meteorological report No. 6, United Kingdom Meteorological Office, London
Schwartz E, Kroo IM (2009) Aircraft design: trading cost and climate impact. 47th AIAA Aerospace Sciences Meeting, Orlando
Shaw T (2005) Airline crew recovery: an integrated pairing and roster recovery approach. INFORMS Annual Meeting, San Francisco
Shebalov S (2008) Multi-objective gate assignment. AGIFORS Airline Operations Study Group Meeting, Amsterdam
Shebalov S, Klabjan D (2006) Robust airline crew pairing: Move-up crews. Transp Sci 40(3):300–312
Smith BC, Johnson EL (2006) Robust airline fleet assignment: Imposing station purity using station decomposition. Transp Sci 40(4):497–516
Sorensen JA, Goka T (1985) Design of an advanced flight planning system. American Control Conference 663–668
Stojković G (2008) Airline operations process modeling. Annual AGIFORS Symposium, Montréal
Stojković M, Soumis F (2001) An optimization model for the simultaneous operational flight and pilot scheduling problem. Manage Sci 47(9):1290–1305
Stojković M, Soumis F (2004) Simultaneous aircraft and crew recovery system. AGIFORS Airline Operations Study Group Meeting, Washington
Stojković M, Soumis F (2005) The operational flight and multi-crew scheduling problem. YUJOR 15(1) 25–48; also published in Les Cahiers du GERAD G-2000-27. École des Hautes Études Commerciales, Montréal
Stojković G, Soumis F, Desrosiers J, Solomon M (2002) An optimization model for a real-time flight scheduling problem. Transp Res Part A 36:779–788
Stojković M, Soumis F, Desrosiers J (1998) The operational airline crew scheduling problem. Transp Sci 32(3):232–245
Tang, B., A. Mukherjee (2000) System and method for generating optimal flight plans for airline operational control. US Patent 6,134,500
Teodorović D, Guberinić S (1984) Optimal dispatching strategy on an airline network after a schedule perturbation. Eur J Oper Res 15:178–182
Teodorović D, Stojković G (1990) Model for operational daily airline scheduling. Transp Plan Technol 14(4):273–285
Teodorović D, Stojković G (1995) Model to reduce airline schedule disturbances. J Transp Eng 121:324–331
Thengvall BG, Bard JF, Yu G (2000) Balancing user preferences for aircraft schedule recovery during irregular operations. IEE Trans 32:181–193
Thengvall BG, Yu G, Bard JF (2001) Multiple fleet aircraft schedule recovery following hub closures. Transp Res Part A 35:289–308

Thengvall BG, Bard JF, Yu G (2003) A Bundle Algorithm approach for the aircraft schedule recovery problem during hub closures. Transp Sci 37(4):392–407

Tino CP, Ren L, Clarke J.-PB (2009) Wind forecast error and trajectory prediction for en-route scheduling. AIAA Guidance, Navigation, and Control Conference, Chicago

Tiourine S, Chiraphadhanakul V (2009) Understanding uncertainty in the airline operations. AGIFORS Airline Operations Study Group Meeting, Montreal

US Department of Transportation, Bureau of Transportation Statistics (2004) Report, Pocket Guide to Transportation, Washington DC, http://www.bts.gov/publications/pocket_guide_to_transportation/2004/pdf/entire.pdf

US Department of Transportation (2010a) Bureau of Transportation Statistics. http://www.bts.gov

US Department of Transportation, Bureau of Transportation Statistics (2010b) Long Tarmac Delays in August Show Steep Drop from Last Year,. http://www.bts.gov/press_releases/2010/dot186_10/html/dot186_10.html

US Department of Transportation (2010c) Office of Aviation Enforcement and Proceedings, Washington, DC http://airconsumer.dot.gov/rules/FAQ%20on%20Consumer%20Rule%20April%2028%202010.pdf

Vaaben B, Marla L, Barnhart C (2010) Improving passenger service by extending disruption management with flight planning. INFORMS Annual Meeting, Austin

Vanderstraetan G, Bergeron M (1988) Automatic assignment of aircraft to gates at a terminal. Comput Ind Eng 14:15–25

Vasquez-Marquez A (1991) American Airlines Arrival Slot Allocation System (ASAS). Interfaces 21(1):42–61

Wei G, Yu G, Song M (1997) Optimization model and algorithm for crew management during airline irregular operations. J Combin Optim 1(3):305–321

Weide O, Ryan D, Ehrgott M (2009) An iterative approach to robust and integrated aircraft routing and crew scheduling. Comput Oper Res 32(6):1451–1476

Wikipedia (2009) February 2007 North American Winter Storm. http://en.wikipedia.org/wiki/February_2007_North_America_Winter_Storm

Wilson J, Wright C, Couluris GJ (1997) Advanced free flight planning and dispatcher's workstation: Preliminary design specification. NASA Technical Report, under contract NAS2-14289

Wolf C (2010) US tarmac-delay rule coming for foreign airlines. Bloomberg.com News, Decembe 31

Xu J, Bailey G (2001) The airport gate assignment problem: Mathematical model and a Tabu Search algorithm. Proceedings of the 34th Annual Hawaii International Conference on System Science, IEEE, Piscataway, NJ, p 3032

Yan S, Lin C (1997) Airline scheduling for the temporary closure of airports. Transp Sci 31:72–82

Yan S, Tu Y (1997) Multifleet routing and multistop flight scheduling for schedule perturbations. Eur J Oper Res 103:155–169

Yan S, Yang D (1996) A decision support framework for handling schedule perturbations. Transp Res Part B: Methodol 30:405–419

Yan S, Young H (1996) A decision support framework for multi-fleet routing and multi-stop flight scheduling. Transp Res-A 30(5):379–398

Yu G, Argüello M, Song G, McCowan SM, White A (2003) A new era for crew recovery at continental airlines. Interfaces 33(1):5–22

Chapter 7
Air Traffic Flow Management

Thomas W. M. Vossen, Robert Hoffman, and Avijit Mukherjee

7.1 Introduction

Air transportation systems are some of the most complex logistical systems imaginable. The world's airlines transported over 2.2 billion passengers in 2008, and transported approximately 40% of world trade (measured by value). There are nearly 2,000 airlines worldwide, which have a total fleet of nearly 23,000 aircraft and serve some 3,750 airports through a route network of several million miles managed by around 160 air navigation service providers.

Like other critical infrastructures, the air transportation network has an enormous impact on local, national, and international economies. In a recent report, the Air Transportation Action Group (ATAG) noted that air transport directly employs five million people (a number that increases to 29 million if indirect and induced jobs are included) and generates 400 billion in output. In addition, ATAG noted that the total world output generated by air transport (through its direct, indirect, and induced impacts) was nearly 1.1 trillion, or 2.3% of global ground delay program (GDP; Air Transport Action Group 2008).

The sheer volume and sustained growth of the aviation sector have put an enormous strain on the air transportation system. This is evidenced by a worldwide increase in flight delays and airport congestion. In 2007, approximately one out of every 4 flights in the United States was delayed or canceled (http://www.transtats.bts.gov). The resulting delays have a significant economic impact.

T. W. M. Vossen (✉)
Leeds School of Business, University of Colorado, Boulder, CO 80309, USA
e-mail: Vossen@Colorado.EDU

R. Hoffman
Metron Aviation, Inc., Herndon, VA 20170, USA

A. Mukherjee
University of California, Santa Cruz, Santa Cruz, CA 95064, USA

C. Barnhart and B. C. Smith, LLC (eds.), *Quantitative Problem Solving Methods in the Airline Industry*, International Series in Operations Research & Management Science 169,
DOI: 10.1007/978-1-4614-1608-1_7,

The US Senate Joint Economic Committee has estimated that system delays cost passengers, airlines, and the US economy more than 40 billion in 2007 (US Senate Joint Economic Committee 2008). In Europe, a similar situation exists: approximately 11% of flights were delayed in 2007, and these delays were estimated to have an annual (direct) cost of 2 billion (http://www.eurocontrol.int/).

A continued growth of the aviation sector will therefore require significant investments in the system's infrastructure, and numerous modernization and airport expansion efforts are currently underway or planned. In the US alone, runways are being added at 7 of the largest airports, and ten other new runway projects are planned. It is unlikely, however, that efforts to increase the physical capacity alone can accommodate the projected growth in air traffic. Although effective, such initiatives are usually very costly and take a long time to complete. Moreover, they can be difficult to implement due to space restrictions and other constraints. As a result, capacity enhancements will also require increased efficiencies in the management processes of the air traffic system. The broad term "air traffic management" (ATM) is commonly used to represent the overall collection of these processes, and can be defined as the composite of services that support the ultimate goal of safe, efficient, and expeditious aircraft movement.

It is common to distinguish two basic ATM components: air traffic control (ATC) and air traffic flow management (ATFM). ATC refers to processes that provide tactical separation services, that is, real-time separation procedures for collision detection and avoidance. ATC is usually performed by human controllers who watch over three-dimensional regions of airspace, called sectors, and dictate local movements of aircraft. Their aim is to maintain separation between aircraft while moving traffic as expeditiously as possible and presenting the traffic in an orderly and useful manner to the next sector. Each sector can only be occupied by a limited number of aircraft; the limit is determined by a controller's ability as well as the complexity of traffic patterns. As such, ATC actions are of a more tactical nature and primarily address immediate safety concerns of airborne flights. ATFM, on the other hand, refers to processes of a more strategic nature. ATFM procedures detect and resolve demand-capacity imbalances that jeopardize safe separation. By keeping the workload of air traffic controllers to a manageable level, traffic flow management can be viewed as the first line of defense in maintaining system safety. Whereas ATC generally controls individual aircraft, ATFM usually adjusts aggregate traffic flows to match scarce capacity resources. Accordingly, ATFM actions have a greater potential to address system efficiency.

For the majority of the previous century, the coordination of air traffic proceeded largely through tactical ATC procedures. This was deemed sufficient, as the demand for air traffic was generally well within the capacity limits. Periodic congestion was usually resolved by procedural changes or technologic advances (see Nolan 1999) for a comprehensive review of the evolution of ATM). In the US, it was not until the aftermath of the air traffic controllers' strike of 1981 that the Federal Aviation Administration (FAA) first implemented a systematic form of flow management known as ground holding. Under ground holding, aircraft departures are restricted until it is determined that sufficient airspace is available for each aircraft. Initially, the use of ground holding was primarily instituted to reduce workload for the inexperienced controllers that were hired in the wake of the mass firings that accompanied the strike. However, the continued growth in air traffic that followed the airline deregulation act of 1978, as well as changes in traffic patterns such as the "hub and spoke" scheduling practices used by airlines, gradually increased the scope of ATFM practices. In Europe, the evolution of flow management proceeded in a similar fashion. When initial schedule coordination efforts proved insufficient in dealing with the first signs of congestion in the late sixties, countries started establishing flow management units in the early seventies. Eventually, this culminated in the establishment of a central flow management unit (CFMU) under the auspices of EUROCONTROL (Leal de Matos and Ormerod 2000). Over time, the use of ATFM initiatives has therefore become increasingly important, and will undoubtedly play an even more important role in the future.

7.1.1 Air Traffic Flow Management Objectives

The objective of ATFM is to match the capacity of the air transportation system with the demand for it, so as to ensure that aircraft can flow through the airspace *safely* and *efficiently*. In the long term, this implies efforts to prevent structural imbalances by reducing demand or increasing capacity. Typically, this requires changes in policy or regulations, such as changes in separation standards or airport access restrictions. A notable example of such efforts is the imposition of slot lotteries at New York LaGuardia Airport (LGA) in 2001, which aimed to relieve the congestion that resulted when access to airport was opened due to the enactment of the so-called Air-21 legislation in May 2000.

In the short term, however, ATFM aims to avoid congestion and delays. When delays must be imposed, the objective is to reduce their impact on airspace users as much as possible. Generally speaking, there is little need for ATFM intervention under nominal operating conditions. Apart from a few brief periods at the busiest airports, demand is less than the nominal capacity and aircraft can flow through the system freely.

As a result, it is important to recognize that the need for ATFM stems from the fact that nominal operating conditions are (increasingly) rare. The fundamental challenge for ATFM, therefore, arises when the system is disrupted. Fluctuating weather conditions, equipment outages, and demand surges cause significant capacity-demand imbalances. Adverse weather conditions, in particular, frequently cause temporary and substantial reductions in airspace and airport capacity. Because these disruptions are highly unpredictable, ATFM will need to resolve the resulting capacity-demand imbalances in a dynamic fashion. However, instead of using local measures (e.g. holding aircraft in the airspace), ATFM attempts to balance the system and prevent local overloading by adjusting the flows of aircraft on a national or regional basis. This is further complicated by the fact that airlines' flight schedules are usually highly interconnected. The aircraft, crews, and passengers that compose the flight schedule might all follow different itineraries, thus creating a complex interaction between the airline's flight legs. As a result, delays of a single flight leg can propagate throughout the network and local disruptions might have a global impact. Changes in traffic patterns over time, such as the recent growth in unscheduled air traffic, further complicate ATFM.

When system disruptions require the use of ATFM procedures, safety is an overriding objective. Typically, safety concerns are addressed by temporarily establishing appropriate constraint levels for the airspace elements that are impacted. That is, ATFM will establish minimum separation standards, limit the number of operations at an airport, etc., to ensure that aircraft can progress safely through the system. Although paramount, safety cannot be the only objective of TFM, as this is vacuously achieved by grounding all aircraft. This in fact occurred on 12 September 2001, when all aircraft in the US were grounded in response to the terrorist attacks that took place the day before. Clearly, the air transportation system exists to facilitate the movement of passengers and goods, and users must be allowed to conduct their operations.

Second only to safety, the objective of ATFM is to ensure that the system operates as efficiently as possible. Given the complexity of the system, as well as the large number of stakeholders involved, it is difficult to define an appropriate notion of efficiency. Traditionally, performance of the system has been measured in terms of schedule deviations, in that ATFM procedures aimed to minimize the delay between actual and scheduled operations (Barnhart et al. 2006). While this provides aggregate performance indicators that are valuable to the air traffic service provider, they do not necessarily reflect the extent of the service provided to users. Due to the complexity of their schedules, aircraft operators have unique objectives when faced with system disruptions, which aggregate performance measures fail to account for.

From a user's point of view, therefore, two important objectives that can be recognized are *flexibility* and *predictability*. Flexibility can be defined as the amount of operational latitude granted to the carriers in meeting their individual objectives (e.g. on-time arrival, network preservation, profit) when disruptions occur. Through flexibility, users are best able to maintain a reasonable level of efficiency given their internal operational constraints, which are generally only understood by the user on the day of operations and largely invisible to the service provider. The notion of predictability is closely related, and can be defined as the reduction of uncertainties in the implementation of ATFM initiatives. Admittedly, most traffic flow management initiatives are the result of inherently unpredictable events. Nevertheless, ATFM initiatives should provide the user with time to react, and the provider's intent should be communicated as clearly and as far in advance as possible. Note, however, that the resulting notion of predictability should not be confused with flight-specific predictability in the form of guaranteed trajectories and arrival slots.

In a system with a multitude of users, the distribution of capacity (or alternatively, delays) that results from ATFM initiatives is also of paramount concern. As such, a final objective of ATFM is to ensure equity of access to airports and other system resources. Basic notions for equity, such as first-come first-served and first-ready first-served, have long been used and endorsed. However, the increasing complexity of ATFM has exposed the limitations of these concepts, in that they can create systematic biases and fail to provide incentives for desirable user behavior (Vossen 2002), such as the timely provision of accurate flight information.

7.1.2 Collaborative Decision-Making

At the heart of the objectives and challenges of ATFM is the fact that decision-making responsibilities are shared between a number of stakeholders. The actions performed by these stakeholders are highly interdependent, and therefore necessitate a significant degree of coordination. It is therefore no surprise that the coordination and cooperation between air traffic service providers (e.g. the FAA) and the airspace users has become increasingly important. To implement appropriate ATFM actions, the service provider needs an accurate picture of flight status and intent. Airspace users, on the other hand, need the flexibility to adjust their schedules, and can only provide accurate information if they know the actions planned by the provider. Given the relatively short response times, the real-time exchange of information between the service provider and users is therefore a critical component of ATFM functionality. In addition, it has become increasingly clear that the service provider should not be solely responsible for determining the delays, reroutes, etc. required to resolve congestion. While both the service provider and users can possibly delay or reroute flights, certain actions that can alleviate congestion are only available to airlines. In the US, for example, only an airline can decide to cancel flights or to reassign passengers, crew, and aircraft. Consequently, any successful attempt at flow management will require significant involvement from airlines and other users. Such decisions involve economic tradeoffs that the air traffic service provider is not in a position to make.

The need for increased cooperation between service providers and users has not gone unrecognized. In the US, for instance, nearly all efforts to improve ATFM during the last decade have been guided by a collaborative decision-making (CDM) philosophy.[1] Under CDM, the management of traffic flows and the associated resource allocation decisions are conducted in a way that gives significant decision-making responsibilities to airspace users. The overall objectives of CDM can be summarized as:

- generating better information, by merging flight data from the airspace system with information generated by airspace users;

---

[1] CDM has since been renamed collaborative air traffic management (CATM). In this chapter, we will use the term CDM, however, to maintain continuity with the numerous CDM references. For an overview of CDM, see Ball et al. (1999, 2000) and Wambsganss (1996).

- creating common situational awareness by distributing the same information both to traffic managers and to airspace users;
- creating tools and procedures that allow airspace users to respond directly to congestion and to collaborate with traffic flow managers in the formulation of flow management actions.

CDM was conceived in the mid-1990s within the FAA airline data exchange (FADE) project, which originally was created as a short-term experiment to see whether up-to-date airline schedule information would result in improved flow management decisions (Wambsganss 1996). The issues revealed during extensive human-in-the-loop experiments eventually led to the initial implementation of CDM, which primarily focused on the development of new operational procedures and decision support tools for implementing and managing GDPs, which are commonly used by the FAA to delay flights that are bound for an airport that is experiencing bad weather conditions. The general idea behind a GDP is to delay these flights prior to their departure (hence the term ground delay), both to increase safety and reduce cost.

The initial implementation of CDM, known as GDP enhancements (GDP-E) began its prototype operations at San Francisco and Newark airports in January of 1998. In GDPs under CDM, airlines send operational schedules and changes to schedules to the central air traffic control systems command center (ATCSCC) on a continual basis. The schedule information includes flight delay information, cancelations, and newly created flights. The ATCSCC uses this information to monitor and possibly implement GDPs, using a decision support tool called flight schedule monitor (FSM). It is important to note that this information is shared with all users (e.g. airlines also have access to FSM), creating a common picture of current and projected airport conditions. Essential to these procedures is the use of certain resource allocation procedures, which have removed previously existing disincentives for airlines to provide accurate information. The effects of these procedures has been significant: it has been stated that since their initial implementation in January of 1998, over 50,000 h of assigned ground delay have been avoided (http://www.metronaviation.com/cdm.php) through improved utilization of the available capacity.

The success of these initial CDM efforts highlighted the potential benefits of increased collaboration in ATFM, and has been the starting point for a large number of CDM-inspired ATFM improvement initiatives that continue today. The scope of these projects has been broad and varied. There have been a number of of initiatives, for instance, that aim to improve information sharing and dissemination. Examples include improvements in database management through the so-called enhanced traffic management system (ETMS), as well as improvements in the way airlines can submit schedule updates using so-called simplified substitutions (Ball et al. 2005b). Other initiatives have aimed to increase airline control over scheduling decisions, within the slot credit substitutions effort (Ball et al. 2005b). In addition, there have also been efforts toward the implementation of more comprehensive flow management procedures. An example of these can be found in the recently instituted airspace flow programs (AFPs; Krozel et al. 2006), which aim to extend GDP procedures to the en-route environment. Moreover, numerous other CDM-based projects are currently under development. As such, it has become increasingly clear that the CDM philosophy provides a blueprint for almost all efforts to improve ATFM, and that communication and coordination between stakeholders has become a basic requirement in proposals for ATFM modernization.

7.1.3 Chapter Outline

This chapter provides a broad overview of air traffic flow management, both from a practical and theoretical perspective. Throughout this chapter, we emphasize communication and interaction between the various stakeholders in the air transportation system, following the tenets outlined in the CDM philosophy.

We start in Sect. 7.2 by describing the current practices of ATFM. While we focus on the practice of ATFM in the US, key differences with European practice will be outlined. The complexity and relevance of air transportation systems has motivated a considerable amount of research. In Sect. 7.3, we present an overview of this large and varied body of research. We start with a broad taxonomy, so as to organize the models that have been proposed along their different characteristics. We outline the main lines of research, and discuss a few particularly relevant models in greater detail. We conclude by discussing the challenges that arise in achieving user adoption of these models. Finally, Sect. 7.4 assesses the state of research in this area, by considering current trends, needs, as well as research opportunities.

7.2 Air Traffic Flow Management in Practice

In this section, we present an overview of the ATFM practice in the US. First, we introduce the various stakeholders, and discuss their roles and responsibilities, as well as the manner in which they interact and coordinate their decisions. Subsequently, we present an overview of the existing controls used by air traffic service providers to mitigate the impact of system disruptions, and discuss the manner in which users respond. Finally, we summarize the main systems and tools that are used to support ATFM decisions.

7.2.1 Stakeholders

Flow management in the National Airspace System (NAS) is performed by a distributed system in which decision-making responsibilities are shared by a multitude of stakeholders. The stakeholders in the NAS are both the air traffic service provider and users. In the United States, air traffic services are provided by the FAA. The primary users of the NAS are airlines; other users are general aviation and the military. The roles and responsibilities of the FAA and the airlines can be outlined as follows.

7.2.1.1 Federal Aviation Administration Roles and Responsibilities

In the US, the FAA is responsible for the coordination of air traffic. Its primary task is the enforcement of proper separation requirements in the controlled airspace. To carry out this function, the FAA has divided the airspace in the continental US into 22 air route traffic control centers (ARTCCs), each of which is responsible for separation of aircraft within its boundaries. Each ARTCC is further divided into 20–80 three-dimensional sectors of airspace. Within each sector, one or two air traffic controllers track and control aircraft movements via voice communication with pilots. As an aircraft moves from one sector to the next, responsibility for the aircraft is handed off from one controller to another, until it arrives within roughly 50 nautical miles and 10,000 feet of its destination airport, at which point control of the aircraft is assumed by terminal radar approach control facilities (TRACONs). Finally, airport towers control aircraft while they taxi to and from runways and during takeoffs and landings. Accordingly, the ATC functions performed by the FAA form a highly distributed process. Air traffic controllers (cf. TRACON/control tower representatives) are responsible only for the movement of aircraft within their region of airspace, and their decisions are mainly based on local and near real-time information about the flights entering their sectors.

At this tactical level, controllers have only minutes to react to an overage of aircraft. Traffic volume is curtailed within a sector or ARTCC by requesting that aircraft being handed off to them be spaced at minimum intervals, usually expressed as minimum distance between aircraft. For instance, 15 miles-in-trail would mean that each aircraft in a given stream must be at least 15 miles behind the aircraft in front of it. Thus, flow restrictions are imposed by downstream facilities on upstream facilities.[2]

The (strategic) ATFM functions performed by the FAA, on the other hand, are coordinated and administered primarily by traffic management specialists at the FAA's ATCSCC, or command center. The ATCSCC continuously monitors current and projected demand within the NAS, and identifies system constraints or other conditions (e.g. weather) that affect the capacity limits. Whenever it is predicted that demand will exceed capacity limits within a 15min interval, FAA

[2] Occasionally though, controllers can also be concerned with downstream effects, so as to prevent the simultaneous operation of too many aircraft in an area (Hall 1999).

regulation mandates a response. In that case, the ATCSCC generates and implements strategies to resolve the projected congestion.

At each ARTCC, traffic management coordinators (TMCs) within a traffic management unit (TMU) perform ATFM activities within their geographical borders. The ATCSCC, on the other hand, coordinates the activities among the ARTCCs. For instance, requests for traffic flow restrictions made by one ARTCC that would affect flow in other ARTCCs must be cleared with the ATCSCC. Thus, while some of the ATFM duties are distributed among the TMUs, the ATCSCC sits at the apex of a hierarchical control structure.

7.2.1.2 Airline Roles and Responsibilities

An airline's operational objectives are usually markedly different from those that underly the FAA's ATFM initiatives: whereas the FAA is concerned with aggregate flows and capacity limits, the ultimate goal of airline operational control is to preserve its published flight schedule. An airline's flight schedule represents its primary product, and often reflects its competitive strategy.

Airlines typically coordinate their daily operations at centralized airline operational control centers (AOCs), which interact with airport and maintenance stations and with individual pilots. Schedule preservation needs to consider both individual flights and schedule interdependencies. Therefore, airline operations require greater consideration of cascading effects of flight delays (see for instance the examples in Sect. 7.2.2.3). This presents a challenge in particular when airlines face so-called *irregular operations*, that is, when they need to respond to ATFM restrictions imposed by the FAA or to other schedule disruptions.

Important functions that need to be performed by airline operational control include the following (also see Airline Dispatchers Federation and Seagull Technology, Inc. 1995 and Grandeau et al. 1998):

- *Schedule adjustment.* On a daily basis, unforeseen events, such as delays or mechanical problems, might disrupt an airline's flight schedule. To prevent the cascading effects these disruptions can have, the AOC will make schedule adjustments that allow a return to a more balanced condition. Schedules might be adjusted in several ways. One option is to delay selected flights. Other possibilities are to reallocate the resources needed to operate flights (e.g. aircraft, crews, but also airport arrival slots), or even to cancel flights to reduce the demand on those resources. In addition, airlines sometimes create flights to balance the schedule. It should be noted that balancing the schedule might be interpreted differently by individual airlines: For one airline the objective might be the ability to return to the normal schedule by the next day, while for another it might mean flying as many of its scheduled flights as possible (cf. Grandeau et al. 1998).
- *Flight planning and dispatch.* An important aspect of airline operations is to determine flight routes and payload that minimize costs and meet the overall airline flight objectives. Winds, aircraft type, and restrictions all affect the

choice of route, which involves a complex tradeoff between speed, altitude, payload, and fuel load. In addition, flight planning has to take into account that regions of airspace can be congested or temporarily inaccessible.
- *Flight monitoring.* This includes monitoring all aspects of flights in progress, such as ensuring that the flight stays within safe and legal limits, assessing weather conditions en route and at destination and alternate airports, and assisting crews in solving problems that might arise. Thus, AOCs are in constant communication with crews during flight.

Schedule planning is usually performed by dedicated coordinators. Flight planning, dispatch, and monitoring are performed by flight dispatchers, which are licensed personnel responsible for individual flights. By law, the responsibility for the safety and control of flights is shared between the dispatcher and pilot; thus, dispatchers at the AOC maintain frequent contact with pilots prior to and during the flight. Other tasks of airline operational control include crew scheduling and tracking, aircraft maintenance operations, and gate management. Typically, these tasks are performed by separate departments that interact with the AOC. For instance, airport stations manage gate allocations and other ground-based resources (e.g. passenger and baggage handling), while maintenance stations handle the coordination of required aircraft maintenance checks (e.g. ensure that aircraft are routed through the maintenance stations).

7.2.1.3 Interaction and Coordination

Both on the side of the FAA and on the side of the airlines, decision-making responsibilities are shared between a number of stakeholders. The actions these stakeholders perform are of course highly interdependent, and therefore necessitate a significant degree of coordination. On the FAA's side, operational processes are essentially distributed among three organizational levels. At the first level, we find the ATCSCC. The ATCSCC oversees aggregate traffic flows and monitors the current and projected capacity limits and demands. Major flow management actions, such as GDPs and rerouting around severe weather, are usually initiated by the ATCSCC. The ATCSCC coordinates these ATFM initiatives with TMUs at the various ARTCCs, TRACONs, and Towers, which form the second organizational level. The entities at this level are responsible for coordinating air traffic in their assigned regions of the airspace. Besides their interaction with the ATCSCC, adjacent centers at this level also interact to coordinate the air traffic between their regions. ARTCCs, TRACONs, and Towers further delegate responsibilities to the individual air traffic controllers, which form the third organizational level. The primary interaction at this level is between controllers at adjacent sectors to transfer control of aircraft. On the airlines' side, on the other hand, daily operations are primarily coordinated at centralized AOCs. Specific tasks, such as gate assignments and maintenance are coordinated with various stations, and flight

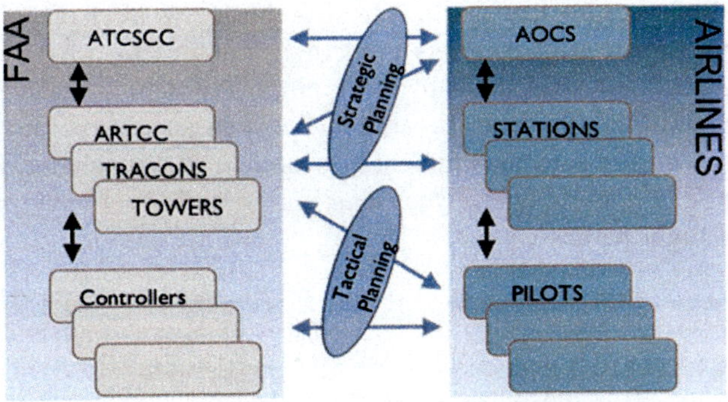

Fig. 7.1 Operations-level interactions between the FAA and airlines

dispatch is of course in constant communication with pilots to monitor and control the progress of individual flights.

The interaction between the FAA and airlines during daily ATFM operations can be separated according to interactions at the strategic and the tactical level, as shown in Fig. 7.1. At the strategic level, interactions occur primarily between the ATCSCC, ARTCCs, and the airlines' AOCs. When the ATCSCC predicts a sustained period of congestion, it responds with an ATFM initiative (e.g. a GDP or a Severe Weather Avoidance Plan), which is communicated to the airlines' AOCs. Typically, these plans are formulated 2–4 h in advance. In turn, airlines communicate the schedule adjustments they intend to make in response to these disruptions. It should be noted that this flow of information is highly important, as the FAA's decisions are partly based on the information they receive. At the tactical level, the interactions occur primarily between controller, pilots (to ensure separation), centers, and stations. Typically, these interactions concern ATC initiatives (e.g. ensuring the separation standards), and other near real-time initiatives.

7.2.2 ATFM Initiatives

Air transportation distinguishes itself from most other forms of transportation in one important respect: vehicles cannot be stopped en route, and therefore cannot allow traffic jams to develop. As a result, traffic managers implement various traffic management initiatives (TMIs) to curtail demand before congestion occurs. Each TMI can be viewed as a combination of the following control techniques:

- Ground holding controls, that is, the selective assignment of delays to flights prior to their departure. These take the form of GDPs that control flights bound for a common location in a common time period, or ground stops that control flights departing a common region in a common time period.
- Rerouting controls, which impose constraints or exercise options on the flight paths that an aircraft can fly.

- Airborne holding controls, which result in flight delays after takeoff. Airborne delays can be applied using a variety of methods, ranging from *Spacing* to *Speed controls* and *Vectoring*. Spacing, between aircraft traveling in the same direction, specifies and controls the separation between successive aircraft. Speed control aims to ensure safe and efficient flow of aircraft by selectively increasing or decreasing their speed, while vectoring corresponds to minor spatial deviations from a flight path.

Generally speaking, ground holding and rerouting techniques are used to support strategic activities, in that they are applied proactively hours in advance. Airborne holding controls, with the exception of spacing, are commonly used for tactical flow management and are initiated reactively (minutes in advance, as opposed to hours).

In the remainder of this section, we discuss the major ATFM initiatives that are used to implement these basic control options. Because ATFM is more of a strategic activity, we direct our attention to initiatives that apply ground holding, rerouting, and spacing.

7.2.2.1 National Rerouting and Spacing Programs

The National System Strategy Team (NSST) is a functional unit within the ATCSCC that coordinates NAS-wide flow strategies. Its functions include planning teleconferences, coordinating and implementing AFPs, managing reroutes, and facilitating international operations. The NSST is divided into distinct areas of geographical responsibility, as shown in Fig. 7.2. Domestic responsibilities are assigned to specific Regional Airspace Managers who oversee geographical areas that coincide with major traffic flows: North Atlantic, Pacific and Polar, Central and South America, and Caribbean. Responsibility for international operations resides with an International Operations Manager.

In mid-January 2006, a National En Route Spacing Position (NESP) was created within the NSST to monitor the NAS, promote system balance, and mitigate the impact of flow constraints in the system.[3] The full spectrum of TMIs is available to the NESP for managing overall NAS constraints. The NESP coordinates, implements, and manages AFPs. Other TMIs, however, are normally coordinated and implemented by NSST members and the Terminal Area National Traffic Management Officer (NTMO). The development and implementation of rerouting and spacing TMIs is largely based on the expertise and judgment of the traffic managers involved, with little use of systematic planning techniques or decision support systems.

---

[3] Prior to January 2006, NESP functionality was provided by a Severe Weather Unit at the ATCSCC.

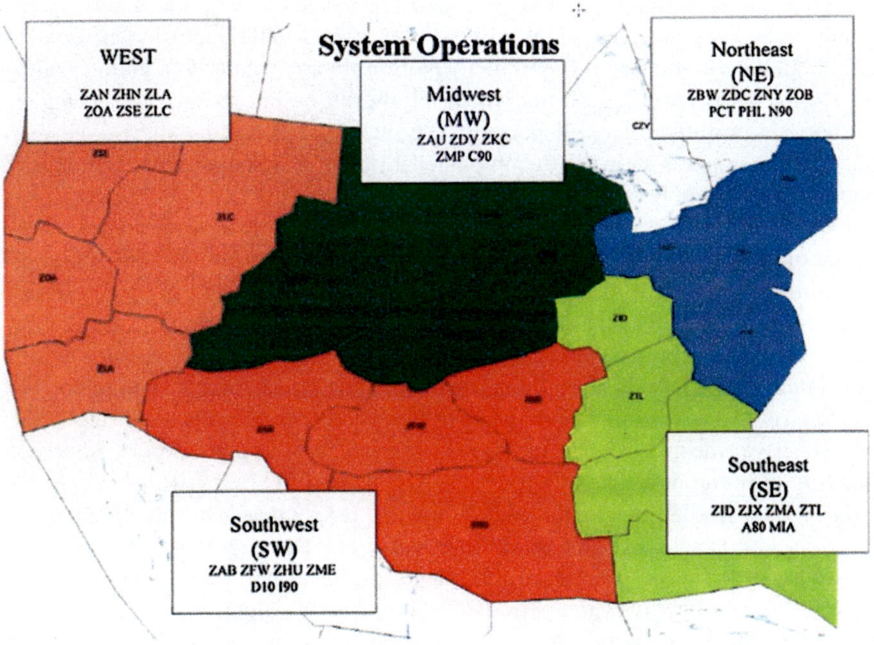

Fig. 7.2 Regional airspace managers oversee the Northeast, Southeast, Midwest, Southwest, and West regions of the country

7.2.2.2 Ground Holding Initiatives

Airports frequently become bottlenecks to traffic flow within the air transportation system. One of the primary responsibilities of air traffic managers is to formulate and apply strategic initiatives to alleviate anticipated demand-capacity imbalances at airports. Traffic managers are required to take action whenever forecasted demand exceeds forecasted capacity for any 15-min period of time. Minor imbalances can be addressed simply by concluding that the excess demand can be handled tactically. More serious demand overloads for an airport necessitate TFM intervention in the form of a ground holding initiative.

In a ground holding initiative, arriving aircraft are held at their respective origin airports to reduce the inflow of aircraft at the affected airport. While the source of a demand-capacity imbalance can be related to both arrivals and departures, the focus on throttling back the arrival flows has considerable merit. The reason for this is that arrival flow reductions will also (implictly) reduce the departure flows. This a natural consequence of the fact that in the air transportation network, airports act as sources and sinks for *flights* but act as through-nodes for *aircraft*: overnight stops and mechanical servicing aside, the vast majority of airframes touch down, reload passengers or cargo, and takeoff within an hour or two.

Ground holding initiatives take one of two forms: (1) Ground Stops or (2) Ground Delay Programs. These initiatives differ mainly in the inclusion criteria for aircraft and the abruptness of the implementation. In a *Ground Stop*, an interval of time $T = [t_1, t_2]$ is specified during which all aircraft bound for an affected region or airport are prohibited from departing. The interval T is usually short-lived, e.g. 30min. This temporal scope is accompanied by a spatial scope, which limits the control policy to aircraft within a geographical region, usually ATC centers near the affected airport. Ground stops are usually imposed with little or no notice, as they are often a reaction to sudden, unexpected events that prompt traffic managers to shut off all arrivals immediately (such as a violent thunderstorm or an aircraft stuck on a runway). Because this action does not affect airborne aircraft, the net effect is to severely reduce (rather than stop) the airport's arrival flow, but not until n min after t_1, where n is the minimum travel time of all aircraft bound for the affected airport that are on the ground at time t_1. Nonetheless, a ground stop is the most effective means of shutting down arrival flow to an airport, short of grounding airborne aircraft.

In a GDP, a similar interval of time $T = [t_1, t_2]$ is specified. During this interval, each flight that is scheduled or estimated to arrive at the airport is subjected to a flight-specific period of ground holding at its origin airport. The objective is to reduce the arrival flow over the period T, which usually coincides with a period of anticipated capacity reduction due to adverse weather conditions. As in a ground stop, traffic managers specify a spatial scope which limits the delays to certain ATC centers, or to those flights whose origin airport is within a number of miles or minutes of flying time of the affected airport. Unlike a ground stop, a GDP generally has significant lead time; the difference between t_1 and the time t_0 at which the GDP is imposed is usually on the order of hours rather than minutes. A GDP provides predictability over an extended planning period; though the delays imposed are often on the order of 30–90 min per flight, airlines have a chance to mitigate the ensuing schedule disruptions. This, in turn, admits a rich variety of resource allocation policies, which we will discuss in greater detail below. In contrast, a ground stop applies a near-universal delay disruption served by flights over a common time period. Another important distinction between a ground stop and a GDP is that inclusion within a ground stop is determined by flight departure time, whereas inclusion within a GDP is determined by flight arrival time. Because a period of degraded arrival capacity at an airport usually provides the impetus for the ground holding initiative, it seems that the GDP inclusion criteria are more appropriate. Moreover, because a GDP can always be created with zero-lead time and a targeted arrival rate of zero flights per hour, it would appear that a GDP subsumes the functionality of a Ground Stop. A GDP, however, tends to have much more procedural overhead associated with its implementation and design. Thus, Ground Stops are preferred when time is of the essence. In fact, a ground stop can be issued by central traffic managers with a single phone call.

In practice, GDPs are far more prevalent than ground stops. In 2005, there were over 1,350 GDPs implemented in the US, applying delays totaling more than 16.8 million min, distributed over 530,000 flights. Because GDPs admit a much richer set of control options and airline responses, we use the remainder of this section to discuss GDPs in greater detail. As stated before, a GDP is a traffic management initiative that reduces arrival flow to an airport by holding inbound aircraft at their points of origin. A GDP is characterized by five elements, each of which we discuss in turn.

1. *Temporal scope*. Because a GDP disrupts flight schedules, users want to know when the initiative will begin and how long it will last. For this reason, each GDP has associated with it

   - a time of implementation, t_0, called a file time (or report time), and
   - a GDP time horizon, $T = [t_1, t_2]$, where t_1 is the start time and t_2 is the end time. The interval T is used to determine the aircraft that are affected. Each aircraft that is scheduled or estimated to arrive during this interval is issued a controlled time of arrival, which is no earlier than the time the flight would have arrived in the absence of the GDP. Usually, T coincides with the period during which airport arrival capacity is expected to be low. However, T might extend into a period of restored capacity to allow the cumulative demand to fall below cumulative capacity.
   In hindsight, calling t_1 the 'start' time of the program is somewhat misleading. With respect to departure control, the effect of the program is first felt at the file time, t_0. For instance, given a file time of $t_0 = 12:00$ for a program starting at $t_1 = 14:00$, a 2-h flight whose scheduled arrival time of 14:00 is delayed until 14:15 will serve its ground delay time between 12:00 and 12:15. So, between t_0 and t_1, departures are being controlled but the effects are not felt at the GDP airport until after t_1. Conversely, the end time is the latest time at which a flight arrives whose departure was delayed by the GDP. Therefore, the last controlled departure took place earlier than the end time.
   If the file time is too close to the start time, there will be a loss of control during the GDP time horizon. In that case, ground delays alone no longer bring demand back in line with capacity because too many aircraft bound are already airborne. At the same time, however, the file time should be postponed long enough to give a clearer picture of future airport conditions.
   Once a GDP is announced, it is uncommon for the file time or start time to be revised. However, the end time is frequently revised once the program is in place, with the most common revisions being GDP cancelation (flights revert to their normal operating schedule) and GDP extension (end time of the program is moved later in time). This variability in the end time (duration) of the program reflects the variability of the anomalous event that triggered the GDP, e.g. uncertain weather forecasts).

2. *Capacity forecasts*. In the current practice, airport capacity is treated as a deterministic value. For each 15-min time period over the GDP time horizon, an estimated airport acceptance rate (AAR) is provided. The length of each slot is computed as the inverse of the AAR. For instance, an AAR of 30 flights per hour would yield 30 slots of 2-min lengths each. The deterministic rate must be

estimated from forecasted airport conditions. This is not as daunting as it might seem, because the aggregate airport arrival capacity can assume one of a limited number of values. Airports operate under a small set of possible runway configurations; each configuration has associated with it a limited set of viable AARs, which is known through operational experience by airport controllers and traffic managers. So, the forecasted weather conditions need only be mapped into one of the possible runway configurations. There is some level of negotiation between airport personnel and traffic managers in the setting of these rates, as airport capacity is affected by human performance of air traffic service providers and by the natural tradeoff for airport resources between arrivals and departures (Gilbo 1993; Hall 1999). The GDP AAR serves both as a capacity estimate and a target arrival rate. A primary measure of GDP performance is the degree to which the hourly airport acceptance rate matches the planned AAR (see Hoffman and Ball 2001 for a general discussion).

The prediction of weather conditions complicates this capacity estimation process. Weather conditions are highly stochastic and difficult to predict. The consequence of being wrong in the planned AAR is either (1) airport underutilization, which takes the form of excess assigned ground delay, or (2) airport over-utilization, which takes the form of airborne holding or, in severe cases, airborne holding plus a ground stop. The uncertainty in capacity forecast is generally addressed dynamically by program revisions.

3. *Arrival flow models*. In order to determine when each flight should arrive during a GDP, some model of airport arrival flow must be adopted. As shown in Fig. 7.3, airport arrival flows consist of the confluence of aircraft over several (usually four) arrival fixes, or corner posts, near the boundary of the TRACON airspace. The arrival fixes provide structure by aligning aircraft into a single flow. An arrival fix usually feeds one runway, depending on the airport runway configuration. Thus, one can view airport arrival flow as a queueing system with multiple servers (one for each arrival fix). Because all aircraft also share resources at the airport itself, all aircraft are served by a common airport server. Thus, a given aircraft requires available capacity at one arrival fix and at the airport.

   In operational practice, however, it generally suffices to model arrival flow as a single-server queueing system. There is some loss of efficiency in adopting the single-server, single queue model, as it does not allow the ability to plan for capacity constraints that occur at some but not all of the fixes, or rerouting of aircraft between arrival fixes. Nevertheless, there are several practical advantages to the aggregation of arrival flows. First, it avoids the need to predict which arrival fix each aircraft will use—something which is generally not known until flights are near departure.[4] In addition, the aggregation avoids the need to make arrival fix-specific capacity predictions. Currently, weather forecasts cannot accurately predict fix capacities hours in advance, when most GDPs are planned. Finally, this

[4] Most airlines are in the habit of filing flight plans no more than 90 min prior to departure, which is generally later than the time that traffic managers would like to assign ground delay.

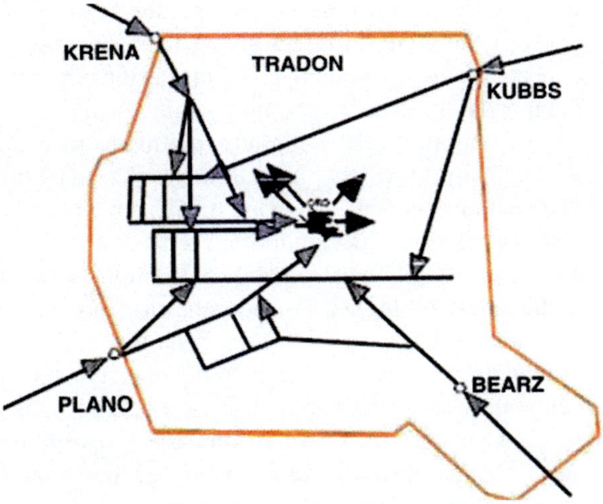

Fig. 7.3 Terminal area arrival patterns at Chicago O'Hare airport; Krena, Kubbs, Plano, and Bearz represent arrival fixes

allows the collection of inbound aircraft to be modeled as a single commodity, thereby enabling substitution of assigned arrival times between flights.

Another simplifying assumption within the current TFM model of arrival flow is that every flight requires the same amount of arrival resource (e.g., 1 min to land for an AAR of 60). It is well known that the spacing between two successive arrivals depends, at least in part, on wake vortex created by the lead flight (see Beasley et al. 2000; Carr et al. 1998 for more details). Ideally, therefore, aircraft should be sequenced during GDP planning so that wake vortex effects are minimized. Variability in en-route travel times, however, tend to scramble the planned arrival order (Ball et al. 2001), and GDP planning therefore assumes uniform landing time intervals.

4. *Flight filtering criteria*. Once the time horizon of a GDP has been determined, the set of flights to be *included* in the program is the set of all flights scheduled or estimated to arrive during the horizon. Each such flight is given a controlled time of arrival (CTA) and a controlled time of depature (CTD). Those flights not meeting this criteria are said to be *excluded* from the program and do not receive control times. Although the ultimate objective of a GDP is to have each flight arrive at its CTA, the CTD is the actual control mechanism. Each flight is expected to depart within 5 min (plus or minus) of its CTD, or else the pilot must contact the FAA traffic manager managing the program for a new CTD and CTA.
   However, not all flights can be issued a ground delay. Flights departing from abroad, for instance, cannot be controlled.[5] Emergency equipment and VIPs constitute another class of exempt flights. Therefore, a class of *exempted* flights is created within the set of included flights. These flights are exempt from FAA-assigned ground delay. Each exempt flight still receives a CTA and a CTD.

[5] With the exception that flights departing certain Canadian airports are subject to US control.

However, the CTA is set to the time the flight would like to have arrived at the onset of the GDP, in effect giving them no delay. The exempt flights are then expected to depart at their CTD, just as the non-exempt flights are.In addition to the mandatory delay exemptions, traffic managers often add an arbitrary class of delay exemptions: those flights departing from an airport outside a radius *r* about the GDP airport, where *r* is a parameter chosen by the traffic managers each time a GDP is implemented. There are many subtle reasons (not germane to our current discussion) why this exemption radius is applied. For our purposes, a primary reason for establishing the radius of exemption is to shy away from applying delay to the long-haul flights. Consider the case in which a flight departs from the west coast of the US for a GDP airport on the east coast. Clearly, in order to apply ground delay to the flight at the point of origin, the appropriate amount of ground delay must be set (and applied) 5 or 6 h in advance, depending on the speed and route of the aircraft. Weather forecasting several hours in advance is not highly accurate. Often times, traffic managers find that the airport conditions unexpectedly improve while the flight is en route, thereby negating (in hindsight only) the need for whatever ground delay was applied to the flight. So, one way of hedging against lack of confidence in weather/capacity forecasting is to apply little or no delay to the long-haul flights. The exemption radius *r* is one way to segregate the long-haul flights from the short-haul flights (short-haul flights being those departing within the radius and long-haul being those flights departing beyond the radius).

As an alternative to setting a radius of exemption, a traffic manager can run a *tier-based GDP*, in which (discretionary) exemption status is determined by the air route traffic control center (ARTCC, or center) from which the flight was departing. There are 20 such centers in the continental US. For instance, a first-tier program includes all flights departing from traffic centers adjacent to the traffic center that contains the GDP airport. When geographical proximity to the GDP airport (measured in miles or minutes of flying time) really is the filtering criteria sought, a distance-based GDP is a more precise filtering criterion than a list of traffic centers, because the traffic centers often have irregularly shaped boundaries (see Ball and Lulli 2004) for a discussion of pros and cons of distance-based GDPs).

5. *Allocation policy*. The final component of a GDP is the allocation policy that assigns arrival resources to each flight. Over time, this allocation has become quite intricate—arrived at by trial and error and by consensus with users from government–industry working groups (Chang et al. 2001).

The current rationing policies for GDPs are a blend of automated algorithms and accepted practices of FAA traffic managers. The primary algorithm, *Ration-By-Schedule* (RBS) is housed in a decision support tool known as the FSM. The details of the algorithm are highly involved, as it must address a wide range of anomalous cases and operational situations. Here, we provide only the level of detail necessary to understand the allocation principles at work. Generally speaking, FSM logic creates a virtual queue of arrival slots in accordance with

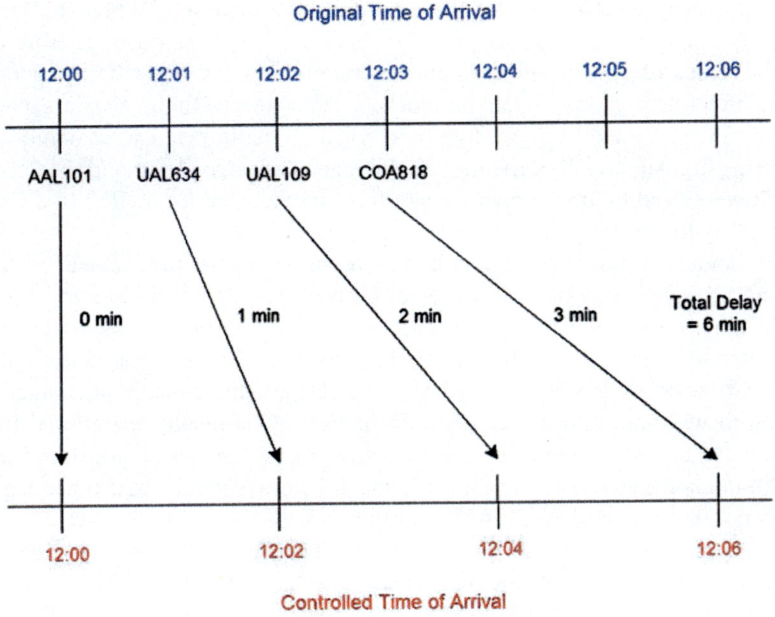

Fig. 7.4 The net effect of ground delays is to stretch out the flow of arrivals over time

the forecasted airport acceptance numbers for future hours (as discussed before). The GDP horizon nearly always has more flights vying for arrival slots in each hour than there are arrival slots (hence the need for the GDP). Each flight is given a CTA corresponding to the time stamp of its allotted slot. Because en route travel time can be estimated by the FAA or provided by carriers, a controlled time of departure (CTD) is also computed by subtracting the en route travel time from the CTA. The difference between the CTD of the flight and its desired departure time is the amount of FAA-imposed ground delay.

Because the anticipated arrival capacity of the airport is implicit in the number of virtual slots created, airport arrival flow can be reduced to a manageable level by allocating each flight to exactly one slot. The net effect is to stretch the anticipated queue of flights out over time, as seen in Fig. 7.4. The question is how to do this in a manner that users will find acceptable.

Prior to 1998, flights were allocated to arrival slots based on a first-come, first-served (FCFS) priority rule, known as *Grover-Jack*. The flight with the earliest estimated time of arrival (ETA) was awarded the earliest slot it could use, then the flight with the next earliest ETA was awarded the earliest slot it could use, subject to the first slot assignment, and so on. It was soon discovered that this seemingly innocent extension of FCFS practices already in use by air traffic controllers created a disincentive for carriers to submit accurate information. The problem was that any flight delay reported by an airline to the FAA just prior to slot allocation was effectively magnified by the queue stretch. When the airport capacity is cut in half, for instance, the delay factor for the queue (before allocation versus after allocation) is approximately 2. Each flight delay report of x min effectively yields $2x$ min of *additional* delay to the flight. Carriers quickly learned that they could avoid this (perceived) 'double penalty' by simply not reporting flight delay. This, in turn, created inaccurate arrival demand predictions, which were used by traffic managers to forecast the need for a GDP (see Chang et al. 2001 for further details).

As of January 1998, the FCFS rule was replaced by the first-scheduled, first-served (FSFS) priority rule known as ration-by-schedule (RBS). Under RBS, the earlier slots are awarded to the flights scheduled to arrival earlier, rather than those estimated to arrive earlier. By making the priority rule dependent on static flight information created prior to the day of operations, the carriers no longer had reason to withhold real-time flight information. As a result, operational information and demand counts became much more accurate. As an added incentive for voluntary flight data submission, the airlines were rewarded with other degrees of operational freedom (see also Wambsganss 1996).

RBS is applied to the set of flights included in the program, which are divided into two classes: exempt flights and non-exempt flights. Because the RBS algorithm must honor allocations already made in a prior execution of the algorithm (e.g. when the GDP is revised or extended), there is a third class of flights that consists of those flights that have already been assigned a slot. As such, the resulting three priority classes are as follows:

- Class 1. Exempt flights, sorted by increasing estimated arrival time;
- Class 2. Non-exempt flights with no control time, sorted by scheduled arrival time;
- Class 3. Non-exempt flights with a control time (CTD and CTA), sorted by increasing CTA. RBS first assigns the Class 1 flights first to slots, then Class 2 flights, and finally Class 3 flights. The "ration-by-schedule" philosophy only applies to the flights in Class 2. Exempt flights (Class 1) are processed by an FCFS principle so that they can receive the slots they want; this is especially important for airborne flights. Previously controlled flights (Class 3) are processed by an earliest CTA principle that ensures the CTA order is preserved. CTA order might not match order by scheduled arrival time because the airlines reserve the right to substitute slots among their own flights at any time during a GDP.

  The allocation process is almost always followed by a *Compression algorithm*, which fills open slots created by flights that are canceled or delayed beyond their scheduled arrival times; this restores operational feasibility of the allocation process. The combined process, RBS plus Compression, is formally called RBS^{++}. As RBS is virtually never run in the absence of compression, however, we drop reference to the "++" and informally refer to the overall algorithm simply as RBS.

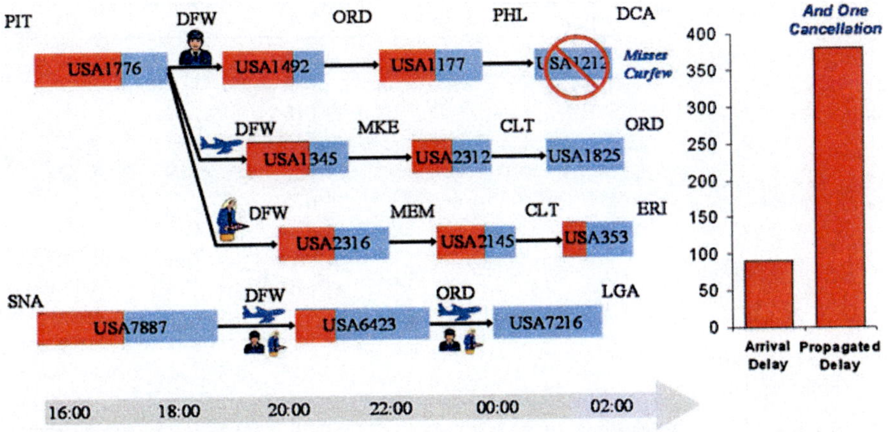

Fig. 7.5 Delay of a single flight (USA1776) bound for DFW triggers massive downstream delays throughout the air carrier operation (*red* portion of flight icons). The same delay imposed on USA7887 has much less impact

7.2.2.3 Airline Responses to Ground Delay Programs

A major lesson learned from recent R&D efforts such as collaborative decision making (CDM) is that efficiency should be measured relative to the standards and values of the NAS users, rather than by using generic efficiency metrics such as throughput, which tend to treat all flights as interchangeable. Air carriers have a strong desire to maintain the integrity of their published schedules, as this provides the best possible service to their passengers. The following example shows the impact that FAA-imposed delays can have on an airline network.

Example 1 Figure 7.5 shows the downstream carrier connectivity of airframe, pilot, and flight crew for two flights, USA1776, and USA7887, both bound for Dallas Fort-Worth Airport (DFW). For instance, the crew of USA1776 connects with flight USA1492 at DFW, which connects with USA1177 at ORD, which connects with USA1212 at PHL. Suppose that a 90-min delay has been imposed on USA1776 as a result of a ground delay program (GDP) at DFW. This induces downstream delays on all subsequent flights of the airframe, pilot, and crew chains (red-shaded portions of flight icons). The final flight in the airframe chain, USA1212, must be canceled because the inbound crew will have operated beyond its legal flying time. In all, the original 90-min ground delay leads to 380 min of delay throughout the air carrier operation, plus one cancelation.

Contrast this with the case of flight USA7887, also bound for DFW, which is exempt from the ground delay program and assigned no delay. The downstream connections for USA7887 have more time built into the schedule to absorb the delays. Ideally, USA would like to transfer the delay from USA1776 to USA7887, thereby minimizing the impact on their operations. To accommodate this type of transaction, sophisticated auxiliary procedures have been developed for FAA

Table 7.1 Initial flight-slot allocation

Airline	Flight	ETA	CTA	Delay
A	1	1,000	1,000	0
B	2	1,005	1,010	5
A	3	1,010	1,020	10
C	4	1,015	1,030	15
A	5	1,020	1,040	20
B	6	1,025	1,050	25
A	7	1,030	1,100	30
A	8	1,035	1,110	35
C	9	1,040	1,120	40
A	10	1,045	1,130	45

ground delay programs to allow air carriers to transfer delay between flights. The 90 min of delay is transferred from USA1776 to USA7887, as in the lower portion of Fig. 7.5. This reduces USA's operational delay from 385 to 100 min, even though it has no apparent effect at the GDP airport. Because the delay transfer is performed by trading arrival slots between the two flights, the traffic flow management objective of reducing DFW arrival flow to a certain number of flights per hour is preserved.

This simple example shows two important things: (1) the need to have mechanisms in place for carriers to mitigate potential damages of unavoidable delays; and (2) a fixed amount of delay imposed on a single aircraft can have dramatically different effects, depending on the flight's place in an operational context. Under existing airport arrival slot swapping rules, an airline can perform any of the following operations:

- cancel a flight;
- move a flight into an arrival slot vacated by one of its flights; and
- swap the arrival slots (arrival times) of two of its flights.

The following example shows how an airline might apply these operations.

Example 2 Suppose that arrival slots have been allocated to the flights of airlines A, B, and C, as in Table 7.1. Let us adopt the vantage point of airline A and determine their best substitution and cancelation strategy based on the following information: (1) The cost of delaying A3 is relatively low (e.g., it has a small number of passengers and the aircraft has a long layover); (2) The cost of delaying A5 is quite high; (3) Passengers of flight A7 can be easily rerouted onto flight A8. One strategy airline A might use is to:

1. swap flights A3 and A5;
2. cancel A7 (transfer passengers to A8);
3. move A8 into slot vacated by A7;
4. move A10 into slot vacated by A8;
5. hold onto the slot vacated by A10.

Table 7.2 Flight-slot allocation after subs/swaps

Airline	Flight	ETA	CTA	Delay
A	1	1,000	1,000	0
B	2	1,005	1,010	5
A	5	1,020	1,020	15
C	4	1,015	1,030	20
A	3	1,010	1,040	30
B	6	1,025	1,050	25
A	8	1,035	1,100	25
A	10	1,045	1,110	25
C	9	1,040	1,120	40
A	Hold	–	1130	–

This substitution process reduces the overall delay for airline A from 135 to 80 min, as seen in Table 7.2. Similarly, airlines B and C would make substitutions and cancelations beneficial to their operations. We will refer to this reallocation procedure on the part of the airlines as substitution–cancelation. The substitution–cancelation process is vital for the mitigation of damages incurred by delays in a ground delay program.

The substitutions performed by the airline in the prior example happened to minimize their overall delay but this is not necessarily their objective. Overall delay minimization assumes that delay costs are quantifiable entities that are the same for every flight and that they increase linearly with delay. Substitution and cancelation decisions for a given airline are highly peculiar to its flight schedule, flight connectivity, and operational methodology. Substitution decisions are generally made in real time by operational personnel [as opposed to operations research (OR) experts who form the original airline schedules]. Virtually all airlines use vendor-supplied software that admits selective substitutions, which is convenient for resolving issues with the most critical flights, and global substituions, which are solved as an assignment problem.

The need for Compression within the flight substitution procedure was available to the air carriers since the early days of GDPs (as of 1987). One of its perceived inequities, however, was the loss of access to airport resources when a flight cancelation was made. This was overcome by modifying the Compression algorithm to give preference for filling vacated slots to the carrier that previously owned the slot. See Wambsganss (1996) for a detailed discussion and examples that illustrate the need to preserve an airline's claim to resources during a GDP and how the interests of one air carrier can conflict with the desire for overall efficiency.

7.2.2.4 Airspace Flow Programs

The initial implementation of collaborative traffic flow management has been aimed at airport GDPs. However, it has become increasingly evident that very significant delays and throughput degradations have arisen from en route airspace problems and limitations, particularly from convective weather activity and demand surges. Work is currently underway within the aviation community to apply CDM technology and concepts to the management of en-route traffic (Collaborative routing).

An essential component in the successful application of CDM to GDPs was the development of consensus around a fair allocation scheme for landing time slot assignment [ration by schedule, using the official airline guide (OAG) schedule as a baseline]. No such baseline exists for the en route airspace. Although there are several concepts for resource allocation before the CDM community, the feasible implementation of these concepts has yet to be demonstrated. Current practices that are used to manage en-route constraints, primarily miles-in-trail restrictions, are quite rudimentary and lead to significant inefficiencies.

Recently, an airspace flow program (AFP) concept has been developed and put into operation on a prototypical basis (summer of 2006). Essentially, an AFP is a ground delay program run for a designated area of constrained airspace. Under the AFP concept, FAA traffic managers can designate:

- a flow constrained area (FCA);
- a time interval, *T*;
- flight filtering criteria; and
- a demand percent reduction, *r*.

The AFP then reduces the flow rate into the FCA by the percent *r* over the interval *T*. This is done by applying ground holds to flights meeting the filtering criteria. The FCA can be a three-dimensional polygon of airspace or just a line, which then acts as a trip wire for catching certain traffic classes. This is convenient, for instance, when traffic must be slowed down into the New York City area.[6] So, the specified airspace need not be constrained itself. Rather, an AFP can be created for other strategic reasons.

Prior to running AFPs, traffic managers had no systematic way of slowing flow to a region of the country. In the convective weather season, GDPs were often put into place at several airports simultaneously as a means of controlling en route traffic. The practice was known as *GDPs in support of severe weather avoidance programs, or GDP-SWAP*. Because GDPs are non-directional, this adversely affected aircraft bound for the GDP airports that were not contributing to the en route volume problem. Although their application is still in an early stage, AFPs have so far proven to be an effective replacement for GDP-SWAP.

Flight Schedule Monitor software has been adapted to run AFPs. For each (filtered) flight estimated to arrive in the FCA, a pseudo-arrival time to the FCA boundary is computed. Subsequently, the ordinary GDP rationing logic is applied (see Sect. 7.2.2.2). Ideally, when the FCA is a volume of airspace, some consideration should be made for the dwell time of each aircraft in the FCA, as this

---

[6] This is sometimes needed when severe weather blocks departure routes from the New York City area. If arrivals were not curtailed, the NYC airports would soon become saturated with surface aircraft.

dictates the amount of resource that the aircraft requires. AFPs, however, have not reached that level of functionality yet. Estimating capacity is an area of outstanding research for arbitrary regions of airspace, especially in the face of adverse weather conditions. As a result, traffic managers currently set a demand reduction rate based on their past experience and strategic goals. For instance, they might specify that the arrival flow to the FCA be reduced by 50%.

A key difference between a GDP and an AFP is that for aircraft caught in a GDP, some amount of FAA-assigned ground delay is inevitable (short of manipulating ground delay amounts through substitution procedures). This is because most flights are operating under published schedules, and therefore cancelations and diversions are quite costly. When ground holds are applied to a region of airspace, however, many flights can reroute around the troubled airspace and remove themselves from the AFP. The issue of who should, or can, reroute out of the AFP and what incentives might go with this is a topic of active research. For now, the filters that traffic managers can apply provide tremendous flexibility in how the FCA resource is allocated. Usually, their criteria for flight filters are completely driven by operational needs (e.g. slow flow over a certain ARTCC boundary). This might not lead to equitable allocation for FCAs, which is also a topic of ongoing research.

Another difference between a GDP and an AFP is that, in principle at least, an AFP can be run for any region of airspace. However, GDPs tend to be run at major airports, due in part to operational constraints and operational tendencies. As a result, general aviation traffic has historically been less affected by GDPs than mainstream commercial traffic.[7] As AFPs become more prevalent, more and more general aviation traffic will be affected by AFPs, thereby increasing their need to become involved in traffic flow management issues.

7.2.3 Decision Support Systems

The ATFM initiatives outlined above heavily rely upon the real-time flow and dissemination of information. The implementation of CDM efforts, in particular, has required a significant enhancement to the data management and distribution capabilities that support traffic flow management. In the remainder of this section, we briefly discuss the key elements of this information infrastructure, and describe the systems that have been developed to support ATFM decision-making.

7.2.3.1 Information Infrastructure

The sheer scale of air transportation systems requires an extensive flight information database to support its operations. In the US, the ETMS operates this

---

[7] Another contributing factor is that general aviation can dynamically change their destination airport in reaction to TFM programs, an option that is frequently exercised.

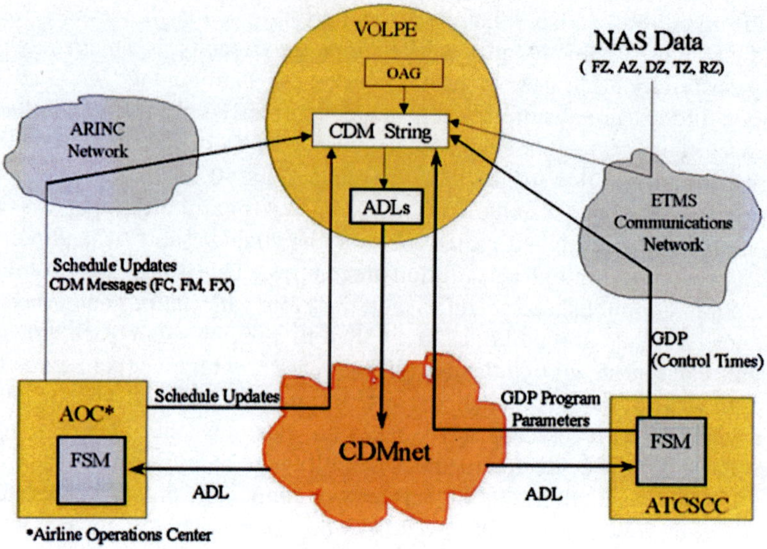

Fig. 7.6 CDMnet overview; Source: http://www.metronaviation.com/cdm.php

database, and acts as the primary interface with traffic flow managers The ETMS acquires airspace, flight intent and position, and other TFM data from a variety of sources, and integrates this data into a national picture of aviation activity. Subsequently, the system redistributes this information to FAA facilities, National Airspace Users, and other entities. The ETMS is composed of two major subsystems: a Hub and Field Sites. The ETMS Hub is located in Cambridge, MA, and represents the system's main computing and communications center. The field sites are located at FAA facilities, and consist of a file server for gathering and processing data together with workstations for dissemination of data.

The ETMS is a unique system that has evolved over a long period of time. Traditional database components have been (and are) integrated with newer technologies and systems in an incremental fashion. As a result, the ETMS consists of many different hardware and software systems that have been added over the years. This legacy often limits the ability to implement recent technologies, and newer systems are usually adopted slowly while the original architecture retains most of its functionality.

The CDM efforts present another step in this evolution, in that they required a new architecture to allow a rapid information exchange between both the airlines and the FAA. Given the limitations of previous communication links, this led to the creation of a new private extra-net that connects the ATCSCC, the participating AOCs, the ETMS hub site, as well as certain other parties. A high-level overview of this network, which is commonly known as CDMnet or AOCnet, is shown in Fig. 7.6.

The use of CDM has produced new information by combining the data from the FAA with the data provided by airlines that have implemented data feeds from their operations systems into this network. Airlines provide information about flight cancelations, mechanical delays, and other events that impact the demand on the NAS; this is done by sending messages either over a pre-existing network (the ARINC network) or over the newly created CDMnet. At the same time, the FAA provides a variety of airspace system status information (NAS data) that is generated in the above-mentioned ETMS, such as flight plans, position updates, arrivals, departures, etc. A separate CDM Hub site at the Volpe National Transportation Systems Center then merges these information streams into a real-time data feed, which is provided to the FAA and the airlines in the form of an aggregate demand list (ADL).

Through the CDMNet, the CDM-enhanced information has been distributed in an unprecedented fashion. In fact, probably the most significant aspect of the new CDM information infrastructure is that the airline operations centers receive the same information as the FAA ATCSCC specialists. Such information is critical in enabling airline operations specialists to plan responses to changing conditions and possible FAA control actions. Previously, such information was not available to airline operations planners or was only available "after-the-fact", when it could no longer be used to influence decision-making. Analysis (Ball et al. 2000) has shown that the information generated by the CDMnet can yield substantial improvements in the information quality.

7.2.3.2 Visualization

The information infrastructure outlined above gives traffic managers the data and tools they need to anticipate and alleviate problems in the air transportation system. The traffic situation display (TSD), for example, is the main user interface to ETMS and allows FAA traffic managers access to the ETMS data. The TSD displays a variety of data and provides numerous commands for manipulating the data as well as the display. In particular, the TSD provides a graphical display of a geographic area with overlays chosen by the user; for example, flights, weather, alert information, flow-constrained areas, reroutes, airports, sector boundaries, and navigational aids.

The TSD, however, is only available at the ETSM field sites; as a result, the resulting information is not available for airlines or FAA facilities that do not have direct access to the ETMS. To improve information sharing and common situational awareness, the common constraint situation display (CCSD) was created. The CCSD is a web-based tool that can easily be accessed over the CDMnet, and allows airlines to view a graphical display of information that they can use to monitor the state of the NAS and to manage their operations. The data shown on the CCSD are largely the same as the data on the TSD; certain information, such as data on sensitive flights, lightning data, and aircraft icons is omitted.

7.2.3.3 Prediction and Planning

In addition to monitoring systems such as TSD and CCSD, the information infrastructure also provides a number of predictive and planning tools to support

TFM decision-making. The ETMS Hub, for example, uses flight intent information to predict the number of aircraft that will arrive at airports, depart from airports, and occupy adapted volumes of airspace for 15-min intervals over the next several hours. If the predicted traffic counts exceed adapted thresholds, an alert is generated. The prediction of traffic counts and the generation of alerts are a part of what is known as the Monitor Alert capability, and the resulting information can be displayed in the TSD and CCSD.

CDM efforts have also introduced an alternative way to forecast convective activity to aid in strategic weather-related TFM initiatives. The so-called collaborative convective forecast product (CCFP) is different from any other weather forecast product currently available to TFM decision makers. The CCFP is a consensus forecast in which CDM-participating airlines, Center Weather Service Units, and meteorologists from the Aviation Weather Center all contribute to its production. The underlying objective of creating a consensus forecast is to improve convective forecasting and alleviate the problem of conflicting forecasts by producing a single source of convective weather information for use by all traffic planners. The CCFP forecasts have become a vital component used in TFM decision-making: analysis has shown that the use of CCFP does lead to better weather-driven flight plan decisions. Improvements in route planning have reduced weather-impacted airspace congestion, and flights are less likely to incur airborne tactical reroutes.

Without a doubt, however, the FSM is the main decision support system introduced under CDM. FSM houses the logic for initial resource allocation and for subsequent manipulation during GDPs, including both the RBS schedule and compression algorithms. The major features of FSM are monitoring of demand and departure time compliance, generation of flight delay statistics, planning GDPs (including a "Power-Run" feature that allows the specialist to review the statistical consequences of the various alternatives for parameter settings in forming a GDP), and the issuance of ground delays and GDP parameter settings for implementation of a GDP. One of the innovations of CDM is that FSM software is available and used by the AOCs as well as the FAA. This provides airlines with the same situational awareness as the FAA, which is one of the cornerstones of successful distributed decision-making.

7.3 Models for Air Traffic Flow Management

The explosive growth in air traffic has motivated a considerable amount of research that considers the application of operations research models to ATFM. It is perhaps not surprising that the interest in operations research applications to support flow management has largely coincided with the problems caused by increased congestion in the air transportation system. The use of decision models to support ATFM received relatively little attention prior to the 1980s, and most of the literature dates after the emergence of formal flow management procedures that followed the air traffic controllers' strike in 1981. This section provides a comprehensive classification and overview of operations research models that have been proposed in support of ATFM.

To provide a more coherent overview of the models we consider, we distinguish between *prescriptive* and *descriptive* models. Prescriptive models are usually optimization models, which suggest a course of action for the decision maker(s). These are perhaps the most widely studied and best-known applications of operations research for ATFM. Descriptive models, on the other hand, are models that aim to analyze and/or predict the overall behavior and key characteristics of the air transportation system. Usually, this entails the use of simulation or queueing models; alternatively, this also encompasses models that are used to support the development of performance metrics.

7.3.1 Prescriptive Models

Prescriptive models can be classified according to the time horizon of the flow management application they address. A systematic description of the resulting categories is given by Odoni (1987), who classifies ATFM initiatives as long-, medium-, or short-term:

- Long-term approaches typically focus on increasing capacity. Examples include the construction of additional airports (which could take 10–15 years), the introduction of new technologies (e.g., satellite-based navigation tools), and the addition of runways to existing airports. Although effective, such initiatives are usually very costly and difficult to implement. Airport expansions, for instance, frequently encounter the resistance of local communities and other special interest groups, who might be concerned with noise, real estate depreciation, and other factors; moreover, they are usually subject to strict environmental regulations.
- Medium-term approaches are mostly administrative or economic in nature, and try to alleviate congestion by modifying spatial or temporal traffic patterns. At some airports, for example, flight schedules are coordinated bi-annually according to IATA guidelines (IATA 2000). In other cases, the aviation authorities negotiate a schedule reduction with carriers (Aviation Week 2004). Similar medium-term approaches include the use of slot lotteries at New York's LaGuardia Airport (DeCota 2000), as well as current proposals for slot auctions and/or congestion pricing (DeCota 2000).
- Short-term approaches consider the strategic adjustment of air traffic flows to match available capacity, and typically span a planning horizon that is about 12 h in the United States and 48 h in Europe. These operational ATFM initiatives attempt to mitigate the unavoidable congestion that arise from unforeseen and unpredictable disruptions as efficiently as possible. Such periods of congestion arise frequently when bad weather causes sudden capacity reductions.

Without a doubt, most of the Operations Research methods have focused on optimization models that address short-term initiatives. Prescriptive models dealing with medium- and long-term initiatives have received considerably less attention, and have primarily considered the use of auctions and congestion pricing for demand management in the medium term. Optimization models that address medium- and long-term initiatives other than demand management are scarce, and concentrate on the design of airspace and route networks. Barnier and Brisset (2004) have considered the problem of flight level allocation to aircraft flows in order to ensure their separation, which they represent as a graph-coloring problem. Other research (Fondacci et al. 1998; Mehadhebi 2000) has focused on the design of a network of predefined routes such that ATC conflicts are minimized.

In the remainder of this section, we begin with a brief overview of models that address demand management. Subsequently, we discuss models that address short-term initiatives in more detail. First, we consider airport capacity allocation models, which are local in nature. Subsequently, we present an overview of airspace allocation models, which are more global in nature in that they explicitly address the interconnectivity in the air transportation system.

7.3.1.1 Demand Management

Within the context of ATFM, demand management refers to any set of administrative and/or economic policies that aim to constrain the demand for access to airspace elements during certain times when congestion would otherwise occur (Barnhart et al. 2006). The practice of demand management has been more common outside of the United States; within the United States, only four airports (New York's LaGuardia and Kennedy airports, Chicago O'Hare airport, and Washington Reagan National Airport) in the United States have limits on the number of hourly flight movements under the so-called "high-density rules" (HDR). The administrative policies that have been used to impose schedule restrictions have been subject to frequent and widespread criticism. The criteria for allocating airport capacity (which are predominantly based on historical precedent) have been said to stifle competition. Moreover, the difficulty in establishing appropriate capacity limits creates the possibility that valuable capacity is wasted, while the absence of economic criteria hides the true value that users place on airport capacity (see Ball et al. 2005a) for a comprehensive discussion). Auctions and congestion pricing provide an alternative to administrative options, and research in this area has attempted to address these issues.

The use of auctions to allocate airport landing slots was first considered by Grether et al. (1979) [later published as Grether et al. (1989)], who proposed a sealed-bid auction mechanism followed by an automated after-market. Rassenti et al. (1982) explored the use of combinatorial sealed-bid package auctions, where airlines would be able to bid for combinations of airport landing and departure slots. This initial research was motivated by the passage of the Airline Deregulation Act in 1978 and the resulting expectation that, due to new entrants, airline slots might become an increasingly scarce resource. Similarly, recent research has been motivated by the so-called AIR-21 legislation of 2000, which has called for the gradual elimination of HDR-imposed slot limits. Ball et al. (2005a) present a framework for airport slot auction design in the United States, which is further analyzed in Ball et al. (2007). Other recent studies (DotEcon 2001; NERA 2004) propose and analyze the use of slot auction schemes at European airports. It is also interesting to note that auction-based approaches have also been considered for the real-time allocation of airport landing slots (Hall 1999; Cholankeril 2003).

Congestion pricing presents a more reactive approach to demand management. There are no predefined slots, and users can land at any time as long as they pay the appropriate congestion fee. The use of congestion pricing, and the resulting problem of determining landing fees, was first proposed by Levine (1969) and by Carlin and Park (1970). Daniel (1995) models and estimates equilibrium congestion prices at hub airports, while more recent proposals for the implementation of congestion fees can be found in Fan (2003) and Ball and Hoffman (2005). Other research is of a more descriptive nature; Fan and Odoni (2002) and Hansen (2002) use queuing models to estimate external delay costs at different airports, and show that these costs are substantially higher than the fees that are currently charged. Another line of such research (Pels and Verhoef 2003; Fan 2003; Bruckner 2003) considers complications that arise when carriers operate multiple flights. If this occurs, a certain amount of the external delays are internal to the carrier, who should not be charged fees for the delay it already internalizes. For a review we refer to Schank (2005), who also examines why congestion pricing has never been effectively implemented. Through a number of case studies, Schank highlights the political and social equity problems that arise: pricing schemes often discriminate against a particular group of aircraft, which will lead to substantial opposition including legal challenges. A particular concern is that congesting pricing may exclude these groups without providing alternatives, which leads Schank to argue that congestion pricing cannot be considered in isolation, without considering substitutes in the broader transportation system.

7.3.1.2 Airport Capacity Allocation Models

Two consecutive operations on one airport runway, such as aircraft landings and takeoffs, require a minimum amount of time gap in between them. The time gap, which is imposed due to safety considerations, mainly depends on the type of aircraft involved in the two operations, meteorological conditions at the time operations take place, and the type of operations—whether they are landings, takeoffs, a landing followed by a takeoff, or a takeoff followed by a landing (see Horonjeff and McKelvey 1994) for details on separation standards). The time separation standards impose an upper bound on the number of operations that can be handled by each runway at an airport. The upper bound on the number of

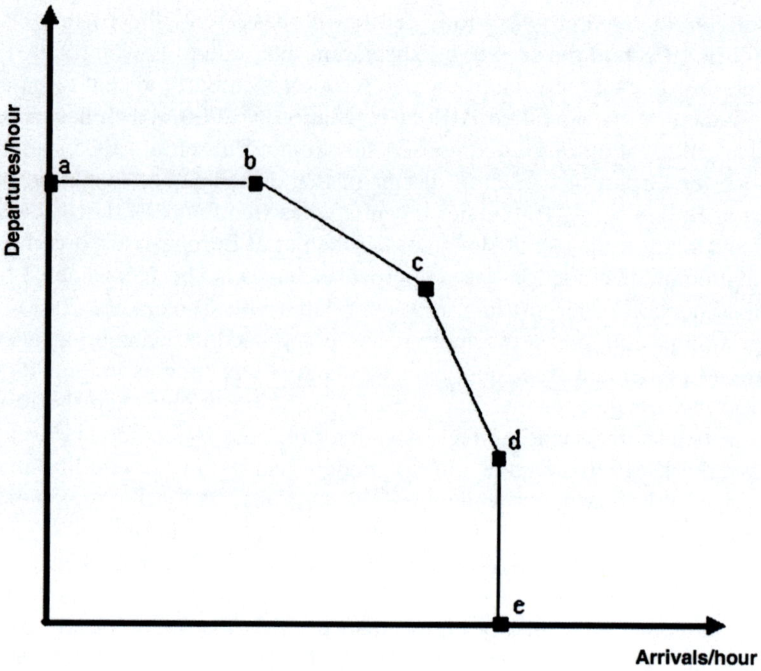

Fig. 7.7 Capacity envelope for a single runway

operations that can occur simultaneously on all available runways at an airport is defined as the instantaneous airport capacity. Sometimes, if two runways are close to each other, such as those at the San Francisco International Airport (SFO), the separation standards on the operations on both runways are not independent, and hence impose additional limitations on the total airport capacity.

In theory, the capacity of an airport at any time is the instantaneous rate of operations that the airport can handle. In practice, however, airport capacity is measured as the upper bound on the total number of operations in discrete time intervals, typically in units of an hour. Airport capacity is further divided into arrival and departure capacities. In many airports, in which runways are close to each other, or there are crossing runways, the arrival and departure capacities are interdependent, and are typically connected through a piecewise linear convex functional relationship (Barnhart et al. 2006; Gilbo 1993). Each point on such a frontier gives the combination of arrival and departure capacities at which an airport can operate during any time interval. Figure 7.7 shows such a relationship for operations on a single runway. Points "a" and "e" indicate the runway capacity when it is used only for departures and only for arrivals respectively. Point "d" represents a higher efficiency point where certain numbers of departures take place without hindering the arrival capacity, i.e., without increasing inter arrival spacing. Similarly, point "b" indicates the conditions when arrivals can take place without reducing departure capacity. Between points "b" and "d" are those where the runway capacity is shared among arrivals and departures. Here, increasing the arrival throughput can only be attained by reducing departures. Gilbo (1993) provided a methodology to estimate such a capacity envelope for an airport, using historical data reporting the number of arrivals and departures during various time intervals.

Meteorological conditions, particularly convective weather, significantly impact the capacity of an airport. Other forms of adverse weather such as fog, snow/ice, low cloud ceiling, and poor visibility can also reduce airport capacities significantly. Reduction in airport capacity can induce delays among flights. A flight can face delay on the ground before it departs from its origin airport, which is commonly known as ground delay. A flight can also face airborne delay while it is en-route to its destination. Airborne delays typically occur in the form of rerouting aircraft, increasing aircraft spacing, reducing aircraft speed, and airborne holding. Airborne delay, although quite common in the NAS during adverse weather conditions, is undesired mainly due to expensive fuel consumption and safety issues related to controlling aircraft and maintaining separation standards while they are airborne. This has motivated researchers to develop optimization models and algorithms to decide on the delays that must be imposed on various flights when there is capacity shortfall at one or more airports.

Optimization models for allocating airport capacity can be classified into two categories. The first set of models focuses on sequencing arriving (or departing) flights to improve runway utilization. Typically, the scope of the runway sequencing models is in the order of few minutes. In the second category of models, the focus is towards allocating airport capacity, during discrete time intervals, among aircraft scheduled to arrive at, and depart from, the airport. The airport capacity values, which these models require as input, are the estimated maximum number of operations, during various time intervals, without considering the exact sequence and separation constraints of such operations. The scope of such models extends up to a few hours.

In general, the problem of sequencing and scheduling arrivals at a runway is defined as follows: given a set of aircraft F that are in the vicinity of an airport, the earliest and latest times—E_f and L_f respectively—that a flight $f \in F$ could land at a runway, and the minimum gap τ_{ij} required between the landing times of two consecutive aircraft $i, j \in F$, the problem is to assign landing times to individual aircraft. Maximizing runway throughput and minimizing total flight delay are two most common objectives used in these models; although other variants of the objective function exist in the literature. Soomer (2008) provides an excellent survey of models for the sequencing and scheduling runway operations. Dear and Sherif (1991) addressed equity by limiting the number of positions by which an aircraft can deviate from its first-come-first-served (FCFS) position in the original sequence. Beasley et al. (2000) provides mixed-integer LP formulation for sequencing landings at multiple runways. The arrival sequencing problem becomes complicated when the pool of flights changes over time. This happens when new flights arrive at or near the airport TRACON and join the landing queue, or when an aircraft actually lands, and hence no longer requires sequencing. Sequencing arrivals dynamically is considered in Beasley at al. (2004). Computational intractability of large instances of sequencing problem has motivated researchers to propose heuristics to achieve acceptable solutions fast. Erzberger (1995) presented an algorithm for real-time scheduling of landings at a runway. The Center TRACON Automation System (CTAS) software, developed by NASA, implements runway sequencing algorithms in real time (see Davis et al. 1997). In recent years, researchers have focused on scheduling departure operations at a runway. Rathinam et al. (2009) extended the dynamic programming formulation for aircraft arrival sequencing problem (Psaraftis 1980) to schedule departures. Gupta et al. (2009) proposed a MILP for the deterministic departure scheduling problem, while considering multiple objectives, and providing strong valid inequalities to enhance computational efficiency.

In the remainder of this section, however, we focus on the second class of problems, which allocate airport capacity over a longer time horizon. This problem is known as the Ground Holding Problem (GHP), and aims to assign ground delays to various flights given the airport capacity constraints. Typically, the objective is to minimize the sum of ground and airborne delays weighted by their relative costs. Past research has focused on a special case (or sub-category) of the GHP—the single-airport ground holding problem (SAGHP). In this problem it is assumed that only one airport in the system faces capacity reduction, and that all other resources in the NAS have either unlimited capacity or do not impose any capacity constraints. Moreover, only the arrival demand-capacity imbalance at the destination airport is addressed in this problem. The goal in the SAGHP is to assign ground delays to flights that are bound for the destination airport, in order to mitigate the arrival demand-capacity imbalance.

Deterministic Models. In a deterministic setting, where the airport capacities during future time intervals are known in advance, the SAGHP can be formulated as a minimum-cost network flow problem (Terrab 1990; Terrab and Odoni 1993; Richetta and Odoni 1993; Hoffman and Ball 2000). The problem formulation is given below.

Let F be the set of flights bound for an airport that faces capacity shortfall. Let a_f be the scheduled arrival time of a flight $f \in F$. Let the time horizon be discretized into time periods of equal duration $t \in 1, \ldots, T$. Each time period could represent, say, a 15-min interval. Let M_t be the capacity of the airport for each time period t. As mentioned before, in the deterministic setting, it is assumed that the value of M_t is known in advance. The decision variables, represented by X_{ft}, are defined as follows:

$$X_{ft} = \begin{cases} 1, & \text{if flight } f \text{ is assigned to time interval } t; \quad \text{and} \\ 0, & \text{otherwise.} \end{cases}$$

The amount of ground delay faced by a flight f is given by the expression $\sum_{t=a_f}^{T} (t - a_f)X_{ft}$. Non linear ground delay cost functions can be easily incorporated, and, for each flight, are given by the expression $\sum_{t=a_f}^{T} g(t - a_f)X_{ft}$, where

$g(\cdot)$ is any arbitrary cost function. For example, a quadratic ground delay cost function is given by $\sum_{t=a_f}^{T}(t-a_f)^2 X_{ft}$. The ground delay costs can also be weighted by flight-specific constants C_f, as defined by Hoffman and Ball (2000). The generalized objective function for the deterministic ground holding problem is given by:

$$\text{Min.} \sum_{f \in F} \sum_{t=a_f}^{T} C_f g(t-a_f) X_{ft}. \tag{7.1}$$

Note that the objective function remains linear although the ground delay cost is non linear. Therefore the problem can be solved as a linear integer program. The set of constraints is defined as follows:

$$\sum_{t=a_f}^{T} X_{ft} = 1 \quad \forall f \in F \tag{7.2}$$

$$\sum_{f \in F : a_f \leq t} X_{ft} \leq M_t \quad \forall t \tag{7.3}$$

$$X_{ft} \in \{0,1\} \quad \forall f, t. \tag{7.4}$$

Constraint set (7.2) ensures that each flight is assigned to exactly one time interval, while constraint set (7.3) imposes the capacity restriction on the number of aircraft that arrives during each time period. The constraint matrix of the above formulation is totally unimodular, and hence the LP relaxation of the integer program guarantees integer solutions.

Variants. There are several variants of the deterministic single-airport ground-holding problem. Hoffman and Ball (2000) added a set of *banking constraints* to the above formulation. Some airlines schedule *banks* of operations at their hub airports, meaning a group of flights whose arrival (or departure) times fall within a specified time window. Such temporal grouping of flights facilitates the transfer of passengers, baggage, and airline crews in a hub-and-spoke system. For example, suppose the passengers from three flights A, B, and C are scheduled to transfer at the hub to another flight D, of the same airline, with a common destination to them all. For this to happen A, B, and C must arrive prior to D, and preferably form an arrival bank so that there is a short waiting time for passengers during the transfer. The banking constraints are added to the ground-holding problem in order to keep flights of each bank temporally grouped. Banking constraints worsen the computational performance—i.e. time taken to obtain integer optimal solutions—because the LP relaxation of the problem rarely yields integer solutions. In their paper, Hoffman and Ball present and analyze several alternative sets of banking constraints, and show that the resulting models can be solved efficiently. The input requirements of this model, however, complicate its application within the CDM framework.

Most of the optimization models for the SAGHP address only the arrival capacity shortfall at an airport, and decide on the amount of ground delay to impose on various incoming flights. However, adverse weather also affects the departure capacity of an airport. Furthermore, as mentioned before, arrival and departure capacities are usually interdependent. Gilbo (1993) presented an optimization model to assign ground delays to both arrival and departure traffic at an airport. His model not only solves for ground delays of aircraft, but also optimal allocation of airport capacity to arrival and departure operations. Dell'Olmo and Lulli (2003) also addressed a similar problem, and provided a dynamic programming algorithm to enhance the efficiency in achieving optimal solutions. Gilbo (1997) extended his previous formulation to consider arrival and departure fix capacities at an airport along with that of its runways.

Exempting long-haul flights from a ground delay program (GDP) raises equity issues in slot allocation, and might result in a systematic bias towards airlines operating long-haul flights. Flights are exempted to hedge against uncertainty in the airport capacity. (Sometimes exemptions also occur if certain flights are already airborne when a ground delay program is initiated.) Delaying long-haul flights can lead to unnecessary delays in the system if weather clears earlier than what was forecast. Therefore, flight exemptions during ground delay programs can be seen as a technique to increase efficiency (or throughput) of the system. We will present models for the ground holding problem that explicitly incorporate uncertainty later in this section. Here, we discuss a deterministic optimization model, presented by Vossen et al. (2003), to mitigate exemption bias during ground delay programs. In essence, they minimize the deviation between the slot assigned to each flight and the ideal slot of a flight. The ideal slots are determined by applying the ration-by-schedule (RBS) algorithm, which is based on the first-scheduled-first-served principle, to all flights without any exemptions. The model and some results from their paper are discussed below.

Let A be the set of airlines, and b_a be the number of slots owned by each airline $a \in A$. Let there be n slots to be allocated. It is assumed that the total number of slots available is at least equal to the total number of slots to be assigned to all airlines, i.e. $n = \sum_{a \in A} b_a$. Each slot $j \in \{1, 2, \ldots, n\}$, has a time stamp associated with it, and is denoted by s_j. (Note that $s_i < s_j$ if $i < j$.) Each airline must be assigned b_a slots, denoted by $j'(a,1), j'(a,2), \ldots, j'(a,b_a)$. The kth slot of an airline is denoted by $j'(a,k)$. This slot ordering is consistent with the slot time ordering, i.e. $s_{j'(a,i)} < s_{j'(a,k)}$ if $i < k$. Let the ideal slot time for the slot $j'(a,k)$, which can be determined by executing the unconstrained ration-by-schedule algorithm, be denoted by a binary indicator p defined as: $p_{jak} = 1$ if it is the ideal slot for $j'(a,k)$; 0 otherwise. Furthermore, let $e(a, k)$ and $l(a, k)$ be the earliest and latest slots that can be assigned to $j'(a,k)$. Due to exemptions, certain slots of an airline a, denoted by the set $FA(a)$, are allocated a-priori. The decision variables are defined as follows:

$$x_{jak} = \begin{cases} 1 & \text{if slot } j \text{ is the } k\text{th slot allocated to airline } a; \quad \text{and} \\ 0 & \text{otherwise.} \end{cases}$$

The objective of the assignment model is to allocate slots to airlines so that the squared deviation between the slots allocated and their ideal locations is minimized. The objective function and the set of constraints are defined as follows:

$$\text{Min.} \quad \sum_{jak} \left(j - p_{jak}\right)^2 x_{jak} \tag{7.5}$$

subject to

$$\sum_{ak} x_{jak} \leq 1 \quad \text{for all slots } j \text{ where } j \notin FA(a) \text{ for some } a \tag{7.6}$$

$$\sum_{k} x_{jak} = 1 \quad \text{for all } a \text{ and } j \in FA(a) \tag{7.7}$$

$$\sum_{j:e(a,k) \leq j \leq l(a,k)} x_{jak} = 1 \quad \text{for all } a \text{ and } k \tag{7.8}$$

$$x_{jak} \in \{0, 1\} \quad \text{for all } a, k \text{ and } j. \tag{7.9}$$

Constraints (7.6) and (7.7) ensure that each slot is assigned to at most one pair and also that each fixed (or exempt) slot is allocated to the appropriate airline. Constraint (7.8) imposes lower and upper bounds on the slot time assigned to all non-exempt slots. Vossen et al. tested how their model reduced exemption bias in eight ground delay programs during the first 4 months in 2001 at Boston's Logan Airport (BOS). The leftmost graph in Fig. 7.8 shows for eight different airlines, the difference between the average delay under unconstrained ration-by-schedule allocation and under RBS with exemptions (a negative number on the vertical axis means that an airline received more delays due to exempting flights of other airlines than it should under unconstrained RBS—i.e., if exemptions were not made). Clearly, flight exemptions have significant impact on the distribution of delays. For example, US Airways (USA) received, on an average, 11.7 min of excess delay due to flight exemptions. The rightmost graph in Fig. 7.8 shows a similar plot after the assignment model to mitigate exemption bias was applied to all eight GDPs. The effect is evident from the plot. The average excess delay per flight faced by the US Airways was reduced to 3.4 min.

Stochastic Models. Deterministic models for ground holding are applicable when airport capacities in future hours are known in advance with perfect information. In reality this is rarely the case. As pointed out by Odoni (1987), the problem of assigning ground delays is both stochastic, because capacity forecasts have a certain degree of uncertainty, and dynamic, because the forecasts update frequently and provide new information about how the weather conditions at an airport are changing. The problem of assigning ground delays to aircraft with uncertain airport capacity is known as the stochastic ground holding problem. Research on the stochastic ground holding problem has been mostly limited to the single airport case, although Vranas et al. (1994a, b) present some treatment of the multi-airport case. Moreover, most of the research has focused on capacity

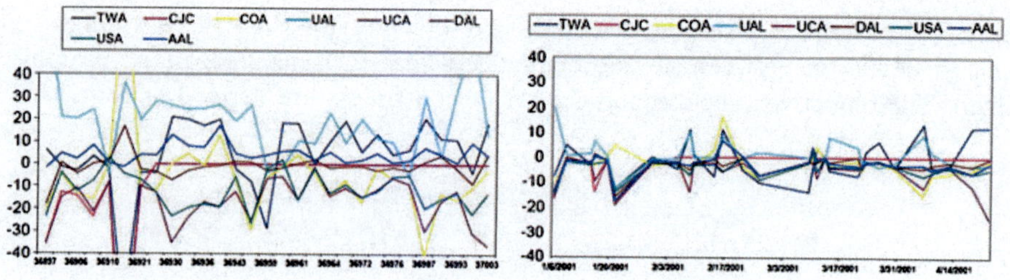

Fig. 7.8 Impact of the optimization model at Boston Logan Airport (source: Vossen et al. 2003)

uncertainty. There is a limited amount of literature (Ball et al. 2001; Willemain 2002) that focuses on the uncertainty in arrival demand, rather than capacity, in solving the ground holding problem.

To our knowledge, Andreatta and Romanin-Jacur (1987) were the first to propose a math-programming model for the SAGHP under uncertainty in airport arrival capacity. However, their formulation was limited to a single time period case. The problem they addressed is defined as follows. A set of flights is scheduled to arrive during a specified time interval at the destination airport. The airport arrival capacity during that time period is uncertain, and the probability distribution of possible capacity values is given. Also given are the unit ground and airborne delay costs of each flight. It is assumed that the airport has infinite capacity in the next time period, and therefore any flight can face a ground delay of at most one time period. The objective is to minimize the expected delay costs, which consist of both the deterministic ground delays and the uncertain airborne delays, by determining which flights are assigned a ground delay. In their paper, the authors also propose a recursive algorithm to assign ground delays to flights when a vector of priority numbers, representing the order in which the flights are to be processed, is given. They also propose an optimal priority rule, which depends only on the airborne delay costs of flights, that minimizes the expected delay cost. Terrab and Odoni (1993) address the same problem and propose a closed form solution for the strategy of assigning ground delays that minimizes the total expected delay costs. They also extend the dynamic programming algorithm for the single-period case, proposed by Adreatta and Romanin-Jacur (1987), to the multi-period case. Because the dynamic programming approach was computationally burdensome, Terrab and Odoni proposed several heuristics to achieve approximate solutions to the problem.

Richetta and Odoni (1993) first proposed an integer programming model to solve the multi-period static stochastic SAGHP. In their model, uncertainty in airport capacity is represented by a finite set of scenarios, each of which represents a time-varying profile of the airport capacity that is likely to occur. The goal is to assign ground delays to flights, given uncertainty in airport capacity, in order to minimize the total expected delay cost. The model formulation is given below.

As in most of the discrete optimization models, the planning horizon is divided into equal time periods. Let there be Q capacity scenarios, each scenario depicting a possible evolution of airport arrival capacity over the planning period with the scenario $q \in \{1, \ldots, Q\}$ having a probability of occurrence equal to p_q. Let M_t^q denote the capacity at time period t under the scenario q. In order to ensure that all flights that are scheduled to land get assigned a landing slot during a time-period, let there be a time-period $T + 1$ with unlimited capacity under all scenarios. In their model, Richetta and Odoni classified the flights that are scheduled to arrive during each time-period into K cost classes. Let N_{kt} denote the number of flights, belonging to cost category k, that are scheduled to arrive at the airport during the time period $t \in 1, \ldots, T$. The cost of ground holding a flight of class k for i time units is denoted by the cost function $G(k, i)$. As will be shown later, this cost function is capable of capturing non-linear ground delay costs for flights, while keeping the objective function linear in decision variables. Let C_a denote the unit cost of airborne holding for all flights.

The decision variables are denoted by X_{ktj}^q, which indicate the number of flights in class k scheduled to arrive during time-period t that are reassigned to arrive during time-period j under capacity scenario q. Let W_t^q denote the number of aircraft that are unable to land during time-period t under scenario q, and hence face airborne holding during that time period. The objective function minimizes the total expected cost of ground and airborne delays. The integer program is given as follows:

$$\text{Min.} \quad \sum_{k=1}^{K} \sum_{t=1}^{T} \sum_{j=t}^{T+1} C_g(k, j-t) X_{ktj} + C_a \left\{ \sum_{q=1}^{Q} p_q \sum_{t=1}^{T} W_t^q \right\} \tag{7.10}$$

subject to

$$\sum_{j=t}^{T+1} X_{ktj} = N_{kt} \quad k = 1, \ldots, K; \quad t = 1, \ldots, T \tag{7.11}$$

$$\sum_{k=1}^{K} \sum_{j=1}^{t} X_{kjt} + W_{t-1}^q - W_t^q \leq M_t^q \quad q = 1, \ldots, Q; \quad t = 1, \ldots, T+1 \tag{7.12}$$

$$W_0^q = W_{T+1}^q = 0 \quad \forall q = 1, \ldots, Q \tag{7.13}$$

$$X_{ktj} \geq 0 \quad \text{and integer}, \quad k = 1, \ldots, K; \quad t = 1, \ldots, T; \quad j = 1, \ldots, T+1 \tag{7.14}$$

$$W_t^q \geq 0 \quad \text{and integer}, \quad t = 1, \ldots, T; \quad q = 1, \ldots, Q. \tag{7.15}$$

Constraint set (7.11) ensures that all flights scheduled to arrive during any time period get rescheduled to land before the end of planning horizon. Constraint set (7.12) imposes an upper bound on the number of aircraft that can land during a time period under different scenarios.

Kotnyek and Richetta (2006) showed that the constraint matrix of the above formulation is not totally unimodular, and in some cases the LP relaxation to the above IP will not yield integer solutions. However, if all flights belong to only one cost category, and if the ground delay cost function is monotonically increasing, the Richetta–Odoni model guarantees integer solutions [see Kotnyek and Richetta (2006) for details]. Monotonically increasing cost functions are of particular interest because they tend to give priorities to flights based on their schedule. This is consistent with the ration-by-schedule (RBS) slot allocation policy, which is widely accepted as an equitable method.

For practical implementation of a static stochastic model for the SAGHP under CDM, it is desirable to obtain a set of planned arrival numbers, which are the number of flights rescheduled to arrive in each time period, under different scenarios. Kotnyek and Richetta (2006) showed that the optimum set of planned arrivals can be easily obtained from the decision variables in the Richetta and Odoni (1993) model. Ball et al. (2003) proposed a simple model, formulated as an integer program, to achieve these numbers, which they name as the Planned Airport Arrival Rates (PAARs). The advantages of their model are as follows: (1) one can use the PAARs and assign slots to flights by using any algorithm of choice. For example, one can apply pure RBS to all flights, or, exempt certain flights and then apply RBS to the remaining flights. The objective function remains invariant to the slot allocation method, which is not the case in the Richetta and Odoni (1993) model; (2) the objective function also remains unchanged after flight substitutions that can take place post the slot assignment process; and (3) the constraint matrix of their formulation is totally unimodular, and hence does not pose a computational burden in solving problems of realistic size. Below, we present the Ball et al. model (wherever possible, we use the same notations as that in the Richetta and Odoni model presented above).

Let N_t be the number of flights scheduled to arrive during time interval $t \in 1, \ldots T+1$. Let λ denote the cost ratio between unit airborne and ground delay of any flight. The PAARs are the decision variables, and are denoted by A_t.Let G_t denote the number of flights that are delayed on the ground from time-periods t to $t+1$.The objective function and the set of constraints are defined as follows:

$$\text{Min.} \quad \sum_{t=1}^{T} G_t + \sum_{q=1}^{Q} p_q \sum_{t=1}^{T} \lambda W_t^q \tag{7.16}$$

subject to

$$A_t - G_{t-1} + G_t = N_t \quad t = 1, \ldots, T+1, \quad (G_0 = G_{T+1} = 0) \tag{7.17}$$

$$A_t + W_{t-1}^q - W_t^q \leq M_t^q \quad t = 1, \ldots, T+1, \quad q = 1, \ldots, Q\left(W_0^q = W_{T+1}^q = 0\right) \tag{7.18}$$

$$A_t,\ G_t,\ W_t^q \geq 0 \quad \text{and integer} \quad t = 1, \ldots, T+1, \quad q = 1, \ldots, Q. \tag{7.19}$$

One can easily solve the above integer program, and obtain the optimal PAARs, which can then be used to assign slots to individual flights based on a certain preference rule. Under the CDM paradigm, airlines can then use the slots assigned

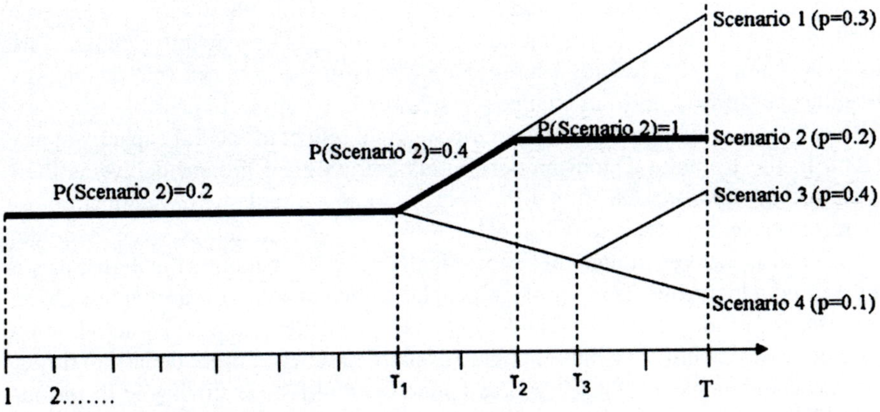

Fig. 7.9 Scenario tree of evolving airport arrival capacity

to them as their assets and perform cancelations and intra-airline substitutions. After a round of cancelations and substitutions, airlines can provide an updated schedule of their flights, and at that point in time, the Ball et al. model can be re-applied with possibly updated probabilities and capacity scenarios. Such repetitive application of the static stochastic model is a partially dynamic approach to solving the SAGHP.

Richetta and Odoni (1994) first developed a multi-stage stochastic integer program with recourse for the SAGHP. Such a model not only accounts for uncertainty, but also utilizes updated information about capacity changes in the decision-making. As in the static models, uncertainty in airport capacity is represented by a finite set of scenarios. The scenarios are arranged in a probabilistic decision tree, which is commonly known as a scenario tree. Each branch of the tree represents a capacity scenario or a group of scenarios realized as the time of day progresses. A capacity scenario corresponds to a possible time—varying arrival capacity profile. For example, Fig. 7.9 shows a scenario tree for which there are four possible scenarios of airport arrival capacities. The probabilities of occurrence are shown in parentheses. The time of day is represented by the horizontal axis, which is divided into T equal time periods. All scenarios are possible at the beginning of the day, but at time-period T_1, either scenarios 1 and 2 or scenarios 3 and 4 are eliminated as possibilities. Similarly at time T_2, scenario 1 or 2 is realized if the upper branch evolves. If scenarios 3 or 4 evolve, no new capacity information becomes available until time T_3. The marginal probabilities of the scenarios change with time, as specific branches of the scenario tree are realized, but the conditional probabilities remain the same. For example, the bold line in Fig. 7.9 represents evolution of Scenario 2. Prior to T_1, the marginal probability is 0.2. At time T_1, this probability changes to 0.4 if the upper branch of the tree is realized, and goes to 0 if the lower branch is realized. Similarly, if the upper branch is realized, the probability of Scenario 2 goes to 1 or 0 at T_2.

The scenario tree reveals the availability of information about airport operating conditions. The information is based on forecasts, so that capacity changes are anticipated before they occur. If branching points in the scenario tree occur only when the operating conditions change physically (for example, at possible fog burn-off times) the active branch of the scenario tree will reflect the actual capacity at any instant. In the Richetta–Odoni model, flights are assigned ground delays as their departure time approaches, so that the decisions can be made with the most up-to-date information about capacity forecasts. However, ground delays, once assigned, cannot be revised, even though this is technically possible so long as the flight has not yet departed. On the one hand, this results in less efficient solutions as there might be unnecessary ground delays that can be recovered by appropriate revision, while on the other hand it results in a higher degree of predictability of flight departure times. Mukherjee and Hansen (2003, 2004) extended the multi-stage stochastic IP formulation of Richetta and Odoni, and proposed a dynamic model for the SAGHP that can revise ground delays of flights in response to updated information on airport capacity. Their formulation is discussed below (wherever possible, we will use the same notation in the static models presented above).

Let Φ be a set of flights that are scheduled to fly to an airport for which a ground holding program is necessary. As in the static model, the time of day is divided into a finite set of time periods of equal duration. The scheduled departure and arrival times of a flight f are denoted by d_f and a_f respectively. The scenario tree is input to the model through the following variables. Let B $(B \geq Q)$ be the total number of branches in the scenario tree. Each branch corresponds to a set of scenarios. The scenarios corresponding to branch $b \in \{1,...,B\}$ are given by the set $\Omega_b = \{S_1^b,...,S_k^b,...,S_{\pi_b}^b\}$; $S_k^b \in \{1,...,Q\}$. We assume that each branch has start and end nodes corresponding with the beginning of time periods. The time periods corresponding to start and end nodes of branch b are denoted by o_b and μ_b, $b \in \{1,...,B\}$. The decision variables in the model are binary variables defined as follows:

$$X_{ft}^q = \begin{cases} 1 & \text{if flight } f \text{ is planned to arrive during time-period } t \text{ under scenario } q\text{; and} \\ 0 & \text{otherwise.} \end{cases}$$

Corresponding to the X_{ft}^q is a set of corresponding auxiliary variables for the departure time period. Specifically we define:

$$Y_{ft}^q = \begin{cases} 1 & \text{if flight } f \text{ is released for departure during time-period } t \text{ under scenario } q\text{; and} \\ 0 & \text{otherwise.} \end{cases}$$

The departure release variables track the planned arrival times but are displaced earlier in time by the amount $a_f - d_f$. Hence the variables Y_{ft}^q are related to X_{ft}^q as follows:

$$Y_{ft}^q = X_{ft+a_f-d_f}^q \in \{d_f,...,T\}, q \in \{1,...,Q\}. \tag{7.20}$$

The objective function and the set of constraints are given as follows:

$$\text{Min.} \quad \sum_{q=1}^{Q} p_q \left\{ \sum_{f\in\Phi} \sum_{t=a_f}^{T+1} (t - a_f) X_{ft}^q + \sum_{t=1}^{T} \lambda W_t^q \right\} \tag{7.21}$$

subject to

$$\sum_{t=a_f}^{T+1} X_{ft}^q = 1 \quad f \in \Phi, \quad q \in \{1, \ldots, Q\} \tag{7.22}$$

$$\begin{aligned} &\underset{t-1}{\overset{q}{W}} - W_t^q + \sum_{f\in\Phi:a_f \le t} X_{ft}^q \le M_t^q \quad t \in \{1, \ldots, T+1\}; \quad q \in \{1, \ldots, Q\}; \\ &\left(W_0^q = W_{T+1}^q = 0\right) \end{aligned} \tag{7.23}$$

$$\begin{aligned} &Y_{ft}^{S_1^b} = \cdots = Y_{ft}^{S_k^b} = \cdots = Y_{ft}^{S_{\pi_b}^b}; \quad f \in \Phi, t \in \{1, \ldots, T\}; \\ &S_k^b \in \Omega_b : \pi_b \ge 2, o_b \le t \le \mu_b \end{aligned} \tag{7.24}$$

$$\begin{aligned} &X_{ft}^q, Y_{ft}^q \in \{0, 1\}; \quad W_t^q \ge 0 \quad \text{and integer} \quad f \in \Phi, \quad q \in \{1, \ldots, Q\}, \\ &t \in \{1, \ldots, T\}. \end{aligned} \tag{7.25}$$

Constraint set (7.21) implies that all flights are planned to arrive by the end of planning horizon $T + 1$. Constraint set (7.22), which is similar to (7.12) and (7.17) in the static model presented above, ensures that the number of arrivals during any time-period is limited by the scenario-specific airport arrival capacity for that time period. The number of arrivals in a time period t is the sum of the reduction in the size of the arrival queue between the end of t and the end of the previous time period $t - 1$, and the number of flights whose planned arrival time is in t. If the number of planned arrivals during a time period exceeds the arrival capacity, then the excess flights are subject to airborne delay and added to the arrival demand for the next time period. Constraint set (7.23) is a set of coupling constraints, sometimes known as non-anticipatory constraints in the literature (see Birge and Louveaux 1998), on the ground holding decision variables. These constraints force ground delay decisions to be made solely on information available at time t. For a given time-period t, it is required that the ground holding decisions are the same for all scenarios associated with the same scenario tree branch b (in other words the scenarios belonging to the set Ω_b) in that time period.

Scenario and scenario tree generation is a vital issue in application of multi-stage stochastic optimization problems. While scenario generation has mostly been studied in the context of stochastic programming problems in finance, Liu (2006) has recently applied statistical clustering techniques to develop capacity scenarios and scenario trees for the stochastic SAGHP from historical airport capacity data.

One of the primary shortcomings of the scenario-based models for SAGHP is that they assume a limited number of capacity profiles can occur. However, in reality the set of possible scenarios can change with time. Furthermore, scenario-based models impose a decision tree structure when in reality improved information about future capacity can be obtained continually rather than at discrete branching points. In light of these shortcomings, Liu and Hansen (2007) proposed a "scenario-free" sequential decision making model, based on dynamic programming techniques, for the stochastic SAGHP. In order to reduce the computational complexity associated with large-scale problems, they proposed several prioritization-based heuristics. Another body of research (Hoffman et al. 2007) showed that a simple heuristic, which is based on giving priority to flights based on their flight distances, minimizes overall delays under the assumption that the airport capacity can increase earlier than anticipated. Hoffman et al. (2007) also proposed an algorithm, which they termed as "equitable ration-by-distance", which performs better, with respect to both equity and efficiency, than the distance-based flight exemptions currently used in practice by the FAA when implementing a GDP.

In the SAGHP, the network effect (or delay propagation) of arrival delays of flights is not considered. Typically, on any given day, an aircraft performs multiple flight operations. Therefore, when an aircraft arrives late at an airport, the subsequent flight performed by the same aircraft might face departure delay. Moreover, at a hub airport, arrival delay of an aircraft can translate into departure delays of several other aircraft due to passenger, baggage, and crew transfer issues discussed earlier. There exists a body of literature, within the domain of the GHP, which considers a network of airports and optimizes the ground delay assignment to various flights, so that delay on a given flight segment can propagate to downstream segments flown by the same aircraft. This problem is commonly known as the multi-airport ground holding problem (MAGHP). In general, these models account for arrival and departure capacities at several airports, but ignore en-route airspace capacity constraints. (Later, we will discuss the existing literature that addresses both airport and airspace capacity constraints.) As pointed out by Bertsimas and Odoni (1997), the existing optimization models for the MAGHP are applicable mainly under the conditions prevailing in the United States, where most of the congestion and delays occur at the airports. But in the context of European airspace, where en-route sectors attribute heavily towards congestion, such models might not be directly applicable. Another commonality of the existing models for the MAGHP is that they ignore flights whose origin or destination, or both, do not belong to the set of airports that are being considered. However, one can increase the set of airports, and include all flights that are important.

Most of the optimization models for the MAGHP are deterministic, i.e. an assumption is made that the airport capacities are known. Even under the deterministic setting, the models, which are integer programs in formulation, pose a severe computational burden. In general, these optimization models attempt to minimize the overall cost of ground and airborne delays, subject to the airport capacity constraints as well as the flight connectivity constraints. Vranas et al. (1994b) first proposed such a model to solve the MAGHP. Later, Bertsimas and Stock (1998) proposed an alternative formulation that performed better computationally (see Andreatta and Brunetta 1998 for details). We will discuss the

Bertsimas and Stock (1998) formulation for the deterministic MAGHP later when we present their model for airspace capacity allocation problems. Andreatta et al. (1995) proposed a heuristic, based on a set of priority rules in assigning ground delays to flights, to solve the deterministic MAGHP. Vranas et al. (1994a) provided some treatment of the stochastic MAGHP. They presented an integer program formulation of the static stochastic MAGHP, and then proposed a heuristic by which one can reapply the static formulation repeatedly in order to revise ground holding decisions as the time of day progresses.

Like the SAGHP, there are several extensions of the deterministic models for the MAGHP. Bertsimas and Stock (1998, 2000) proposed constraints that can capture the interdependence between arrival and departure capacities at airports. With such an extension, it is possible to determine the optimal ground holding strategy and the allocation of runway capacity between arrivals and departures simultaneously. In all the above models for the MAGHP, a set of constraints, commonly known as flight connectivity constraints, are imposed between successive flights performed by the same aircraft, i.e. single connections are accounted for. Navazio and Romanin-Jacur (1998) proposed an alternative formulation for the deterministic MAGHP in which they addressed the delay propagated by an aircraft onto several other flights; a situation that commonly arises at hub airports where the arrival and departure times of several flights are interlinked. In their paper, Bertsimas and Stock (1998) also discussed possible extensions of their model to capture multiple connections among flights.

7.3.1.3 Airspace Capacity Allocation Models

Optimization models and algorithms that address en route capacity constraints treat the airspace system as a multiple origin-destination network on which traffic flow must be assigned. Deterministic optimization models addressing en route capacity constraints were formulated as a multi-commodity network flow problem by Helme (1992), and more recently by Bertsimas and Stock (2000). These models deal with aggregate flows instead of individual flights. Also, compared to single-commodity flow network formulations, these models are more computationally intensive. One of the assumptions made by Helme is that each aircraft route is determined prior to its departure. The Bertsimas and Stock model, on the other hand, addresses routing as well as scheduling decisions. Their model is formulated as a dynamic, multi-commodity, integer network-flow problem with certain side constraints. Aggregate flows are generated by solving a Lagrangian relaxation of the LP, in which the capacity constraints are relaxed into the objective function. Subsequently, a randomized rounding heuristic is applied to decompose the aggregate flows into a collection of individual flight paths. Finally an integer packing problem is solved to obtain feasible, and near-optimal, individual flight routes.

Disaggregate deterministic 0-1 integer programming models for deciding ground and airborne holding of individual flights when faced with airport and airspace capacity constraints were formulated by Lindsay et al. (1994). The proposed model, which was named the Time Assignment Model (TAM), decides on the temporal and spatial location of each aircraft, given a set of capacity constraints on NAS resources. The input parameters are the origin and destination airports, and a set of en route fixes each aircraft must fly over. The time-varying capacities (deterministic) of each airport and each fix are also provided as input to the TAM model. The decision variables are the times at which aircraft are to be at each fix on their predetermined flight path. The objective function minimizes the sum of ground and airborne delay costs. Usually, the cost of airborne holding is greater than the same amount of ground holding, and the cost differences are provided as input parameters. There are five sets of constraints in the TAM model. The first set of constraints ensures that the capacities of any resource—airport or en route fix—are not violated during any time interval. The second set specifies a minimum flight time between two fixes for a given aircraft. The third set of constraints ensures that an aircraft passes over an en route fix on its route only once. The fourth set specifies the earliest departure time of each flight, and the fifth set specifies the minimum time an aircraft must be at its airport before it is ready to depart its next flight.

More recently, Bertsimas and Stock-Patterson (1998) presented a deterministic 0-1 IP model to solve a similar problem. For each aircraft, a predetermined set of en route sectors is specified as the route between its origin and destination. The model decides on the departure time and sector occupancy time of each aircraft. Bertsimas and Stock-Patterson showed that their formulation is NP-hard. In many practical cases the LP relaxation of their IP yields integer optimal solutions, and hence their model is considered to be computationally efficient in practice. The Bertsimas–Patterson model is presented here.

Let K denote a set of airports and F be the set of flights scheduled between those airports, and J denote the set of en route sectors. Let Ψ denote the set of pairs of flights that are continued, i.e., $\Psi := \{(f, f') : f' \text{ is continued by } f\}$. Let the planning horizon be divided into T time intervals of equal duration. For a given flight f, let N_f denote the number of resources (i.e., sectors and airports), and $P(f,i)$, $1 \leq i \leq N_f$, denote the ith resource along flight f's path. Note that $P(f,1)$ and $P(f, N_f)$ represent the departure and arrival airports respectively. Depending on the trajectory, each flight is required to spend a minimum number of time units, l_{fj}, in a sector j that lies along its flight path. Let the capacity of resources during a time-interval t be denoted as follows: $D_k(t)$ equals the departure capacity of airport $k \in K$, $A_k(t)$ the arrival capacity of k, and $S_j(t)$ the sector capacity (i.e., the number of aircraft allowed to be present) in sector $j \in J$. The flight-specific scheduled times and delay costs are denoted as follows: d_f, a_f, and s_f are the scheduled departure, arrival, and turn-around times (the minimum ground time for an aircraft between flights) respectively, while c_f^g and c_f^a denote the unit costs of delaying a flight on the ground and in the air.

The binary decision variables, which are non-decreasing, are defined as follows:

$$X_{ft}^j = \begin{cases} 1 & \text{if flight } f \text{ arrives at sector } j \text{ by time } t; \text{ and} \\ 0 & \text{otherwise.} \end{cases}$$

To reduce the size of the formulation, Bertsimas and Patterson (1998) defined for each flight a feasible time window that establishes when that flight can occupy a resource along its flight path. The feasible time periods, for a flight f to be present in sector j, are represented by a set T_f^j, $j \in P(f), 1 \leq i \leq N_f$. Based on the decision variables, the total ground and airborne delays of a flight are given by the following expressions:

$$g_f = \sum_{t \in T_f^k, k=P(f,1)} t\left(w_{ft}^k - w_{f,t-1}^k\right) - d_f; \text{ and}$$

$$r_f = \sum_{t \in T_f^k, k=P(f,N_f)} t\left(w_{ft}^k - w_{f,t-1}^k\right) - a_f - g_f.$$

The objective function and the set of constraints are defined as follows:

$$\text{Min.} \quad \sum_{f \in F} \left(c_f^g g_f + c_f^a r_f\right)$$

subject to

$$\sum_{f:P(f,1)=k} (w_{ft}^k - w_{f,t-1}^k) \leq D_k(t) \quad \forall k \in K, \quad t \in \{1,\ldots,T\} \tag{7.26}$$

$$\sum_{f:P(f,N_f)=k} (w_{ft}^k - w_{f,t-1}^k) \leq A_k(t) \quad \forall k \in K, \quad t \in \{1,\ldots,T\} \tag{7.27}$$

$$\sum_{f:P(f,i)=j, P(f,i+1)=j'} (w_{ft}^j - w_{f,t}^{j'}) \leq S_j(t) \quad \forall j \in J, \quad t \in \{1,\ldots,T\} \tag{7.28}$$

$$w_{f,t+l_{fj}}^{j'} - w_{ft}^j \leq 0 \quad \begin{cases} f \in F, t \in T_f^j, j = P(f,i), \\ j' = P(f,i+1), \quad i < N_f \end{cases} \tag{7.29}$$

$$w_{ft}^k - w_{f', \; t-s_{f'}}^k \leq 0 \quad \begin{cases} (f',f) \in \Psi, t \in T_f^k, \\ k = P(f,1) = P(f',N_{f'}) \end{cases} \tag{7.30}$$

$$w_{ft}^j - w_{f,t-1}^j \geq 0 \quad \begin{cases} \forall f \in F, j \in P(f,i), \\ 1 \leq i \leq N_f, \quad t \in T_f^j \end{cases} \tag{7.31}$$

$$w_{ft}^j \in \{0,1\} \quad \forall f \in F, j \in J, \quad t \in \{1,\ldots,T\}. \tag{7.32}$$

The objective function minimizes the total cost of flight delays. The set of constraints are classified into two categories: capacity constraints (7.26–7.28) and connectivity constraints (7.29–7.31). The capacity constraints ensure that the flow is bounded by the capacities of each resource in the system—airports and sectors. For example, constraint set (7.28) ensures that the total number of flights within a sector during any time interval does not exceed the sector capacity during that time period. Within the connectivity constraints, there are two sub-categories: sector and flight connectivity. The sector connectivity constraints (7.29) ensure that each flight passes through the proper sequence of sectors in its route between origin and destination airports. The flight connectivity constraints (7.30) ensure that an aircraft must spend a minimum "turnaround" time at an airport before it can depart on its subsequent leg. Constraint set (7.31) ensures that the decision variables are non-decreasing, while (7.32) ensures they are binary.

There are several variants to the Bertsimas–Stock formulation. If sector capacity constraints are removed, the formulation corresponds to a multi-airport ground holding problem. Multiple connections at a hub airport can be addressed by modifying flight connectivity constraints, while interdependence between arrival and departure capacity constraints can be captured using the concept of capacity envelopes discussed before. An interesting alternative is proposed by Lulli and Odoni (2007), who introduce a more macroscopic version of the Bertsimas–Stock formulation that omits some of its details (i.e. speed control, en route airborne holding). Lulli and Odoni argue that the resulting model is particularly appropriate for ATFM in Europe, where congestion on en route sectors is common and much more prevalent than in the United States. Using their model, the authors show that fundamental conflicts may arise between efficiency and equity, and illustrate the potential benefits of selective airborne delay assignments. Another important alternative is proposed by Bertsimas et al. (2008), who extend the Bertsimas–Stock formulation by allowing for rerouting decisions. Specifically, they present an innovative formulation where reroute options are represented using a small set of additional constraints. Their experiments indicate that the resulting models can be solved efficiently for realistic large-scale problem instances.

The models for airspace capacity allocation discussed above prescribe delay and routing for individual flights. Therefore these models can be applied in a setting where one controlling entity is authorized to assign delays and reroutes to individual flights. This is not the case under the CDM paradigm, where users (airlines) actively participate and exchange information during the ATFM decision making process. Goodhart (2000) proposed a centralized procedure for the airspace capacity allocation problem, in which airline priorities can be accommodated while assigning ground delays and reroutes to individual flights. Sherali et al. (2002) proposed a 0-1 IP for deciding ground delays and routes for a set of flights so that a global objective function—typically the overall cost of delay—is minimized. Given a set of flight plans for each individual flight, their model selects one flight plan (a combination of delay and route choice) for each flight so that certain criteria for controller workload, airspace safety, and equity among airlines are satisfied. The authors proposed a polynomial-time algorithm that yields solutions within 0.01% of the optimality in several instances.

Sherali et al. (2003) proposed a model, which is essentially an extension of the model presented in Sherali et al. (2002), to decide on delays and routes of a set of flights subject to efficiency, equity, controller workload, and safety constraints. The resulting Airspace Planning and Collaborative Decision Making (APCDM) model has several subroutines, as described in Fig. 7.10. The first, namely, the

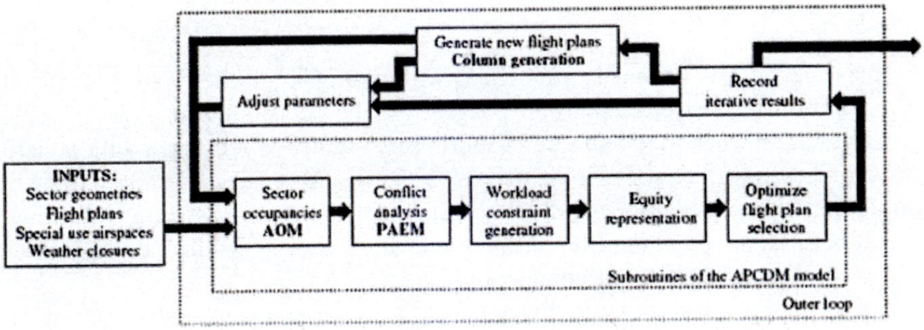

Fig. 7.10 Conceptual framework of the APCDM model (source:)

Airspace Occupancy model (AOM), depicts the time-varying sector occupancy for all given alternative flight plans. The Probabilistic Aircraft Encounter Model (PAEM) conducts a conflict analysis between pairs of aircraft flight-plans. The third subroutine formulates the workload and conflict resolution constraints that are input into a large-scale 0–1 IP. The fourth subroutine defines several equity constraints. Sherali et al. (2006) proposed the development of the cost function, and several parameters of the APCDM model, and presented computational results for realistic problems using Enhanced Traffic Management System (ETMS) data.

So far we have highlighted the research on air traffic flow management under deterministic capacities of en route resources—sectors, fixes, airways, etc.—in the NAS. As discussed before in Sect. 3.3.1, weather, which is a key factor in determining capacity in the NAS, is highly unpredictable. Therefore, although in practice ATFM is performed under the assumption of deterministic evolution of the system, there is a high degree of uncertainty both in terms of demand and capacity NAS-wide. Nilim et al. (2001) addressed the problem of dynamically routing an aircraft under uncertain weather using a Markov Decision Process. Nilim and El Ghaoui (2005) proposed a methodology, based on a robust dynamic programming algorithm, which introduces robustness to the solutions when the components of the transition matrix are stochastic.

Mukherjee (2004) extended the formulation of the stochastic dynamic ground holding problem described in the previous section (Mukherjee and Hansen 2003) to address the problem in which weather impacts the capacity of airport arrival fixes along with that of the airport itself. The decisions made under uncertainty in the capacity of airport and en route arrival fixes are the amount of ground holding assigned to flights at their respective origin airports and the amount of traffic that needs to be rerouted (dynamically) to avoid overload at the weather impacted arrival fixes. Mukherjee (2004) applied this model to problems of a realistic size. However, the applicability of their proposed methodology to large-scale problems involving traffic between multiple origin-destination airports requires further investigation. It is likely that computational complexity is an issue, and developing algorithms that give faster solutions might be necessary.

7.3.1.4 Airline Response

Whereas the FAA is primarily concerned with aggregate traffic flows and overall throughput during periods of airport congestion, the decisions and tradeoffs faced by individual airlines in a GDP are of a different nature. When faced with a GDP, for example, an airline typically responds to the resulting schedule disruptions by trading-off flight cancelations and delays. Such decisions are based on a multitude of factors, such as the disruption to and the cost of crew schedules, the passenger costs of delay, possible flight connections, etc.

The ground delays imposed by a GDP can create severe disruptions to an airline's flight schedule, which not only affect the delayed flights but may also propagate delays to other flights. To mitigate these disruptions, airlines may cancel flights and substitute flight-slot assignments. A decision model to support this slot swapping process was first presented in Vasquez-Marquez (1991), which also describes its application at American Airlines. Other models for resolving schedule disruptions through slot swapping are proposed in Luo and Yu (1998a, b) and in Bard and Mohan (2008). Generally speaking, these models can be viewed as assignment problems (that is, they assign an airline's flights to arrival slots) with potential side constraints. These additional constraints depend both on the objective function that is considered (i.e. minimize the number of delayed flights, the passenger delay, the maximum delay, etc.) and the extent to which downstream impacts are taken into account. Similar models are also considered in Hoitomt et al. (1999) and Niznik (2001), who describe their application at both United Airlines and American Airlines; given the real-time nature of the decision process, the problems are often solved heuristically. Another approach, which explicitly considers the connection dependencies of hub operations but leads to less efficient algorithms, is proposed in Milner (1995) and further extended in Carlson (2000). More recently, Abdelghany et al. (2007) also propose decision support models that can be used to mitigate the impact of the disruptions caused by airspace flow programs. In addition to slot swaps, these models also consider rerouting decisions, in that an airline may opt to reroute its flights to avoid congested areas. It should be emphasized, however, that none of these models fully reflect the complexity of airline decision-making during GDPs. For instance, these models typically do not incorporate the decision to cancel a flight, which is one of the most important decisions during a GDP.

Another family of decision models for resolving schedule disruptions may be found in Bard et al. (2001), Cao and Kanafani (1997a, b), Teodorovic and Guberinic (1984, 1990), and Thengvall et al. (2000, 2001). While the details of these models are beyond the scope of this survey, this class of models attempts to find an operable, system-balanced flight schedule when aircraft shortages disrupt an airline's flight schedule (that is, they consider an airline's entire network of flights and disruptions that are not necessarily due to ATFM restrictions). The application of these models at United Airlines is described in Jarrah et al. (1993) and Rakshit et al. (1996). These models, however, typically do not incorporate arrival slot constraints. Their use is primarily in schedule recovery after the disruptions from a GDP have occurred.

7.3.2 Descriptive Models

The prescriptive models outlined in the previous section generally aim to develop plans and/or procedures to support the service provider's decision-making. Descriptive models, on the other hand, aim to analyze or predict key characteristics of the air transportation system. The applications of descriptive models are widespread and varied, and their objectives range from evaluating capacity and delays in certain airspace elements and understanding major design changes to developing performance measures. Generally speaking, however, descriptive models are more strategic in nature and consider long term and more aggregate components of the air transportation system.

As with prescriptive models, we distinguish between airspace models and airport models. Airspace models are more global in nature and explicitly consider the interconnectivity in the system, while airport models consider local characteristics of their facilities (runways, taxiways, etc.). In addition, one can further make a distinction between analytical models and simulation models, and also classify descriptive models according to their ultimate objective.

7.3.2.1 Airspace Models

Descriptive airspace models can be used to assess airspace capacity, understand the system impact of disruptive events, and evaluate the effectiveness of potential responses or policy changes. These models have substantial differences in scope, methodology, and objectives.

The most comprehensive approaches rely on detailed simulation models to analyze the entire airspace system (or large segments thereof). The dominant models in this are the Airport and Airspace Simulation Model (known as SIMMOD; Federal Aviation Administration 1993) and the Total Airport and Airspace Modeler (TAAM; Marietta 1993). They are often referred to as microscopic models, in that they allow for a high level of detail: users can input data to configure airspaces and airport environments from gates, terminals, pushback, taxiways, runways, and terminal and en route airspace to flight schedules, routing options, resource availabilities, and operational procedures. These models have modules to capture the interaction between airspace and airport operations, and support conflict detection and resolution (Garcia 2003). Another example is the Reorganized ATC Mathematical Simulator (RAMS), which is used by Eurocontrol to analyze airspace operations and procedures (Leal de Matos and Ormerod 2000). These systems have become increasingly sophisticated over time, and are often used in complex design and assessment studies (see also Odoni et al. 1997 for a review). Example applications include environmental assessments for airport improvement projects, evaluation of airport runway configuration options, and investigations of rerouting alternative options. While microscopic large-scale simulation models can provide a powerful tool, the level of detail also implies that they require extensive user expertise (both of inputs and modeling components) that can complicate a broader modeling effort. Moreover, these models are generally less applicable when their inputs are subject to significant uncertainty, because it can be hard to develop scenarios at the level of detail required (Barnhart et al. 2003).As a result, these models are perhaps best suited for detailed analysis and preliminary design evaluations.

To support more aggregate analyses that consider a broader scope, more macroscopic models have been developed. Such models are more appropriate for traffic flow analysis, cost-benefit studies, or policy evaluation. Some of the models in this class still contain significant levels of detail, but have a more narrow focus. One such example is the Detailed Policy Assessment Tool (DPAT; Wojcik (2001) developed by the MITRE corporation. DPAT is a discrete-event simulation model that captures individual flights and itineraries through a sequence of constrained resources. It has been used to analyze the system-wide effects of weather, and TFM initiatives such as ground delay programs and miles-in-trail restrictions. Another example is the Future ATM Concepts Evaluation Tool (FACET; Bilimoria et al. 2000) developed by NASA. FACET considers en route airspace operations over the contiguous United States, and is used to analyze the system-wide impact of new air traffic flow management initiatives. Other approaches rely on analytical queueing models, which generally require less detailed inputs and execute quite fast. As a result, these models are more applicable for system-wide studies. An example is LMINET (Long et al. 1998), a queueing network model of the NAS developed by the Logistics Management Institute (LMI). LMINET estimates delays at airports and en route sectors, allowing consideration of congestion issues. The strengths of these approaches naturally imply a limited amount of detail in certain model aspects. LMINET, for instance, does not consider the interdependencies that occur within a flight schedule, and therefore cannot capture the resulting delay propagation that occurs.

Large-scale simulation models have also been used to analyze the impact of TFM disruptions on airline operations. SIMAIR (Rosenberger et al. 2002), for instance, is a stochastic model of airlines operations that includes an event generator to sample random events such as ground delays and unscheduled maintenance. A dedicated module simulates an airline's operational control center to evaluate recovery procedures and plans under operational conditions. Closely related is the MIT Extensible Air Network Simulation (MEANS, Clarke et al. 2005), which can be used to predict the impact of air traffic management disruptions as well as airline operations control and scheduling actions.

In addition to these general-purpose models, a considerable amount of research has also addressed the modeling of individual sectors within the airspace. Generally speaking, these models aim to estimate ATC sector capacities, evaluate sector traffic complexity, or analyze the probability of aircraft conflicts. They can be used to assess and quantify the workload of air traffic controllers, to measure en route safety, and to monitor airspace system conditions. Alternatively, they can also be incorporated within the large-scale simulation and optimization models outlined before.

Sector capacity is commonly measured in terms of the number of simultaneously present aircraft or the number of aircraft traversing the sector per unit time (Sridhar et al. 1998). Establishing sector capacities presents a challenge due to the heavy reliance on human controllers. Mathematical models to calculate the theoretical capacity of an ATC enroute sector have been proposed by Janic and Tosic (1991), who compute expected sector capacity based on sector geometry, traffic conditions, and ATC separation rules. Subsequent models (Janic 1997) have also aimed to incorporate human factors, that is, the conflict solving efficiency of air traffic controllers. Other, somewhat more general, approaches have concentrated on quantifying sector complexity (Sridhar et al. 1998; Masalonis et al. 2003). These approaches aim to develop metrics that account for the factors (and relationships between them) which determine the difficulty of the controllers' tasks. Examples (Hinston et al. 2001) include airspace factors (sector dimensions, spatial distribution of airways and navigational aids, coordination with neighboring sectors, etc.), traffic factors (number and spatial density of aircraft, homogeneity of aircraft performance, etc.), and operational factors (limitations of communications systems, procedural flow restrictions, etc.). In addition, stochastic models have been proposed to capture the probability of aircraft conflicts along airways (Erzberger et al. 1997; Kuchar and Yang 2000; Prandini et al. 2000). Aircraft conflict probabilities can be used as an alternate measure of complexity, or to assess the risk of aircraft collisions. Recent work (Barnett 2000; Willemain 2003), for instance, has applied collision risk modeling to analyze the impact of moving toward a "free flight" paradigm, where pilots are given additional authority to fly directly from one point to another without restricting movement to the established airways. This would require advanced navigational capabilities (e.g. area navigation, or RNAV) that allow aircraft to operate independent of ground-based radio navigation devices. Free flights are being introduced incrementally in select regions of the country.

Statistical estimation methods have been applied to assess the impact of weather on air traffic, and hence delays. Research in this direction is useful to benchmark performance of the ATM, predict future performance, and to establish benefits of improving weather and traffic demand forecasts. Callaham et al. (2001) first introduced the concept of Weather Impacted Traffic Index (WITI). WITI is computed as the product of instantaneous traffic count and an indicator of severe weather in a region of airspace. The results can be aggregated over time or space to produce a daily or hourly WITI of specific Centers or the entire NAS. Sridhar et al. (2006, 2007) and Klein et al. (2007) further the research on WITI by characterizing its time series over a day using as reference the traffic counts from a weather-free day. Sridhar et al. (2007) also analyze the impact of Center-based WITI on the total NAS delay, and identify some specific Centers in the NAS wherein the weather impact is most severe. Hansen and Xiong (2007) estimated the impact of WITI on a different NAS performance metric, the Daily Average Flight Time (DAFT); the DAFT metric provides an attractive alternative to the use of reported flight delays, which are often distorted as airlines pad their schedules (by adding buffers to the scheduled flight times) to give the appearance that flights arrive on time more frequently than they actually do.

7.3.2.2 Airport Models

In addition to the more general airspace models, a significant amount of research has also considered the analysis of airport facilities. The earliest work in this area considers the estimation of airport runway capacities. Blumstein (1959) was the first to propose analytical models for determining the capacity of a single runway, which he defined as the number of possible movements (landings or takeoffs) per period in the presence of separation constraints. Subsequent work extended this model to cases where multiple runways operate simultaneously (Swedish 1981), and have proven to be valuable in airport planning (see also Barnhart et al. 2003 for a more detailed review). The use of queueing models to estimate airport delays has also received considerable attention. The challenge with these models arises due to the fact the airport queues are non-stationary, given that both demand and service vary during the course of the day because of demand patterns and weather conditions. This has led to the development of numerical approaches for calculating airport delays, which was first analyzed by Koopman (1972). Subsequent work (Kivestu 1976; Malone 1995) extended this model to yield power approximation methods for the analysis of dynamic airport queues. They have been used, among others, to study gridlock conditions at New York LaGuardia Airport (Fan and Odoni 2002), and to calculate the airport capacities published in the FAA's benchmark report of 2001 (Federal Aviation Administration 2001).

Another stream of research has focused on predicting the delays that flights incur due to airport ground operations, that is, their delays in taking off. These models aim to assist both in the management of airport resources as well as broader flow management initiatives. Better predictive ability, for instance, will improve the *Monitor Alert* module that is used by the FAA to predict airspace traffic levels and indicate when sector overloads occur. It is common to distinguish between push-back (or departure) delays and taxi-out delays. Push-back delays measure the discrepancy between the scheduled departure time and the actual departure time from the gate (the push-back time). Taxi-out times, on the other hand, measure the difference between the takeoff time and the push-back time. Push-back and Taxi-out delays depend on many factors ranging from ground operations (luggage handling, airline policy, etc.) to weather conditions, runway configuration, and downstream restrictions. Odoni et al. (1994) developed a non-homogeneous queueing model to analyze the takeoff delays at congested airports. Shumsky (1997) extended this model and estimated takeoff times under non-steady state conditions, using a regression model that attempts to predict taxi-out times using airline, departure runway, and departure demand as the explanatory variables. Idris et al. (2002) further analyzed the causal factors that determine taxi-out delays, and developed a queueing model for taxi-out time estimation. Another recent paper by Tu et al. (2005) develops a statistical model for estimating push-back delays based on three components: seasonal trend, daily propagation pattern, and random residuals.

7.4 Research Opportunities

While the current body of research in ATFM is certainly large and varied, it is nevertheless important to note that OR models have generally not seen a widespread adoption within the ATFM practice. Whereas the field of OR has had a critical impact within other areas of the airline industry (e.g. revenue management, airline schedule planning, and crew rostering), its applications to flow management have been more isolated and, generally speaking, of a more limited scope. Certainly, important streams of research have found their way into practice. The body of work on descriptive models to analyze airspace and/or airport capacities and delays, for instance, has yielded commercial software packages and proven to be indispensable in numerous strategic design studies. On the other hand, however, prescriptive optimization models have rarely—if ever—been adopted in practice. Given the attention these models have received over the last 20 years, this is perhaps somewhat surprising.

While somewhat speculative, we nevertheless believe it is worthwhile to consider the causes for this discrepancy between theory and practice. It has been argued, for instance, that optimization models are often regarded with suspicion by users (Leal de Matos and Ormerod 2000) who view their mathematical content as obscure and distrust the "black-box" nature of the results. In addition, optimization models can be very time-consuming, which is a significant limitation in a near real-time ATFM environment. This is exacerbated further by the oftentimes intensive data requirements, which complicates the implementation of optimization models in practice. The large body of work on stochastic Ground Holding Problems, for instance, generally uses a scenario tree as an input. The development of appropriate decision trees, however, presents a formidable task in and of itself, and has received attention only recently (Innis and Ball 2004; Hansen et al. 2006). Moreover, the models' users might find it hard to determine appropriate values for models parameters that—while mathematically convenient—do not correspond well with the manner in which they make their decisions. Models from the ground holding problem have traditionally used parameters to represent the relative "cost" of ground and airborne holding. It can be difficult for the service provider to determine values for such parameters, or understand how varying these parameters will impact the results. On a broader level, the advent of the CDM paradigm has illustrated the importance of distributed decision-making responsibilities, information sharing, and incentives. While many ideas of the traditional, large-scale, aggregate optimization models could potentially be transferred to a CDM environment (Barnhart et al. 2003), addressing the resulting issues presents a major challenge.

At the same time, however, it is important to note that the context in which TFM operates makes it difficult to adopt OR models on a large scale. Because TFM is concerned with safety and day-to-day operations, the environment will naturally be more conservative and slower to adopt new models than a for-profit business operating in a competitive environment. In a system this complex, an incremental and evolutionary approach to adopting new technologies has long been preferred. Thus, although we believe that the research on ATFM has made significant progress and yielded important new ideas, it is safe to say that the field as a whole is still in its early stages. As such, we believe that there is a clear need and opportunity for further ATFM research, and in the remainder of this chapter we discuss the potential research that we believe to hold considerable promise. To frame our discussion, however, we start with an overview of the important developments surrounding ATFM.

7.4.1 Trends

Over the next two decades, the demand for air traffic in the US is expected to grow to two or three times its current level (http://www.jpdo.gov). Given that the air transportation system can barely manage the current demand levels, stakeholders in the system are actively pursuing ways to accommodate this future growth. These developments are wide-ranging, and promise to shape the direction of ATFM—in practice as well as in research—for the foreseeable future.

Broadly speaking, these developments or trends fall into three categories. First, one can recognize trends at the policy and program level. In addition, there are developments that aim to improve ATFM technology and procedures. Finally, we can also distinguish a number of trends in the ATFM research.

Policy and Programs

In 2003, the US Congress passed the Vision 100-Century of Aviation Reauthorization Act (Public Law 108-176), which mandates the design and deployment of an air transportation system that will be able to meet the needs for aviation services in 2025. Recognizing that an inability to meet these needs would curtail growth of the economy, the Vision 100 Act launched several programs and organizations that aim to transform the air transportation system. Chief among those is the establishment of a Joint Planning and Development Office (JPDO), which oversees and coordinates the research and development efforts of seven different government agencies[8] and serves as a liaison between the government and the private sector (http://www.jpdo.gov).

The main responsibility of the JPDO is to create and carry out an integrated plan for what is known as the Next Generation Air Transportation System (NextGen). The NextGen concept envisions a fundamental departure from the current approach to air transportation operations that provides a common framework for safety, efficiency, security, and environmental concerns. Moreover, it aims to align technology improvements and procedural advancements to create

[8] The Federal Aviation Administration, Department of Transportation, Department of Defense, Department of Commerce, Department of Homeland Security, National Aeronautics and Space Administration, and White House Office of Science and Technology Policy

a system that can handle the anticipated increases in air traffic demand. As such, the NextGen initiative has established a clear trend toward more integrated approaches (Pearce 2006).

The NextGen vision might appear to be a departure from the CDM paradigm that has been the driving force behind most policy changes and technology deployments for ATFM in the last decade. Whereas NextGen represents an overarching top-down initiative, CDM initiatives have traditionally evolved along an incremental bottom-up approach. Nevertheless, the basic tenets of the CDM philosophy play a key role in the NextGen concept, and provide the foundation for most of the technology and procedural improvements that are envisioned within NextGen. If anything, the notions of information sharing, common situational awareness and decentralized decision-making responsibilities will only become more prevalent under NextGen.

It is important to note that similar developments are also underway in Europe. In 2004, the European union adopted the Single European Sky ATM Research (SESAR) initiative to transform its air transportation system (Ky and Miaillier 2006). While the European system faces somewhat different problems (there is, for instance, a considerably larger fragmentation of technologies and policies among its member states), the overall objective of the SESAR concept is similar to NextGen. Specifically, the SESAR initiative calls for a European air transportation system that will be synchronized, both in a geographical sense and from a system perspective, and that will take an integrated approach from research to operations.
Technology and Procedures

The NextGen concept requires significant increases in the safety, security, capacity, and efficiency of air transportation operations, which can only be achieved by a combination of both technological and procedural advances. The NextGen vision organizes these needs by identifying eight key capabilities that will help achieve these goals (Concept of operations for the next generation air transportation system 2007; Arbuckle et al. 2006).

Performance-Based Operations and Services (*PBO*) enables services to be tailored to individual user needs. This represents a fundamental paradigm shift away from today's system, which offers uniform services and thereby caters to the least-equipped aircraft. PBO offers multiple service levels, allowing users to equip their aircraft in accordance with the operational benefits they would receive. This is in marked contrast with today's system in which advanced aircraft capabilities are unilaterally conceived, imposed, and regulated by the FAA. PBO enables more cost-effective service provision and encourages user innovation.
The ramification for TFM is that multiple aircraft capabilities must be considered in models for allocating TFM resources and services. A subsequent area of research is to study the tradeoffs between service levels. For instance, reserving airspace for aircraft able to fly in close proximity to each other or along precise pathways (required navigation performance, RNP) can displace lesser equipped aircraft. Distribution of benefits must be considered across user groups and the system as a whole.

Network-Enabled Information Access requires that information will be available, securable, and usable in real-time for different communities and air transportation domains. Real-time aggregation and access to information is a central component of NextGen, and aims to improve the speed and quality of TFM in a distributed fashion.

Positioning, Navigation, and Timing Services (*PNT*) in the NextGen concept, will be provided where and when needed to assure reliable aircraft operations in nearly all conditions. The objective is to eliminate the restriction of aircraft movements to ground-based navigational aids (NAVAIDs), by the introduction of satellite-based (GPS) navigation systems that will allow operators greater flexibility and efficiency in choosing their flight paths. As such, this component of NextGen subsumes the well-known "Free Flight" initiative and programs that were started in the mid-1990s.

The possible use of these technologies has a variety of implications for TFM, ranging from in incre effective capacity to a greater consideration of user preferences in routing.

Aircraft Trajectory-Based Operations (*TBO*) present another fundamental change, which proposes a transformation of today's approach to flight planning and ATM. TBO envisions a system that manages operations based on 4D aircraft trajectories (time-based paths from gate to gate that include ground segments), regularly adjusts airspace structure to best meet user needs, and relies on automation for trajectory analysis and separation assurance. Real-time exchange of the planned trajectories will allow service providers to assess their impact, understand the impact of demand, and evaluate potential resource allocation plans.

The need to consider and manage trajectories as entities unto themselves requires a substantial change in TFM procedures. The incorporation of surface movements into the trajectories, for example, will create a significant linkage with en route operations. Largely for convenience, today's system tends to separate the two.

Super Density Operations also relies on the other capabilities to improve the overall flow management in and out of busy metropolitan airspace. This will allow maximum use of the busiest airports, by introducing new procedures to improve airport surface movements, and reduce spacing and separation requirements.

Weather Assimilated into Decision Making relies on network-enabled information access to provide a common weather picture for decision support. The overall objective is a single, integrated system that aggregates weather information from multiple sources and combines analysis and visualization tools with the ability to apply probabilistic weather information to ATM decision support systems directly.

Equivalent Visual Operations relies upon a combination of the aforementioned capabilities to allow aircraft to operate as well under poor or low visibility conditions as they can under high visibility conditions. This could have significant implications for TFM, in that it yields more predictable and efficient operations and an increase in the system's effective capacity.

Layered, Adaptive Security presents an integrated approach to system security, by constructing "layers of defense" that help reduce the overall risk of a threat reaching its objectives while minimally affecting efficient operations. While these capabilities should not be viewed as a definite blueprint of a future air transportation system, they do provide a set of end-state goals to work toward that will therefore shape ATM research and development along both the technological and procedural dimensions.

The NextGen vision outlines a large variety of technological advances, ranging from the introduction of new navigation systems to the development of decision support tools and predictive methods. Nevertheless, one can argue that all of these proposals are meant to support an infrastructure that emphasizes information sharing and common situational awareness to allow for a high level of responsiveness. Moreover, they recognize that a high level of automation will be needed to create a system that is robust in the presence of uncertainty and scalable to accommodate the growth in operations.

At the same time, the procedural changes advocated under NextGen share a common user focus that emphasizes distributed decision-making. Initiatives such as Performance-Based Services and Aircraft Trajectory-based Operations aim to provide more flexibility to users while reducing the need for service provider intervention and control of resources. The NextGen vision further expands the CDM paradigm where, as much as possible, decisions are made at the local level with an awareness of global implications. Overall, these procedures intend to better align the service provider's processes and practices with customer needs.

We note, however, that one important issue related to NextGen vision lies in how it is to be implemented. In recent years there have been calls to commercialize the provision of air navigation services, by creating an altogether different organization that is physically separated from the FAA. Proponents of ATC privatization point to the experience of other countries (such as Canada, the United Kingdom, and Australia) to argue that this will improve service quality, reduce costs, and improve modernization (McDougall and Roberts 2009). Critics, on the other hand, argue that these benefits are not as clear-cut, and that modernization may have increased user fees and come at the expense of safety (Sclar 2003). Given the differences in countries' approaches toward commercialization and the diversity in traffic strictures, however, it is perhaps difficult to draw firm conclusions about the effects of such reforms (Button and McDougall 2006).

7.4.2 Research Opportunities

While the NextGen vision will undoubtedly undergo changes and revisions in the years to come, we believe that its key capabilities and fundamental characteristics offer numerous opportunities for OR analyses and modeling. Indeed, the high level of automation that will be needed to support this system indicates that decision support systems that rely on OR methods will become indispensable. At the same time, the notion of a distributed, information-centric air transportation system significantly expands the scope of potential OR analyses and models.

To conclude this chapter, we discuss a number of these possible research areas and propose concrete research problems. To organize our discussion, we distinguish between the technology and the procedural improvements that have been proposed under NextGen.

Technology

The current system's infrastructure essentially dates back to the 1950s, relying on radar and analog radio to navigate air traffic. This infrastructure limits the system's ability to grow and adapt, which has motivated the large number of technological advances proposed within the NextGen concept. The inability to grow the current system, however, also implies that these technologies present a radical transformation that will require substantial investments: initial estimates by the JPDO suggest 15–22 billion in federal spending will be needed to support NextGen through 2025 (not including operations or airport improvements), with an additional cost of 14–20 billion to equip aircraft with NextGen technologies (http://www.jpdo.gov).

As such, the development of descriptive models, whether analytical or simulation, to assess and measure the benefit of new technologies presents a first research opportunity (see, for example, Miller and Clarke 2007). Clearly, such models will have to consider not only the system-wide impact, but also the impact on individual users or user groups. A related area of interest (and contention) is the development and analysis of funding mechanisms to support the cost of the system. Alternatives to the current system of taxes and fees have been and continue to be proposed and discussed (Dillingham 2006; Yamanaka et al. 2006), due to concerns that the current system creates a disconnect between the revenues contributed by users and the costs they impose on the system. OR models and analysis could add to these discussions, providing valuable insights about issues such as revenue adequacy and equity.

Similar models can also be useful to evaluate the impact of other changes to the ATM environment. The General Accounting Office (GAO; United States General Accounting Office 2007), for example, recently analyzed how Very Light Jets (VLJs) might impact the NAS. VLJs represent a new type of small jet aircraft that are equipped with advanced avionics and can be certified for single-pilot operations, at a price that is below other business jets. The primary concern for ATFM is having a non-uniform collection of aircraft (e.g. slower and faster) operating in close proximity of each other. The GAO study indicated that the introduction of this type of aircraft might increase delays by approximately 1–10%, with little impact on safety. The results are inconclusive, since the rate of VLJ introduction and routing choices are difficult to forecast. The rate of introduction is most likely to be gradual. For this reason, ATFM has no specific plans for addressing VLJs, but separation of airspace by performance capability is one possibility. Exactly how to segregate the airspace is an outstanding research problem.

Another area of potential research is related to the development of technologies and methods to support the real-time information sharing and common situational awareness that characterize NextGen. This builds upon the CDM efforts in this area, and aims to extend these further to a system-wide environment. Because weather conditions have a profound impact on system operations, we expect these efforts to focus on reducing uncertainty by improved forecasting, information aggregation, and visualization methods.

Along with tools to increase the prediction accuracy of unfolding events, however, this introduces new possibilities for the development of models that integrate weather information within both strategic and tactical ATFM decision making. While a considerable amount of progress has been made in this area (see Sect. 7.3), we believe the emphasis on real-time information sharing can lead to more dynamic models that enable a rapid response to changing conditions. One promising area is the development of real-time simulation models that integrate with optimization techniques to support efficient and equitable ATFM decisions. The resulting models could be used to increase common situational awareness by analyzing the impact of possible decision alternatives in real-time (the FACET tool developed by NASA presents a step in this direction, Sweet et al. 2002). Alternatively, such models might also automate the overall decision-making process with limited intervention by air traffic controllers and managers. Dynamically adjusting/revising decisions when provided with updated information on system conditions (Ball et al. 2005c) will be pivotal to increasing the efficiency in the system, in particular when full-scale contingency planning is not possible.

Given the potential cost of information gathering and accuracy improvements, we also believe there to be opportunities to study the *value* of information within NextGen at a more strategic level. Initial research (Willemain 2002; Ball et al. 2001), for example, evaluates how information sharing under CDM impacts ground holding decisions by considering the tradeoff between airborne and ground holding given the accuracy of demand forecasts. Understanding how information impacts decisions, both from a system and user perspective, can guide efforts to create common situational awareness and enforce compliance.

Procedures

The NextGen vision requires significant procedural advances that will have a major impact on TFM operations. Concepts such as Performance-Based Operations and Services and Aircraft Trajectory-Based Operations present a fundamental change in paradigm, and introduce a new approach to the management of system resources and capacity. This generates a number of possibilities for the development and evaluation of policy alternatives, which can address important questions with respect to the appropriate definitions and access to the air transportation system's resources.

A central aspect of these models will be the ability to account for user behavior and response. One of the proposals within the Performance-Based Services initiative, for example, is to prioritize access for aircraft that are equipped with advanced technologies. Along with the previously mentioned investment models that evaluate the benefits of adoption from the users' perspective, this necessitates models which capture user behavior to evaluate whether policies generate the desired incentives. Another potential area of research is the application of market-based mechanisms, both in the medium and short term, for managing the air transportation system's resources. In the medium term, for instance, airlines could bid for and own a certain proportion of system resources. Subsequently, they might be able to trade the resources they own with other users on a daily basis. Such an approach would require users to develop models that valuate their resources, and support decisions related to resource trading on a secondary market. The service provider, on the other hand, would have to design this "marketplace", and provide a platform for resource trading. Examples of initial research in this area include the use of auctions to assign airport arrival slots (Ball et al. 2005a) and the use of slot trading during GDPs (Vossen and Ball 2006).

In addition, the use of new models for disruption management offers several possibilities. The growth in air traffic demand implies an increasing impact of disruptions, and new initiatives will be necessary to maintain stability in the system. Given that most disruptions are due to bad weather conditions, the above-mentioned decision models that integrate uncertainty form one important research direction in this area.

Another promising area is the development of decision support models that will facilitate contingency planning, to increase responsiveness under changing conditions. From the service provider's perspective, this will require robust scheduling models that incorporate user-specified contingency options so that recovery times can be reduced vastly. Such a framework also needs corresponding models at the user side, to allow for fully integrated airline recovery methods that can evaluate and establish user preference under the various potential scenarios. Other opportunities might concentrate on the coordination of multiple ATFM initiatives. Currently, most ATFM decisions aim to resolve local problems. When GDPs are implemented at several airports, for example, each of those is treated as an isolated event and little is done to coordinate them. This is likely to change, and efforts are underway to develop decision support tools that will facilitate strategic decision making when multiple ATFM initiatives are needed to mitigate congestion in various parts of the air transportation system. An interesting possibility in this area is the development of procedures that can configure airspace capacity dynamically, to best match the available resources with projected demand. An interesting possibility in this area is the development of procedures that can dynamically configure the airspace to maintain safety, efficiency, and equity in the traffic flow. Some of the emerging concepts in this area are the development and assessment of a "tube network" connecting major cities in the US (Sridhar et al. 2006)), and the dynamic reconfiguration of sectors so that traffic and controller workloads are more-or-less equitably distributed (Yousefi and Donohue 2004; Gianazza 2007).

Finally, we note that there are also a number of opportunities at a more tactical level. In recent years, for example, several studies have centered on the concept of continuous descent approach (CDA), in which aircraft descend and decelerate continuously (Clarke et al. 2004; Ren and Clarke 2004; Tong et al. 2006; Stell 2009). The CDA concepts present a departure from conventional approaches, where aircraft approach a runway in a step-down vertical profile with extended level segments at low altitudes during landings. Since level segments are avoided in a CDA, the vertical profile of an aircraft is usually higher than in a conventional approach procedure. Also, in a CDA, deceleration is delayed, resulting in lower engine thrust. According to Ren and Clarke (2004) CDA reduces noise, emissions, and fuel burn of aircraft during landing operations. Predicting aircraft trajectory in a CDA is, however, associated with higher level of uncertainty than the conventional approach procedure. While recent studies addressed this issue (Tong et al. 2006; Stell 2009), further research and field testing is needed before CDA can be made fully operational.

References

Abdelghany K, Abdelghany A, Niznik T (2007) Managing severe airspace ow programs: the airlines side of the problem. J Air Transp Manag 13:329–337

Air Transport Action Group (2008) The economic and social benefits of air transport. http://www.atag.org

Airline Dispatchers Federation and Seagull Technology, Inc (1995) Airline operational control overview. Prepared by Airline Dispatchers Federation and Seagull Technology, Inc., Cooperative Research and Development Agreement 93-CRDA-0034, September 1995

Andreatta G, Brunetta L (1998) Multiairport ground holding problem: a computational evaluation of exact algorithms. Oper Res 46:57–64

Andreatta G, Romanin-Jacur G (1987) Aircraft flow management under congestion. Transp Sci 21:249–253

Andreatta G, Brunetta L, Guastalla G Multi-airport ground holding problem: a heuristic approach based on priority rules. In: Bianco L, DellOlmo P, Odoni AR (eds) Modeling and simulation in air traffic management. pp 71–89 Springer, Berlin, (1995)

Arbuckle D, Rhodes D, Andrews M, Roberts D, Hallowell S, Baker D, Burleson C, Howell J, Anderegg A (2006) US vision for 2025 air transportation. J Air Traffic Cont 48(1):15–21

Aviation Week (2004) United, American agree to cut peak schedule by 5% at O'hare. Aviation Week, October 2004.

Ball M, Hoffman K (2005) Nextor congestion management project: interim report. Technical report, NEXTOR, The National Center of Excellence for Aviation Operations Research

Ball MO, Lulli G (2004) Ground delay programs: optimizing over the included flight set based on distance. Air Traffic Cont Q 12:1–25

Ball M, Chen C-Y, Hoffman R, Vossen T (1999) Collaborative decision making in air traffic management: current and future research directions. In: Proceedings of ATM'99, workshop on advanced technologies and their impact on air traffic management in the 21st century

Ball M, Hoffman R, Knorr D, Wetherly J, Wambsganss M (2000) Assessing the benefits of collaborative decision making in air traffic management. In: Proceedings of 3rd USA/Europe air traffic management R&D seminar

Ball M, Vossen T, Hoffman R (2001) Analysis of demand uncertainty effects in ground delay programs. In: Proceedings of 4th USA/Europe air traffic management R&D seminar

Ball M, Hoffman R, Odoni A, Rifkin R (2003) A stochastic integer program with dual network structure and its application to the ground-holding problem. Oper Res 51:167–171

Ball M, Donohue G, Hoffman K (2005a)Auctions for the safe, efficient, and equitable allocation of airspace system resources. In: Cramton P, Shoham Y, Steinberg R (eds) Combinatorial auctions. MIT Press, Cambridge, pp 507–538

Ball M, Hoffman R, Lovell D, Mukherjee A (2005b) Response mechanisms for dynamic air traffic flow management. In: Proceedings of the 2005 air traffic management research and development seminar, Baltimore, MD

Ball MO, Hoffman R, Lovell D, Mukherjee A (2005c) Response mechanisms for dynamic air traffic flow management. In: Proceedings of the 6th USA/Europe air traffic management R&D seminar

Ball MO, Ausubel LM, Berardino F, Cramton P, Donohue G, Hansen M, Hoffman K (2007) Market-based alternatives for managing congestion at New York's LaGuardia Airport: optimal use of scarce airport capacity. In: Proceedings of AirNeth annual conference, The Hague, The Netherlands

Bard J, Mohan DN (2008) Reallocating arrival slots during a ground delay program. Transp Res Part B 42:113–134

Bard J, Yu G, Arguello MF (2001) A time-band optimization model for the irregular operations problem. IIE Trans Oper Eng 33:931–947

Barnett A (2000) Free-flight and en route air safety: a first-order analysis. Oper Res 48(6):833–845

Barnhart C, Belobaba P, Odoni A (2003) Operations research in the air transport industry. Transp Sci 37:368–391

Barnhart C, Ball M, Nemhauser G, Odoni A (2006) Air transportation: irregular operations and control. In: Laporte G, Barnhart C (eds) Transportation handbook II (to appear)

Barnier N, Brisset P (2004) Graph coloring for air traffic flow management. Ann Oper Res 130:163–178

Beasley JE, Krishnamoorthy M, Sharaiha YM, Abramson D (2000) Scheduling aircraft landings—the static case. Transp Sci 34:180–197

Beasley JE, Krishnamoorthy M, Sharaiha YM, Abramson D (2004) Displacement problem and dynamically scheduling aircraft landings. J Oper Res Soc 55:54–64

Bertsimas D, Odoni A (1997) A critical survey of optimization models for tactical and strategic aspects of air traffic flow management. Technical report, NASA

Bertsimas D, Stock Patterson S (1998) The air traffic flow management problem with enroute capacities. Oper Res 46:406–422

Bertsimas D, Stock Patterson S (2000) The traffic flow management rerouting problem in air traffic control: a dynamic network approach. Transp Sci 34:239–255

Bertsimas D, Lulli G, Odoni A (2008) An integer optimization approach to large-scale air traffic flow management. IPCO, pp 34–46

Bilimoria K, Sridhar B, Chatterji G, Sheth K, Grabbe S (2000) Facet: future atm concepts evaluation tool. In: Proceedings of the 3rd USA/Europe air traffic management R&D seminar

Birge JR, Louveaux F (1998) Introduction to stochastic programming. TIMS Institute Of Management Sciences

Blumstein A (1959) The landing capacity of a runway. Oper Res 7(6):752–763

Bruckner JK (2003) Airport congestion when carriers have market power. Am Econ Rev 92:1357–1375

Button K, McDougall G (2006) Institutional and structure changes in air navigation service-providing organizations. J Air Transp Manag 12:236–252

Callaham M, DeArmon J, Cooper A, Goodfriend J, Moch-Mooney D, Solomos G (2001) Assessing nas performance: normalizing for the effects of weather. In: Proceedings of the 4th USA/Europe air traffic management R&D seminar, Santa Fe, USA

Cao J-M, Kanafani A (1997) Real-time decision support for integration of airline flight cancelations and delays, part I: mathematical formulation. Transp Plan Tech 20:183–199

Cao J-M, Kanafani A (1997) Real-time decision support for integration of airline flight cancelations and delays, part II: algorithm and computational experiments. Transp Plan Tech 20:201–217

Carlin A, Park RE (1970) Marginal cost pricing of airport runway capacity. Am Econ Rev 60:310–319

Carlson P (2000) Exploiting the opportunities of collaborative decision making: a model and efficient algorithm for airline use. Transp Sci 34:381–393

Carr GC, Erzberger H, Neuman F (1998) Airline arrival prioritization in sequencing and scheduling. In: Proceedings of the 2nd USA/Europe air traffic management R&D seminar

Chang K, Howard K, Oiesen R, Shisler L, Tanino M, Wambsganss MC (2001) Enhancements to the FAA ground-delay program under collaborative decision making. Interfaces 31:57–76

Cholankeril E (2003) Evaluation of an auction mechanism for allocating airport arrival slots. Master's thesis, Massachusetts Institute of Technology

Clarke JPB, Ho NT, Ren L, Brown JA, Elmer KR, Tong K-O, Wat JK (2004) Continuous descent approach: design and flight test for louisville international airport. J Aircraft 41:1054–1066

Clarke JP, Melconian T, Bly E, Rabbani F (2005) Means: the MIT extensible air network simulator. In: Simulation: transactions of the society international for computer simulation. Society for Computer Simulation International, San Diego

Concept of operations for the next generation air transportation system (2007). http://www.jpdo.gov, June 2007

Daniel J (1995) Congestion pricing and capacity at large hub airports: a bottleneck model with stochastic queues. Econometrica 63:327–370

Davis TJ, Isaacson DR, Robinson III JE, Den Braven W, Lee KK, Sanford B (1997) Operational test results of the passive final approach spacing tool. In: Proceedings of the 8th IFAC symposium on transportation systems, Chania, Greece

Dear RG, Sherif YS (1991) An algorithm for computer-assisted sequencing and scheduling of terminal area operations. Transp Res Part A Gen 25(2–3):129–139

DeCota W (2000) Matching capacity and demand at LaGuardia airport. Presented at the NEXTOR workshop on airline and national strategies for dealing with airport and airspace congestion

Dell'Olmo P, Lulli G (2003) A dynamic programming approach for the airport capacity allocation problem. IMA J Manag Math 14:235–249

Dillingham G (2006) Observations on potential FAA funding options. Technical report, US Government Accountability Office (GAO)

DotEcon Ltd (2001) Auctioning airport slots: a report for HM treasury and the department of the environment, transport and the regions. Technical report, DotEcon Ltd

Erzberger H (1995) Design principles and algorithms for automated air traffic management. AGARD lecture series, number 200. Wiley, New York

Erzberger H, Paielli RA, Isaacson DR, Eshow MM (1997) Conflict detection and resolution in the presence of prediction error. First USA/Europe air traffic management R&D seminar, Saclay, France, June, pp 17–20

Fan TP (2003) Market-bases airport demand management—theory, model and applications. PhD thesis, Massachusetts Institute of Technology

Fan TP, Odoni AR (2002) A practical perspective on airport demand management. Air Traffic Cont Q 10:285–306

Federal Aviation Administration (1993) SIMMOD reference manual, 2nd edn. Federal Aviation Administration, Washington, DC

Federal Aviation Administration (2001) Airport capacity benchmark report, Federal Aviation Administration

Fondacci R, Goldschmidt O, Letrouit V (1998) Combinatorial issues in air traffic optimization. Transp Sci 32(3):256–267

Garcia E (2003) Development of a framework for the assessment of capacity and throughput technologies within the national airspace system. PhD thesis, Georgia Institute of Technology

Gianazza D (2007) Airspace configuration using air traffic complexity metrics. In: Proceedings of the 7th USA/Europe air traffic management R&D seminar

Gilbo E (1993))Airport capacity: representation, estimation, optimization. IEEE Trans Cont Syst Tech 1:144–154

Gilbo E (1997) Optimizing airport capacity utilization in air traffic flow management subject to constraints at arrival and departure fixes. IEEE Trans Cont Syst Tech 5:490–503

Goodhart J (2000) Increasing airline operational control in a constrained air traffic system. PhD thesis, University of California at Berkeley

Grandeau S, Clarke M, Mathaisel D (1998) The processes of airline system operations control. In: Yu G (ed) Operations research in the airline industry. Kluwer, Dordrecht, pp 312–369

Grether D, Isaac M, Plott C (1979) Alternative methods of allocating airport slots: performance and evaluation. Technical report, Polynomics Research Laboratories, Inc, January 1979

Grether D, Isaac M, Plott C (1989) The allocation of scarce resources: experimental economics and the problem of allocating airport slots. Westview Press, Boulder

Gupta G, Malik W, Jung YC (2009) A mixed integer linear program for airport departure scheduling. In: Proceedings of the 9th AIAA aviation technology, integration, and operations conference, Hilton Head, SC

Hall WD (1999) Efficient capacity allocation in a collaborative air transportation system. PhD thesis, Massachusetts Institute of Technology

Hansen M (2002) Micro-level analysis of airport delay externalities using deterministic queueing models: a case study. J Air Transp Manag 8:73–87

Hansen M, Xiong J (2007) Weather normalization for evaluating nas performance. In: Proceedings of the 7th USA/Europe air traffic management R&D seminar

Hansen M, Liu B, Mukherjee A (2006) Scenario-based air traffic flow management: from theory to practice. Technical report, University of California, Berkeley

Helme MP (1992) Reducing air traffic delay in a space-time network. IEEE Int Conf Syst Man Cybernetics 1:236–242

Hinston JM, Aigoin G, Delahaye D, Hansman RJ, Puechmorel S (2001) Introducing structural considerations into complexity metrics. In: Proceedings of the 4th USA/Europe air traffic management R&D seminar

Hoffman R, Ball MO (2000) A comparison of formulations for the single-airport ground-holding problem with banking constraints. Oper Res 48:578–590

Hoffman R, Ball MO (2001) The rate control index for traffic flow management. IEEE Trans Intel Transport Syst 2:55–62

Hoffman R, Ball MO, Mukherjee A (2007) Ration-by-distance with equity guarantees: a new approach to ground delay program planning and control. In: Proceedings of the 7th USA/Europe air traffic management R&D seminar

Hoitomt D, Kraay D, Tang B (1999) Flight sequencing under the FAA's simplified substitution rules. Presentation: AGIFORS operations, joint study group meeting, control study and ground resources, April 1999

Horonjeff R, McKelvey FX (1994) Planning and design of airports. McGraw-Hill Professional, New york

IATA (2000) Worldwide scheduling guidelines, 3rd edn. International Air Transport Association, December 2000

Idris H, Clarke JP, Bhuva R, Kang L (2002) Queuing model for taxi-out time estimation. Air Traffic Cont Q 10(1):1–22

Innis T, Ball MO (2004) Estimating one-parameter airport arrival capacity distributions for air traffic flow management. Air Traffic Cont Q 12:223–252

Janic MA (1997) Model for air traffic control sector capacity based on air traffic controlller workload. Transp Plan Tech 20:311–335

Janic M, Tosic V (1991) En route sector capacity model. Transport Sci 25:299–307

Jarrah A, Yu G, Krishnamurthy N, Rakshit A (1993) A decision support framework for airline flight cancelations and delays. Transport Sci 27:266–280

Kivestu P (1976) Alternative methods of investigating the time-dependent M/G/K queue. SM thesis, Department of Aeronautics and Astronautics, Massachusetts Institute of Technology, Cambridge, MA, USA

Klein A, Jehlen R, Liang D (2007) Weather index with queuing component for nas performance assessment. In: Proceedings of the 7th USA/Europe air traffic management R&D seminar

Koopman BO (1972) Air-terminal queues under time-dependent conditions. Oper Res 20(6):1089–1114

Kotnyek B, Richetta O (2006) Equitable models for the stochastic ground-holding problem under collaborative decision making. Transp Sci 40:133–146

Krozel J, Jakobovits R, Penny S. (2006) An algorithmic approach for airspace flow programs. Air Traffic Cont Q 14(3):203–229

Kuchar JK, Yang LC (2000) A review of conflict detection and resolution modeling methods. IEEE Trans Intel Transp Syst 1(4):179

Ky P, Miaillier B (2006) Sesar: towards the new generation of air traffic management systems in Europe. J Air Traffic Cont 48(1):11–14

Leal de Matos P, Ormerod R (2000) The application of operational research to european air traffic flow management—understanding the context. Eur J Oper Res 123:125–144

Levine ME (1969) Landing fees and the airport congestion problem. J Law Econ 12:79–108

Lindsay K, Boyd E, Burlingame R (1994) Traffic flow management modeling with the time assignment model. Air Traffic Cont Q 1:255–276

Liu B (2006) Managing uncertainty in the single airport ground holding problem using scenario-based and scenario-free approaches. PhD thesis, University of Calinfornia, Berkeley

Liu B, Hansen M (2007) Scenario-free sequential decision model for the single airport ground holding problem. In: Proceedings of the 7th USA/Europe air traffic management R&D seminar

Long D, Lee DA, Johnson JP, Gaier EM, Kostiuk PF (1998) Modeling air traffic technologies with a queueing network of the national airspace system. Technical report CR-2001-210875, NASA

Lulli G, Odoni A (2007) The European air traffic management problem. Transp Sci 41:431–443

Luo S, Yu G (1998a) Airline schedule perturbation problem: ground delay program with splittable resources. In: Yu G (ed) Operations research in the airline industry. Kluwer, Dordrecht, pp 433–460

Luo S, Yu G (1998b) Airline schedule perturbation problem Landing and takeoff with non-splittable resource for the ground delay program. In:Yu G (ed) Operations research in the airline industry. Kluwer, Dordrecht, pp 404–432

Malone KM (1995) Dynamic queueing systems: behavior and approximations for individual queues. PhD thesis, Operations Research Center, Cambridge University, Cambridge, MA

Marietta GA (1993) TAAM solutions, 2nd edn. Preston Aviation Solutions Inc, Preston

Masalonis AJ, Callaham MB, Wanke CR (2003) Dynamic density and complexity metrics for realtime traffic flow management. In: Proceedings of the 5th USA/Europe air traffic management R & D seminar, Budapest, Hungary, 23–27 June 2003

McDougall G, Roberts A (2009) Commercializing air traffic control: have the reforms worked. Legal studies paper series, research paper 09–11, Suffolk University Law School, February 2009

Mehadhebi K (2000) A methodology for the design of a route network. In: Proceedings of 3rd USA/Europe air traffic management R&D seminar

Miller B, Clarke JP (2007) The hidden value of air transportation infrastructure. Tech Forecasting Soc Change 74(1):18–35

Milner J (1995) Dynamic slot allocation with airline participation. PhD thesis, Massachusetts Institute of Technology

Mukherjee A (2004) Dynamic stochastic optimization models for air traffic flow management. PhD thesis, University of California at Berkeley

Mukherjee A, Hansen M (2003) Dynamic stochastic model for single airport ground holding problem (submitted for publication)

Navazio L, Romanin-Jacur G (1998) Multiple connections multi-airport ground holding problem: models and algorithms. Transp Sci 32:268–276

NERA (2004) Study to assess the effects of different slot allocation schemes. Technical report, National Economic Research Associates

Nilim A, El Ghaoui L (2005) Robust control of Markov decision processes with uncertain transition matrices. Oper Res 53:780–798

Nilim A, El Ghaoui L, Duong V, Hansen M (2001)Trajectory-based air traffic management (tb-atm) under weather uncertainty. In: Proceedings of the 4th USA/Europe air traffic management R&D seminar

Niznik T (2001) Optimizing the airline response to ground delay programs. In: Proceedings of AGIFORS airline operations meeting

Nolan MS (1999) Fundamentals of air traffic control, 3rd edn. Brooks/Cole Publishing Company, Pacific Grove

Odoni A (1987) The flow management problem in air traffic control. In: Odoni AR, Bianco L, Szego G (eds) Flow control of congested networks. Springer, Berlin, pp 269–288

Odoni AR, Rousseau JM, Wilson NHM (1994) Models in urban and air transportation. In: Operations research and the public sector: handbooks in operations research and management science, vol 6. Elsevier, Amsterdam, pp 107–150

Odoni AR, Deyst J, Feron E, Hansman RJ, Khan K, Kuchar JK, Smipson R (1997) Existing and required modeling capabilities for evaluating atm systems and concepts. Technical report, International Center for Air Transportation, Massachusetts Institute of Technology

Pearce RA (2006) The next generation air transportation system: transformation starts now. J Air Traffic Cont 48(1):7–10

Pels E, Verhoef E (2003) The economics of airport congestion pricing. Discussion paper 03-083/3, Tinbergen Institute, October 2003

Prandini M, Hu J, Lygeros J, Sastry S (2000) A probabilistic approach to aircraft conflict detection. Intel Transp Syst IEEE Trans 1(4):199–220

Rakshit A, Krishnamurthy N, Yu G (1996) System operations advisor: a real-time decision support system for managing airline operations at united airlines. Interfaces 26:50–58

Rassenti S, Smith V, Buffin R (1982) A combinatorial auction mechanism for airport time slot allocation. Bell J Econ 13:402–417

Rathinam S, Wood Z, Sridhar B, Jung YC (2009) A generalized dynamic programming approach for a departure scheduling problem. In: Proceedings of the 9th AIAA guidance, navigation, and control conference, Illinois, USA

Ren L, Clarke (2004) JPB Flight-test evaluation of the tool for analysis of separation and throughput. J Aircraft 45:323–332

Richetta O, Odoni A (1993) Solving optimally the static ground holding policy problem in air traffic control. Transp Sci 24:228–238

Richetta O, Odoni A (1994) Dynamic solution to the ground-holding problem in air traffic control. Transp Res Part A Pol Practice 28:167–185

Rosenberger J, Schaefer AJ, Goldsman D, Johnson EL, Kleywegt AJ, Nemhauser GL (2002) A stochastic model of airline operations. Transp Sci 36:357–377

Schank JL (2005) Solving airside airport congestion: why peak runway pricing is not working. J Air Transp Manag 11:417–425

Sclar E (2003) Pitfalls of air traffic control privatization. Report commissioned by National Air Traffic Controllers Association, February 2003

Sherali HD, Smith JC, Trani AA (2002) An airspace planning model for selecting flight-plans under workload, safety, and equity considerations. Transp Sci 36:378–397

Sherali HD, Staats RW, Trani AA (2003) An airspace planning and collaborative decision-making model: part I-probabilistic conflicts, workload, and equity considerations. Transp Sci 37:434–456

Sherali HD, Staats RW, Trani AA (2006) An airspace-planning and collaborative decision-making model: part II-cost model, data considerations, and computations. Transp Sci 40:147–164

Shumsky RA (1997) Real-time forecasts of aircraft departure queues. Air Traffic Cont Q 5(4):281–308

Sridhar B, Swei S (2006) Relationship between weather, traffic, and delay, based on empirical methods. Proceedings of the 6th AIAA Aviation Technology, Integration, and Operations Conference, Wichita, USA

Sridhar B, Swei S (2007) Computation of aggregate delay using center-based weather impacted traffic index. Proceedings of the 7th AIAA Aviation Technology, Integration, and Operations Conference, Belfast, Ireland

Sridhar B, Seth K, Grabbe S (1998) Airspace complexity and its application in air traffic management. In: Proceedings of the 2nd USA/Europe air traffic management R&D Seminar

Sridhar B, Grabbe S, Sheth K, Bilimoria KD (2006) Initial study of tube networks for flexible airspace utilization. In: Proceedings of AIAA guidance, navigation, and control conference, Keystone, CO

Stell L (2009) Flight management system prediction and execution of idle-thrust descents. In: Proceedings of the 28th digital avionics systems conference

Swedish W (1981) Upgraded FAA airfield capacity model supplemental user's guide

Sweet DN, Manikonda V, Aronson JS, Roth K, Blake M (2002) Fast-time simulation system for analysis of advanced air transportation concepts. In: Proceedings of the AIAA modeling and simulation technologies conference, pp 5–8

Teodorovic D, Guberinic S (1984) Optimal dispatching strategy on an airline network afer a schedule perturbation. Eur J Oper Res 15:178–182

Teodorovic D, Guberinic S (1990) Model for operational daily airline scheduling. Transp Plan Tech 14:273–285

Terrab M (1990) Ground holding strategies for air traffic control. PhD thesis, Massachusetts Institute of Technology, Department of Civil Engineering

Terrab M, Odoni A (1993) Strategic flow management for air traffic control. Oper Res 41:138–152

Thengvall B, Bard JF, Yu G (2000) Balancing user preferences for aircraft recovery during airline irregular operations. IIE Trans Oper Eng 32:181–193

Thengvall B, Bard J, Yu G (2001) Multiple fleet aircraft schedule recovery following hub closures. Transp Res Part A 35:289–308

Tong KO, Boyle DA, Warren W (2006) Development of continuous descent arrival (CDA) procedures for dual-runway operations at houston intercontinental. In: Proceedings of the 6th AIAA aviation technology, integration, and operations conference, Wichita, USA

Tu Y, Ball M, Jank W (2005) Estimating flight departure delay distributions—a statistical approach with long-term trend and short-term pattern. Technical report, University of Maryland

US Senate Joint Economic Committee (2008) Your flight has been delayed. http://jec.senate.gov/

United States General Accounting Office (2007)Very light jets: several factors could influence their effect on the national airspace system. Technical report, United States General Accounting Office, GAO-07-1001, August 2007

Vasquez-Marquez A (1991) American airlines arrival slot allocation system. Interfaces 21:42–61

Vossen TWM (2002) Fair allocation methods in air traffic management. PhD thesis, University of Maryland

Vossen TWM, Ball MO (2006) Slot trading opportunities in collaborative ground delay programs. Transp Sci 40(1):29–54

Vossen T, Ball M, Hoffman R, Wambsganss M (2003) A general approach to equity in traffic flow management and its application to mitigating exemption bias in ground delay programs. In: Proceedings of the 5th USA/Europe air traffic management R&D seminar

Vranas P, Bertsimas D, Odoni A (1994a) Dynamic ground-holding policies for a network of airports. Transp Sci 28:275–291

Vranas P, Bertsimas D, Odoni A (1994b) The multi-airport ground holding problem in air traffic control. Oper Res 42:249–261

Wambsganss M (1996) Collaborative decision making through dynamic information transfer. Air Traffic Cont Q 4:107–123

Willemain T (2002) Contingencies and cancelations in ground delay programs. Air Traffic Cont Q 10:43–64

Willemain TR (2003) Factors influencing blind conflict risk in en route sectors under free-flight conditions. Transp Sci 37(4):457–470

Wojcik L (2001) DPAT: a tool for system-level simulation of airports and airspace. Technical report, The Mitre Corporation

Yamanaka S, Karlsson J, Odoni A (2006) Aviation infrastructure taxes and fees in the United States and the European union. Transp Res Rec 1951(1):44–51

Yousefi A, Donohue GL (2004) Temporal and spatial distribution of airspace complexity for air traffic controller workload-based sectorization. In: Proceedings of AIAA 4th aviation technology, integration, and operations (ATIO) forum, Chicago, IL

Index

C. Barnhart and B. C. Smith, LLC (eds.), *Quantitative Problem Solving Methods in the Airline Industry*, International Series in Operations Research & Management Science 169, DOI: 10.1007/978-1-4614-1608-1, © Springer Science+Business Media, LLC 2012